D1718815

Wolfgang Pompe
Gerhard Rödel
Hans-Jürgen Weiss
Michael Mertig

Bio-Nanomaterials

Related Titles

Niemeyer, C.M., Mirkin, C.A.

Nanobiotechnology

Concepts, Applications and Perspectives

2004

Print ISBN 978-3-527-30658-9

Kumar, C.S.

Biofunctionalization of Nanomaterials

2005

Print ISBN 978-3-527-31381-5

Ruiz-Hitzky, E, Ariga, K, Lvov, Y.M.

Bio-inorganic Hybrid Nanomaterials

Strategies, Syntheses, Characterization and Applications

2008

Print ISBN 978-3-527-31718-9

Mano, J.F.

Biomimetic Approaches for Biomaterials Development

2012

Print ISBN 978-3-527-32916-8

Fratzl, P, Harrington, M.J.

Introduction to Biological Materials Science

2014

Print ISBN 978-3-527-32940-3

Taubert, A, Mano, J.F., Rodríguez-Cabello, J.C.

Biomaterials Surface Science

2014

Print ISBN 978-3-527-33031-7

Wang, J

Nanomachines

Fundamentals and Applications

2013

Print ISBN 978-3-527-33120-8

Zhang, Q, Wei, F

Advanced Hierarchical Nanostructured Materials

2014

Print ISBN 978-3-527-33346-2

Reich, S, Thomsen, C, Maultzsch, J

Carbon Nanotubes

Basic Concepts and Physical Properties

2004

Print ISBN 978-3-527-40386-8

Yakushevich, L.V.

Nonlinear Physics of DNA

2 Edition

2004

Print ISBN 978-3-527-40417-9

Jung, R

Biohybrid Systems

Nerves, Interfaces, and Machines

2012

Print ISBN 978-3-527-40949-5

Wolfgang Pompe
Gerhard Rödel
Hans-Jürgen Weiss
Michael Mertig

Bio-Nanomaterials

Designing Materials Inspired by Nature

WILEY-VCH Verlag GmbH & Co. KGaA

The Authors

Prof. Dr. Wolfgang Pompe
Technische Universität Dresden
Inst. f. Werkstoffwissenschaft
Dresden
01062
Germany

Prof. Dr. Gerhard Rödel
Technische Universität Dresden
Institut für Genetik
Dresden
01062
Germany

Dr. Hans-Jürgen Weiss
Rabenau
Hainsberger Str. 22
01734
Germany

Prof. Dr. Michael Mertig
Max Bergmann Zentrum
Technische Universität Dresden
Inst. f. Physikalische Chemie
Dresden
01062
Germany

Left: Atomic force microscopy image of a self-assembled artificial 4-arm DNA junction; image size: 350 nanometer × 350 nanometer. (Courtesy of Alexander Huhle, Ralf Seidel, and Michael Mertig, Technische Universität Dresden.)

Middle: Human hematopoietic stem cell growing in a fibrillar network of collagen Type I on a silicone substrate with a reactive polymer coating. (Courtesy of Ina Kurth, Leibniz Institute of Polymer Research Dresden e.V., Germany.)

Right: High-resolution TEM image of a trabecula of rat bone showing a mineralized collagen bundle (below left to above right), about 15nm across, composed of triple helical collagen fibrils, about 1.5nm across. The fine striation pattern corresponds to the lattice constant of the *c*-axis of hydroxyapatite, 3.44 Å. The pattern of large stripes running parallel to the fibril axis corresponds to the *a*-lattice planes. (See also D. Grüner, T. Kollmann, C. Heiss, R. Schnettler, R. Kniep, P. Simon, "Hierarchical structure of spongeous tibia head of rats: SEM studies and HR-TEM of FIB prepared trabecula", to be published. (Courtesy of Paul Simon, Max Planck Institute for Chemical Physics of Solids Dresden, Germany.)

■ All books published by **Wiley-VCH** are carefully produced. Nevertheless, authors, editors, and publisher do not warrant the information contained in these books, including this book, to be free of errors. Readers are advised to keep in mind that statements, data, illustrations, procedural details or other items may inadvertently be inaccurate.

Library of Congress Card No.: applied for

British Library Cataloguing-in-Publication Data
A catalogue record for this book is available from the British Library.

Bibliographic information published by the Deutsche Nationalbibliothek
The Deutsche Nationalbibliothek lists this publication in the Deutsche Nationalbibliografie; detailed bibliographic data are available on the Internet at <http://dnb.d-nb.de>.

© 2013 Wiley-VCH Verlag GmbH & Co. KGaA, Boschstr. 12, 69469 Weinheim, Germany

All rights reserved (including those of translation into other languages). No part of this book may be reproduced in any form – by photoprinting, microfilm, or any other means – nor transmitted or translated into a machine language without written permission from the publishers. Registered names, trademarks, etc. used in this book, even when not specifically marked as such, are not to be considered unprotected by law.

Print ISBN: 978-3-527-41015-6
ePDF ISBN: 978-3-527-65529-8
ePub ISBN: 978-3-527-65528-1
mobi ISBN: 978-3-527-65527-4
oBook ISBN: 978-3-527-65526-7

Cover Design Adam-Design, Weinheim, Germany

Typesetting Thomson Digital, Noida, India

Printing and Binding Markono Print Media Pte Ltd, Singapore

Contents

Preface

The various challenges that we are confronted with today require novel solutions that will influence future developments in the field of materials science worldwide. This concerns the necessity to master the transition to regenerative energies. Also, the foreseeable exhaustion of essential resources necessitates developing new materials strategies, such as to use renewable raw materials, to exploit low-grade ores, or to establish widespread materials recycling. In view of this situation, the attitude toward nature has changed: in the past, the progress of mankind was based on extending its domination over nature. Now consensus is growing that future progress has to be achieved in close accordance with nature. Such attitude gave rise to the concept of biologically inspired materials engineering. It includes the development and production of novel materials, such as living tissue for regenerative bone therapy, and novel materials processing techniques, such as biologically controlled mineralization via microorganism–silica hybrid composites. "Bio-inspired" relates to inspiration by some mechanisms or processes present in the organic world, and the attempt to adapt them to technology. According to this nomenclature, "bio-inspired approach" denotes the following: The richness of biomolecular structures and processes serves as basis for the creation of nano-structured materials with novel functionalities, commonly summarized under the term "bionanotechnology." Here we follow the definition of bionanotechnology as proposed by Ehud Gazit in his book "Plenty of Room for Biology at the Bottom: An Introduction to Bionanotechnology" (Gazit, 2007). In the following, we will focus on materials or processes where adaptation includes the use of biomolecules or living cells, and hence on biologically inspired materials development in a narrower sense. The enormous progress in molecular biology and microbiology over the past 50 years has generated a huge knowledge as the basis to tackle such tasks. Genetic engineering allows the generation of tailored recombinant proteins or microorganisms and thus provides a large "toolbox" for the implementation of biological structures in a technical environment.

   Progress of synthetic biology will probably provide a further qualitative leap. In the paper entitled "Creation of a bacterial cell controlled by a chemically synthesized genome," J. Craig Venter and coworkers reported the creation of an artificial bacterial chromosome and its successful transfer into a bacterium, where it replaced the native DNA (Gibson *et al.*, 2010). Under the control of the synthetic genome, the cell started to produce proteins, eventually leading to DNA replication and cell division. The creation of this self-replicating synthetic bacterial cell called *Mycoplasma mycoides* JCVI-syn1.0 can be regarded as a milestone on the way from molecular genetics to synthetic biology. An old dream of biologists may become reality soon: engineering organisms designed for specific technological use, such as the efficient production of particular medical drugs or of biofuels via photosynthesis. Close interdisciplinary cooperation of biologists, materials scientists, chemists, physicists, and computer scientists is required to develop this research area successfully and to further public acceptance of the novel products, possibly including even artificial organisms in the future. Interdisciplinary approaches are also necessary regarding ethics and biosafety problems that require thorough assessments of the risk potential on the basis of profound and broadly oriented scientific work.

Based on our experience to teach biologically inspired materials science in various courses at the Technische Universität Dresden, our book aims at providing the basics of this scientific field for students of biology, biotechnology, bioengineering, materials science, chemistry, and physics and thus to lay the ground for interdisciplinary research. The already existing knowledge basis in bio-inspired materials science allows us to arrange practical results around a few general principles identified in the living world. Thus, we have organized the book in seven main chapters coauthored by two or three colleagues: Chapter 1 "Molecular units" by M. Mertig, W. Pompe, and G. Rödel; Chapter 2 "Molecular recognition" by W. Pompe and G. Rödel; Chapter 3 "Cell adhesion" by T. Pompe and W. Pompe; Chapter 4 "Whole-cell sensors" by W. Pompe and G. Rödel; Chapter 5 "Biohybrid silica-based materials" by W. Pompe, H.-J. Weiss, and H. Worch; Chapter 6 "Biomineralization" by M. Gelinsky, W. Pompe, and H.-J. Weiss; and Chapter 7 "Self-assembly" by M. Mertig and W. Pompe. It is recommended that one should begin with more biologically oriented subjects and later turn to those with a stronger materials science focus. The selection and the explanation of general principles have been motivated by particular biological case studies. Every chapter devoted to one such principle is introduced by a few subjectively selected biological case studies. These examples provide the background for elucidating the particular principle in the second section. In the third part of every chapter, examples for materials processing in engineering, medicine, and environmental technologies are given. We are aware that the subject of every chapter could be extended into a whole monograph. However, we see that students of materials science as well as of biology prefer to get an introduction to the whole field allowing them to initiate deeper studies of special topics. Therefore, we try to develop the basic principles as a kind of focusing and connecting part. In addition to biological principles, basic physical and chemical laws have been included since they are likewise essential for successful bio-inspired materials processing. Preferably, we chose a heuristic approach to the various topics. Occasionally, small tasks for quantitative estimates or simple modeling are formulated, including hints for the solutions. We hope that it will motivate the reader to address more complex calculations in the related original literature.

Acknowledgments

The engagement with bio-inspired materials science at the Technische Universität Dresden dates back to an elucidating and exciting discussion between one of us (WP) and Arthur Heuer of Case Western Reserve University at Cleveland 20 years ago. Just at that time, Arthur Heuer, together with a group of other well-known American materials scientists, issued a *Science* paper on "Innovative materials processing strategies: a biomimetic approach" (Heuer *et al.*, 1992), where he emphasized the great potential of mimicking biological processing strategies. He generously shared his ideas on what could possibly be done by materials scientists in this interdisciplinary research field. Later on, we repeatedly benefited from his personal engagement, as well as from that of Manfred Rühle at the Max-Planck Institute for Materials Science, Stuttgart, by establishing a research group for bio-inspired materials science at the Max-Bergmann Centre at the Technische Universität Dresden. We thank Arthur Heuer deeply for his great visionary advice and permanent support. We would also like to thank the many students and colleagues who supported us with valuable contributions of their research work and by reading drafts of particular chapters. Special thanks go to Michael Ansorge, Annegret Benke, Anne Bernhardt, Anja Blüher, Manfred Bobeth, Martin Bönsch, Horst Böttcher, Lucio Colombi Ciacchi, Florian Despang, Hermann Ehrlich, Angela Eubisch, Christiane Erler, Annett Groß, Katrin Günther, Thomas Hanke, Sascha Heinemann, Klaus Kühn, Mathias Lakatos, Lynne Macaskie, Sabine Matys, Iryna Mikheenko, Martin and Msau Mkandawire, Kai Ostermann, Ralf Seidel, Paul Simon, and Ulrich Soltmann, as well as to many colleagues for providing figures from their work. We also thank the staff of Wiley-VCH, in particular Ulrike Fuchs and Nina Stadthaus, whose engaged work and manifold advices during the extended preparation of the manuscript enabled us to finally complete it.

Dresden, Germany
March 2012

Wolfgang Pompe
Gerhard Rödel
Hans-Jürgen Weiss
Michael Mertig

References

1 Gazit, E. (2007) *Plenty of Room for Biology at the Bottom: An Introduction to Bionanotechnology*, Imperial College Press, London.

2 Gibson, D.G., Glass, J.I. *et al.* (2010) Creation of a bacterial cell controlled by a chemically synthesized genome. *Science*, **329** (5987), 52–56.

3 Heuer, A.H., Fink, D.J., Laraia, V.J., Arias, J.L., Calvert, P.D., Kendall, K., Messing, G.L., Blackwell, J., Rieke, P.C., Thompson, D.H., Wheeler, A.P., Veis, A., and Caplan, A.I. (1992) Innovative materials processing strategies: a biomimetic approach. *Science*, **255** (5048), 1098–1105.

1
Molecular Units

1.1
Case Studies

Living beings are "open" systems whose sustained existence requires fluxes. Since even very simple open systems such as a candle flame or a water jet form their own shape and restore it after a disturbance, one may readily accept the idea that more complex open systems are able to form and sustain more complex structures in space and time. Apparently, there are open systems involving chemical reactions with a tendency toward the formation of substances and reaction cycles with increasing complexity, ending up in the formation of life. The molecular processes of life are usually confined to enclosed (but not closed) spaces, the cells and their internal compartments. Eukaryotic cells, which are discriminated from prokaryotic cells by the presence of a nucleus harboring the vast majority of genetic information, are equipped with a variety of such functionally defined compartments, collectively summarized as organelles (Figure 1.1). This type of confinement is realized by membranes shielding the interior and controlling the flow of substances, energy, and information in and out. The information flow is facilitated by the membranes' capability of signal detection and transduction. Structural flexibility of the plasma membrane is a necessary precondition for cell motility and division. The cell is filled with cytoplasm, an assembly of functional entities and filamentous networks immersed in an aqueous solution, the cytosol.

Since there are good reasons for the assumption that all organisms have descended from a hypothetical common progenitor, their relationship has the topology of a tree and hence is usually visualized as a graph known as the phylogenetic tree of life (Figure 1.2). One can be sure that the tree of life obtained with a particular advanced technique, as the one in Figure 1.2, does not much differ from the real one and thus can serve as a basis for considerations as if it were the real one. As seen in the figure, the tree of life consists of three major domains. The vast majority of organisms are unicellular. Multicellular species are only found in a few branches of the Eukaryota, which are distinguished by the presence of a nucleus containing most of the genetic information. Autotrophic and heterotrophic organisms are present in every major domain. These terms refer to the source of the energy-rich organic substances (nutrients) required to drive the metabolism.

Bio-Nanomaterials: Designing Materials Inspired by Nature, First Edition. Wolfgang Pompe, Gerhard Rödel, Hans-Jürgen Weiss, and Michael Mertig.
© 2013 Wiley-VCH Verlag GmbH & Co. KGaA. Published 2013 by Wiley-VCH Verlag GmbH & Co. KGaA.

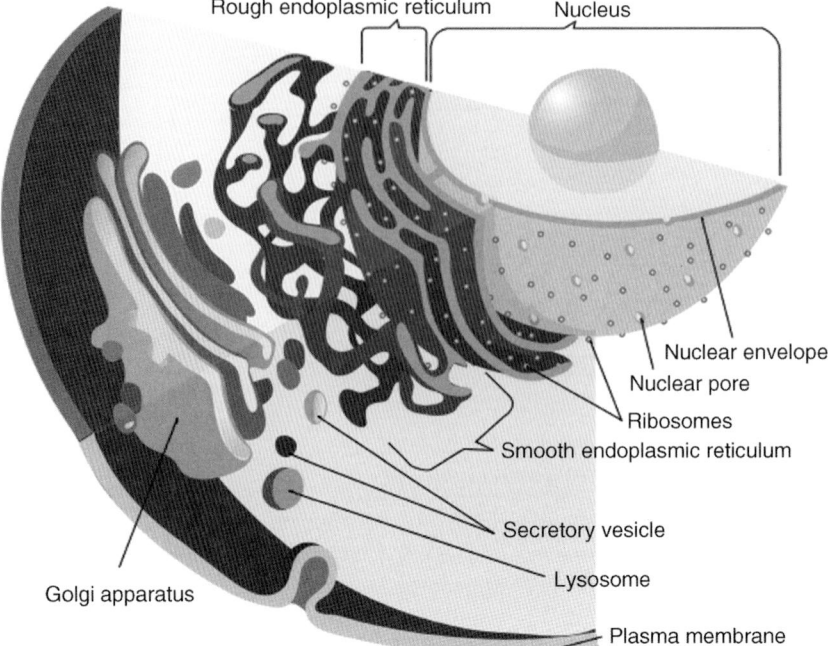

Figure 1.1 Eukaryotic cell structure with the endomembrane system. (*Source*: Wikimedia Commons; author Mariana Ruiz.)

Autotrophic organisms are able to produce the nutrients by themselves, starting from inorganic substances. In contrast, heterotrophic organisms are unable to synthesize their nutrients and hence have to acquire them by consuming organic substances. Besides the well-known photoautotrophy of plants, an alternative form of autotrophy, the so-called chemoautotrophy, is widespread among the Prokaryota. In the presence of oxygen, chemoautotrophic organisms make use of the energy released by oxidation, notably of inorganic substances, enabling them to live in extreme habitats such as salt lakes, hot springs, deep sea floors, and so on. This property makes chemoautotrophs interesting for bioengineering. Photoautotrophic cyanobacteria have recently been considered with respect to their suitability for biofuel production. Eukaryotes are especially valuable for biotechnology, bio-inspired materials development, and medical engineering. Fermentation by means of yeast, for example, has been applied for millennia. Recently, animal stem cells have been widely used in tissue engineering developments. The huge variety of organisms offers a wealth of objects with structures differing on the molecular level that may be suitable as building blocks in biotechnology and biologically inspired materials science. Today, we are still in a very early stage of exploring their potential. Our responsibility for the protection of life on Earth implies that progress in this field of research should always be complemented with adequate risk assessment.

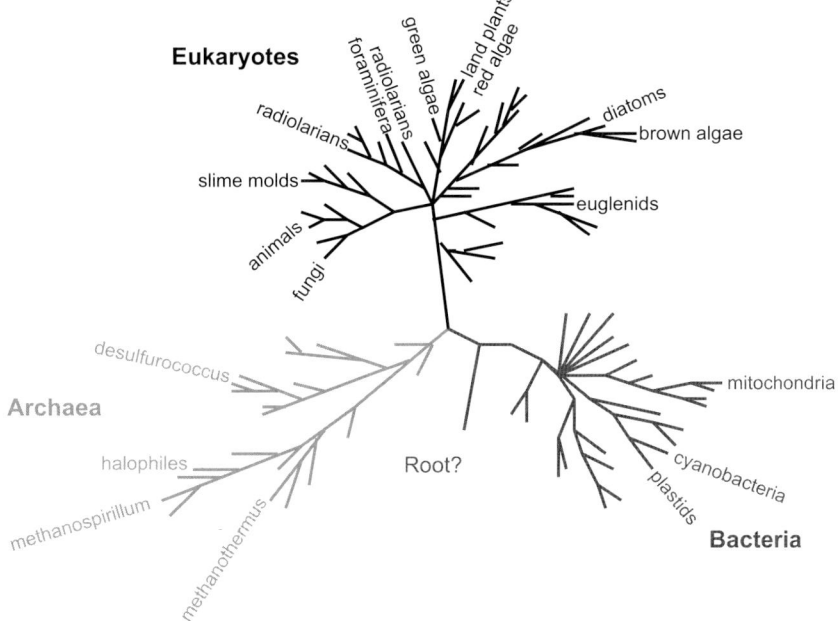

Figure 1.2 The phylogenetic tree of life based on the comparison of ribosomal RNA. (Bacteria and Archaea are also called Prokaryota. The names of most taxa are omitted here for simplicity.)

The sizes of the bimolecular structures investigated as potential building blocks range from molecular (0.2 nm) to cell size (0.1 mm) (Figure 1.3). Remarkably, organisms utilize only a small fraction of the chemical elements. Obviously, they are sufficient to form the large variety of organic compounds required for sustaining the processes of life. Let us consider the composition of the bacterium

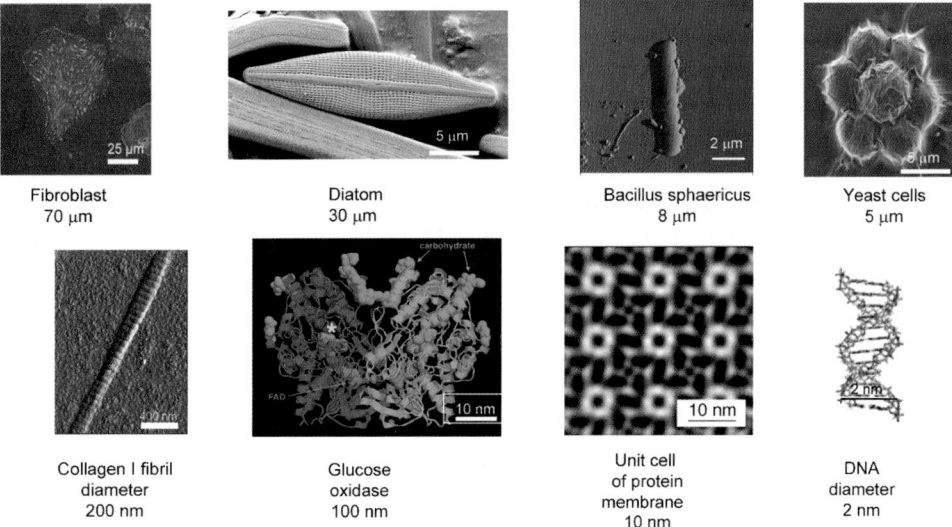

Figure 1.3 Size variation of biological components.

Table 1.1 Molecular composition of *E. coli* according to Nelson and Cox (2008).

	Percentage of total cellular weight	Approximate number of different molecular species
Water	70	1
Proteins	15	3000
Nucleic acids		
DNA	1	1–4
RNA	6	>3000
Polysaccharides	3	10
Lipids	2	20
Monomeric subunits and intermediates	2	500
Inorganic ions	1	20

Escherichia coli, with size is about $2\,\mu m \times 1\,\mu m$ (Table 1.1). Its cytoplasm contains the nucleoid usually with one DNA chain, eventually a few small circular DNA molecules called plasmids, about 15 000 ribosomes (the sites of protein synthesis), 10 to several hundred copies of about 1000 different enzymes, about 1000 smaller organic compounds with a molecular weight less than 1000 (metabolites or coenzymes), and various inorganic ions. The cytoplasm is surrounded by the cell envelope, which consists of an outer and an inner membrane composed of lipid bilayers and peptidoglycans. Connected to the envelope are specific protein structures such as flagellae for cell propelling, pili providing adhesion sites, and surface layer proteins for mechanical stabilization and acting as filter and ion transport structures.

Eukaryotic cells with a size of about $5–100\,\mu m$ show a higher structural complexity. The essential differences to bacteria are the presence of a nucleus, a number of membrane-enclosed organelles (e.g., mitochondria, endoplasmic reticulum, Golgi complexes, peroxisomes, and lysosomes), and the cytoskeleton, a highly structured network of protein filaments (microtubules, actin filaments, and intermediate filaments) organized by numerous proteins that regulate the assembling and disassembling of the various filaments. Characteristic components of plant cells are chloroplasts and vacuoles. A concise overview of the structure and properties of the main groups of biomolecules available for a bottom-up design of nanostructured materials – nucleic acids, proteins, carbohydrates, and lipids – is provided below.

1.1.1
Nucleic Acids

The storage, replication, and transfer of genetic information in living organisms is mediated by chain-like macromolecules called nucleic acids, the deoxyribonucleic acid (DNA) and several types of ribonucleic acid (RNA) (Figure 1.4).

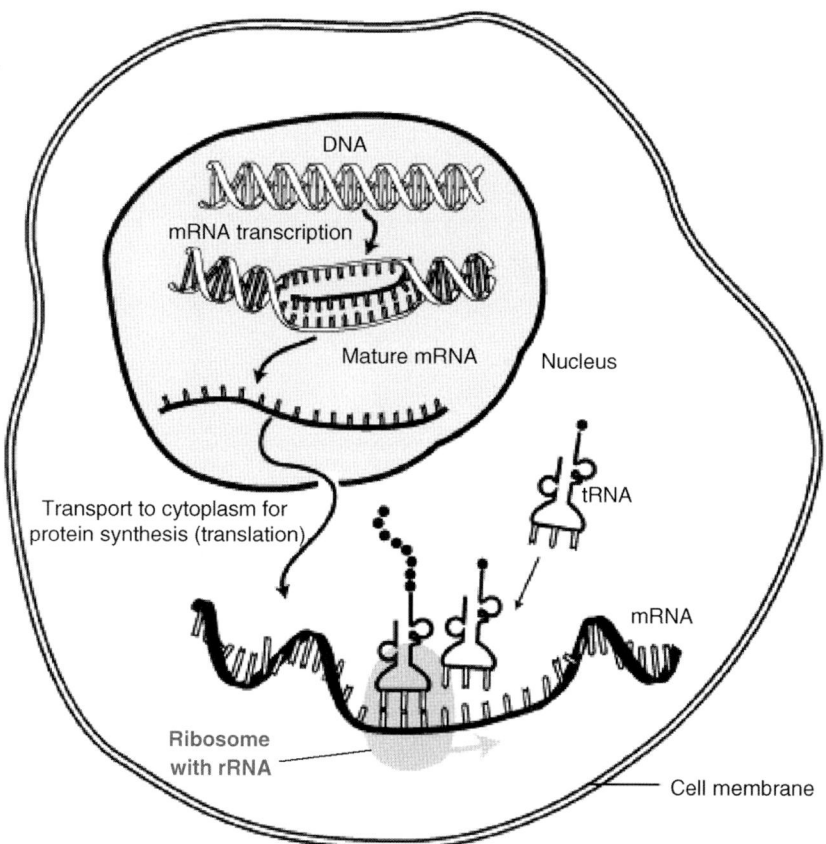

Figure 1.4 Functions of the three major types of RNA in protein synthesis: DNA serves as a template for the synthesis of messenger RNA. The respective genetic information can be translated into a specific sequence of amino acids by ribosomes that are composed of proteins and ribosomal RNA. The amino acids are provided by transfer RNA. (*Source*: Wikimedia Commons; author Sverdrup; http://www.genome.gov/Pages/Hyperion/DIR/VIP/Glossary/Illustration/mrna.shtml.)

The nucleotide sequence of the DNA encoding the genetic information is transcribed into messenger RNA (mRNA), which is released from the nucleus into the cytoplasm, where it associates with one of the ribosomes, the sites of protein synthesis. Groups of three bases of the nucleotide sequence of the mRNA serve as the codons for one of the 20 amino acids found in proteins. Transfer RNAs (tRNAs) are the carriers of the amino acids, which constitute the raw material for protein synthesis. Each tRNA carries at its 3'-end a particular amino acid that is incorporated in the growing polypeptide chain at a specific site of the ribosome, when a sequence of three bases (anticodon) of the tRNA is bound to the complementary sequence (codon). Every nucleotide of DNA and RNA consists of a phosphate group, a sugar group, and a nitrogen-containing base (Figure 1.5). Natural nucleic acid contains

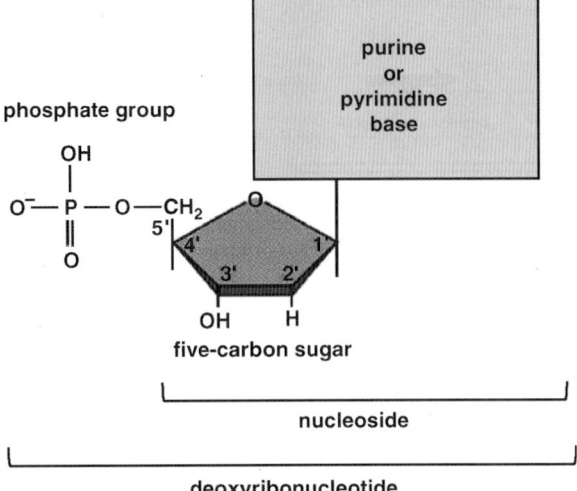

Figure 1.5 Structure of a deoxyribonucleotide. Ribonucleotides possess —OH instead of —H on the 2′-carbon of the pentose ring.

four different bases. The bases are derivatives of two compounds, purine and pyrimidine. The purine bases adenine (A) and guanine (G) and the pyrimidine base cytosine (C) are common to DNA and RNA. The pyrimidine base thymine (T) is only found in the DNA, and uracil (U) only in the RNA (Figure 1.6). The sugar units in a DNA are 2′-deoxy-D-ribose units, and in RNA D-ribose units (Figure 1.7).

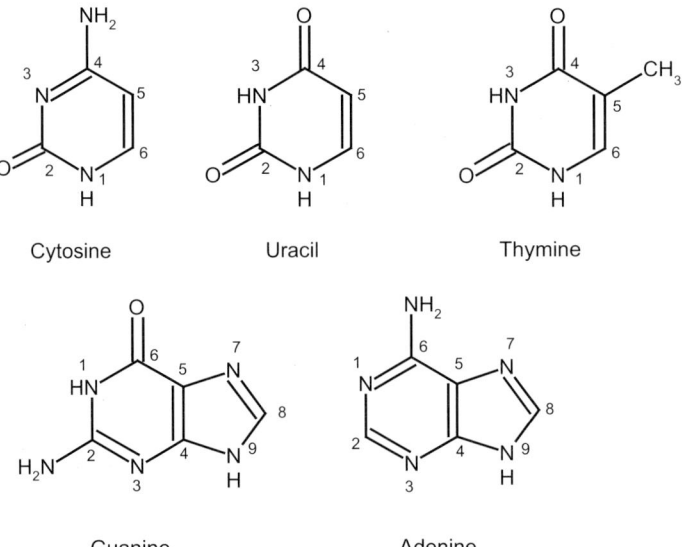

Figure 1.6 The structure-forming bases of DNA and RNA: pyrimidine bases (top) and purine bases (bottom).

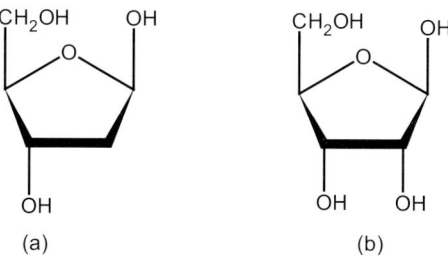

Figure 1.7 2′-Deoxy-D-ribose (a) and D-ribose (b) – the sugar building blocks in DNA and RNA, respectively.

The nucleotides of DNA and RNA are successively linked by phosphate group bridges. The phosphate groups connect the 5′-hydroxyl group of one pentose with the 3′-hydroxyl group of the pentose of the next nucleotide (Figure 1.8). The backbone of DNA and RNA consists of negatively charged phosphate and pentose residues. Like any charge, it polarizes the surrounding medium containing dipoles, such as water in this case. This is equivalent to an attraction of water by the DNA backbone; hence, it is hydrophilic. The purine and pyrimidine bases are nonpolar molecules, which implies hydrophobic behavior. In water, such

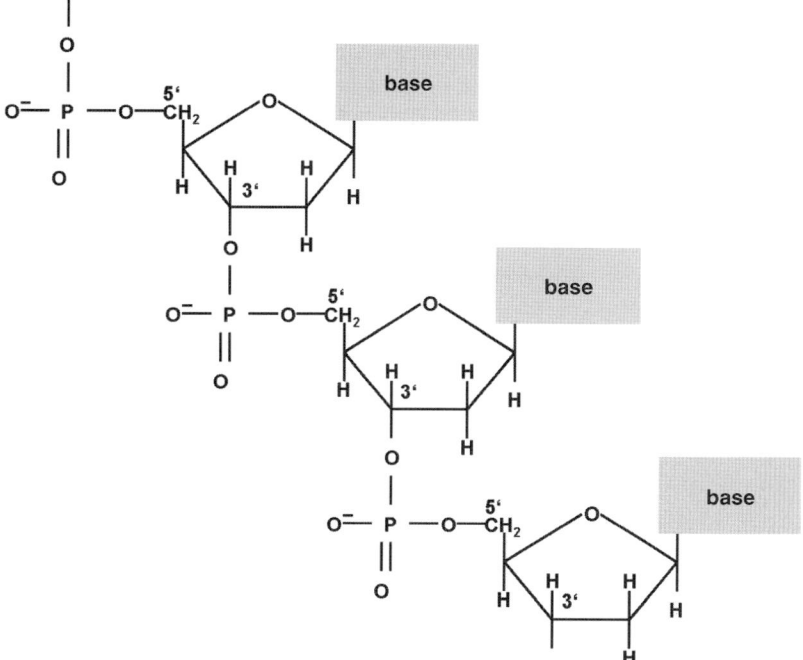

Figure 1.8 Single-stranded DNA or RNA consisting of three nucleotide units linked by phosphate groups.

behavior manifests as an effective interaction between the bases with the tendency to arrange themselves in stacks so that their contact with water is minimal. This hydrophobic base-stacking interaction also stabilizes the structure of nucleic acids. In addition to the hydrophobic interaction, there are binding sites on the bases, provided by ring nitrogens, carbonyl groups, and exocyclic amino groups. Amino and carbonyl groups enable the formation of hydrogen bonds between base stacks of two or more nucleic acids. (Artificial nucleic acids with more than two strands are relevant in biotechnology.) In their fundamental work on the molecular structure of double-stranded DNA and RNA published in 1953, Watson and Crick had revealed that these hydrogen bonds make the base pairings G–C and A–T in DNA (and G–C and A–U in RNA) (Watson and Crick, 1953) (see Figure 1.9).

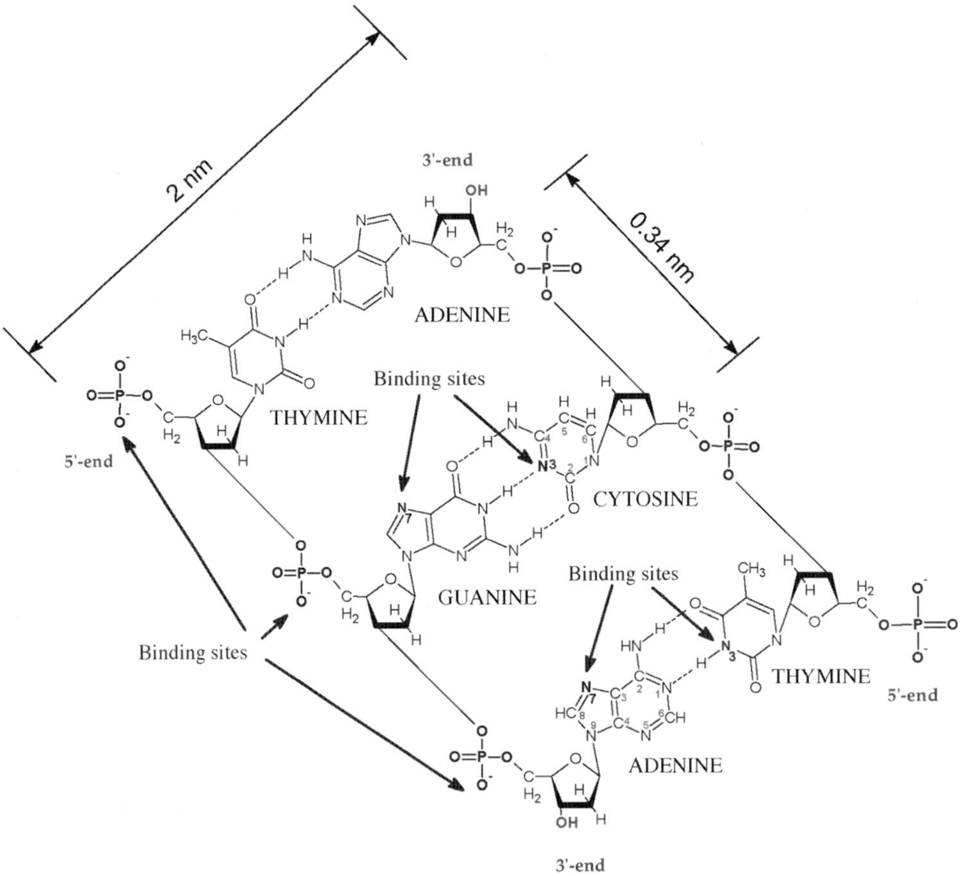

Figure 1.9 Chemical structure of a short sequence of a DNA double helix. The hydrogen bonds between the bases are indicated as dashed lines. There are two classes of binding sites of interest for DNA-templated nanowire fabrication: negatively charged phosphate groups in the backbone, N7 atoms of bases G and A, and N3 atoms of bases C and T acting as electron donors.

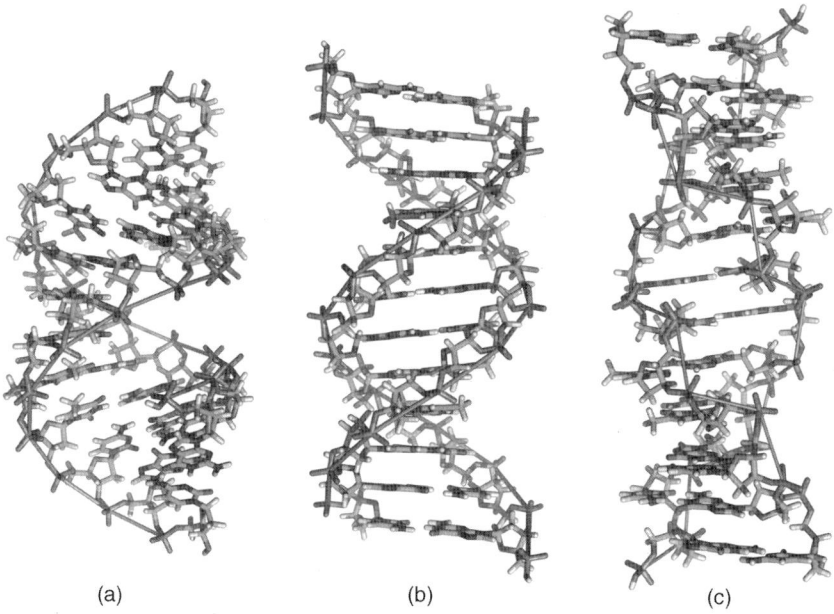

Figure 1.10 Structures of double-stranded DNA: (a) A-DNA; (b) B-DNA; (c) Z-DNA (*Source*: Wikimedia Commons; author Richard Wheeler (Zephyris).)

The double helix consisting of two DNA strands wound around a common axis, with the hydrophobic purine and pyrimidine bases arranged inside and the hydrophilic backbone of each strand facing the surrounding water, has an energetically favorable configuration. As seen in Figure 1.9, the oligonucleotides are not symmetrical with respect to reversal of direction. Note the opposite orientations of the nucleotide sequences, aligned from the end with the 5′-pentose site to the end with a 3′-pentose site. The base sequence TGA of the left strand matches the TCA sequence of the right strand only if the two are arranged in an antiparallel manner. There are three different conformations of the double helix structure. Under *physiological conditions*, the most stable form of DNA is the so-called B-form (Figure 1.10b). It is a right-handed double helix with a diameter of ∼2.0 nm and 10.5 base pairs per one helical turn. With *decreasing humidity*, the B-DNA can change its structure into A-DNA, a right-handed double helix, which is slightly more densely packed. It is not known whether the A-form DNA occurs in cells. The third conformation, the Z-DNA, is a slender left-handed double helix with a zigzag-shaped backbone. Particular nucleotide sequences, as alternating pyrimidine–purine sequences, fold preferentially into Z-DNA. Short strands of Z-DNA have been found in bacteria and eukaryotes. Characteristic parameters of the three forms are given in Table 1.2. Usually, the B-DNA is applied in bionanotechnology.

Double-stranded DNA (dsDNA) is stable under conditions near pH 7.0 and room temperature. Above 80 °C or at extreme pH, the viscosity of an aqueous solution of dsDNA is much lower than that under normal conditions, which is the result of a

Table 1.2 Characteristic geometric parameters of the DNA configurations.

	A-form	B-form	Z-form
Helical sense	Right-handed	Right-handed	Left-handed
Diameter	2.6 nm	2.0 nm	1.8 nm
Base pairs per helical turn	11	10.5	12
Helix rise per base pair	0.26 nm	0.34 nm	0.37 nm
Tilt of the bases normal to the helix axis	20°	6°	7°

reversible structural transition called *denaturation*. The denaturation consists in splitting of the double-stranded DNA into single-stranded DNA (ssDNA), which arrange themselves in random coils (Figure 1.11). The UV absorption (near 260 nm) of ssDNA is higher than that of dsDNA. By setting the conditions back to normal, the denaturation is reversed. The midpoint temperature T_m of the transition range is called the *melting point* of the DNA (Figure 1.12). It depends on the ratio of G–C and A–T base pairs. As a consequence of the higher effective binding energy of G–C pairs with three hydrogen bonds compared to the A–T pairs with two hydrogen bonds, the melting point of DNA increases with increasing G–C content.

The reversibility of the dissociation of double-stranded DNA and the association of the respective single-stranded DNA (the so-called hybridization) is the key for the replication of the genetic information in living organisms. This simple structural design in connection with the high mechanical stability of the information inscribed

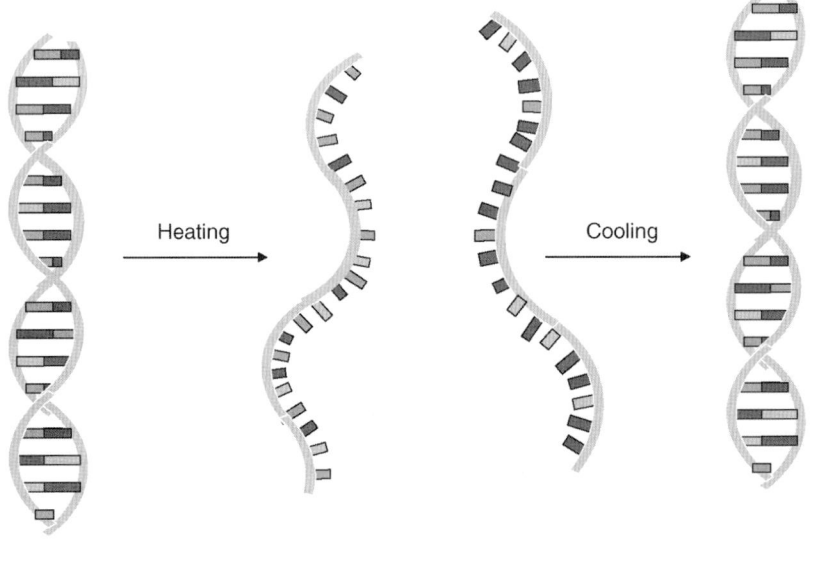

Native DNA   Denatured DNA   Renatured DNA

Figure 1.11 Denaturation (melting) and renaturation (hybridization) of a double-stranded DNA.

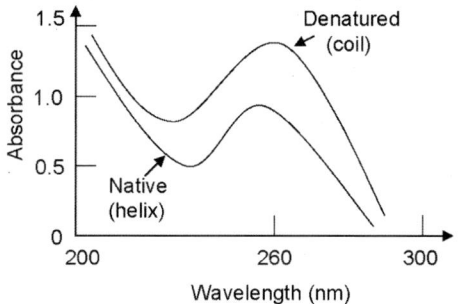

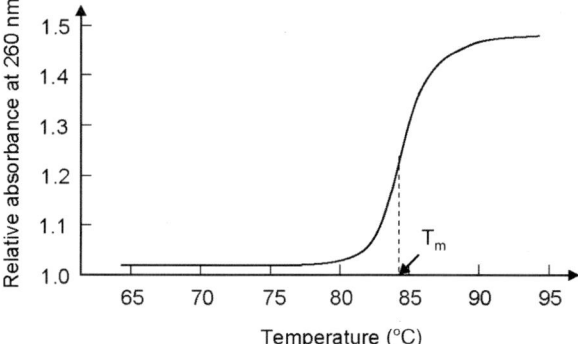

Figure 1.12 Change of the optical absorbance caused by melting of double-stranded DNA (T_m = melting temperature).

in the double-stranded chain molecule is crucial for minimizing errors in the replication process. However, rare alterations in the DNA sequence (mutations) are also stably replicated. Clearly, the molecular structure of the DNA is one essential precondition for the evolution of life with sufficient stability. At the same time, the ability to form double-stranded DNA segments by hybridization of complementary single-stranded oligomers and the possibility to reversibly disassemble the resulting double-stranded DNA into the constituent single-stranded DNA make DNA oligomers an ideal "brick" in soft nanotechnology, as we will see in the following.

1.1.2
Proteins

Proteins can be subdivided into structural and functional proteins can be. *Structural proteins* (e.g., collagen in soft and hard tissues, surface layers of many bacteria, and the cage-forming ferritin protein) are responsible for the molecular (pre)organization of structural units. *Functional proteins* (*enzymes*) catalyze the primary life processes of the organism, such as oxygen transport, sugar cycles, regulation of fuel supply for cellular machinery (e.g., synthesis of ATP and GTP), formation of the new structural proteins, transport processes via cell membrane,

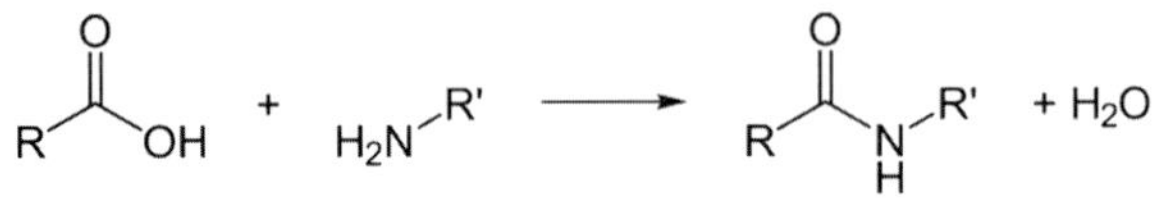

Figure 1.13 Formation of a peptide bond by condensation.

remodeling of tissue, and motion of muscles. Therefore, they can be regarded as the "functional molecules of life" (Lezon, Banavar, and Maritan, 2006).

The *primary structure* of a protein is quite simple. It is assembled as a chain of a limited number of 20 α-amino acids that differ in their side chains. These short side chains, or residues (R groups), determine the physical and chemical properties of the different amino acids, including their solubility in water. The backbone of a protein is formed by the so-called peptide bonds. They are formed by the removal of OH^- from the α-carboxyl group of one amino acid and H^+ from the α-amino group of a second amino acid. The result of this condensation reaction is a dipeptide and water (Figure 1.13).

The terminal amino acid residue with a free α-amino acid group is called the amino-terminal (or N-terminal) residue, whereas the other end with the α-terminal carboxyl group is the carboxyl-terminal (or C-terminal) residue. Peptide bonds are very stable with an average half-life of about 7 years under intracellular conditions excluding enzymatic cleavage. Condensed amino acids are called *oligopeptides* or *polypeptides*, depending on the number of amino acid residues. Polypeptides with a molecular weight higher than 10 000 are referred to as proteins. Typically, proteins are chains of few hundred up to few thousand amino acid residues. The 20 *natural amino acids* can be subdivided into seven groups of similar functionality, which they fulfill in the chain sequence of a particular protein (see Table 1.3).

Proteins usually consist of one long single polypeptide chain but there are also *multimeric proteins* consisting of two or more mostly noncovalently bonded polypeptide chains. For example, *E. coli* RNA polymerase, the enzyme that mediates the transcription of the genetic information of double-stranded DNA into RNA strands, is composed of five polypeptide chains.

Proteins often perform a specific task million-fold with high reliability. Therefore, proteins must be stable under changing conditions. However, stability alone is not the only property that accounts for the utmost importance of proteins in all living organisms (Lezon, Banavar, and Maritan, 2006). In order to realize the essential feature of living matter – cooperation – on all scales from a small organelle inside a cell up to a complex organism, the proteins exhibit properties such as high specificity of interaction and functional diversity, self-organization in higher ordered networks, and high sensitivity with respect to external signals. High specificity and diverse functionality can be realized by a combination of rigid subunits that can be arranged with high flexibility. An arrangement of the subunits in a chain is a possible solution for that. However, the variety of a longer chain must be restricted to a limited number of structures. It should be excluded that each mutation leads to a significant change of the phenotype. Otherwise, the number of phenotypes to be displayed should be large enough that evolution is not precluded when the

Table 1.3 Structure and functionality of natural amino acids classified by the R groups (for abbreviation a three-letter or one-letter code is used).

Name/short notation	Structure	Functionality
Control of protein backbone properties		
Glycine (Gly, G)		Smallest amino acid without side chain, making the protein chain more flexible
Proline (Pro, P)		With two covalent bonds to the backbone, forming a kink in the chain
Carbon-rich side chains driving protein folding		
Leucine (Leu, L)		Hydrophobic domains forming closely packed clusters inside the protein
Isoleucine (Ile, I)		Hydrophobic domains forming closely packed clusters inside the protein
Alanine (Ala, A)		Hydrophobic domains forming closely packed clusters inside the protein
Valine (Val, V)		Hydrophobic domains forming closely packed clusters inside the protein
Aromatic rings driving protein folding		
Phenylalanine (Phe, F)		Hydrophobic domains, aromatic rings often form stacks of one upon another
Tryptophan (Trp, W)		Hydrophobic domains, aromatic rings often form stacks of one upon another

(*continued*)

Table 1.3 (*Continued*)

Name/short notation	Structure	Functionality
Tyrosine (Tyr, Y)		Hydrophobic domains, aromatic rings often form stacks on top of one another, hydroxyl group acts as possible binding site
Diverse hydrogen bonding groups		
Histidine (His, H)		Imidazole group can be neutral or charged depending on pH, can fulfill catalytic tasks, coordinates strongly with metal ions
Serine (Ser, S)		Very often at protein surfaces, offering hydrogen bonding groups
Asparagine (Asn, N)		Very often at protein surfaces, offering hydrogen bonding groups
Threonine (Thr, T)		Very often at protein surfaces, offering hydrogen bonding groups
Glutamine (Gln, Q)		Very often at protein surfaces, offering hydrogen bonding groups
Carboxylic acid binding groups		
Aspartate (Asp, D)		At neutral pH negatively charged; at protein surfaces, metal ion binding

Table 1.3 (Continued)

Name/short notation	Structure	Functionality
Glutamate (Glu, E)		At neutral pH negatively charged; at protein surfaces, metal ion binding
Positively charged surface binding sites		
Lysine (Lys, K)		Amine end group ionized under neutral pH, carbon-rich chain plays a role in interaction with other carbon-rich molecules
Arginine (Arg, R)		Guanidinium group ionized under neutral pH, carbon-rich chain plays a role in interaction with other carbon-rich molecules
Sulfur-containing groups		
Cysteine (Cys, C)		Most reactive amino acid, forms covalent disulfide cross-links, coordinates strongly with metal ions
Methionine (Met, M)		Has a hydrophobic sulfur atom, promotes protein folding, sulfur atom coordinates with metal ions

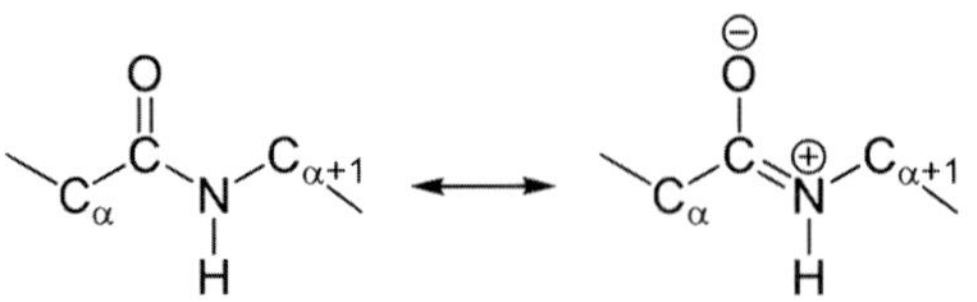

Figure 1.14 Peptide bond. It exists in two resonance forms, an uncharged (about 60%) and a charged (about 40%) form.

environmental conditions are changing. These demands to be fulfilled by a "molecule of life" explain why we observe the folding of the protein chains into more compact structures with pronounced stability and specificity (Lezon, Banavar, and Maritan, 2006). Primary, secondary, tertiary, and quaternary structures distinguish the various structure levels of folded proteins.

The *primary structure* of a protein is determined by the sequence of amino acids. Peptide bonds and disulfide bonds link the amino acids in a polypeptide chain. The covalent bonds in the polypeptide backbone cause essential constraints on the protein structure. The peptide group forming that backbone group consists of six atoms arranged in one plane, where the oxygen atom of the carbonyl group stays in *trans* position to the hydrogen atom of the amide nitrogen (Figure 1.14). The α carbons of adjacent amino acid residues are separated by three covalent bonds, aligned as C_α—C—N—$C_{\alpha+1}$. The C—N bond has a partial double bond character (about 40%). Therefore, the oxygen has a partial negative charge and the nitrogen a positive charge, which leads to a small electric dipole. These dipoles can line up into large dipoles. The regular assembly of electric dipoles of the peptide bond along the protein backbone is the basis of the observed piezoelectric behavior of proteins (Lemanov, 2000). There is an ongoing discussion whether piezoelectricity could play a role in the biomineralization of collagen (see also Section 6.3.2.1).

The *secondary structure* describes stable spatial arrangements of amino acids in a particular segment of the whole protein, without regarding the arrangement of the side chains of the amino acids or their relation to other segments. Only a few types of stable secondary protein structures are realized. Partial sequences of the backbone are often organized as α-helices or β-sheets (Figures 1.15 and 1.16). Already in the 1950s, Pauling and Corey have pointed out that the short-ranging *weak hydrogen bonds* between the amino acids are the reason for the formation of the α-helices and β-sheets (Pauling and Corey, 1951). In α-helices, the protein backbone is wound around an imaginary axis. The side groups of the amino acids are arranged outside the helix. The step height of the helix is about 0.54 nm. Each helical turn includes 3.6 amino acids. Hydrogen bonds between the hydrogen attached to the nitrogen atom of the peptide bond of the nth amino acid in the chain along the backbone and the oxygen atom of the carboxyl group of the $(n + 4)$th amino acid stabilize the helix. Every peptide bond along the helix involves a hydrogen bond. The β-sheets are composed of aligned strands connected by hydrogen bonds. The strands of the β-sheets are arranged in two-stranded ribbons, multistranded sheets, or barrel-like structures. The amino-terminal to carboxyl-terminal orientations of adjoining chains can be the same (parallel β-sheets) or opposite (antiparallel β-sheets). A

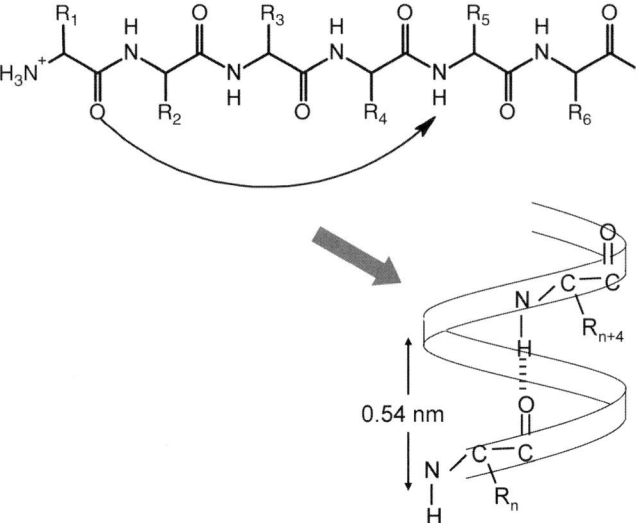

Figure 1.15 α-Helix. Each turn of the helix consists of 3.6 amino acids, which places the C=O group of amino acid n exactly in line with the H—N group of amino acid $n + 4$.

survey of all known natural proteins reveals that about one-third of all amino acids are found in α-helices and one-fifth in β-sheets.

The *tertiary structure* is the three-dimensional structure of a polypeptide chain, as defined by the atomic coordinates. There are filamentous and globular tertiary structures. Functional properties of proteins such as molecular recognition, catalytic activity, and preorganization of mineralized tissue are to a large extent regulated by

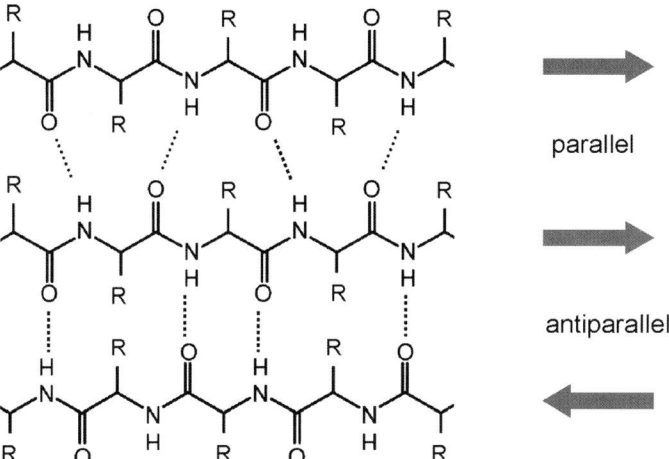

Figure 1.16 β-Sheet. Extended strands are lined up side by side, and H-bonds bridge from strand to strand. Identical or opposed strand alignments make up parallel or antiparallel β-sheets.

Figure 1.17 Structure of human hemoglobin. The protein α and β subunits are in red and blue and the four iron-containing heme groups in green. Binding of one oxygen molecule to one of four possible binding sites leads to a conformation change resulting in enhanced oxygen binding efficiency. (*Source*: Wikimedia Commons; author Richard Wheeler (Zephyris).)

the tertiary structure. Therefore, the successful use of proteins in biohybrid man-made materials requires that the tertiary structure of the protein is reproduced with high precision.

Quaternary structures are arrangements of multiple polypeptide subunits. Often these subunits can also fulfill regularity tasks. Binding of small molecules to subunits can cause large changes in the quaternary structure. A well-studied example is the tetrameric hemoglobin, the carrier of oxygen in the blood, which consists of four subunits (Figure 1.17). Binding of oxygen or carbon dioxide causes essential conformational changes that allow an optimized uptake and release of oxygen in the various tissues.

Another example is the iron storage protein ferritin (see also Section 6.2.6), which consists of 24 monomers. The quaternary structure, a spherical cage formed by 12 dimers, contains reaction centers for the catalytic oxidation of Fe(II) to Fe(III), and specific reaction sites for the deposition of ferrihydrite ($Fe_2O_3 \cdot nH_2O$).

Capsids Capsids are higher order protein structures constituting the shell of viruses. Their high regularity and stability make them suitable for various applications in nanobiotechnology. The capsids form a cage for the genetic material of the virus, either DNA or RNA. The coat proteins are arranged in the capsids mainly in an *icosahedral* or *helical* way. The icosahedral shape is composed of 20 equilateral triangular faces. Three examples of icosahedral viruses are shown in Figure 1.18.

The helically shaped capsids form filamentous cylinders. Capsids can consist of one or more different proteins. The filamentous capsid of bacteriophage M13, which

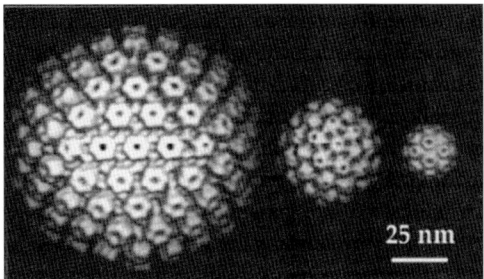

Figure 1.18 Image reconstructions of spherical (icosahedral) viruses: equine herpes virus, cauliflower mosaic virus cowpea, and chlorotic mottle virus (left to right). (Reproduced with permission from Douglas and Young (1999). Copyright 1999, Wiley-VCH Verlag GmbH.)

is often used in bioengineering, is a flexible rod of about 1 μm length. It consists of five different proteins: the major protein gp8 with about 2700 copies per phage, 5 copies of the minor proteins gp3 and gp6 that are placed at one end of the rod, and 5 copies of the proteins gp7 and gp9 at the other end (Figure 1.19).

The tobacco mosaic virus (TMV) is a very stable filamentous virus. The capsid is 300 nm long with 18 nm outer and 4 nm inner diameter (see Figure 1.20). A RNA is

p3   p6   p7   p8   p9

Figure 1.19 Structure of the filamentous M13 phage.

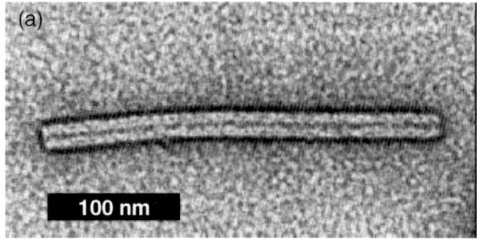
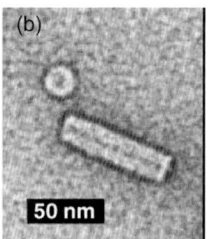

Figure 1.20 Transmission electron microscopy (TEM) images of TMV at 400 kV. (a) Intact virus, stained with uranyl acetate. The central channel and the exterior surface appear black, that is, uranium-rich. (b) Virus fragments: the circular shape is a very short fragment standing upright on the substrate. (Reproduced with permission from Knez *et al.* (2004). Copyright 2004, Wiley-VCH Verlag GmbH.)

surrounded by a helical assembly of 2130 coat proteins. The TMV exhibits a surprising stability at temperatures up to 90 °C and pH values from 3.5 up to 9. These properties make TMV a useful template in bionanotechnology.

Some virus capsids can undergo structural transitions induced by chemical switching, a property of interest for potential applications in bionanotechnology. The plant virus cowpea chlorotic mottle virus (CCMV), for example, shows a pH-dependent structural change (Douglas and Young, 1999). Its capsid is composed of 180 identical coat protein subunits assembled to an empty cage with an outer diameter of 26 nm and an inner diameter of about 20 nm. CCMV shows a reversible swelling of 10% at pH >6.5. This increase in diameter is connected with the formation of 60 pores in the protein shell (diameter 2 nm). At pH <6.5, the pores are closed. Thus, the reversible shape transition enables a pH-dependent switching of entrapment and release of substances.

The so-called bacteriophages (or phages), members of the group of viruses that infect bacteria, are of major interest in nanotechnology. The above-mentioned M13 virus and tobacco mosaic virus are examples. Bacteriophages are assumed to be the most abundant entities in the biosphere. About 10^9 phages per milliliter are found at the surface of biofilms. The fast replication cycles of phages in the bacterial host cells are an essential advantage for potential nanotechnological applications. The virus attaches to receptors on the surface of the bacterium and injects genetic material through the cell wall, whereupon the bacterium interrupts its normal synthesis processes and churns out virus components. Within a few minutes, new phages are assembled and released via cell lysis or budding. Filamentous phages can also be released continuously. These are rapid processes that enable phage libraries with typically 10^9 or more phages to be set up rather quickly (see also Section 2.3).

1.1.3
Carbohydrates

The term "carbohydrates" indicates that this large group of organic compounds can be formally regarded as hydrates of carbon, as they seem to be mainly composed of a

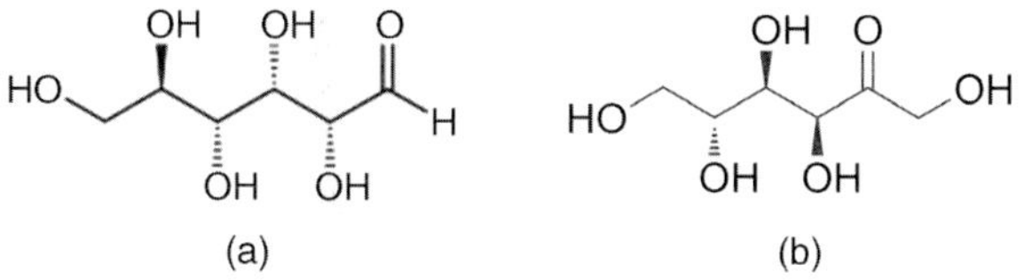

Figure 1.21 Chain structure of D-glucose (a) and D-fructose (b).

backbone of carbon to which hydroxyl groups and hydrogen atoms are attached. Carbohydrates such as cellulose, sugar, or starch are the biomolecules with largest abundance on Earth. The photosynthesis is the unlimited process for production of cellulose in plant cells. Sugar and starch are important energy deposits in biological structures. According to the number of combined monosaccharide units, carbohydrates are usually subdivided into the following: monosaccharides, disaccharides, oligosaccharides, polysaccharides, and glycoconjugates.

Monosaccharides have an unbranched backbone. The backbone can be an open chain or a cyclic structure. They exhibit two or more hydroxyl groups. In the open chain, one of the carbons is double bonded to one oxygen atom to form a carbonyl group. This carbonyl group can be either at the end of the open chain (*aldose*) or at any other position of the chain (*ketose*). Hydroxyl groups are attached to all other carbon atoms. The aldohexose D-glucose and the ketohexose D-fructose are the most frequent natural monosaccharides (Figure 1.21). The carbonyl group can be oxidized to a carboxyl group, which implies that glucose and other sugars of similar structure are reducing molecules. The carbonyl group can form a covalent bond with a hydroxyl group further down the chain resulting in a cyclic structure (Figure 1.22). In aqueous solution, saccharides with four or more carbon atoms form preferentially cyclic structures.

Disaccharides (e.g., maltose, lactose, and sucrose) consist of two monosaccharides covalently joined by an O-glycosidic bond. This bond can be hydrolyzed under acidic conditions (Figure 1.23).

Oligosaccharides are short chains of n monosaccharides ($3 \leq n \leq 20$) joined by glycosidic bonds. Oligosaccharides are often joined to lipids or proteins forming the so-called glycoconjugates. *Polysaccharides* containing only one carbohydrate species are called *homopolysaccharides*, while *heteropolysaccharides* are composed of different carbohydrates. Examples of homopolysaccharides are starch (Figure 1.24) and glycogen (animal starch), the storage forms of monosaccharides,

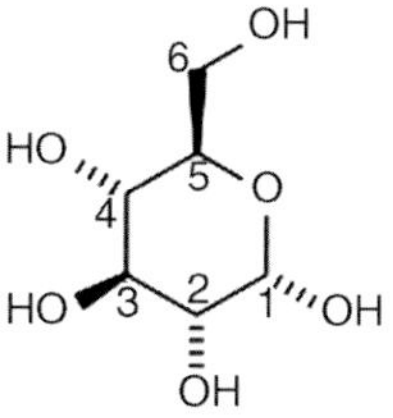

Figure 1.22 Cyclic structure of D-glucose.

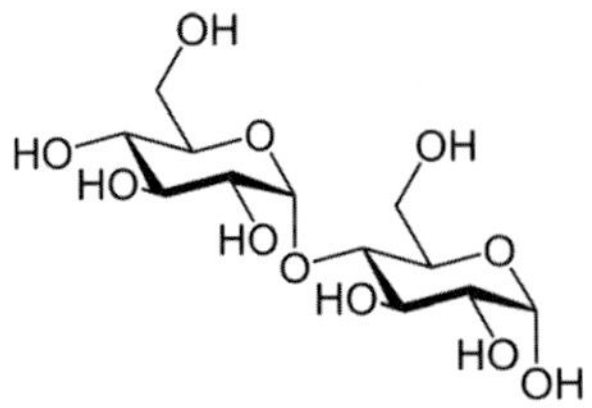

Figure 1.23 Structure of the disaccharide maltose, which consists of two D-glucose units.

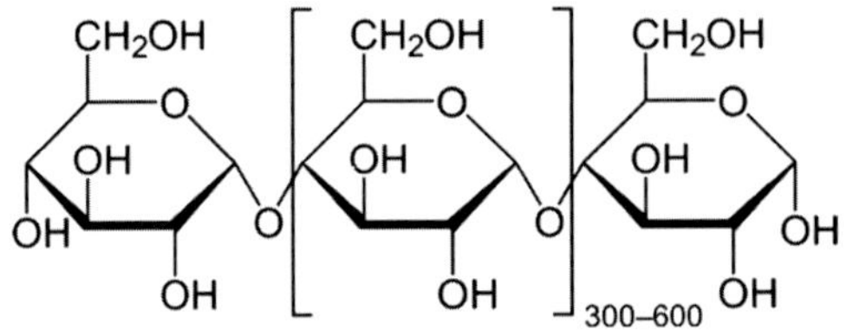

Figure 1.24 Starch or amylose consists of a large number of α-D-glucose units joined together by glycosidic bonds. It is produced by green plants as energy store.

and structure-forming molecules such as cellulose (Figure 1.25), chitin, and chitosan (Figure 1.26). Examples for structural heteropolysaccharides that are of interest for the development of bio-inspired materials are hyaluronic acid (or hyaluronan) and alginate (Figures 1.27 and 1.28).

Glycoconjugates are hybrid structures of polysaccharides and proteins or lipids (see Table 1.4 and Figure 1.29). They are signal carriers for the intramolecular assembling of biomolecules. Usually, oligosaccharides are covalently linked with proteins or lipids to form glycoconjugates, which are subdivided into proteoglycans, glycoproteins, and glycolipids.

Oligo- and polysaccharides can be important information carriers, for example, in the communication between cells and the surrounding media. Examples are recognition sites on the cell surfaces. Often a thick layer of oligosaccharides, the glycocalyx, covers the plasma membrane of eukaryotic cells. This layer contains molecular information inscribed in the carbohydrates. Oligosaccharides and polysaccharides offer multiple binding sites for amino acids and

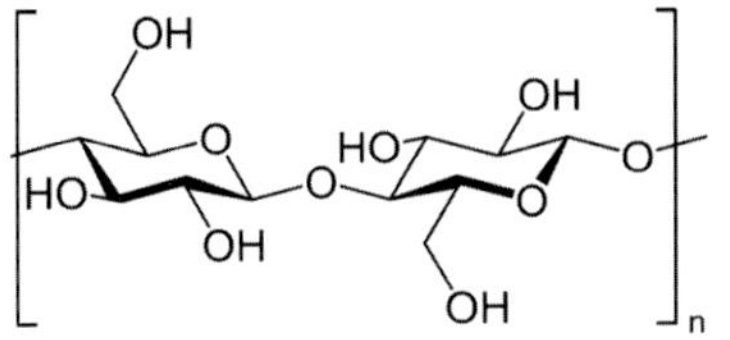

Figure 1.25 Cellulose consisting of a linear chain of several hundred to more than 10 000 β-D-glucose units. Cellulose is the structural component of the primary cell wall of green plants, many forms of algae, and oomycetes. Some bacteria secrete it to form biofilms.

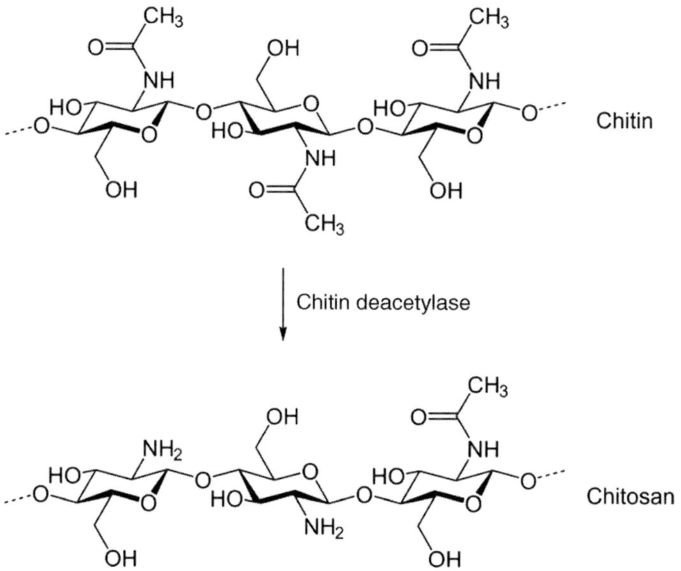

Chitin

Chitin deacetylase

Chitosan

Figure 1.26 Chitin and chitosan. Chitin, a chain polymer of *N*-acetylglucosamine, is the main component of cell walls of fungi, the exoskeletons of arthropods (crustaceans and insects), or the radula of mollusks. Chitosan is produced by deacetylation of chitin. It is composed of randomly distributed glucosamine (deacetylated unit) and *N*-acetylglucosamine (acetylated unit). The biocompatible chitosan is an interesting scaffold material for tissue engineering.

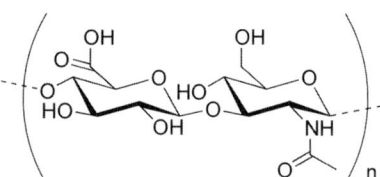

Figure 1.27 Hyaluronic acid, an anionic glycosaminoglycan, is composed of D-glucuronic acid and D-*N*-acetylglucosamine. It is one of the main components of the extracellular matrix distributed widely throughout connective, epithelial, and neural tissues.

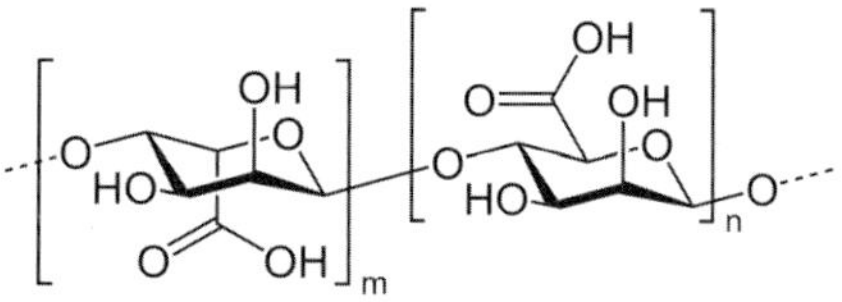

Figure 1.28 Alginate, an anionic polysaccharide, is widely distributed in the cell walls of brown algae. It is a linear copolymer with homopolymeric blocks of β-D-mannuronic acid (right block in the figure) and α-L-guluronic acid residues (left block in the figure).

Table 1.4 Subgroups of glycoconjugates.

Glycoconjugates	Structures
Proteoglycans	The basic unit is a core protein. One or more glycosaminoglycan chains are covalently joined to membrane proteins or proteins of the extracellular matrix. The glycosaminoglycan chain can bind via electrostatic interaction to extracellular proteins. Proteoglycans organize the extracellular matrix or are integrated in the cellular membrane.
Glycoproteins	One or more oligosaccharides (glycans) are covalently joined to a protein outside or inside cells. The glycans are smaller and more branched than the glycosaminoglycans of proteoglycans. By the binding of negatively charged oligosaccharides, the polarity and solubility of the proteins can be changed. This can lead to rod-like structures. Furthermore, it can protect the proteins from the attack of proteolytic enzymes. Glycoproteins form highly specific recognition sites, for example, for lectins.
Glycolipids	The head groups of some membrane lipids can be complex oligosaccharides. Thus, the outer membrane of Gram-negative bacteria such as *E. coli* and *Salmonella typhimurium* is composed of lipopolysaccharides (Figure 1.29). These structures are targets of antibodies produced by the immune system in response to bacterial infection.

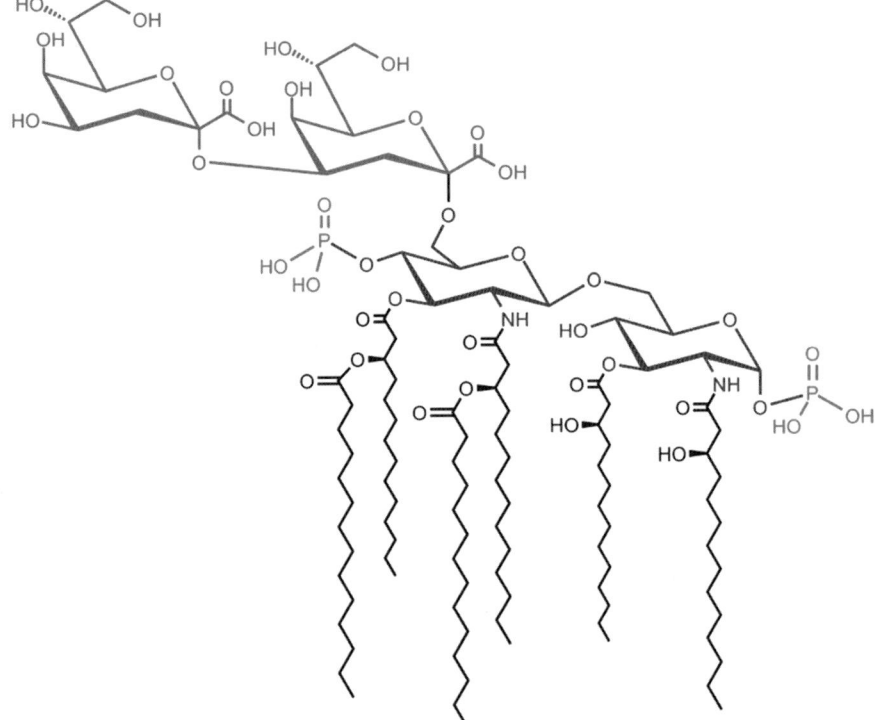

Figure 1.29 Structures of a bacterial lipopolysaccharide, the (3-deoxy-D-manno-octulosonic acid)$_2$–lipid A endotoxin from *E. coli* K-12. (*Source*: Wikimedia Commons; author Tim Vickers.)

nucleotides. To get an impression of the enormous wealth of information, we can make a rough estimate of the number of combinations of a hexameric oligosaccharide with the various amino acids or nucleotides. For conjugations with the 20 natural amino acids, we get 20^6 possibilities, whereas there are 4^6 possibilities for linkages with nucleotides. Despite stereochemical restrictions that will reduce these numbers, the estimate demonstrates the enormous information density that could be realized in glycoconjugates. Lectins, a ubiquitous group of proteins, bind with high specificity to carbohydrates. Lectins are found in plants (e.g., concanavalin A), animals (e.g., galectin-1 and mannose binding protein A), viruses (e.g., hemagglutinin), and bacteria (enterotoxin). It is reasonable to expect that further progress in glycobiology will significantly increase the relevance of the *sugar code* for application in bionanotechnology.

1.1.4
Lipids

Lipids are a large group of insoluble biomolecules with diverse functions. The insolubility in water is the group defining property. Depending on the functions, three subgroups of lipids are distinguished: storage lipids (fats and oils) acting as energy reservoirs, structural lipids (mainly phospholipids and sterols) constituting the major structural components of membranes, and a larger group of functional lipids such as cofactors, electron carriers, light absorbing pigments, chaperones, and emulsifying agents. Compared to nucleic acids or proteins, lipids are small molecules. With a charged head group and a carbon–hydrogen tail they are amphiphilic, which means they combine the hydrophilicity of the polar head and hydrophobicity of carbon–hydrogen tail in one molecule. As we will discuss in Chapter 7, this combination provides the ability of forming self-organized structures of a particular type. The chemical diversity of the head groups permits the realization of a large variety of different functions in biological structures. Several structural characteristics of the three subgroups are compiled in Table 1.5 (see Figures 1.30–1.32).

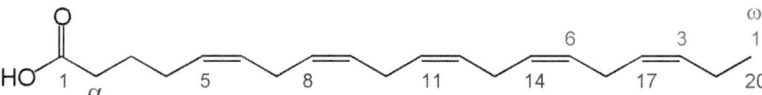

Figure 1.30 Example of a polyunsaturated fatty acid: 20:5 ($\Delta^{5,8,11,14,17}$) eicosapentaenoic acid (EPA), an omega-3 fatty acid. Two nomenclatures are used for fatty acids. In the standard version, the number 1 denotes the carboxyl carbon (C-1), and α the carbon next to it. The positions of double bonds are assigned by Δ followed by numbers indicating the lower numbered carbons in the double bonds. In an alternative version, the carbons are numbered in the opposite direction: number 1 is the methyl carbon at the end of the chain (ω carbon). The positions of double bonds are indicated relative to the ω carbon.

Table 1.5 Classification and structural characteristics of lipids.

Lipids	Structures
Fatty acids	Fatty acids are carboxylic acids with hydrocarbon chains with saturated (single) and unsaturated (double) carbon bonds (number of carbons in the chain n_{Carbon} from 4 to 36). The fatty acids exhibit a very low oxidation state. Therefore, they are particularly suited for energy storage. The melting point is higher for longer hydrocarbon chains and smaller number of unsaturated carbon bonds (unsaturated fatty acids). Example: see Figure 1.30
Structural lipids	Double layers of structural lipids constitute the main structural element of biological membranes. Their general structure with a polar head group (hydrophilic) and hydrophobic tails (fatty acids) explains the amphiphatic character of membrane lipids. Depending on the head groups, five general types of membrane lipids are distinguished: • glycerophospholipids • galactolipids and sulfolipids • archaeal tetraether lipids • sphingolipids • sterols. Example: see Figure 1.31
Functional lipids	Functional lipids are essential components of the cell metabolism. They can act as signal molecules (e.g., hormones and growth factors), enzyme cofactors (e.g., for electron-transfer reactions in chloroplasts and mitochondria), light absorbing pigments, or volatile lipids in plant communication. Example: see Figure 1.32

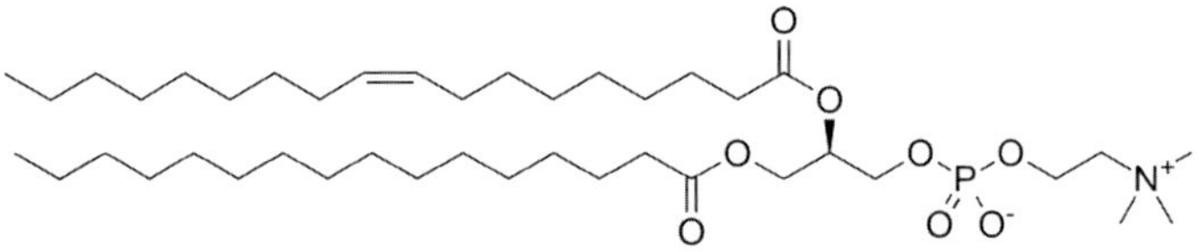

Figure 1.31 Chemical structure of the glycerophospholipid 1-palmitoyl-2-oleoylphosphatidylcholine (POPC).

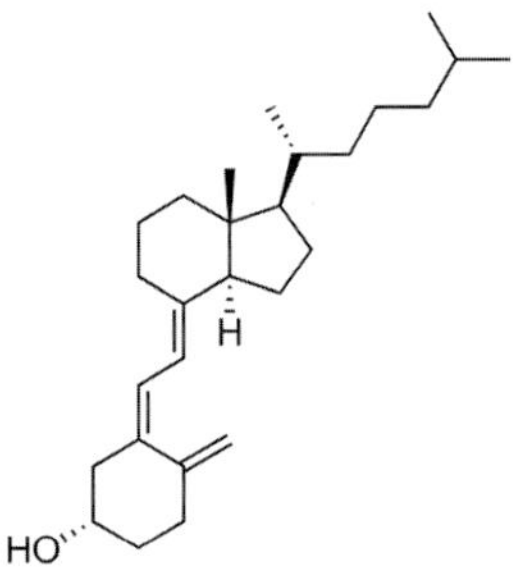

Figure 1.32 Structure of cholecalciferol (vitamin D$_3$). This hormone regulates the Ca^{2+} metabolism in kidney, intestine, and bone.

1.2
Basic Principles

In this chapter, a number of principles is explained which are the reasons for more general structure–property relations observed with biomolecules. They are also important for the understanding of the following chapters. While some properties result from the polymer structure, others reflect unique features directly connected with the particular biochemical properties of the molecules.

1.2.1
The Persistence Lengths of Biopolymer Chains

It is a typical structural characteristic of many biomolecules that their basic chemical entities are arranged in a chain structure. Obviously, the chain is the geometrically best solution to inscribe the information needed for a specific function in a molecular structure. In addition, the arrangement of the chain in the space offers ideal options for controlling readout processes of the information. Therefore, a more detailed geometrical characterization of deformed biomolecular chains is of high interest. Phenomenologically, a deformed molecular chain can be described in a continuum model. Let us consider a long thin biopolymer such as a double-stranded DNA or an actin filament, one of the main components of the cytoskeleton. To describe a point at the nondeformed chain, we introduce the distance s from one end along the chain. When the chain is deformed, this point is shifted to the position with the distance $s' = s + \Delta s$ from the chain end, where Δs denotes the displacement of the position s. The complete biopolymer can be described as a chain of small elastic rods of length ds with a tangential vector $\vec{t}(s)$ (see Figure 1.33). Due to the deformation, the length of the small elastic rod is changed to $ds' = ds + d\Delta s \equiv ds(1 + u(s))$. Here $u(s) = d\Delta s/ds$ denotes the tensile

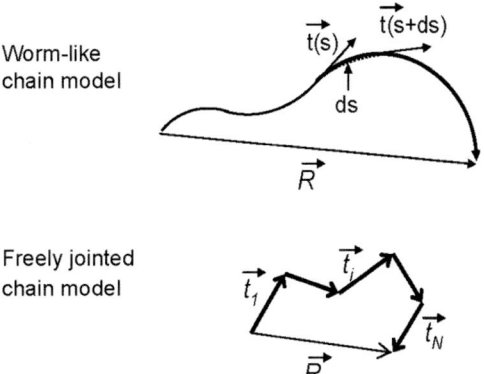

Worm-like chain model

Freely jointed chain model

Figure 1.33 Description of a bent polymer chain in the worm-like chain model and in the freely jointed chain model. $\vec{R}$ denotes a vector pointing from the initial to the end point of the chain.

strain of the chain at the position s. For simplicity, we assume that the non-deformed polymer is a straight rod of length L. The change of the contour length ΔL of the deformed chain now can be expressed by $\Delta L = \int_0^L d\Delta s = \int_0^L u(s)ds$. For a complete description of the deformation of the chain, we have to introduce two more deformations: the curvature or bending deformation $\vec{\beta}(s) = d\vec{t}/ds$, where $\vec{t}(s)$ is the unit tangent vector changing along the chain, and the torsional deformation or twist density $\omega(s) = d\Phi/ds$, with $\Phi(s)$ as the twist angle. The total twist $\Delta\Phi$ is given by $\Delta\Phi = \int_0^{L_{tot}} \omega(s)ds$. If all deformations are elastic, the energy $dW_{el}(s)$ stored in the segment ds can be approximated by a quadratic dependence on the deformations $\vec{\beta}$, u, and ω:

$$dW_{el} = \frac{1}{2}k_B T(A\vec{\beta}^2 + Bu^2 + C\omega^2 + Du\omega)ds. \tag{1.1}$$

As the energy is a scalar, mixed terms including the vector of the bending deformation do not contribute to the energy. We scale the energy coefficients in units of the thermal vibration energy per degree of freedom, $(1/2)k_B T$, with k_B as the Boltzmann constant. $k_B T A$, $k_B T B$, and $k_B T C$ are denoted as bending stiffness, stretch stiffness, and twist stiffness of the chain, respectively, whereas $k_B T D$ is called twist–stretch coupling. A dimensional analysis shows that A and C have the dimension of a length $[L]$, B has the dimension $[L^{-1}]$, and D is dimensionless. A and C are termed as *bend persistence length* and *twist persistence length*, respectively. A stiff biomolecule is characterized by large persistence lengths. In the case of nearly free rotation of the monomers in the chain, the contributions of twist to the energy can be neglected so that C and D can be set equal to zero. In the often realized case where the chain is easily bent by applied forces f so small that the related strain u is negligible, the constant B, too, can be set equal to zero (see also Task 1.1). Thus, the model is reduced to a one-parameter description of the deformation, the so-called *simplified elastic rod model*, among the polymer community also known as the *worm-like chain model* (WLC) or *Kratky–Porod model*:

$$W_{el} = \frac{1}{2}k_B T \int_0^{L_{tot}} A\vec{\beta}^2 \, ds. \tag{1.2}$$

Let us consider a short straight molecule of length l that has been bent with a radius of curvature R (Figure 1.34).

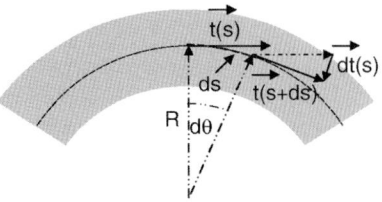

Figure 1.34 Definition of geometric parameters describing a bent polymer chain. R denotes the local radius of curvature.

Table 1.6 Examples of bend persistence lengths of biopolymers.

Biopolymer	Filament radius	Bend persistence length
Double-stranded DNA	$r = 1$ nm	50 nm
Amyloid protofibril		
With 100% ordered β-sheets	$r = 0.7–1.8$ nm	4 µm
With 50% ordered β-sheets	$r = 1.8–2.5$ nm	0.1 µm
Actin filament	Ellipse: $r_{max} = 3$ nm; $r_{min} = 2$ nm	15 µm
Microtubule	Tube: $r_{out} = 14.2$ nm; $r_{in} = 11.5$ nm	6 mm

From Eq. (1.2) for the elastic energy of a thin bent fiber follows $W_{el} = (1/2)k_B T (Al/R^2)$ (see also Task 1.2). Bending into a quarter of a circle, for example, which means $R = 2l/\pi$, requires the energy

$$W_{el}(\pi/2) = \left(\frac{\pi A}{4R}\right) k_B T. \tag{1.3}$$

Obviously, the ratio of bend persistence length, A, and radius of curvature, R, governs the elastic energy in a bent molecule. For small A/R, which means thermal energy fluctuations exceeding the elastic energy required for bending, the average conformation of the polymer is governed by entropy. Then the orientations of distant segments on the chain are not correlated but random. Indeed, biomolecules with lengths $L \gg A$ appear as a random tangle. In other words, the persistence length A is a measure for the length of the statistically independent segments, L_{seg}. This result leads to a second simplified model, in which the overall chain shape is modeled as a freely jointed chain of segments. In this so-called freely jointed chain (FJC) model, the segment length L_{seg} is a phenomenological parameter in the size range of the bend persistence length (for more details see the following section). Several examples of bend persistence lengths are given in Table 1.6. In comparison to these values, common polymers such as polyethylene can be described with persistence lengths of less than 1 nm. Therefore, the stiffer biopolymers such as DNA or fibrous proteins of the cytoskeleton are also called *semiflexible polymers*. The high persistence length of these biopolymers is one essential advantage for their application as building blocks for manufacturing microscopic machine systems or as templates for electronic circuitries.

The bend persistence length provides also a measure for Young's modulus E of the polymer chain. When we describe the polymer chain as a continuous isotropic elastic material, the bend persistence length can be expressed as

$$A = \frac{EI}{k_B T}. \tag{1.4}$$

As derived in Appendix B, here I is the second moment of inertia of the cross section,

$$I = \iint\limits_{\text{cross section}} y^2 \, dx dy, \tag{1.5}$$

where y is the distance perpendicular to the neutral axis and $dxdy$ is an area element of the cross section. For a cylindrical wire with radius r, we get $I_{cylinder} = (\pi/4)r^4$ (for other shapes see solution of Task 1.3). Equation (1.4) shows that the bend persistence length decreases with decreasing fiber radius and increasing temperature.

1.2.2
Equilibrium Shape of a Semiflexible Polymer Chain

Let us now address in more detail the problem of the equilibrium shape of a biopolymer chain under the influence of thermal fluctuations in a solution. We are interested in a mesoscopic description of the polymer geometry. The contour length L should be much larger than the bend persistence length. Thus, we can idealize the polymer as a chain of perfectly straight segments of length L_{seg} connected by perfectly free links (FJC model). The segments, whose number is L/L_{seg}, are assumed to be randomly arranged along the chain. In the freely jointed chain model, the orientation of the segment i does not depend on the orientation of segments $i-1$ and $i+1$. As expected, like the path of a random walker, the polymer shows a coiled equilibrium shape for large numbers of segments. The diameter of the coil can be defined as the average end-to-end distance of the chain (see Figure 1.33).

For the FJC model, we can express the mean square value of that distance as

$$\langle \vec{R}\cdot\vec{R}\rangle = \sum_{i,j=1}^{N}\langle L_{seg}\vec{t}_i \cdot L_{seg}\vec{t}_j\rangle = L_{seg}^2\left(\sum_{i=1}^{N}\vec{t}_i^2 + 2\sum_{i<j=1}^{N}\vec{t}_i\cdot\vec{t}_j\right) = N(L_{seg}^2)$$

$$= L_{seg}L_{tot}. \tag{1.6}$$

The coil diameter of the semiflexible biopolymer D_{coil} scales with the square root of its length:

$$D_{coil} \approx L_{seg}^{1/2}L_{tot}^{1/2}. \tag{1.7}$$

The value for the segment length can be related to the bend persistence length by comparing the result of the FJC model with calculations using the WLC model. For the worm-like chain model, the mean square distance is given as

$$\langle \vec{R}\cdot\vec{R}\rangle = \left\langle\int_0^{L_{tot}}\vec{t}(s)ds\int_0^{L_{tot}}\vec{t}(s')ds'\right\rangle = \int_0^{L_{tot}}ds\int_0^{L_{tot}}ds'\langle\vec{t}(s)\vec{t}(s')\rangle, \tag{1.8}$$

where the correlation function $\langle\vec{t}(s_1)\vec{t}(s_2)\rangle \equiv f(s_1,s_2)$ is governed by the elastic interaction of neighboring segments. For positions well away from the ends of the chain, the orientation correlation function depends only on the distance along the contour of the chain, $\Delta s_{12} = |s_1 - s_2|$; hence, $\langle\vec{t}(s_1)\vec{t}(s_2)\rangle = f(\Delta s_{12})$. The function $f(\Delta s_{12})$ has to fulfill two conditions describing the strong correlation for short distance, $f(\Delta s_{12}) \to 1$ with $\Delta s_{12} \to 0$, and the missing of any correlation for very large distances, $f(\Delta s_{12}) \to 0$ with $\Delta s_{12} \to \infty$. The orientation changes along differ- ent segments of the polymer chain due to thermal fluctuations are statistically independent. When we consider a sequence of three points on the chain, 1, 2, and 3, with $\Delta s_{13} = \Delta s_{12} + \Delta s_{23}$, then the free enthalpy change connected with the orienta- tion change along the segment, Δs_{12}, is not involved in the free enthalpy change of the following segment. This means that the thermal average of the orientation correlation can be written as

$$f(\Delta s_{1s} + \Delta s_{23}) = f(\Delta s_{12})f(\Delta s_{23}). \tag{1.9}$$

The only function that fulfils this relation is $f(\Delta s) = \exp(q\Delta s)$. The unknown parameter q can be determined evaluating the elastic energy of a chain segment with help of the principle of equipartition of energy in thermal equilibrium (Nelson, 2008). It leads to an exponential decay of the correlation function

$$\langle \vec{t}(s)\vec{t}(0)\rangle_{\text{WLC}} = \exp(-|s|/A), \tag{1.10}$$

where the decay length is given by the persistence length A.

Evaluation of the integral in Eq. (1.8) now gives for large polymer length $L_{\text{tot}}/A \gg 1$:

$$\langle \vec{R} \cdot \vec{R} \rangle = \int_0^{L_{\text{tot}}} ds \int_0^{L_{\text{tot}}} ds' \exp(-|s - s'|/A) = 2 \int_0^{L_{\text{tot}}} ds \int_s^{L_{\text{tot}}} ds' \exp(-(s' - s)/A)$$

$$\cong 2AL_{\text{tot}}. \tag{1.11}$$

This means, the diameter D_{coil} of the coiled molecule can be expressed as

$$D_{\text{coil}} \approx (2A)^{1/2} L_{\text{tot}}^{1/2}. \tag{1.12}$$

By comparing the results of the freely jointed chain model and the worm-like chain model, it appears that they coincide if the effective segment length is chosen twice the bending persistence length: $L_{\text{seg}} = 2A$.

1.2.3
The Load–Extension Diagram of a Semiflexible Polymer Chain

Basic Phenomena As explained in the previous section, the equilibrium shape of a long semiflexible biopolymer chain is a random coil. Therefore, untangling a DNA coil by stretching it with the aim to deposit it on a substrate as a long straight molecule would require an external force, f, pulling at the ends. One has to overcome the elastic entropic force caused by the thermal motion of the molecule. Force–displacement measurements on individual DNA molecules can be measured by means of subtle experimental setups (Smith *et al.*, 1992; Cluzel *et al.*, 1996; Clausen-Schaumann *et al.*, 2000; Bockelmann *et al.*, 2002; Günther *et al.*, 2010). Soft microneedles, atomic force microscopes (AFMs), and optical and magnetic tweezers have been used for this purpose. In optical or magnetic tweezers, one end of the molecule is bound at a dielectric or a superparamagnetic microbead, whereas the other end is fixed on a substrate. The load needed to characterize the deformation behavior of a single DNA covers a wide range, from the sub-pN regime of entropic forces acting in coiled molecules up to the rupture force of overstretched molecules in the range of about 600–800 pN. A high stiffness of loading device (relatively to the base pair stiffness) and high load resolution are needed. For instance, in measurements of unzipping a single double-stranded DNA molecule, the molecular stiffness is in the range of $k_{dsDNA} = 0.01$ pN/nm. The stiffness of soft microneedles is typically in the range of $k_{microneedle} = 0.001 - 0.002$ pN/nm, whereas in an optical trap for silica beads of 1 μm diameter in H_2O with a laser power $P = 700$ mW, a stiffness $k_{trap} = 0.25$ pN/nm has been measured (Bockelmann *et al.*, 2002). Spring constants of atomic force microscopes are about of 1000–10 000 times higher than those of optical tweezers. This means the compliances of optical traps or AFMs are almost negligible compared to the molecular compliance.

In an optical tweezers system, the force caused by the displacement of the dielectric bead in a highly localized electric field of a focused laser beam increases with increasing field gradient (increasing laser power). Therefore, a broad force range can be realized. The system allows a full 3D manipulation during loading. Compared to magnetic tweezers, one restriction is given by the missing option of torque application. Another limitation follows from possible radiation damage of the biomolecule at higher laser power.

In magnetic tweezers, the superparamagnetic bead is pulled by an external magnetic field. As the loading system can act in parallel on many molecules, parallel single-molecule experiments can be performed. One interesting feature of magnetic tweezers is the magnetic polarization anisotropy of the microspheres. Therefore, by rotation of the external magnetic field with respect to the substrate, in addition to a stretching force also a torque is caused. Thus, in addition to the stretching elasticity, the torsional stiffness of molecules can be measured (Kauert *et al.*, 2011). However, the missing option of a true 3D manipulation causes a limited applicability for magnetic tweezers.

In Figure 1.35, an experimental setup of force–displacement measurement using magnetic tweezers is shown schematically. The loading system allows covering of a large force range with high resolution. One example for a sensitive load–displacement measurement at low force is shown in Figure 1.35a. Günther *et al.* (2010) addressed the question, how much the mechanical and structural properties of DNA are changed by staining with YOYO-1, a widely used fluorescent dye for DNA. In the figure, the influence of the binding of YOYO-1 on the force–extension curve of DNA is shown. From fits of the experimental data to the WLC model, it was concluded that the persistence length of DNA remains constant independent of the amount of bound YOYO-1. Furthermore, the size of YOYO binding sites at the DNA (3.2 ± 0.6 bp/dye) and the elongation of the DNA per intercalated YOYO-1 molecule (1.6 ± 0.4 bp/dye) were derived from the experimental data. Upon binding of one YOYO molecule, the DNA untwisting is found to be $24° \pm 8°$. This is in line with values for other intercalators such as ethidium bromide, for which a value of $26°$ has been reported. These results demonstrate the manifold information that can be derived from mechanical single-molecule experiments for characterization of molecular interactions. In Figure 1.35b, another example is

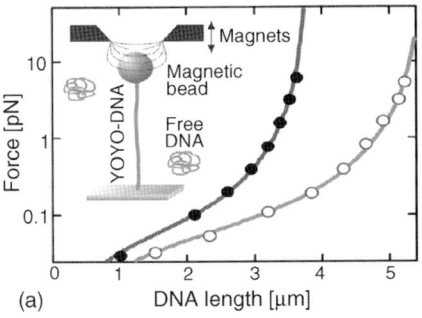

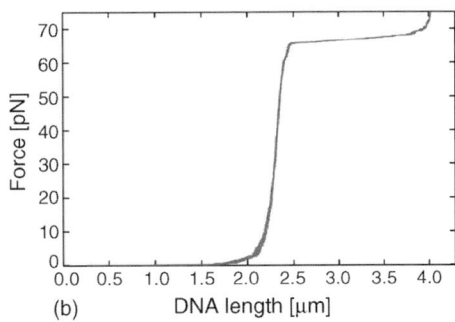

(a)   DNA length [µm]

(b)   DNA length [µm]

Figure 1.35 Force–extension measurements on double-stranded DNA by magnetic tweezers. (a) Force–extension curves of native DNA (filled circles) and YOYO–DNA (empty circles) at 1.4 dye/bp with fits to the WLC model (solid lines) providing persistence lengths of 54 ± 3 and 56 $\pm$ 3 nm and contour lengths of 3.82 ± 0.01 µm and 5.56 ± 0.02 µm, respectively. *Inset*: Sketch of the experimental setup (Günther, Mertig, and Seidel, 2010). (b) Overstretching of a 6.6 kb long DNA. (Courtesy of Daniel Klaue and Ralf Seidel.)

given showing the overstretching of a double-stranded DNA (6.6 kb long) at a high tensile load.

From the above-mentioned extensive studies of the mechanical behavior of single DNA molecules, the general structure of the force–displacement curves of double-stranded DNA has been derived as shown in Figure 1.36. At very low stretching forces of 2–3 pN, the double-stranded DNA is stretched to about 90% of its contour length in the B-form. This range is governed by the entropy change from the random coiled shape to a nearly straight molecule. The higher tensile stiffness of the nearly

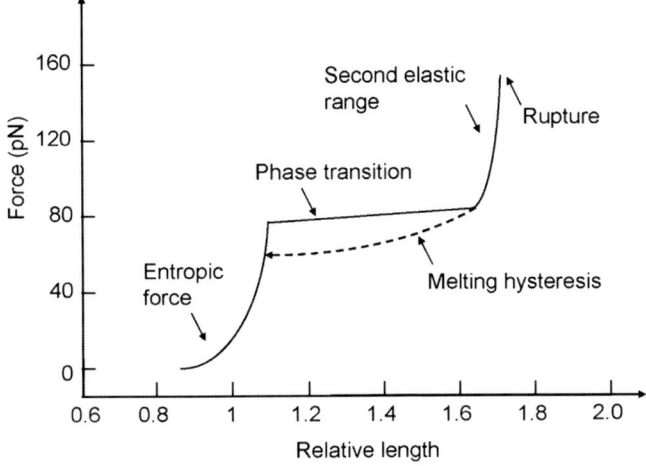

Figure 1.36 Characteristic ranges of the force–displacement diagram of λ-DNA molecules loaded with pulling velocities of about 1–10 µm/s. The displacement is plotted as the relative length normalized to the contour length of 16.3 µm. The first 10 µm of the displacement is not shown because it is indistinguishable from background noise.

straight molecule is represented by the steep slope. At about 65 pN, a remarkable increase of length is observed to about 1.6 of the contour length. Originally, this jump was explained by a reversible transition from the B-form DNA to an S-form DNA. Molecular modeling studies have shown that the DNA can form an over-stretched ladder-like structure, the S-form. The S-form can be stretched elastically until rupture. Additional information can be obtained by recording the response to unloading by reversal of the extension rate. Hysteresis loops, as the one schemati-cally indicated in Figure 1.36, bear witness to the presence of nonelastic components in the stress–strain behavior of the molecule.

A possible explanation of the hysteresis loop has been given by Smith *et al.*, (1996). It has to be assumed that along a double-stranded DNA there can be point-like defects, so-called nicks, at one of the two strands (Figure 1.37). These are over-stressed regions under tensile load. At some critical tension, dsDNA undergoes an overstretching transition connected with partial melting. Recently, Bosaeus *et al.* (2012) have shown by experiments with small 60–64 bp DNA that, if the AT content is high (about 70%), an irreversible rupture including hysteresis is observed. However, GC-rich sequences (about 60%) undergo a reversible overstretching transition into the S-form. Furthermore, the transition depends on salt concentra-tion. Low salt concentration (5 mM NaCl) lowers the stability of dsDNA and favors the irreversible transition. If the overstretched dsDNA remains in low-salt buffer for

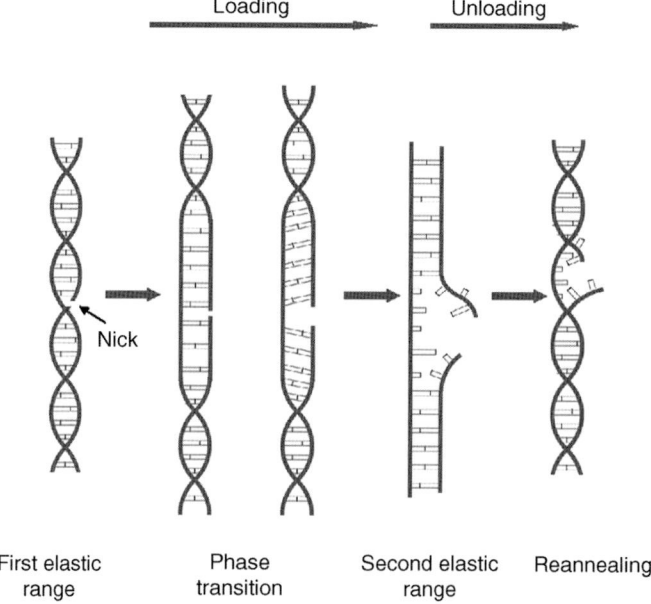

First elastic range   Phase transition   Second elastic range   Reannealing

Figure 1.37 While yielding under tensile load, dsDNA may get into an overstretched state with a "skewed ladder" pattern in the presence of a nick in one of the strands. After unloading, the regions that have remained base paired rapidly contract to B-DNA but the frayed regions take several minutes to reanneal. (Adapted from Smith *et al.* (1996).)

more than a few seconds, then a nicked strand melts and frays back from both sides of the nick. When the tension is released, the regions that have remained base paired rapidly contract to B-DNA but the frayed regions take several minutes to reanneal, perhaps due to secondary structure in the frayed strands. As shown by Clausen-Schaumann *et al.* (2000), melting can also be observed in a defect-free S-DNA. With further stretching of the S-DNA, the double helix melts into single strands. These parallelly arranged strands can be loaded without fracture up to 800 pN. With subsequent unloading, the double helix is restored. Whereas the cooperative B–S transition is not rate-dependent, melting and reannealing are. The observed transition rate of reannealing in the unloaded state is about $(10–20) \times 10^3$ bp/s. The rate-dependence of melting and reannealing reveals itself as a rate-dependent hysteresis.

Basic Models   The extension of a coiled molecule under the action of an external force f can be described in the simplest approximation with a one-dimensional FJC model (Nelson, 2008). In order to extend the force-free model discussed in the previous section, we first consider a one-dimensional version that will be generalized subsequently for 3D. We describe the molecule as a chain of N segments of length $L_{seg} = 2A$. Let an external force f act in the x-direction on a coiled molecule. One end of the chain is fixed at $x = 0$. The mechanical work required for a displacement Δx of the other end of the molecule is $\Delta A_{ext} = f \Delta x$. It contributes to a decrease of the free enthalpy $\Delta G^{\dagger}$ of the molecule with loading system by $\Delta G^{\dagger} = -T\Delta S - \Delta A_{ext}$. Every segment i of the chain can be displaced by thermal fluctuations with a certain probability forward or backward, denoted here by a two-state variable, $\sigma_i = +1$ and $\sigma_i = -1$, respectively. The resulting shift of the free end is given by

$$x = L_{seg} \sum_{i=1}^{N} \sigma_i. \tag{1.13}$$

One set of state variables $\{\sigma_1, \ldots, \sigma_N\}$ describes a particular realization of the stretched chain due to the thermal fluctuations. Its probability in the presence of an external force f is given by

$$p(\sigma_1, \ldots, \sigma_N) = Z^{-1} \exp\left(-\left(-f L_{seg} \sum_{i=1}^{N} \sigma_i\right)/k_B T\right),$$

with

$$Z = \sum_{\sigma_1 = \pm 1} \cdots \sum_{\sigma_N = \pm 1} \exp\left(\left(f L_{seg} \sum_{i=1}^{N} \sigma_i\right)/k_B T\right). \tag{1.14}$$

The mean displacement of the free end is given by the average of Eq. (1.13) over all conformations $\{\sigma_1, \ldots, \sigma_N\}$:

$$\langle x \rangle = L_{seg} \sum_{\sigma_1 = \pm 1} \cdots \sum_{\sigma_N = \pm 1} p(\sigma_1, \ldots, \sigma_N) \sum_{i=1}^{N} \sigma_i. \tag{1.15}$$

As $p(\sigma_1, \ldots, \sigma_N)$ is a product of N identical factors, this expression can be evaluated in a straightforward way (Nelson, 2008), which leads to

$$\langle x \rangle = NL_{\text{seg}} \frac{\exp(fL_{\text{seg}}/k_B T) - \exp(-fL_{\text{seg}}/k_B T)}{\exp(fL_{\text{seg}}/k_B T) + \exp(-fL_{\text{seg}}/k_B T)} = L_{\text{tot}} \tanh(fL_{\text{seg}}/k_B T). \quad (1.16)$$

For small load $fL_{\text{seg}}/k_B T \ll 1$, we get from Eq. (1.16) a linear load–extension relation

$$f = \frac{k_B T}{L_{\text{seg}}} \frac{\langle x \rangle}{L_{\text{tot}}}. \quad (1.17)$$

At low extension, the polymer shows linear elasticity (Hooke's law). The so-called entropic elasticity is characterized by a temperature-dependent effective stiffness $k_B T / L_{\text{seg}}$. For large force or low temperature, the FJC approximation leads to a completely parallel orientation of the single elements:

$$\lim_{f \to \infty} \langle x \rangle = L_{\text{tot}}. \quad (1.18)$$

In the case of large forces where the tensile stiffness $k_B T B$ of the polymer segments becomes relevant, the FJC model of rigid rod segments is no more applicable. The tensile deformation u_{ten} of every element of the chain is $u_{\text{ten}} = f / k_B T B$. This can be simply included in Eq. (1.16):

$$\langle x \rangle = L_{\text{tot}} \left(1 + \frac{f}{k_B T B} \right) \tanh(fL_{\text{seg}}/k_B T), \quad (1.19)$$

where L_{tot} is the contour length of the unloaded polymer. Unlike Eq. (1.16), this equation does not yield a maximum extension but a continuing linear increase.

Equations (1.16) and (1.19) are rough approximations to reality. In order to obtain more realistic theoretical constructs for the overall load–extension relation, the statistic average over all possible orientations of the molecular segments in three dimensions has to be taken into account. The most serious drawback of the FJC model is the assumption of free rotation of the elements of length L_{seg} around the points at which they are assumed to be linked. It is obvious that the shape change of a coiled molecule under load depends also on the bending stiffness of the elements. This has been taken into account by a refined model, the worm-like chain model (Bustamante *et al.*, 1994; Marko and Siggia, 1995). Here, the interaction with adjoining elements has been included by adding the bending energy to the external work. As another modification, the bend persistence length A is substituted for the free parameter L_{seg} of the FJC model (Bustamante *et al.*, 1994). The comparison with experimental data shows that the FJC model describes the elastic behavior of DNA at low extension well (Smith *et al.*, 1992). For larger extension, the tensile deformation has been included in the WLC approach by Odijk (1995). A modified version of the WLC model allows a fairly good description of the load–extension behavior of the double-stranded DNA over the complete range including the elastic entropic force as well as the contributions of the bending energy and the tensile energy (Wang *et al.*, 1997). The corresponding analytical expressions of the mentioned models are compiled in Table 1.7.

Table 1.7 Load–extension models for semiflexible polymer chains.

Model	Formula	Reference
FJC; entropic/tensile strain theory	$\varepsilon = \left(\coth\left(\dfrac{2fA}{k_{B}T}\right) - \dfrac{k_{B}T}{2fA} \right)\left(1 + \dfrac{f}{k_{B}TB} \right)$	Smith *et al.* (1996)
WLC; entropic theory	$f = \left(\dfrac{k_{B}T}{A}\right)\left(\dfrac{1}{4(1-\varepsilon)^{2}} - \dfrac{1}{4} + \varepsilon \right)$	Marko and Siggia (1995)
WLC; entropic/ tensile strain theory	$\varepsilon = 1 - \dfrac{1}{2}\left(\dfrac{k_{B}T}{fA}\right)^{1/2} + \dfrac{f}{k_{B}TB}$	Odijk (1995)
WLC; entropic/ tensile strain theory	$f = \left(\dfrac{k_{B}T}{A}\right)\left(\dfrac{1}{4(1-\varepsilon+f/k_{B}TB)^{2}} - \dfrac{1}{4} + \varepsilon - \dfrac{f}{k_{B}TB} \right)$	Wang *et al.* (1997)

Adapted with permission from Wang *et al.* (1997). Copyright 1997, Elsevier.
$\varepsilon = \langle x \rangle / L_{tot}$ relative extension; $k_{B}TB =$ tensile stiffness.

1.2.4
Cooperativity

Cooperativity occurs in living organisms on all structural levels, beginning with cooperative configuration changes of single macromolecules. The function of hemoglobin already mentioned in Section 1.1.2 as an oxygen binding protein is an example for cooperative binding. Binding of one oxygen at a specific binding site causes the whole tetrameric protein to switch into a well-defined second conformation for binding more oxygens ($n \approx 3$). From this conformation change effected by the binding of one ligand, proteins with such property have got the designation "allosteric proteins," derived from Greek *allos* meaning "other" and *stereos* meaning "shape." Moreover, the affinity of hemoglobin to oxygen can be modulated by binding dissolved carbon dioxide: The related structure change always leads to an accelerated release of oxygen, when it is most needed. Many kinds of allosteric proteins have been formed in evolution to enable cooperative coupling mechanisms as modulating elements of life processes. Another example for cooperativity is the sudden softening of a double-stranded DNA under tensile loading when the load exceeds about 65 pN (Figure 1.36). The overstretching transition from the B-configuration to the S-configuration with an extension by 70% is the result of the cooperative change of the stacking of the base pairs along the DNA chain.

Generally, cooperativity concerns coordinated conformation changes of weakly interacting elements in biopolymers. There are cases where the interacting elements are in close vicinity, but in other cases they are far apart. Cooperative motion of distant subunits is coordinated by signal transmission. However, as we will see in other examples to be discussed later, the structure transitions can be caused also by a change of the temperature or the chemical environment (pH change, for instance). Therefore, it should be possible to describe the phenomenon cooperativity in a more general language. For this purpose, we follow the concept of the famous Ising model of magnetism. Similar to the orientation of magnetic moments in a solid body, we assume that a single structure element i of the biopolymer can occupy discrete energy states $E_i = -\alpha k_B T \sigma_i$. In the previous section, we considered the one-dimensional FJC model with the two states $\sigma_i = \pm 1$. The noninteracting segments L_{seg} can occupy the two energy states $E_i = \pm f L_{seg}$ with their displacement parallel or antiparallel to the direction of the acting external load ($\alpha = f L_{seg}/k_B T$). Cooperativity implies the presence of an interaction energy of adjoining elements:

$$U_{int} = -\gamma k_B T \sum_{i=1}^{N-1} \sigma_i \sigma_{i+1}. \tag{1.20}$$

Hence, adjoining elements with equal orientation, $\sigma_i = \sigma_{i+1}$, contribute a negative amount of energy, $U_{int} = -\gamma k_B T$. Correspondingly, adjoining elements with unequal orientation contribute a positive amount. γ is the so-called cooperativity parameter, which expresses the coupling strength. Extending the FJC model derived above for the average displacement of a stretched one-dimensional chain by adding the interaction energy U_{int} yields a generalized version of Eq. (1.16):

$$\langle x \rangle = N L_{seg} \frac{\sinh \alpha + \left(\sinh \alpha \cosh \alpha / \sqrt{\sinh^2 \alpha + e^{-4\gamma}} \right)}{\cosh \alpha + \sqrt{\sinh^2 \alpha + e^{-4\gamma}}}, \tag{1.21}$$

with $\alpha = f L_{seg}/k_B T$ (Nelson, 2008) (see also Figure 1.38). For small values of the cooperativity parameter, $\gamma \to 0$, Eq. (1.21) reduces to the simple FJC equation (1.16), whereas for strong coupling, $\gamma \to \infty$, we get a step-like jump from $x = 0$ (coiled DNA) to a straight molecule with the load-independent contour length $\langle x \rangle = N L_{seg} = L_{tot}$. The increase of the extension with increasing cooperative interaction can also be seen when we consider the limit case of Eq. (1.21) at low external force:

$$f = \frac{k_B T}{e^{2\gamma} L_{seg}} \frac{\langle x \rangle}{L_{tot}}. \tag{1.22}$$

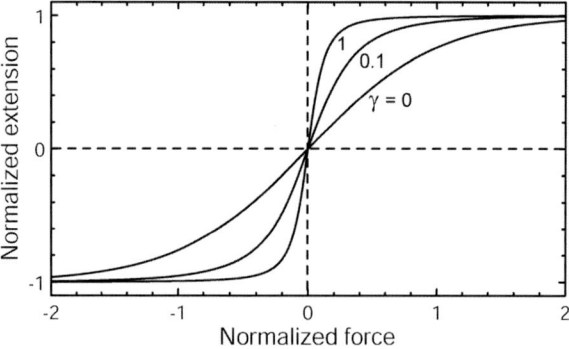

Figure 1.38 Cooperative FJC model. Change of normalized extension $\langle x \rangle / L_{tot}$ in dependence on normalized force $\alpha = f L_{seg}/k_B T$ for various values of the cooperativity parameter γ.

This means the effective "spring constant" of the coil decreases with increasing γ. For larger γ, the transition from the coiled molecule to the straight one is sharper.

1.2.5
Protein Folding

The Levinthal's Paradox and Stable Secondary Structures In order to understand the process of protein folding in more detail, we will reconsider some experimental findings. What is the characteristic timescale to create a functional protein? The time needed to form a functional protein of about 100 amino acids in *E. coli* is typically in the range of 5 s at 37 °C. If we try to explain this process as a random sequential arrangement of all the amino acids of the molecular chain, we come to a paradox first stated by Cyrus Levinthal in 1968. Let us assume that every amino acid residue can realize four different conformations. Then the number of all possible conformations of the protein is $4^{100} \approx 1.6 \times 10^{60}$. This simple rough estimate shows that the real process of protein folding must occur highly parallelly and hierarchically. Obviously, the protein folding cannot be random. Probably in the evolutionary selection process such molecular structures have dominated which developed a kind of cooperativity that favors the formation of stable subunits. From the thermodynamics point of view, folding is a competition between two forces, a driving force due to the decrease of the enthalpy caused by the increasing number of noncovalent bonds (for characteristic interaction energies see also Table 1.8) and a retarding entropic force favoring the randomly coiled chain structure.

The noncovalent weak interactions are larger than the thermal fluctuations by about a factor of 10 (see Table 1.8). Therefore, enthalpy-driven ($\Delta H < 0$) folding of proteins in small stable subunits of a few nanometers in size can occur at room temperature. The retarding contribution of entropy reduction ($T\Delta S < 0$) is about 1–2 kcal/mol for side chains of amino acids. However, this is overcompensated by the many new bonds between the large numbers of amino acids. All together, at room temperature, folding of protein leads to a decrease of the free enthalpy $\Delta G = \Delta H - T\Delta S < 0$. The decrease of the free enthalpy ΔG is in the range of about 4–10 kcal/mol. This value is so large that the folded structure is stabilized with respect to thermal fluctuations at room temperature T_r with $k_B T_r = 0.6$ kcal/mol.

Table 1.8 Interactions that stabilize proteins.

Interaction	Binding energy (kcal/mol)
Covalent bonds	>50
Dispersion forces	<1
Hydrogen bonds	1–7
Electrostatic interaction (low dielectric)	1–6
Hydrophobic interaction	2–3
Average thermal energy (37 °C)	0.6

As we have learned in Section 1.1.2 from the Pauling and Corey arguments, folding of a peptide chain yields *α-helices and β-sheets as stable secondary structures.* The further process is an ongoing compaction of the structure by interaction of the secondary structures. This produces domains of interacting α-helices and β-sheets, which finally leads to the *tertiary structure* of the protein. Up to now, there is no complete theory that can describe the whole process in all details. Folding can be explained only by the interplay of the weak, noncovalent interactions. Whereas the covalent binding along the backbone stabilizes the open-chain structure, noncovalent interactions between the side chains favor a more compact arrangement of the chain. Hydrogen bonds between the main chain groups of the amino acid residues as well as short-range interaction of the side chain groups control the final arrangement of the secondary structure. There are two alternative models explaining main features of the experimental observations. In one of the models, it is supposed that *long-range weak interactions* between the secondary structures direct the formation of the domain structures. Another model favors the assumption that *hydrophobic interactions* cause a spontaneous collapse of the open-chain structure into the so-called *molten globule state*. The existence of hydrophobic regions along the protein chain is the reason why these regions avoid the contact with the aqueous solution in the equilibrium structure. However, the molten globule state cannot be a highly compact state. The side chains of the single amino acid residues are responsible for a self-avoidance of chain segments similar to the self-avoidance of molecules in a real gas described with a finite radius of single atoms. Locally, secondary structures could be formed in the globule (collapsed) state of the protein. It has to be assumed that features of both models can be observed experimentally.

Banavar *et al.* (2004) have shown that the experimentally observed typical ternary folding structures can be simulated already under fairly general assumptions about the protein structure and interaction. They described a homopolymer chain in a continuum approach as a tube of length L and thickness Δ with a range of attractive interaction R. In a Monte Carlo simulation, the thermodynamic equilibrium at zero temperature has been calculated. Elastic bending energy, overall hydrophobic energy, and effective hydrogen bond energy of interacting chain segments have been included in the interaction modeling. A phase diagram of the tubes has been derived, representing the various phases depending on the two parameter ratios Δ/R and L/R (Figure 1.39).

In the limit of large L/R, the model provides two phases. For large Δ/R, a swollen phase with equal weight for all self-avoiding configurations of single tubes is obtained. The possible conformations are essentially equivalent with respect to energy. For large L/R and small Δ/R, a semicrystalline phase with packages of stretched parallel tubes is found. Every stack of tubes is surrounded by six other stacks parallel to each other in a hexagonal array, similar to the nematic phase of liquid crystals (see also Section 7.2.6). For small L/R and small Δ/R, a featureless phase of compact coils is obtained, whereas for thick fibers again the swollen random conformations governed by self-avoiding and elastic energy are favored. Most interesting is the existence of a *marginally compact phase* near the triple point at

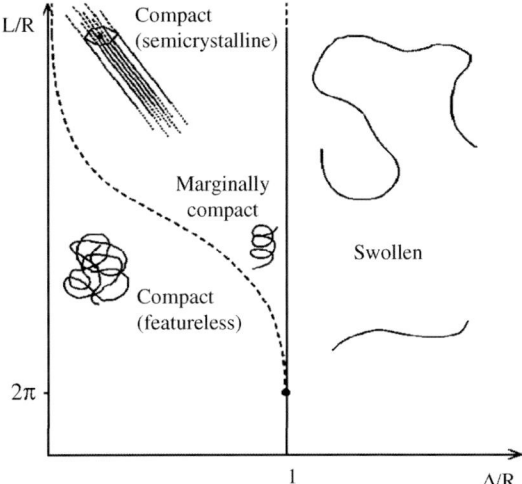

Figure 1.39 Zero-temperature phase diagram of self-attracting tubes depending on the tube length/interaction range ratio L/R and the tube thickness/interaction range ratio Δ/R. (Reprinted with permission from Banavar *et al.* (2004). Copyright 2004, the American Physical Society.)

$\Delta \approx R < L$. In this phase, a *limited number of structures of the discrete homopolymer chains* is obtained: helices, kissing hairpins, regular hairpins, and sheets, with energetic differences among them. The marginally compact phase is sensitive with respect to parameter changes. A slight increase of temperature causes a first-order transition to the featureless compact phase.

Examples of the marginally compact structures are shown in Figure 1.40. These are the results of various modeling runs with a homopolymer of 48 amino acids. The effective range of interaction R was chosen nearly equal to the tube radius, $\Delta/R \approx 1$. Depending on the initially open shape of the chain and the variation of the energy parameters, the simulation led to a limited number of equilibrium configurations.

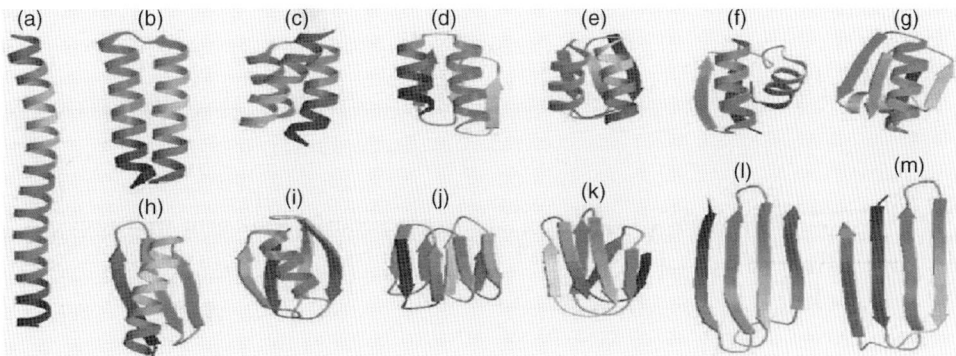

Figure 1.40 A sampling of conformations of a homopolymer modeled as a marginally compact tube of 48 residues. All structures are compositions of helices and strands. (Reprinted with permission from Hoang *et al.* (2006). Copyright 2006, IOP Publishing Ltd.)

This simulation reflects the two characteristic features of marginally compact tube structures. First, they are always composed of stacks of α-helices and β-sheets. Second, the number of different conformations is much smaller than the number of possible random arrangements of the chain. The calculated conformations show a striking similarity to experimentally observed secondary polypeptide structures. The Monte Carlo simulation time yields also reasonable estimates of the real overall folding time.

Universal Features of Protein Folding   The good agreement between the results of model calculations for protein folding and the experimentally observed results leads to the conclusion that the folding of protein chains follows a few rules:

- There are common structural features of globular proteins, the marginally compact structures composed of assemblages of α-helices and β-sheets, as dominating secondary structures.
- The marginally compact structures are phases of minimum free enthalpy of nanometer-sized subunits. The proteins fold reproducibly into the marginally compact structures.
- There exist only a few thousand marginally compact states. The overall tertiary structure of a globular protein shows a pronounced stability with respect to minor changes of the amino acid sequence of the protein chain. This is a favorable precondition for a stable, so-called neutral evolution, which means that the *majority of mutations are not causing essential changes of the phenotype* (Lezon, Banavar, and Maritan, 2006). The formation of life based on the DNA–RNA–protein cycle is relatively stable with respect to mutations of the DNA.
- The folding process proceeds along a series of metastable transition states that are also characterized by assemblages of marginally compact structures of α-helices and β-sheets. The transition can be described in a conformation space as a motion on a *funnel-like energy landscape* along the steep gradient to the minimum of free enthalpy.
- As folding is a cooperative process of local formation of domains of separate α-helices and β-sheets, it proceeds in fractions of a second. Quick and reproducible folding of proteins is a necessary precondition for effective physiological function.
- The tertiary structure of the globular protein composed of well-defined domains of α-helices and β-sheets generates well-defined sites at its periphery for interaction with other proteins and other molecular entities. This is the base for *specificity* as well as *diversity*, two properties essential for the molecular processes of life. Proteins are well suited for molecular *recognition* and *selection*.

Chaperones as an Additional Tool Box for Directed Protein Folding   Chaperones are proteins that interact with only partially folded polypeptides. They assist the folding in the cell in order to direct the folding evolution through the diversity of possible transition states on an optimal path toward the minimum of free enthalpy. They can create microenvironments to facilitate the folding or to correct improperly folded intermediate protein structures. Different kinds of chaperones are available:

- Small containers with hydrophobic interior, called chaperonins, create a temporary hydrophobic environment shielding the protein from contact with interfering cell components and facilitate a well-defined folding.

- The proteins of Hsp70 family (Hsp, heat shock proteins) bind temporarily to regions of unfolded polypeptides that are rich in hydrophobic residues in order to avoid inappropriate aggregation.
- The protein disulfide isomerase (PDI) breaks incorrect cross-links of cysteine amino acids that could fix the folding in an unfavored conformation.
- The peptide prolyl *cis–trans* isomerase (PPI) puts the proline into one of the two possible conformations, which allows a correct fit into the folded protein chain.

The chaperones are essential tools of the self-assembly process, directing and accelerating the whole process of folding.

As discussed above, the helical structure is distinguished among all other protein structures by its particular stability. There is a mechanism called helix–coil transition that facilitates the transition between the two conformations with changing temperature. The helical configuration is stable at low or at high temperatures in water or organic solvents, respectively. The transition can be observed by optical polarization measurements on the translucent protein solution. Circular dichroism (CD) spectroscopy enables sensitive measurements of changes of the secondary structure of proteins (for more details see Appendix C). The chirality of helical molecules affects their interaction with circularly polarized light: The refractive index, n, and the molar extinction coefficient, ε, of solutions of helical polypeptides or proteins differ for left and right circularly polarized light, which is known as circular birefringence (CB) ($n_L \neq n_R$) and circular dichroism ($\varepsilon_L(\lambda) \neq \varepsilon_R(\lambda)$). The latter transforms linearly polarized light into elliptically polarized light. The major axis of the ellipse is tilted by an angle ϕ with respect to the incident linearly polarized light. The *ellipticity* is due to differential absorbance of circularly polarized light by the chiral molecules, whereas the tilt angle ϕ is due to the differential phase velocities (for details see Appendix C). These effects are suitable for monitoring helix–coil transitions of polypeptides and proteins.

As an example, the CD spectra of short alanine-based peptides in aqueous solution at $0\,^{\circ}C$ (Scholtz *et al.*, 1991) are shown in Figure 1.41a. The molar ellipticity $[\theta]$ is a measure for the ellipse axis ratio. Peptides with the generic formula Ac-Y(AEAAKA)$_n$F-NH$_2$ (A: L-alanine; E: L-glutamic acid; K: L-lysine; flanked by an N-terminal acetyltyrsoine and a C-terminal phenylalanyl carboxamide) have been studied. The spectra correspond to chain lengths of 50, 38, 32, 26, 20, and 14 residues, reading from the lower curve at 222 nm to the upper curve. The wavelength region between 210 and 220 nm, which corresponds to the absorption band of peptide bond, yields the main information concerning the length dependence of the helix–coil transition. The measured ellipticity can be converted into fractional helicity of the peptides. This conversion requires the knowledge of $[\theta]_{222}$ for the completely helical and completely coiled configurations (Scholtz *et al.*, 1991). From the observed temperature dependence of the ellipticity, the temperature dependence of the helix–coil transition can be derived as shown in Figure 1.41b. The data show that with increasing peptide length the transition occurs at higher temperature and with a steeper slope. This indicates the increasing cooperativity of the transition with increasing peptide length.

As usual, the driving force for the conformation change is the loss of free enthalpy in the process: $\Delta G = G_{helix} - G_{coil} = \Delta H_{bond} - T\Delta S_{tot}$, where ΔH_{bond}

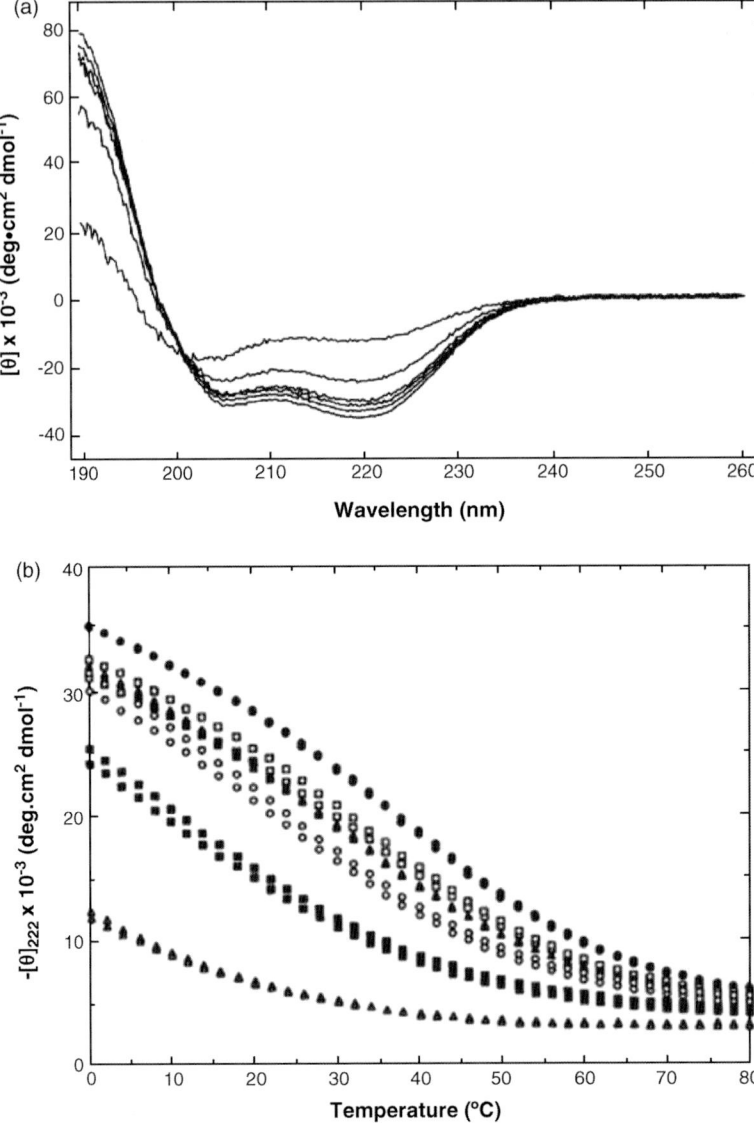

Figure 1.41 α-Helix formation in water versus temperature for short alanine-based peptides. (a) CD spectra of six peptides with chain lengths of 50, 38, 32, 26, 20, and 14 residues, reading from the lower curve to the upper curve, at 222 nm. (b) Thermal unfolding curves of the peptides monitored by CD. The differences between the two curves for every peptide are due to the uncertainty of measurement. Curves are shown for peptides with chain lengths of 50, 38, 32, 26, 20, and 14 residues reading from the upper to the lower curve. (Reproduced with permission from Scholtz *et al.* (1991). Copyright 1991, John Wiley & Sons, Inc.)

and ΔS_{tot} denote the change of the enthalpy and the entropy of the total system consisting of the polypeptide and the solvent. According to Section 1.1.2, an α-helix can be formed by hydrogen bonds between the oxygen atom in the carbonyl group of the ith amino acid and the hydrogen atom in the amide group of the $(i+4)$th amino acid of the polypeptide chain (Figure 1.15). The $(i, i+3)$ spacing of the Glu and Lys residues was selected by Scholtz *et al.* (1991) because side chain interactions are minimal when compared to the $(i, i+4)$ arrangements of Glu and Lys residues. The experiment was meant to be an example for helix–coil transitions associated with the polypeptide backbone and minimal side chain interactions.

From the observation that the helical conformation is stable in aqueous solution at low temperature, we can conclude that $\Delta H_{bond} = H_{helix} - H_{coil} < 0$ should be valid. For the change of the entropy $\Delta S_{tot} = S_{helix} - S_{coil}$, we can assume that there is always a negative contribution of the configuration entropy $\Delta S_{conf} < 0$ due to the limited degree of freedom of the monomers in the α-helix. Furthermore, the degree of freedom of the water molecules is restricted next to the helix–solvent interface. This causes a decrease of the entropy of the surrounding solvent molecules, $\Delta S_{solvent} < 0$, when extended helical structures are formed. Since the total entropy change is negative, $\Delta S_{tot} = \Delta S_{conf} + \Delta S_{solvent} < 0$, there is a transition temperature $T_m = \Delta H_{bond}/\Delta S_{tot}$ beyond which the free enthalpy change ΔG becomes positive. Hence, the coiled configuration is the favored equilibrium structure above this critical temperature.

The driving force for the formation of the helical structure can be expressed by a parameter $\alpha \equiv (\Delta H_{bond} - T\Delta S_{tot})/(-2k_B T)$. Positive α means that the formation of the α-helix is thermodynamically favored. Its growth by one unit leads to a decrease of the free enthalpy by $\Delta G = -2\alpha k_B T$. For the evaluation of experimental data, it is useful to express α by ΔH_{bond} and the transition temperature $T_m \equiv \Delta H_{bond}/\Delta S_{tot}$ (often also called midpoint temperature) with

$$\alpha = \frac{1}{2}\frac{|\Delta H_{bond}|(T_m - T)}{k_B T T_m}. \tag{1.23}$$

The formation of a first bond between the ith and $(i+4)$th amino acid of the polypeptide always leads to some interaction with the neighboring amino acids of the yet coiled positions. This means the cooperativity of the transition has to be included into an extended model. Again we apply the description of the cooperativity, which we have already discussed for the modified FJC model of the load–extension behavior of polymer chains. The binding state of the amino acid at position i is characterized by a state parameter σ_i with $\sigma_i = +1$ for the helix structure and $\sigma_i = -1$ for the coiled structure. For the change of free enthalpy due to formation of a single hydrogen bond, we get $-(2\alpha - 4\gamma)k_B T$, where γ describes the mismatch interaction between neighboring bonded and nonbonded amino acids. A detailed statistical evaluation of the equilibrium distribution yields the average value for the state variable along the polypeptide (Nelson, 2008):

$$\langle \sigma \rangle = \frac{\sinh \alpha}{\sqrt{\sinh^2 \alpha + e^{-4\gamma}}}. \tag{1.24}$$

With increasing cooperativity (large γ), the transition becomes sharper. The comparison with the experimental data shows that the molar ellipticity of the scattered light can be described in good approximation by a linear function of the average state variable, $-[\theta]_{222} = c_1 + c_2\langle\sigma\rangle$.

1.2.6
DNA Melting Transition

As already reported in Section 1.1.1, aqueous solutions of double-stranded DNA are stable at pH 7.0 and room temperature (25 °C). A temperature increase above 80 °C causes the viscosity of the solution to drop sharply and the UV absorption (typically recorded at 260 nm) to increase as a result of a reversible structural transition into two single-stranded molecules, known as melting or denaturation. This reversible transition is very similar to the helix–coil transition of polypeptides. The UV absorption characteristic can be derived from the thermodynamic equilibrium of a cooperative two-stage system as given in Eq. (1.24). Also in this case, weak hydrogen bonds (between the complementary bases) are responsible for the helix stability. The helix is the stable phase at low temperature. When DNA melts, the hydrogen bonds between the bases break. At the same time, the disruption of the base stacks is connected with loss of favorable van der Waals attraction and dipole–dipole interactions. The gain of energy due to reduced Coulomb interaction of the negatively charged backbones is smaller than the energy increase of the above-mentioned interactions. Thus, the total change of the enthalpy is positive, $\Delta H_{bond} = H_{melt} - H_{helix} > 0$, during melting. The overall entropy also increases, $\Delta S_{tot} > 0$. When the two DNA strands are separated, the screening ion clouds around the DNA can relax. The increasing flexibility of the backbones is connected with an increase of the configurational entropy. Furthermore, the unstacked base configurations contribute to this increase. An opposite contribution is caused by exposing hydrophobic regions of the DNA to the surrounding water.

In summary, we get a similar temperature dependence for the free enthalpy change $\Delta G = \Delta H_{bond} - T\Delta S_{tot}$ as for the helix–coil transition of proteins, with the disordered phase at high temperature and the ordered phase at low temperature. Therefore, DNA melting is sometimes called a helix–coil transition. For partially molten DNA, the transition region between double-stranded segments and separated strands is connected with an additional interaction energy that leads to the cooperativity discussed above. This explains the sharpness of the melting transition. A higher content of G–C pairs with three hydrogen bonds in comparison to A–T pairs with two hydrogen bonds leads to a higher melting temperature due to the increase of ΔH_{bond} (Nelson and Cox, 2008). By changing the pH value and the ionic strength, the repulsive electrostatic interaction can be influenced, and therewith also the melting temperature.

1.2.7
Biocatalytic Reactions

Biocatalysts, also called enzymes, mediate the chemical transformations of organic substances inside and outside living cells. The influence of biocatalysis on digestion of food in the stomach has been explored already in the late 1700s. In the 1850s, the fermentation of sugar into alcohol by yeast cells led Louis Pasteur to the assumption that biocatalytic processes are always connected with living cells, a theory known as vitalism. Soon it was found that biocatalytic processes can also occur outside a living cell. (The term *ferment* is restricted to enzymes acting in living organisms.) Many functional proteins and some nucleic acids act as enzymes. Today, enzymes are widely used in food processing, brewing industry, paper industry, textile industry, biofuel industry, drug development, medicine, and molecular biotechnology. The principles that these applications are based on are potentially useful for other novel developments too.

The Reaction Transition State Model In 1946, Linus Pauling formulated a general model of the biocatalytic process based on the concept of *reaction transition states*, first proposed by Michael Polanyi in 1921 and John Haldane in 1930 (Haldane 1930). For simplicity, we consider a reaction where a substrate phase S is transformed by a chemical reaction into a product phase P. The reaction can be described by a reaction diagram.

The driving force for the reaction can be expressed by the difference of the free enthalpy between the S phase and the P phase, ΔG, with

$$\Delta G = \Delta G^0 + RT \ln\left(\frac{[P]}{[S]}\right). \tag{1.25}$$

Here R is the molar gas constant (8.315 J/mol), and [P] and [S] are the concentrations of the product and substrate phases, respectively. For negative $\Delta G^0 < 0$, the product phase is favored in thermodynamic equilibrium $\Delta G = 0$:

$$K_{eq} = \exp\left(-\frac{\Delta G^0}{RT}\right) = \frac{[P]_{eq}}{[S]_{eq}} > 1, \tag{1.26}$$

where K_{eq} is the equilibrium constant of the reaction S $\leftrightarrow$ P. The rate V of any unimolecular reaction S $\rightarrow$ P is determined by the product of rate constant k and substrate concentration $V = k[S]$. Usually a noncatalyzed biochemical reaction is suppressed near room temperature for a long time, as the system has to overcome an activation barrier ΔG_{uncat}. For the rate constant k, the transition state theory yields

$$k = \frac{k_B T}{h} \exp\left(-\frac{\Delta G_{uncat}}{RT}\right), \tag{1.27}$$

where k_B is the Boltzmann constant and h is the Planck constant (see also Task 1.5).

In the presence of an enzyme, the reaction rate is much higher. This must be due to a lower activation barrier in the presence of the enzyme E interacting with the substrate S and the product P. These interactions are included into the formalism by introducing metastable transition states ES and EP (see Figure 1.42), which allows the reaction to be separated into a sequence of thermally activated reactions: E + S $\rightarrow$ ES $\rightarrow$ EP $\rightarrow$ E + P. The binding energies between the enzyme E and the transient species ES and EP are the decisive parameters as they lower the activation barrier by $\Delta G_B = \Delta G_{uncat} - \Delta G_{cat}$. Site-specific catalytic groups of

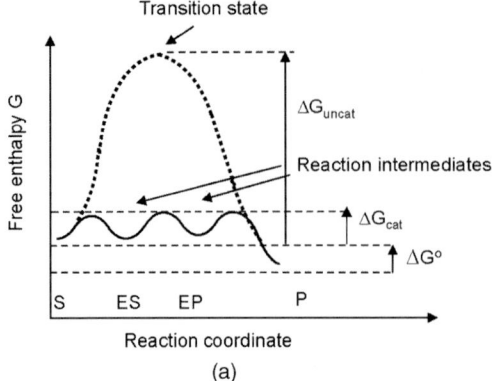

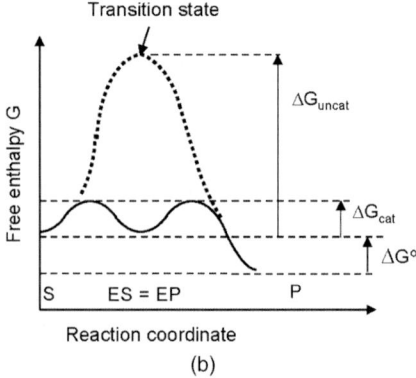

Figure 1.42 Reaction coordinate diagrams. Change of the free enthalpy in an uncatalyzed (dotted line) and catalyzed (full line) reaction in dependence on a reaction coordinate. S and P denote the ground states of the substrate and the product phase, respectively. ES and EP denote less stable intermediate phases of the enzyme-catalyzed reaction (a). Often in a simplified model only one intermediate phase can be assumed to describe a catalytic reaction path (b). This assumption has been used in the Michaelis–Menten model. (Adapted from Nelson and Cox (2008).)

the enzyme, such as amino acid residues, can form transient covalent bonds with the substrate and product phases. In addition, noncovalent bonds, such as hydrogen bonds or hydrophobic and ionic interactions between substrate and enzyme, lower the activation barrier. Some enzymes need additional nonprotein molecules called *cofactors* to be active. These can be inorganic components such as metal ions (e.g., Fe^{2+}, Mg^{2+}, Mn^{2+}, and Zn^{2+}) or *coenzymes*. Coenzymes are small organic groups that transport chemical groups from one enzyme to another. Important examples are NADH (nicotinamide adenine dinucleotide), which provides electrons for reactions in the respiratory chain, and ATP (adenosine triphosphate), which acts as an energy carrier by phosphate ion transfer.

The substrate and the enzyme change their structure during the formation of the intermediate phases ES and EP to realize the energetically optimized reaction path. This differs from the action of inorganic catalysts, and it explains the high efficiency of biocatalysts. Initially, it had been assumed that already a perfect fit between the substrate and the enzyme would explain the high efficiency of the reactions. Such a "lock and key" model explains the high specificity of the enzymatic reaction but not the *existence of a stable transition state*. The latter can be explained by a refined lock and key model taking into account the ability of the enzyme to reshape itself continually during the reaction. In the "induced fit" model, it is assumed that the conformation of subunits is changed during the reaction to provide precise fit with the changing substrate. For an intended use of enzymes as components for biohybrid materials, care has to be taken that the necessary freedom for structural relaxation is not lost by the immobilization or entrapment of the enzyme on or in the carrier material. For accelerating the reaction by a factor of 10, $\Delta G_B = 5.7\,kJ/mol$ is required. The energy contribution of a single weak bond is typically in the range of

4–30 kJ/mol. Therefore, multiple noncovalent interactions will be sufficient to lower the activation barrier by 60–100 kJ/mol. This explains why the reaction rate can be increased with enzymes by 10^5–10^{17} times, which demonstrates the huge potential of enzymes for biochemical engineering.

Enzyme Kinetics It would be highly desirable to evaluate the efficiency of an enzyme by a straightforward experiment. Of particular interest are the probability of transient sticking of the substrate at the enzyme and the maximum turnover rate k_{cat} of the whole process. Stimulated by first studies of Henry in 1903, Michaelis and Menten proposed in 1913 a useful scheme for the evaluation of the kinetics of catalytic reactions. With some simplifying assumptions on the process, an equation can be derived that describes the relation between the reaction velocity of a one-substrate enzyme-catalyzed reaction and the substrate concentration. The beauty of this approach is the possibility to derive the above-mentioned intrinsic enzyme properties in a simple way. Michaelis and Menten have considered the simplest case of the reaction with only one transient state ES (see Figure 1.42):

$$E + S \underset{k_{-1}}{\overset{k_1}{\rightleftharpoons}} ES \underset{k_{-2}}{\overset{k_2}{\rightleftharpoons}} E + P. \tag{1.28}$$

The prerequisite of the model is that only the early stage of the reaction is considered. This means it can be assumed that the concentration [P] of the product phase is small. Therefore, any back reaction can be neglected ($k_{-2} = 0$). Furthermore, it is assumed that the concentration [S] of the substrate is large with respect to the total enzyme concentration [E_{tot}]. In the initial stage, changes of the substrate concentration are negligible in the total reaction balance. Under these assumptions, a reaction rate equation for the transient state can be formulated with [ES] as the concentration of the transient phase ES:

$$\frac{d}{dt}[ES] = k_1[S]([E_{tot}] - [ES]) - (k_{-1} + k_2)[ES]. \tag{1.29}$$

Here [E_{tot}] − [ES] is the concentration of the free (or unbound) enzyme. With the initial condition [ES]($t = 0$) = 0 and the assumption [S] = const, it leads to a very simple solution:

$$[ES](t) = (1 - \exp(-\lambda t))[ES]_{st},$$

with

$$\lambda = k_1(K_m + [S]) \quad \text{and} \quad [ES]_{st} = \frac{[E_{tot}][S]}{K_m + [S]}. \tag{1.30}$$

Here K_m is the *Michaelis constant*:

$$K_m = \frac{k_{-1} + k_2}{k_1}. \tag{1.31}$$

The solution is limited to the early part of the reaction only (see also Figure 1.43). As we see, at very short time (for $t \ll \lambda^{-1}$) there is a linear increase of the occupation of the transition state. The slope increases for higher substrate concentration. For

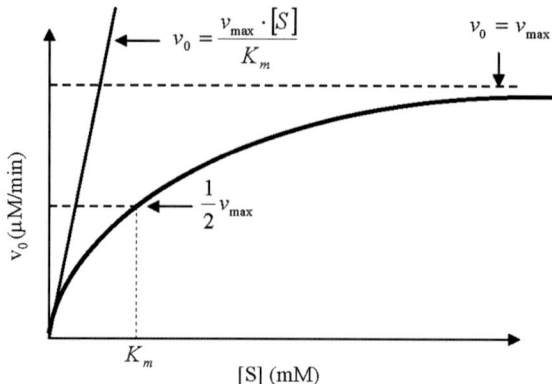

Figure 1.43 The Michaelis–Menten model of enzyme kinetics: the Michaelis–Menten plot. The kinetic parameters can be determined from the asymptotic behavior at low and high substrate concentrations.

longer time, saturation is reached with a stationary value $[ES]_{st}$ for the concentration of the transient phase. This *quasi-steady state* is considered by Michaelis and Menten in their further evaluation. In the quasi-steady state, the reaction velocity v_0 of the product phase is given as

$$v_0 = k_2[E]_{st} = v_{max} \frac{[S]}{K_m + [S]} \quad \text{with} \quad v_{max} = k_2[E_{tot}].\tag{1.32}$$

This *Michaelis–Menten rule* expresses the saturation value of the reaction velocity of the enzyme in the early stage of the reaction depending on the concentrations of the substrate $[S]$ and the enzyme $[E_{tot}]$. v_{max} is the maximal reaction velocity. The maximum turnover number $v_{max}/[E_{tot}] = k_2$ characterizes the internal speed of a single enzyme molecule. The Michaelis constant K_m is a measure for the maximum occupation of the transition state of the enzyme:

$$\text{maximum}([ES]_{st}/[E_{tot}]) = K_m^{-1}.\tag{1.33}$$

A small value of the Michaelis constant corresponds to a high concentration of the transition phase. If the enzyme binds the substrate with a high rate in comparison to the catalysis rate k_2 and the substrate dissociation rate k_{-1}, then we observe a high occupation of the transition state. The Michaelis–Menten model characterizing the early stage of the catalytic reaction allows a universal description of the dependence of the product rate v_0 on the substrate concentration, $[S]$, if the two phenomenological fitting parameters v_{max} and K_m are known (Figure 1.43). The evaluation of the product rate allows a straightforward determination of these parameters by rewriting the Michaelis–Menten equation in the form of the so-called Lineweaver–Burk equation:

$$\frac{1}{v_0} = \frac{K_m}{v_{max}[S]} + \frac{1}{v_{max}}.\tag{1.34}$$

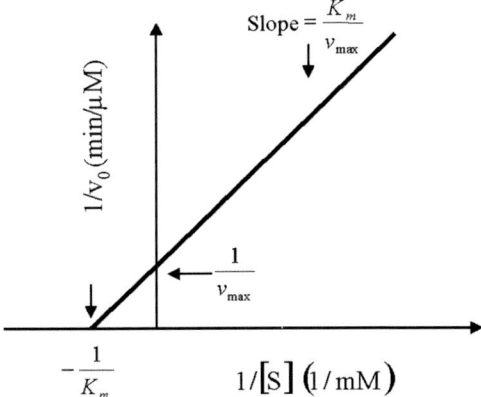

Figure 1.44 The Michaelis–Menten model of enzyme kinetics: the double-reciprocal Lineweaver–Burk plot.

The double reciprocal plot of $1/v_0$ versus $1/[S]$ plot is suitable for fitting experimental data. As one can see in Figure 1.44, the two intercepts on the $1/v_0$ axis and $1/[S]$ axis yield immediately $1/v_{max}$ and $-1/K_m$, respectively.

1.3
Bioengineering

1.3.1
Biointerfacing

Today, the study of biological interfaces with the aim to exploit them for the development of novel materials covers a wide range of sizes lasting from big orthopedic implants down to biofunctionalized nanowires or nanoclusters in advanced biosensors. Accordingly, the topic "biointerfacing of materials" includes a broad field of research activities. It concerns interfaces of artificial materials with (i) single biomolecules, (ii) cellular organelles, (iii) the extracellular matrix of tissue, and (iv) single cells or biofilms. For further discussion, we have a brief view on natural interfaces in biological systems.

Compartmentalization is an essential prerequisite for living systems to generate separate spaces for biochemical reactions. The cell as a whole and the various subcellular organelles of eukaryotic cells are surrounded by membranes, mostly based on lipid bilayers. Lipid bilayers are fluid – in the sense that the molecules can arrange and rearrange themselves more or less easily while staying inside the monolayer or bilayer – thus behaving like a two-dimensional liquid. The fluidity of a membrane mainly depends on the chemical nature of the lipids. Unsaturated lipids make membranes more fluid, while saturated ones and cholesterol have the opposite effect. In order to fulfill their main function of separation, biomembranes are impermeable to most molecules. Only small lipophilic molecules are able to freely diffuse through the membrane. Transport of ions and hydrophilic molecules including water is strictly controlled by specific channels or transport proteins embedded in the membrane. Signal transduction is facilitated by membrane-anchored receptor proteins via binding of specific molecules.

Functionalization of technical surfaces with native biomembranes is extremely challenging because of their complex physical and chemical properties. Another level of complexity is added by the fact that cells permanently remodel the membrane composition according to their physiological needs. Technical surfaces of any chemical constitution will somehow interact with membrane components, and hence tend to produce artifacts. A possible solution to this dilemma is offered by cells whose cell membranes are covered with an additional layer, for example, the cell wall or surface layers, simple membranes that are mainly or exclusively composed of proteins. These rigid structures provide mechanical stability. At the same time, their porous structure allows the access of most molecules to the underlying cell membrane. *Bacteria and yeast cells may therefore be prime candidates* for the application of whole cells to technical devices (see also Chapter 4).

An alternative solution can be the use of the *extracellular matrix as a natural coupling medium* between the living cells and an artificial material. This is an option that has already been successfully used in many developments of biocompatible implants for regenerative therapy in medicine. For example, metallic or ceramic implants for bone surgery can be coated with collagen–hydroxyapatite composites that are the main components of the bone matrix. It is obvious that such coatings substitute the functions of a real bone matrix only to a limited degree. Therefore, it is intended to use cells of the surrounding tissue to rebuild such artificial coatings. With the incorporation of signal molecules, such as cytokines or growth factors, the basic interactions can be controlled (see also Chapter 6).

The sensitivity of the biomolecules with respect to their surroundings may cause serious problems in the case of direct coupling to an artificial material. As explained in Section 1.2, the functionality of proteins depends strongly on their correct folding, which is, for example, difficult to achieve at hydrophobic surfaces. Therefore, it is usually necessary to design multilayered coatings that provide "soft" transitions to the main material.

General Design Rules for Biointerfaces While the functionalization of technical devices with living cells is still in its infancy, a wide variety of methods has been developed to engineer surfaces with biomolecules. Some general rules have to be fulfilled for the successful preparation of biomimetic interfaces by using bio-molecular building blocks.

- *Preservation of the native conformation:* The functionality of biomolecules depends strongly on their conformation in space. The molecular information stored in proteins, DNA, and polysaccharides can only be used if the molecules, assembled at an interface, have the conformational freedom as in their natural environment. This is only possible if the correct three-dimensional structure of a biomolecule is retained during immobilization, which is a difficult challenge for highly complex biomolecules. The conformation of many of chain-like molecules is governed by weak noncovalent bonds. Therefore, uncontrolled adsorption of the biomolecules at surfaces can be critical. Unspecific adsorption can be avoided by an appropriate coating with hydrophilic polymer films such as poly(ethylene oxide) (PEO), leading to a "neutral interface". Hydrophobic molecules like integral constituents of membranes, such as ion channels, exhibit surface-exposed hydrophobic residues. Their native structure can only be attained and preserved in an appropriate *lipophilic environment.*

- *Site-specific binding and controlled patterning:* Usually, biofunctionalization of artificial materials aims at information transfer between living matter and an engineering component. In micro- or nanostructured hybrid constructs such as biosensors, microarrays, hybrid electronic circuitries, or implanted biomedical devices, the interface has to meet specific requirements. Key parameters are the density distribution of molecules which transfer the signals between the living system and the technical device, their uniformity, and often a controllable patterning of the interface. The natural length scale for structure control has to be in the range of a few nanometers, determined by the characteristic sizes of the biomolecular building units. This is preferably realized by *self-assembly* as it occurs in biological systems (see also Chapter 7).

- *Chemical compatibility:* From the chemical point of view, the organic interface has to be compatible with the multiple substrate materials such as silicon, silica, glasses, noble metal films, polystyrene, polydimethylsiloxane (PDMS), and various polymorphic carbon phases (graphite, diamond, and carbon nanotubes). In the case of incompatibility, a suitable transition layer has to be applied.

- *Thickness requirements:* For many sensor applications, such as various optical sensors (e.g., surface plasmon resonance (SPR) spectroscopy and interferometry), and electronic sensors (e.g., field-effect transistors), the highest sensitivity of the hybrid device is obtained with interfaces whose thickness does not exceed a few nanometers. There are other applications, such as whole-cell sensors, that require coatings of several μm thickness providing microcavities for the immobilization of living cells as well as a channel system communicating with a life-sustaining environment.

- *Reusability and regeneration:* Finally, aspects of cost efficiency also have to be considered, for instance, the option of reusability or regeneration of the biological interface in the case of biosensors.

In conclusion, one needs a high versatility of available processing options to meet the multiple demands regarding the design of biological interfaces. This design is based on a modular approach that uses biological components such as DNA, proteins, lipids, membranes, or whole cells as building blocks. To create a stable interface, different interaction mechanisms can be employed that can roughly be divided into *physical absorption, chemical binding,* and *bioaffinity interactions.* As explained in the following sections, there are various technological platforms available that can be applied for the purpose in question. In this chapter, we will preferentially address solutions based on interfaces functionalized with DNA and proteins.

1.3.2
DNA-Based Nanotechnology

The use of DNA for the creation of new materials and processing techniques in nanotechnology has benefited from the rapid methodical progress in life sciences. Advances in high-throughput DNA sequence analysis and expression profiling resulted in the development of diverse techniques for fabrication of DNA microarrays. The biological self-assembly of nucleic acids has been one of the key mechanisms that motivated chemists, physicists, and materials scientists to work with DNA. Due to its restricted number of bases and their unique capability of self-recognition, it has been the most suitable molecule for demonstrating the feasibility of the "bottom-up approach" in the manufacturing of artificial nanostructures. Soon it was found that DNA and RNA are suitable as templates for building nano-structured surfaces of almost any kind of materials. This paved the way for application of DNA-assembled interfaces in materials engineering. The innovative potential of DNA is most obvious in two main areas: (i) biomolecular templates for submicrometer electronic circuitries, and (ii) programmable nanoprobes.

1.3.2.1 Biomolecular Templates for Submicrometer Electronic Circuitries

In 1965, Gordon Moore formulated an empirical observation in the so-called Moore's law that the market for silicon-based microelectronics is characterized by doubling the functionality (bits) per chip surface area every 18 months. Today, this empirical law is still valid in semiconductor industry. The large-scale production of such structures is still realized by using top-down lithography. The microstructures are "carved out" from bulk-like materials by applying sophisticated physical and chemical technologies, such as optical lithography, UV lithography, or electron beam and X-ray lithography. It can be expected that the top-down lithography will come to principal physical limits when the structure sizes will fall below 10 nm. For processing, it means that new technologies summarized as bottom-up approach are needed. Arrangement and interconnection of nanosized electronic components into a functional circuit have to be realized. It has to be processed in parallel with high reliability on a large scale of length (typically tens of micrometers). Therefore, recognition and self-assembly should be essential ingredients of these new techniques. Comparing these demands with the machinery inside living cells evokes the idea of implementing such objects in future nanoelectronic technology. One basic idea behind such developments is the use of DNA for self-assembling nano-structures. The specific base pair recognition, the charged phosphate groups along the backbone, and the existing options for conjugation with additional functional groups are the reasons why DNA can act as a programmable template for the formation of a nanostructured circuit. The whole process of biotemplating can be subdivided into four parts: preparation of the DNA for immobilization, end-specific immobilization, stretching of coiled DNA, and functionalization of DNA templates.

Preparation of DNA for Immobilization   Stable immobilization of the ends of single- or double-stranded DNA (dsDNA) can be achieved by nonspecific physical binding (electrostatic or hydrophobic interactions), covalent binding, site-specific bio-molecular interactions (bioaffinity), or physical embedding of DNA ends in coatings. Whereas physical embedding of the DNA molecule needs no further modification, for covalent binding and bioaffinity the DNA ends have to present specific linkers depending on the substrate chemistry. There are two basic ideas derived from natural DNA replication that can be applied for the end-specific functionalization of a dsDNA (Figure 1.45). During replication, each single strand of an unwound DNA double helix acts in the cell as a template for *immobilization of individual matching nucleotides*. Enzymes catalyze this process. It is possible to use this mechanism for *in vitro* polymerization of nucleotides at so-called single-stranded overhangs of a dsDNA. Figure 1.45a shows how DNA polymerase catalyzes the binding of deoxynucleotide triphosphates dATP, dCTP, dGTP, and dTTP at the sticky end of a dsDNA. In the example, the last nucleotide (dTTP) is functionalized with a reactive

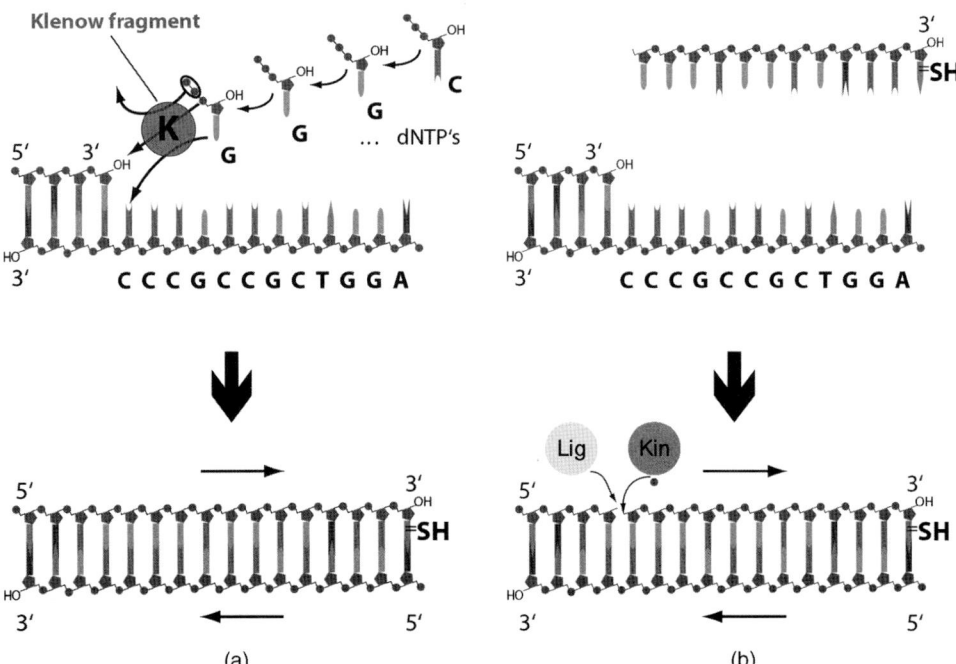

(a)   (b)

Figure 1.45   (a) Sticky-end functionalization. The exonuclease-free Klenow fragment of DNA polymerase incorporates the deoxynucleotide triphosphates dGTP, dCTP, and dATP and the thiol-modified S⁴-TTP, resulting in a functionalized blunt end. (b) Hybridization of a thiol-modified oligomer to the sticky end of a λ-DNA. The single-stranded nick can be closed by a two-step process: phosphorylation of the OH-terminated 5′-end of the oligomer before hybridization using T4 kinase (Kin), and ligation after hybridization using T4 ligase (Lig).

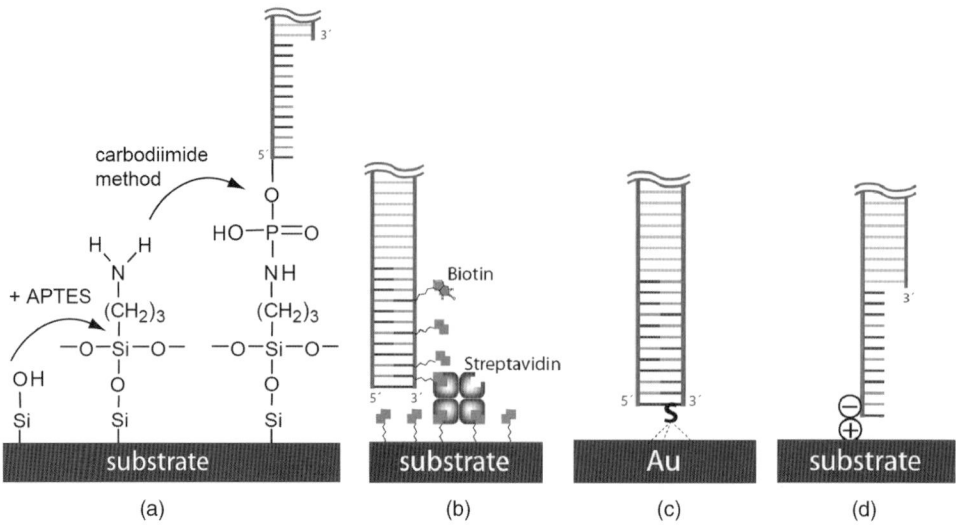

Figure 1.46 Different techniques for end-specific anchoring of DNA to surfaces: (a) At the DNA end, an activated phosphoester formed by the carbodiimide method is covalently coupled with a nucleophilic agent such as an amine formed by silanization to create a stable covalent bond. (b) Affinity reaction forming a streptavidin bridge between the biotinylated DNA end and a biotinylated surface. (c) Specific binding of thiol-functionalized DNA to a gold surface. (d) Electrostatic bonding between positively charged amino groups and twofold negatively charged DNA ends.

SH group. In this way, the blunt end of the dsDNA is modified to bind covalently to a gold surface. Other, often used reactive groups are the coenzyme biotin and the steroid digoxigenin. Alternatively, the sticky end can be filled with an already *functionalized complementary sticky end by hybridization* by means of the enzyme T4 kinase (Kin). Finally, the nick is closed by ligation by means of the enzyme T4 ligase (Lig).

End-Specific Immobilization   The structure of DNA offers many possibilities for coupling to an inorganic material: the amines in the bases, the negatively charged backbone, the phosphodiesters within the backbone, the phosphate at the 5'-end, and the hydroxyl group at the 3'-end. The anchoring of a single end of DNA at a silica substrate or patterned gold contact is an elementary process step for the immobilization of DNA templates for microelectronic circuitries. Various techniques are available for this purpose (Figure 1.46). DNA can be bound covalently to reactive surface sites by various coupling reactions. Besides the NH_2 groups, other reactive groups on the DNA can be generated by activation reactions such as the carbodiimide or the activated ester method. On hydroxyl-terminated substrates such as silica or glass, *DNA can be covalently immobilized by silanization of the substrate with reactive silanes.* 3-Aminopropyltriethoxysilane (APTES), which provides an amino group for chemical reaction, is commonly applied for silanization (Figure 1.46a). The amino group can be connected with an activated phosphoester of the DNA by reaction with carbodiimide. Covalent binding is the preferred technique for DNA immobilization owing to the high bond strength, optional high uniformity, and packing density of the oligonucleotides (provided they do not exceed ~20 bases). Other processing routes to covalent binding are summarized in Table 1.9.

Another approach makes use of *bioaffine coupling*: A biotinylated DNA is anchored to a biotinylated surface by means of a streptavidin bridge (Figure 1.46b). *Biotin and streptavidin* form one of the strongest natural bonds. Streptavidin and the closely related avidin are tetrameric proteins with four binding sites for biotin. Their bond strength to biotin depends on the loading rate. It varies between 5 pN for low loading rate and 170 pN for high loading rate.

The functionalization of gold contacts with oligonucleotides can be accomplished by applying *thiol chemistry* (Figure 1.46c). Thiol groups are used for chemical binding of biomolecules on metals such as gold, silver, platinum, or copper. On gold surfaces, each sulfur atom is coordinatively bonded with three gold atoms. The gold–sulfur bond ruptures at a force of about 1.4 nN (at a loading rate of 10 nN/s).

The fourth option is nonspecific physical binding via *electrostatic interaction* between a positively charged substrate surface (e.g., gold contacts functionalized with amino–thiol groups) and the twofold negatively charged end of the DNA (Figure 1.46d). This method demands an appropriate control of the pH and the ion concentration in the solution to prevent attraction of the whole DNA chain by the positively charged surface. To get preferential binding of the termini of the DNA, the charges should be partially shielded by an appropriate choice of the pH value (about pH 8–9).

As to be discussed in the following subsection, even *hydrophobic interaction* can be used for DNA adsorption on hydrophobic surfaces. In overstretched double-stranded DNA, disturbed hydrogen bonds between complementary bases cause hydrophobic interactions. This mechanism can be applied for transfering oriented stretched DNA on a substrate, which will be discussed in the following subsection. Finally, the immobilization of DNA in a hydrogel offers the possibility of depositing thicker DNA coatings on a substrate. This technique could be relevant if a higher effective surface density of the DNA probes is needed for higher sensitivity of a biosensor. The main processing routes developed for immobilization of DNA on various kinds of materials are compiled in Table 1.9. For more details see Festag *et al.* (2005).

Stretching of Coiled DNA The manufacturing of linear templates for microelectronic circuitries requires to align stretched molecules on the particular substrate. As explained in Section 1.2.3, the stress–strain curve of an individual DNA molecule shows an elastic range followed by an extended yield range where the molecule is stretched without being denatured. This permits the formation of well-aligned templates for nanowires. Various techniques for stretching and parallel alignment of DNA molecules are available: stretching of single molecules by atomic

Table 1.9 Immobilization mechanisms of DNA on solid surfaces.

Interaction	Surface modification	DNA modification	Remarks
Covalent binding		Depending on spacer head group	
Thiol chemistry	Gold substrate with thiol–spacer constructs: thiol–alkyl, multiple thiol anchor, steroid disulfide	Thiol modified	Consecutive probing possible
Amine chemistry	Gold substrate: amine/succinimide, maleimide; silica substrate: 3-aminopropyltriethoxysilane + 1,4-phenylene diisothiocyanate (PDC)	Amino modified	
Epoxy chemistry	Silica substrate with 3-glycidoxypropyltrimethoxysilane (GOPS)	Amino modified	High loading capacity, good uniformity
Bioaffinity interaction	Streptavidin; biotin/streptavidin double-layer S-layer/streptavidin	Biotinylated	High specific binding strength, high binding efficiency and uniformity
Electrostatic interaction	Amino groups	Control of effective charge of the negatively charged backbone	About pH 8–9
Hydrophobic interaction	Hydrophobic substrate	Partially denatured by stress, defects of H-bonds between nucleic acids	Relevant for molecular combing
3D nanoporous hydrogels	Silica substrate with GOPS		Higher loading capacity

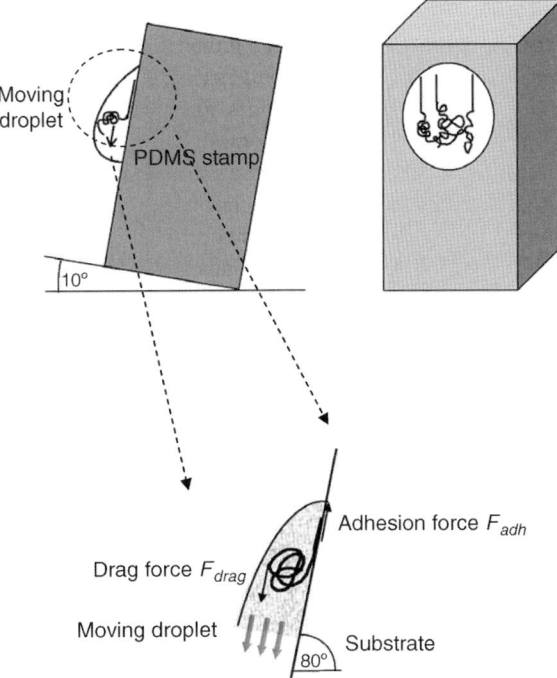

Moving droplet

PDMS stamp

10°

Adhesion force F_{adh}

Drag force F_{drag}

Moving droplet

Substrate

80°

Figure 1.47 Principle of molecular combing with a moving droplet: Coiled DNA suspended in aqueous solution and stuck to the substrate with one end is stretched by the liquid flow.

force microscopy, optical and magnetic tweezers, or simultaneous alignment of many molecules by methods such as molecular combing, hydrodynamic flow, and electrophoretic stretching.

Molecular Combing The principle of molecular combing is depicted in Figure 1.47. A droplet of DNA solution is placed on the surface of a hydrophobic substrate in an inclined position such that it flows down. A favored substrate is polydimethylsilox-ane. This hydrophobic polymer is well suited for the fabrication of stamps enabling the transfer of biomolecules. In order to become stretched by the drag of the liquid flow and adsorbed in the stretched conformation in parallel alignment, the molecules must be stuck to the substrate with one end. The process of parallel alignment of DNA on a substrate is called "molecular combing".

Random sticking can be controlled by the experimental conditions such as pH, the presence of additional cations, and the degree of hydrophobicity of the surface. Figure 1.48 shows the strong influence of the pH value, with the other parameters kept constant.

The sudden change of the aspect of the deposited DNA between pH 3 and pH 4 is due to a transition of the adsorption behavior. At pH 3 and pH 4, the DNA is adsorbed in a coiled shape, whereas beyond pH 4 the DNA is adsorbed at one end only, and then

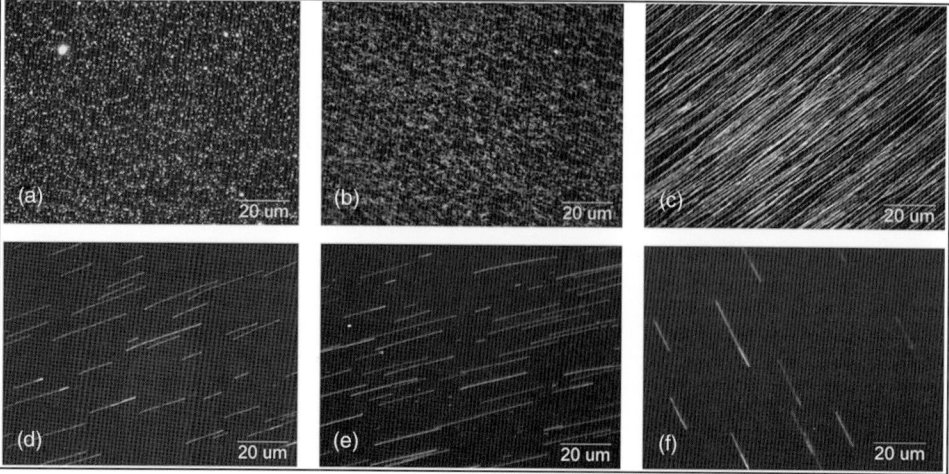

Figure 1.48 pH-dependent behavior of λ-DNA solution in molecular combing experiments on an hydrophobic PDMS surface ((a) pH 2, (b) pH 3, (c) pH 4, (d) pH 6.2, (e) pH 9, (f) pH 10). (Reproduced with permission from Benke *et al.* (2011). Copyright 2011, IOP Publishing Ltd.)

stretched and aligned in the direction of the moving droplet. The stretched molecules are longer than the contour length of λ-DNA, which is 16.2 μm. Apparently, the density of stretched molecules is highest at pH 4. The conspicuously changing behavior of the DNA as a function of pH in this experiment coincides with the change of the surface charge of PDMS. For high acidity, the surface is positively charged by preferential adsorption of hydronium ions (H_3O^+), whereas beyond pH 4 it is negatively charged through adsorbed hydroxide ions (OH^-). This property of PDMS explains why with high acidity the adhesion of coiled DNA molecules with their negatively charged phosphate backbone is so strong that the drag force of the liquid flow is not able to stretch them. In less acidic solutions, only the hydrophobic sticky ends of the dsDNA bind strongly to the substrate so that the coil can be stretched into a straight line. With higher pH, the sticking probability of the DNA is reduced by Coulomb repulsion between the DNA and the surface charge of the PDMS.

The double-stranded λ-DNA possesses 12 unpaired bases on its two single-stranded sticky ends that present hydrophobic binding sites to the PDMS substrate. With one of the sticky ends stuck to the substrate after incidental contact, the main part of the coiled molecule is dragged along by the moving liquid, whereby it becomes unraveled, stretched into straight shape, overstretched to lengths exceeding the contour length by factors of 1.6–2.1, and aligned in the direction of flow (Figure 1.49). Experiments with a controlled stationary flow (Brochard-Wyart, 1995) have shown that the stretching of DNA by the drag force occurs at flow velocities as low as 20 μm/s (see the following subsection). Hence, no "meniscus force" is required to explain the stretching. The overstretching of the natural B-DNA by the moving droplet exposes hidden hydrophobic binding sites that bind to the hydrophobic substrate. Thus, the overstretched DNA is stabilized on the PDMS stamp. On

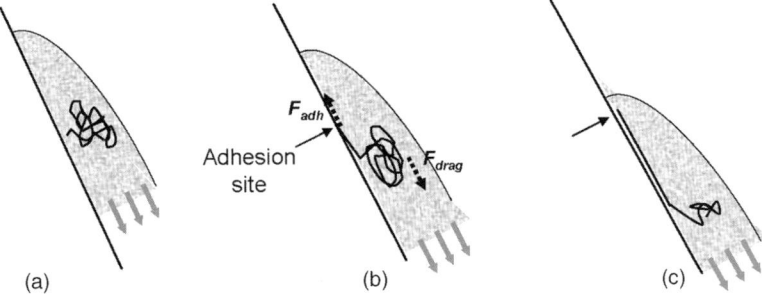

Figure 1.49 Molecular combing of DNA by means of a moving droplet: (a) DNA suspended in solution; (b) one end adhering; (c) partially stretched.

a hydrophilic substrate, such as a PDMS surface with 30° contact angle after treatment with plasma (air), the overstretched DNA relaxes to its contour length and becomes partially unstretched (Figure 1.50).

To sum up, the observed combing can be understood as a three-step process: (i) One end of the coiled DNA suspended in solution gets stuck to the substrate. (ii) The DNA is stretched while the coil becomes unraveled by the drag force of the liquid. (iii) The drag acting on the stretched B-DNA causes overstretching into the S-DNA form or partially molten DNA. After deposition of aligned DNA molecules at the stamp surface, they can be transferred by microcontact printing onto the final substrate. With an appropriate surface profile of the PDMS stamp, structured patterns of well-aligned DNA can be produced (Figure 1.51).

Superposition of DNA patterns can be realized by sequential printing (Figure 1.52).

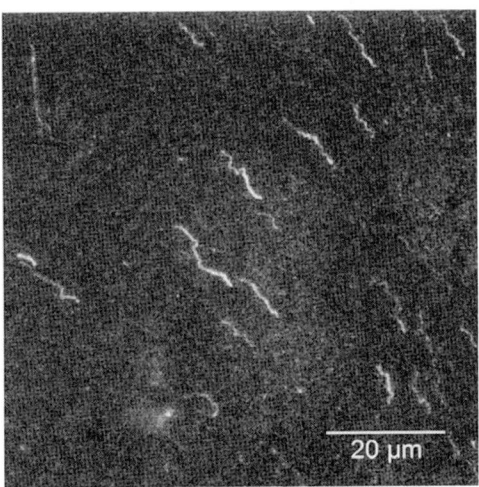

Figure 1.50 DNA relaxed after molecular combing and adsorption on a hydrophilic surface. Adsorption of DNA molecules has been done on plasma-treated hydrophilic PDMS. (Reproduced with permission from Benke *et al.* (2011). Copyright 2011, IOP Publishing Ltd.)

Stamp profile   Stamp loaded with λ-DNA

Figure 1.51 Transfer of λ-DNA, combed at a patterned PDMS surface, onto a glass substrate (Benke, 2007).

Stationary Hydrodynamic Flow The stretching, alignment, and deposition of DNA molecules by means of the drag force in liquid flow, as it has been done simply with a moving droplet of aqueous solution, can be done in a better controllable way by means of a more complex experimental setup (Figure 1.53). An open flow cell is arranged on an inverted microscope. A patterned array of interdigital gold electrodes on a glass slide is chosen as a substrate in this example. It is covered with 200 µl of 100 mM phosphate buffer. A controlled flow velocity is realized as indicated in the drawing. The microcapillaries can be positioned by means of a micromanipulation system. It allows the control of the flow velocity, the flow direction with respect to the electrode array, and the depth profile of the velocity field. For the monitoring of the stretching process, the molecules are labeled with the fluorescent dye YOYO-1. Intercalation of the dye increases the contour length of the labeled DNA. With a staining ratio of 0.72 base pairs per dye molecule, the length of λ-DNA increases from 16.2 to 21.8 µm. In order to facilitate covalent binding of the sticky ends of the double-stranded λ-DNA to the gold electrodes, both ends are thiolated (T–DNA–T)

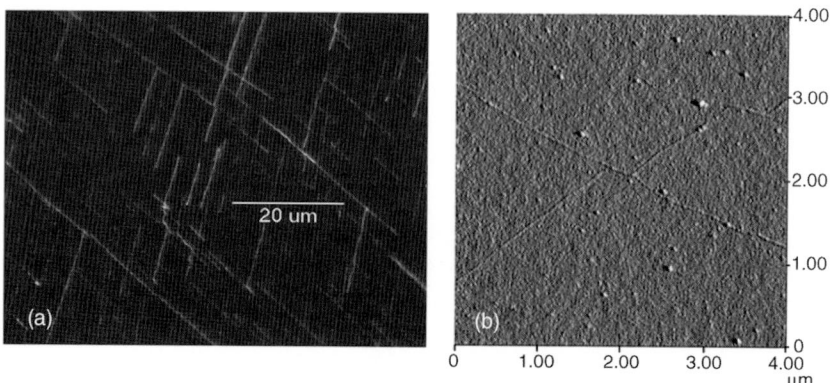

Figure 1.52 Fluorescence image (a) and AFM image (b) of a DNA network of λ-DNA aligned by molecular combing technique on a PDMS stamp and transferred onto a glass surface. The pattern of crossing DNA was produced by two-step stamping. (Reproduced with permission from Benke *et al.* (2011). Copyright 2011, IOP Publishing Ltd.)

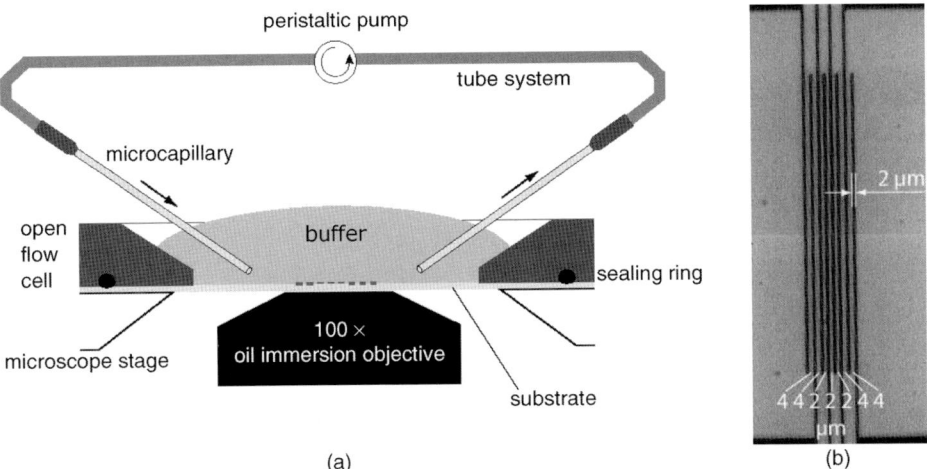

(a)   (b)

Figure 1.53   Setup for hydrodynamic deposition of DNA. (a) Open flow cell arranged on an inverted fluorescent microscope, with microcapillaries positioned above the substrate. (b) Substrate with an interdigital electrode array. (Reprinted with permission from Erler, Guenther, and Mertig (2009) and Erler (2010). Copyright 2009, Elsevier.)

by the Klenow polymerization reaction as described above. The stretching of a coiled DNA molecule tethered at one end to an electrode is controlled by the flow velocity.

In a model experiment with single DNA, Wirtz (1995) has studied the transport behavior of single DNA in liquids. The fluorescently labeled DNA tethered to a small magnetic bead was pulled by a magnetic force through a solution of unlabeled DNA molecules. Three types of conformations have been observed, depending on the flow velocity. Below 3.5 μm/s, a quasi-equilibrated *coil* shape is stable. It changes into a *trumpet* shape for 3.5 μm $< v <$ 20 μm. For higher velocities $v >$ 20 μm/s, a *stem and flower* conformation is observed. As mentioned above, this velocity is nearly three orders smaller than the characteristic velocity of the moving droplet in the molecular combing experiment.

Figure 1.54 shows characteristic patterns of stretched DNA, with the flow direction perpendicular to the electrodes (dark stripes). By varying the concentration of T–DNA–T in the solution, the number of interconnects can be controlled. First, the solutions were flushed in with a flow rate of about 2.5 μl/s to allow single-end binding to the gold electrodes. With a flow rate of 12.5 μl/s, the molecules became stretched, thus being able to form a bridge to the next electrode and get bound there with their free end. By clockwise changing the orientation of the flow direction with respect to the electrodes, the orientation of the interconnects can be changed (Figure 1.55a). The stability and elasticity of the immobilized molecules can be demonstrated by applying an oblique flow of buffer with 210 μl/s (Figure 1.55b). Most of the molecules are bowed out, and a smaller number of molecules are ruptured. Rupture is indicated by the short length of the fragments attached on the left at the electrodes. Probably, this rupture is caused mainly by photoinduced cleavage of the stretched DNA. Failure of the anchorage on the right would leave longer trailing ends.

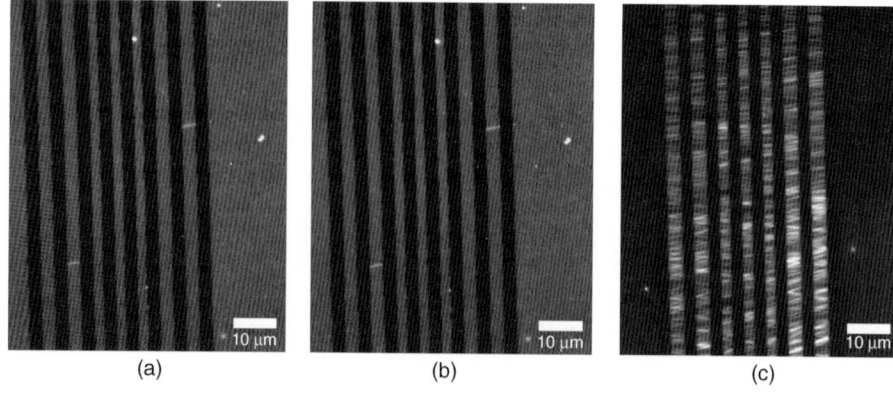

(a)   (b)   (c)

Figure 1.54 T–DNA–T bridging the gaps between electrodes after stretching in liquid flow, depending on DNA concentration of the solution: (a) 0.5 ng/μl; (b) 4 ng/μl; and (c) 10 ng/μl. (Reprinted with permission from Erler, Guenther, and Mertig (2009) and Erler (2010). Copyright 2009, Elsevier.)

DNA Branching for Network Formation Two essential steps of the bottom-up approach for nanoelectronic circuitries by using DNA have already been explained in the previous two sections: site-specific immobilization and controlled alignment of DNA molecules. However, there is still a long way to go from the discussed examples involving single- or double-stranded DNA to complex electronic circuitry. As a further step on this way, the assembling of 2D and 3D network structures has to be considered. In this context, one can try to make use of the unique properties of

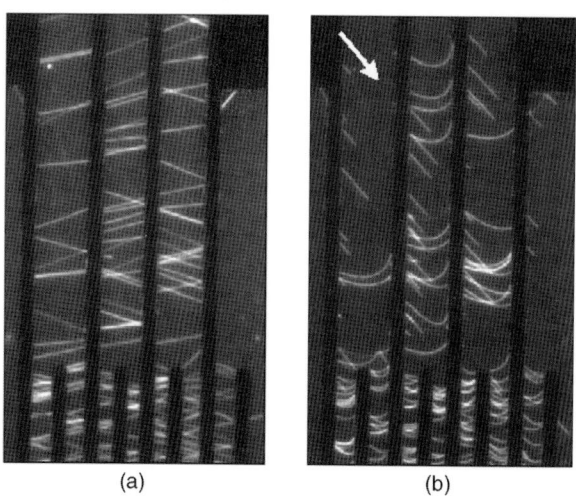

(a)   (b)

Figure 1.55 T–DNA–T bridging the gaps between gold electrodes after stretching in liquid flow: (a) as deposited; (b) affected by an oblique flow of 210 μl/s as indicated by the arrow, suitable for testing the anchoring strength. (Reprinted with permission from Erler, Guenther, and Mertig (2009) and Erler (2010). Copyright 2009, Elsevier.)

DNA, the encoding of information on the nanoscale combined with a reliable and robust recognition mechanism, and the self-assembly of DNA molecules near room temperature, for the *manufacturing of a complex DNA structure in a one-step process by self-assembly.* DNA structures obtained in this way may subsequently serve as templates for the deposition of nanoelectronic functional elements.

DNA Tiles and Origamis The fundament for the design of nanostructured DNA templates has been laid with the so-called tile model by the pioneering work of Nadrian Seeman (Seeman, 1982, 2003; Seeman and Belcher, 2002). The basic idea consists in the preparation of multi-arm junction motifs that are self-assembling tiles for higher order structures. The tiles have to be sufficiently stiff, and they must be equipped with a suitable number of single-stranded overhangs that serve as linkers for the self-assembly of the higher order structure. The tiles are formed by hybridization of single-stranded synthetic oligonucleotides (Figure 1.56a). The stiffness of the tile can be increased by a large number of crossovers between the individual molecules forming one compact tile. Up to now, a large variety of 2D and 3D tiles have been prepared. Such tiles have been successfully used to produce 2D lattices, nanoribbons (Liu *et al.*, 2005), and nanotubes (Liu *et al.*, 2004, 2006). Moreover, various 3D structures have been produced by assembling of oligonucleotides: cubes (Chen and Seeman, 1991), octahedra (Zhang

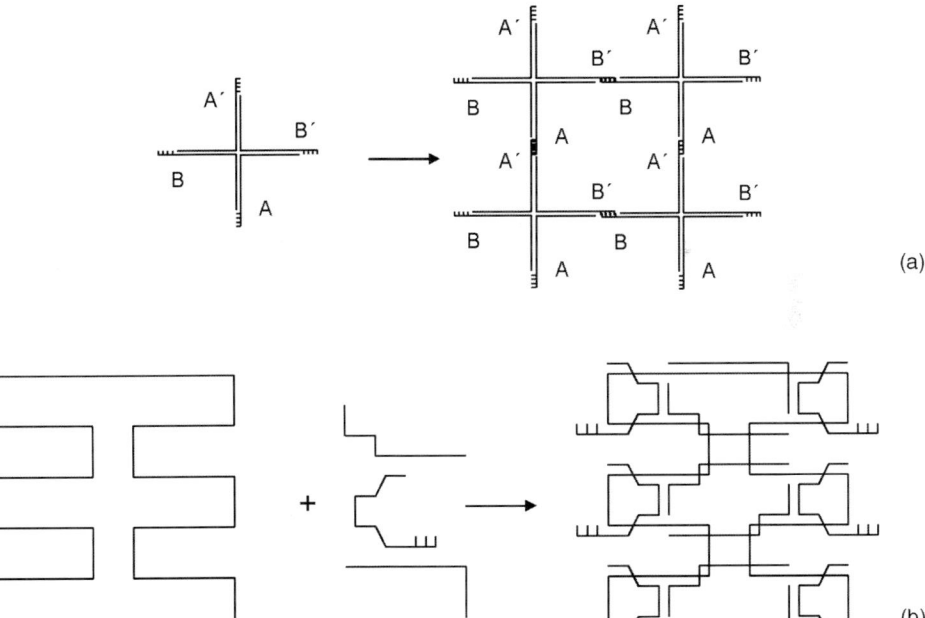

Figure 1.56 Self-assembly of large DNA structures. (a) The tile concept (Seeman, 1982): branched DNA junctions interact via sticky-end hybridization (A–A', B–B') to self-assemble into 2D or 3D lattices. (b) The concept of DNA origami (Shih *et al.*, 2004; Rothemund, 2006): a single-stranded DNA is folded into the intended shape and stabilized by a multitude of short staple strands. Some of the staple strands simultaneously form sticky ends.

and Seeman, 1994), buckyballs (He *et al.*, 2008), and dynamic capsules that can be switched between different states (Aldaye and Sleiman, 2007).

For the design of templates of nanoelectronic circuitries, the Seeman tiles have to be modified. The size of synthetic oligonucleotides is restricted to about 100 bases. This means that the length of single branches is only in the range of a few nanometers. For nanoelectronic networks, this size should be larger by at least a factor of 10. Therefore, the tiles should be enlarged with appropriate elongations of the short arms. This can be achieved by the following strategy (Figure 1.57) (Mertig and Pompe, 2004). In a two-step process, first a small DNA tripod is prepared with three oligonucleotides that have partially complementary sequences. Lines with same gray shades represent complementary DNA duplex strands.

Alternatively to Seeman's tile concept, Shih *et al.* (2004) and Rothemund (2006) have transferred the famous Japanese art of folding complex origamis from one

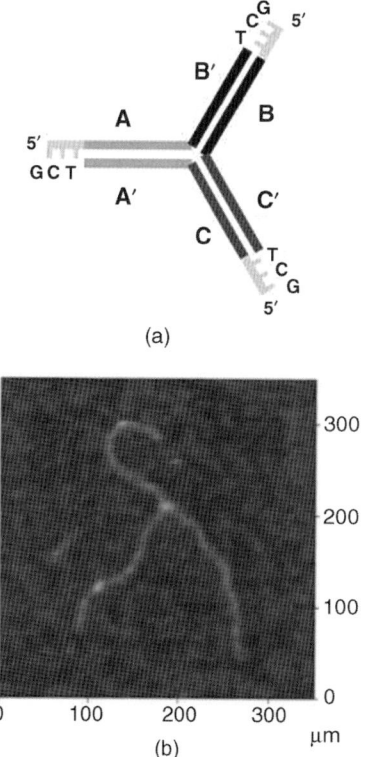

(a)

(b)

Figure 1.57 (a) Schematic viewgraph of a DNA tripod used as central element for a DNA junction. Each duplex arm possesses at its end a GCT 5′ overhang that can be used for ligation of double-stranded DNA fragments of about 500 bp and sticky ends that are complementary to the tripod overhangs. In this way, each arm of the tripod is elongated by about 180 nm long upon ligation. (b) Scanning force microscopy of a three-armed DNA junction. (Reproduced with permission from Mertig and Pompe (2004). Copyright 2004, Wiley-VCH Verlag GmbH.)

piece of paper into the DNA world. To build a complex DNA structure, only one long single-stranded DNA is sufficient if a number of short single strands (staples) are added (Figure 1.56b). Each staple strand has a unique sequence that governs a specific folding action. It allows the additional incorporation of appendices as, for example, loops. Thus, together with the folding, spatially addressable elements can be implemented in the structure. A wide range of structural variation can be obtained in this way. Examples such as a square nut, a 12-tooth gear, and a hollow box that can be opened by an external DNA "key" suggest promising applications in nanotechnology. DNA boxes that can be opened, loaded, and selectively immobilized on cells can be used as nanorobots for targeted transport of molecular payloads in order to stimulate cells in tissue cultures. Douglas, Bachelet, and Church (2012) provided the first example by creating a hexagonal DNA barrel with dimensions of 35 nm × 35 nm × 45 nm (see Figure 1.58). The barrel is formed by two domains that are covalently connected in the rear by single-stranded hinges. The barrel can be

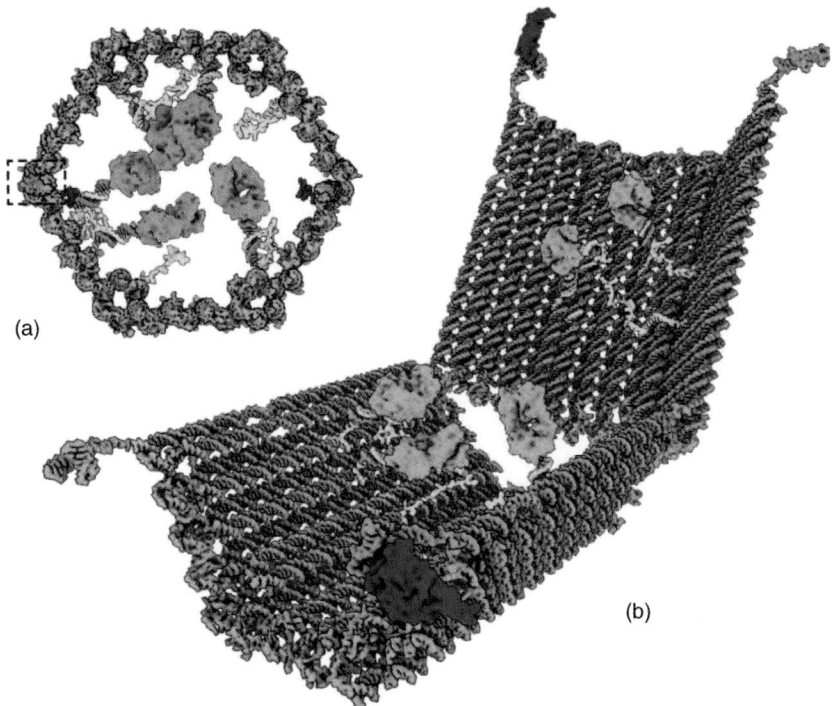

(a)

(b)

Figure 1.58 Design and TEM analysis of an aptamer-gated DNA nanorobot. (a) Schematic front orthographic view of a closed nanorobot loaded with a protein payload. Two DNA aptamer locks fasten the front of the device on the left (boxed) and the right. (b) Perspective view of the nanorobot opened by protein displacement of the aptamer locks. The two domains are constrained in the rear by scaffold hinges. (c) TEM images of robots in closed and open conformations. Left column: unloaded; middle column: robots loaded with 5 nm gold nanoparticles; right column: robots loaded with antibody fragments. Scale bar: 20 nm. (Reprinted with permission from Douglas, Bachelet, and Church (2012) Copyright 2012, AAAS.)

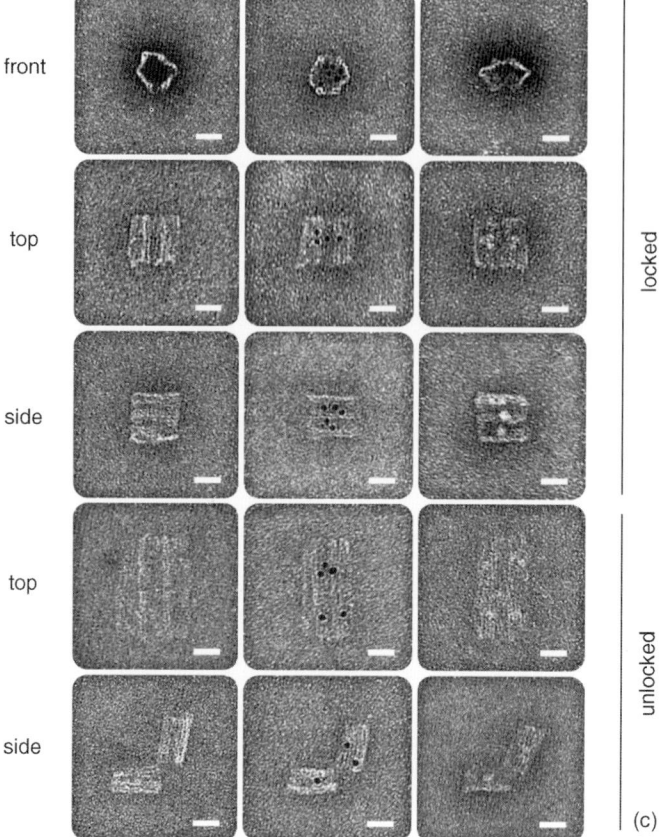

Figure 1.58 *(Continued)*

noncovalently closed in front by staples modified with DNA aptamer-based locks (for aptamers, see also Section 2.1). The locks open in response to binding antigen keys. Each barrel possesses one specific duplex of aptamer locks that undergoes opening when both aptamers recognize their corresponding antigens at the target (cell surface). Then the barrel is opened like an entropic spring. The inner faces of the barrel are exposed, and any payload previously immobilized at the inner barrel faces can interact with surface receptors. In the example, by choosing three different aptamers, six different lock duplexes were prepared. As proof of principle, the nanorobots were loaded with combinations of antibody fragments that stimulated specific signaling pathways in six different cell types leading to inhibition or activation of cellular processes, respectively.

Among the shown examples, assemblies of nanotubes or bundles are of particular interest because of their higher stiffness compared to double-stranded DNA (see also Section 7.3.1). This is useful for the design of well-aligned connections in nanoelectronic circuitries.

Electronic Properties of DNA   For several reasons, DNA molecules seem to be suitable elements for the assembly of nanoelectronic circuitries, among them: (i) synthesis of tailored variants with predesigned length and base order; (ii) controlled self-assembly into well-defined shapes. Since the early 1990s, experimental work addressing the electric conductivity of single- and double-stranded DNA has been performed. The early results were contradictory as the reported *conductivities varied between metal-like and insulating*. Soon it became obvious that the results were affected by temperature, pH, humidity, and presence of counterions and oxygen, and hence did not represent the intrinsic conductivity of the DNA. The latter can be properly measured by integrating an individual DNA molecule into a microscopic solid-state device, such as a molecular field-effect transistor on a silicon chip, or a tunneling barrier between the tip of a scanning tunneling microscope and a gold film.

A favored platform for measuring electrical transport in single DNA molecules is based on nanoscopic contact structures made by single-walled carbon nanotubes (SWCNTs) integrated in a microelectronic structure on a silicon chip (Guo et al., 2008; Roy et al., 2008). Carbon nanotubes and DNA have similar diameters of 1–2 nm. In order to reduce the contact resistance, a strong electronic coupling between the carbon nanotubes and the DNA has to be realized. The DNA can be covalently bound to the carbon nanotubes functionalized with carboxyl groups forming the source and the drain contact. The conducting silicon wafer serves as a back gate. I–V curves have been measured under ambient and vacuum (10^{-5} Torr) conditions for single-stranded and double-stranded DNA (Figure 1.59). The resistance of a very short piece (~6 nm) of double-stranded DNA with well-matched base pairs has been found to be 0.1–5 MΩ (Guo et al., 2008). This is in the same range as the c-axis resistance of highly oriented bulk pyrographite with similar dimensions (about 1 MΩ). This means that the electrical behavior of a stack of DNA base pairs is similar to that of the stack of aromatic graphite planes. The resistance increases with the width of the gap between the source and the drain. Unlike usual conducting materials, where the resistance scales with the length of the conductor, the resistance of DNA increases exponentially with length. For a ~27 nm long double-stranded DNA (80 bp long DNA fragment encoding a portion of the H5N1 gene of the avian influenza A virus), a resistance of 25–40 GΩ was measured under ambient conditions (Roy et al., 2008). At a source–drain voltage of about 1 V in a

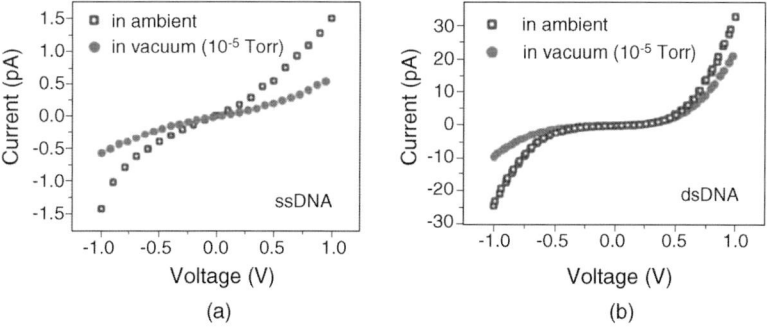

Figure 1.59 I–V characteristics of single DNA: (a) single-stranded DNA; (b) double-stranded DNA. (Reprinted with permission from Roy et al. (2008). Copyright 2008, the American Chemical Society.)

single-stranded DNA of a length ~27 nm, the current is typically only a few pA, whereas in a double-stranded DNA it is about 20 times larger. This is primarily due to the lack of a regular stacking of the nucleotide bases in the single-stranded DNA.

In most experiments, double-stranded λ-DNA has been used. Better information on the electron transport through DNA molecules can be derived from experiments with artificial poly(dA)–poly(dT) and poly(dG)–poly(dC) DNA molecules consisting of identical base pairs. Yoo *et al.* (2001) performed experiments with poly(dA)–poly(dT) and poly(dG)–poly(dC) DNA molecules with average lengths of about 0.5–1.5 μm and about 1.7–2.9 μm, respectively. More than 20 samples were analyzed at ambient conditions and in vacuum and revealed an essential difference between the two molecular species. Current–voltage curves have been measured (Figure 1.60). The

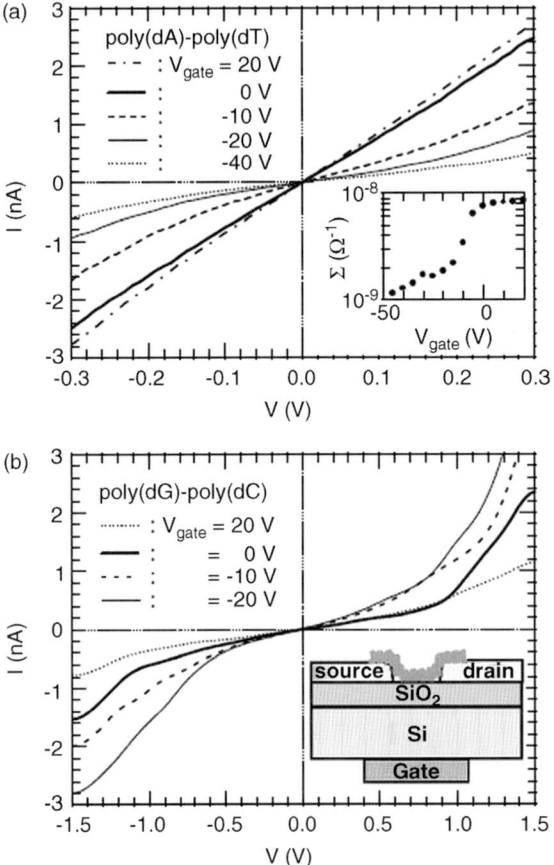

Figure 1.60 *I–V* curves measured at room temperature for various values of the gate voltage (V_{gate}) for poly(dA)–poly(dT) (a) and poly(dG)–poly(dC) (b). In the inset of (a), the conductance at $V = 0$ is plotted as a function of V_{gate} for poly (dA)–poly(dT). The inset of (b) is a schematic diagram of electrode arrangement for gate-dependent transport experiments. (Reprinted with permission from Yoo *et al.* (2001). Copyright 2001, the American Physical Society.)

electron transport between the source and drain electrodes was additionally influenced by an external electric field caused by the constant potential V_{gate} of a gate electrode placed next to the current path (see Figure 1.60b). The evaluation of the conductance Σ (slope of the $I–V$ curve) at $V=0$ reveals a significant difference in the dependence on the gate voltage for the two types of molecules. The inset of Figure 1.60a shows that the conductance decreases when a negative voltage V_{gate} is applied. It means that the charges transported in the poly(dA)–poly(dT) molecules are electrons hindered in their motion by the negative gate potential. Consequently, the molecule behaves like a *n-type semiconductor*. However, in poly(dG)–poly(dC) molecules the conductance Σ increases when the gate voltage gets negative, as can be seen in Figure 1.60b. The charges, transported between source and drain, are positively charged holes. The molecule is like a *p-type semiconductor*. A strong temperature dependence of the conductance has been observed around room temperature with a crossover to a weak temperature-dependent conductance at low temperatures. This dependence can be explained by a stepwise thermal excitation of the vicinity of the charge carriers moving forward as so-called hopping polarons.

Metallization of DNA   One of the conclusions of the previous subsection is that the electronic properties of DNA are not suitable for assembling nanoelectronic circuitry with DNA molecules only. However, their outstanding biochemical prop-erties related to versatile structural transformations such as scission, ligation, and hybridization facilitate various "cut and paste" procedures that may be useful in following a bottom-up approach toward complex structures. In the 1990s the question arose whether it would be possible to modify the DNA chemically in order to improve the electron transport. Generally, the various functional units of the DNA offer options to add further organic or inorganic compounds without loosing its unique ability of self-assembly and information coding. For improving the electronic properties by metallization, the following three concepts have been explored: (i) incorporation of metallic ions into the base pair stack, (ii) DNA-directed assembly of metallic or semiconducting nanoparticles using DNA templates and hybridization of complementary sequences, and (iii) heterogeneous nucleation and growth of metal clusters at the DNA (Figure 1.61).

Incorporation of Metallic Ions into the Base Pair Stack   Already in 1993, a new conformation of DNA, the so-called M-DNA, has been reported (Lee *et al.*, 1993). At pH values above 8, B-DNA undergoes a conformational change in the presence of divalent metal ions Zn^{2+}, Co^{2+}, or Ni^{2+}. NMR studies show that the imino protons with coordination to the N3 position of thymine and the N1 position of guanine are replaced by the divalent ions. This means that the natural hydrogen bonds are replaced by *metal coordination interaction*. The absorption and circular dichroism spectra of the M-DNA differ only slightly from those of B-DNA. In every base pair (T–A or C–G), one divalent metal ion is incorporated while one hydrogen bond is retained. The metal ion is "interchelated". The distance between the metal ions is about 0.4 nm regardless of the sequence of the double-stranded M-DNA. Thus, the normal B-DNA can be reformed without denaturation on removal of the metal ion.

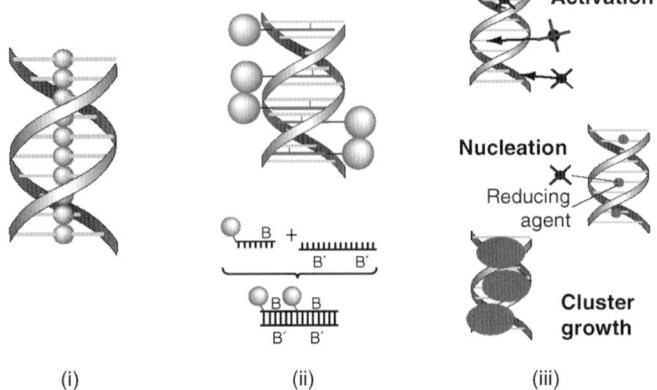

(i)   (ii)   (iii)

Figure 1.61 Possible concepts for improving the electronic properties of DNA: (i) incorporation of metallic ions into the base pair stack; (ii) DNA-directed assembling of metallic or semiconducting nanoparticles, using DNA templates and hybridization of complementary sequences; and (iii) heterogeneous nucleation and growth of metal clusters.

The regular quasiperiodic distribution of the metal ions in the center of the base-paired M-DNA helix implies enhanced electron transport. Resistance measurements in vacuum at room temperature show metal-like conductivity of M-DNA, unlike B-DNA, which behaves as a narrow-bandgap semiconductor (Rakitin et al., 2001). Conductivity measurements were performed on DNA bundles of $\sim 10^2$ molecules. The I–V curves of the B-DNA show a pronounced asymmetry. As a possible explanation, electric fields could build up within the B-DNA bundles, related to double-well potentials associated with the imino protons in the center of the helix with coordination to the N3 of thymine and N1 of guanine in every base pair. In M-DNA, these imino protons are replaced by Zn^{2+} ions leading to a collapse of the double-well potentials (Figure 1.62). The M-DNA behaves as a molecular wire with a diameter of one atom surrounded with a negatively charged insulator. In addition, the wire exhibits all the structural advantages of a natural B-DNA, such as molecular coding of information and the capability of self-assembly.

The substitution of synthetic bases for natural base pairs offers a second route for the incorporation of metal complexes into oligonucleotides of DNA. Tanaka et al. (2006) have developed synthetic nucleosides that form chelator complexes with metal ions such as Pd^{2+}, Pt^{2+}, Cu^{2+}, and Hg^{2+} (Figure 1.63). Two *nucleobases are paired through interstrand metal coordination*. This principle allows a large variety of structures to be generated by incorporation of metal ions into DNA at any position by automated DNA synthesis. Similarly, linear metal arrays can be built by template-directed formation onto preorganized duplexes with appropriate metal binding mismatches. Oligonucleotides have been synthesized that bear salicylic aldehyde mismatches (S–S) as the metal complexing unit and thymine mismatches. The S–S base pair can only complex a metal ion (M = Cu^{2+}, Ni^{2+}, Mn^{3+}, Fe^{3+}, and VO^{2+}) in the presence of an equimolar amount of ethylenediamine (en) (S–M(en)–S). The thymine mismatch (T–T) is used for the Hg^{2+} complexation instead of the pyridine

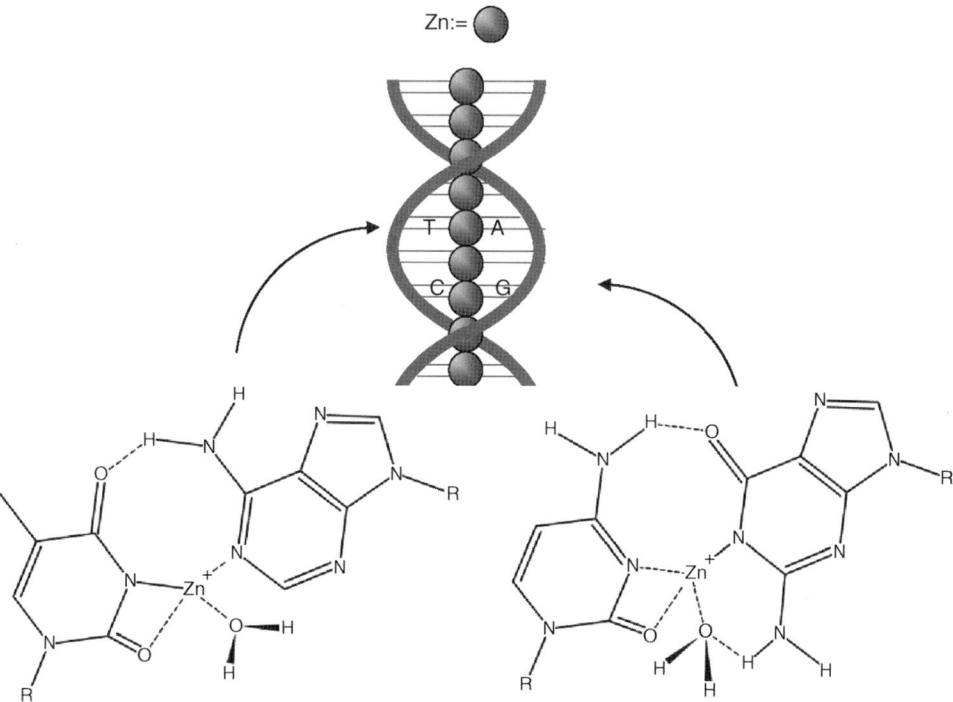

Figure 1.62 Base-pairing scheme for M-DNA with Zn^{2+} (Aich *et al.*, 1999). The imino protons with coordination to the N3 position of thymine and the N1 position of guanine are replaced by the divalent Zn^{2+} ion.

bases (P–P) that are also possible (as shown in the first example). Already Kuklenyik and Marzilli (1996) had demonstrated that two thymine bases in a single-stranded DNA can be cross-linked by a Hg^{2+} ion forming a T–Hg–T pair that can cause conformational changes such as a hairpin–duplex transition. The feasibility of the template concept has been demonstrated by preparing a double-stranded 20-mer with a central cassette of 10 metal complexing units (5 S–S and 5 T–T mismatches) creating a robust bimetallic ion stack inside the DNA double helix (Tanaka *et al.*, 2006). A certain restriction of the Tanaka concept is set by the limited length of synthetic oligonucleotides. The examples show that the development of M-DNA and oligonucleotides with synthetic bases points to a new technology of future molecular electronics in which DNA with *programmable conductivity and magnetism* could be used as nanoelectronic building blocks.

The formation of specific bonds between metal ions and nucleotide bases has found another promising application in nanotechnology for the *development of analytic tools*. Metal ions such as Hg^{2+} and Ag^+ are severe environmental pollutants. Therefore, there is a high demand for a cheap and reliable analytic device for analysis of water and food with respect to such contaminations. The change of the fluorescence of semiconductor quantum dots (QDs) such as CdSe, ZnS, or corresponding alloyed

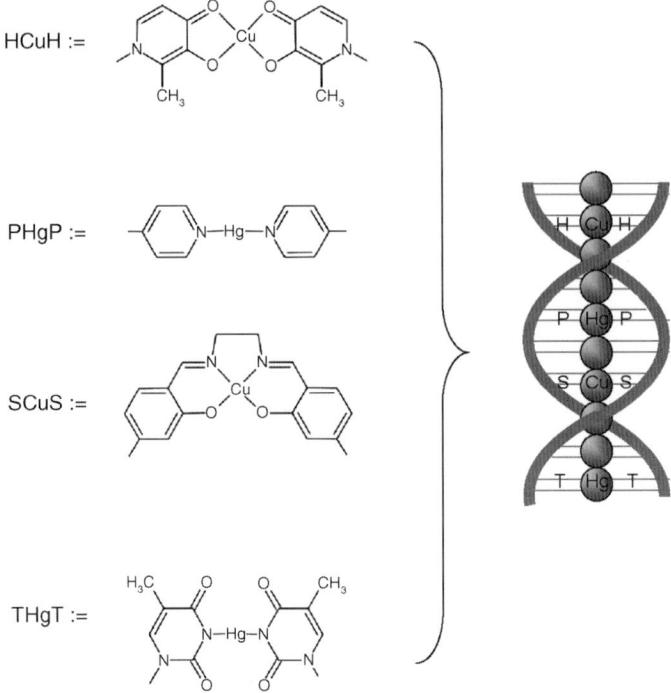

HCuH :=

PHgP :=

SCuS :=

THgT :=

Figure 1.63 Heterogeneous metal assembly through Cu²⁺- and Hg²⁺-mediated duplex formation between two artificial oligonucleotides in which two hydroxypyridone nucleobases (H) and one pyridine nucleobase (P) replace the natural base pairs, and assembly through use of salicylic aldehyde–salicylic aldehyde (S–S) as well as thymine–thymine (T–T) mismatches for heterogeneous metal stacking consisting of S–Cu²⁺(en)–S and T–Hg²⁺–T base pairs (Tanaka *et al.*, 2006).

materials in the presence of metal ions has served as the basic idea for the development of such a sensor system. The fluorescence intensity of the QDs is reduced by electronic energy transfer to metal ions, if there are any nearby. This process, the so-called electron-transfer quenching, can be realized by functionalizing the QDs with oligomers that contain appropriate sequences of thymine and cytosine (see also Figure 1.64). Hg²⁺ ions bridge thymine bases. Similarly, Ag⁺ ions form stable cytosine bridges. By an appropriate arrangement of these bases along the oligomers **1** and **2**, a rigid hairpin structure is formed in the presence of Hg²⁺ and/or Ag⁺ ions, respectively. If QDs of differential sizes, which fluoresce at differential wavelengths, are functionalized with T- or C-modified oligomers, a multiplexed analysis can be realized.

The same principle can be applied to the design of logic gate operations (Freeman *et al.*, 2009). Figures 1.65 and 1.66 show how the two basic states "0" and "1" can be realized by making use of the specific electron-transfer quenching of two QDs of different sizes by Hg²⁺ or Ag⁺ ions as input. Quenching of both wavelengths (in the presence of both ions) may represent the signal "1", whereas quenching of only one

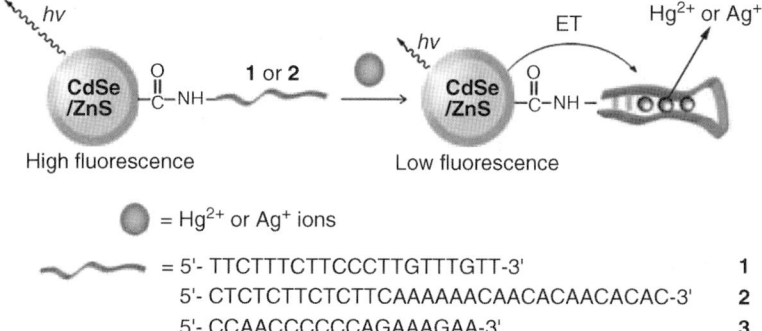

= Hg²⁺ or Ag⁺ ions

= 5'- TTCTTTCTTCCCTTGTTTGTT-3'   1

5'- CTCTCTTCTCTTCAAAAAACAACACAACACAC-3'   2

5'- CCAACCCCCCAGAAAGAA-3'   3

Figure 1.64 Optical analysis of Hg²⁺ and Ag⁺ ions by nucleic acid modified QDs. Examples for possible oligonucleotide sequences are **(1)** 5'-TTCTTTCTTCCCTTGTTTGTT-3' and **(2)** 5'- CTCTCTTCTCTTCAAAAAACAACACAACACAC- 3'. (Reproduced with permission from Freeman *et al.* (2009). Copyright 2009, Wiley-VCH Verlag GmbH.)

wavelength (in the presence of only one of the ion species) may represent the signal "0". The proposed system can be considered as a first step for the implementation of QDs as optical readout signals for logic gate operations.

DNA-Directed Assembly of Metallic or Semiconducting Nanoparticles Using DNA Templates and Hybridization of Complementary Sequences   Prefabricated nanoparticles can be assembled along a DNA by covalent or electrostatic interaction. An

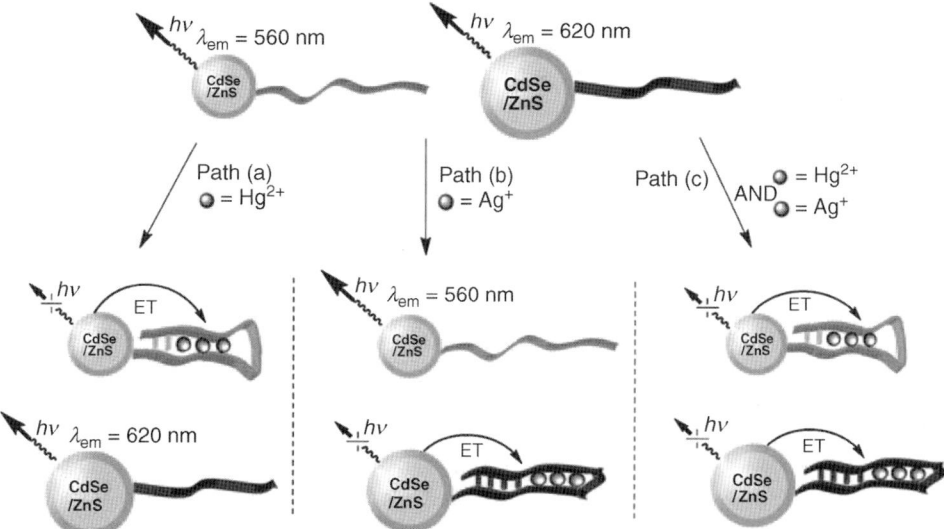

Figure 1.65 The "AND" logic gate system based on the electron-transfer (ET) quenching of **1**- and **2**-QDs by Hg²⁺ and Ag⁺ inputs. (Reproduced with permission from Freeman *et al.* (2009). Copyright 2009, Wiley-VCH Verlag GmbH.)

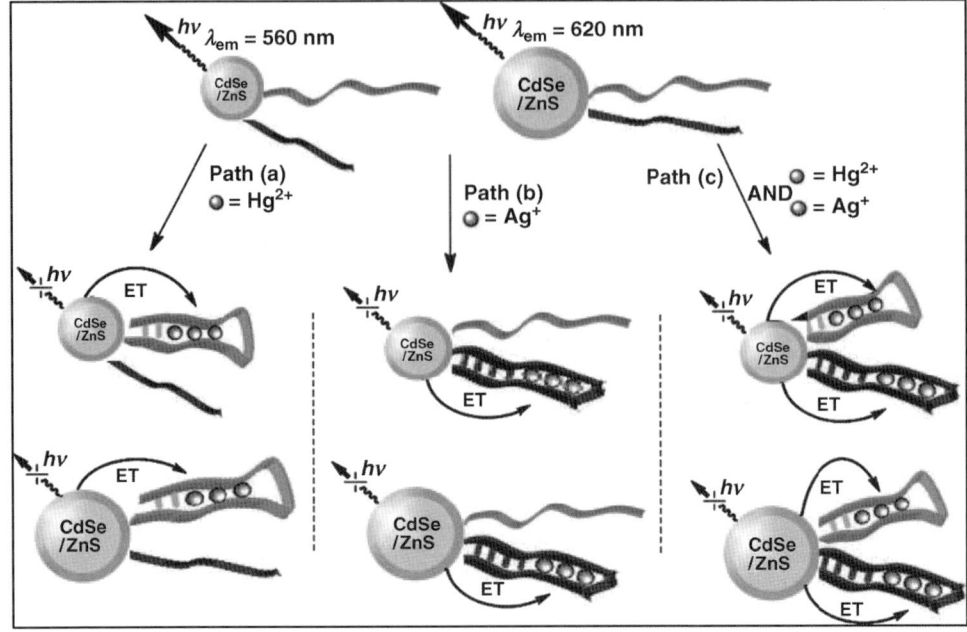

Figure 1.66 The "OR" logic gate system based on the electron-transfer quenching of **1**- and **2**-QD by Hg^{2+} and Ag^+ inputs. (Reproduced with permission from Freeman *et al.* (2009). Copyright 2009, Wiley-VCH Verlag GmbH.)

example for covalent integration has been elaborated by the Willner group in 2002 (Patolsky *et al.*, 2002). Site-specific covalent binding of 1.4 nm Au nanoparticles with thymine bases in a double-stranded poly(A)–poly(T) duplex has been realized in a three-step process. In the first step, the Au particles functionalized with mono-*N*-hydroxysuccinimide are covalently linked with amino psoralen (Figure 1.67), which acts as intercalator in the second step when the modified Au particles are mixed with a solution of poly(A)–poly(T) duplexes of about 900 nm length. After 20 min, the mixture is irradiated with UV light ($\lambda < 360$ nm) for 45 min, which causes a photo-induced covalent attachment of the intercalator at the thymine residues. AFM images prove that a well-organized Au nanoparticle wire is formed in the poly(A)–poly(T) template (Figure 1.68). The average height of 4 nm can be interpreted as the intercalation of single nanoparticles in the helical double-stranded DNA.

Harnack *et al.* (2002) have used Au nanoparticles (1.3 nm) capped with negatively charged tris(hydroxymethyl)phosphine (THP) for the metallization of calf thymus DNA. First, the DNA molecules were spin coated on a silicon substrate in order to stretch the molecules. The solution of THP–Au colloids was then applied to the substrate for 2–15 min, rinsed, and dried. Surprisingly, it was possible to create a dense deposition of the negatively charged particles along the negatively charged DNA. Obviously, the electrostatic interaction was screened by the presence of additional molecules in the solvent (ethanol–water mixture). It is assumed that short-range

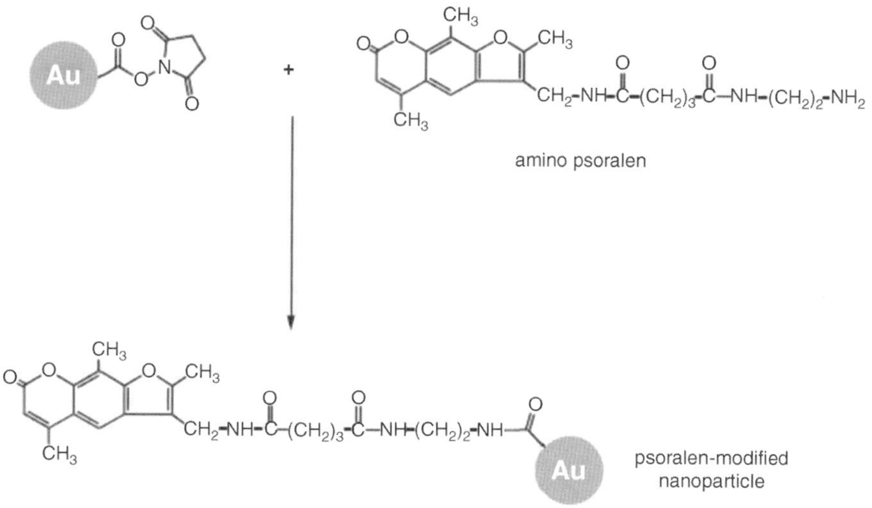

CH₃

amino psoralen

psoralen-modified
nanoparticle

Figure 1.67 Functionalization of Au nanoparticles with amino psoralen, which facilitates intercalation with poly(T)–poly(A) duplexes. (Reproduced with permission from Patolsky *et al.* (2002). Copyright 2002, Wiley-VCH Verlag GmbH.)

interactions such as hydrogen bonds between THP and DNA facilitate the binding. There can also be covalent bonds between THP and amino groups of the DNA bases. In the final process step, the tiny Au nanoparticles were enlarged by electroless gold plating, which provided a nanowire with a diameter of about 30–40 nm. The wires showed ohmic behavior with resistivity of $\sim 3 \times 10^{-5}\,\Omega$ m, which is about 1000 times higher than the bulk resistivity of gold ($2 \times 10^{-8}\,\Omega$ m). A similar strategy has been applied by Sastry *et al.* (2001) exploiting the electrostatic interaction of positively charged Au nanoparticles (4 nm in size) with the negative charges of the phosphate backbones of double-stranded DNA. The gold clusters were capped with lysine

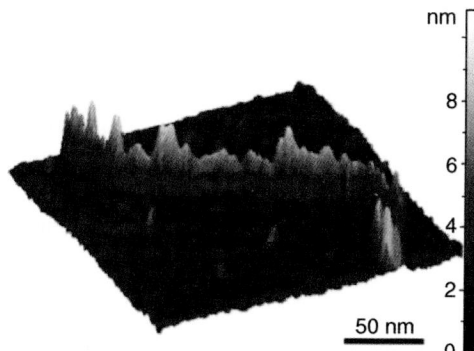

Figure 1.68 AFM image of a chain of Au nanoparticles intercalated in a poly(A)–poly(T) template. (Reproduced with permission from Patolsky *et al.* (2002). Copyright 2002, Wiley-VCH Verlag GmbH.)

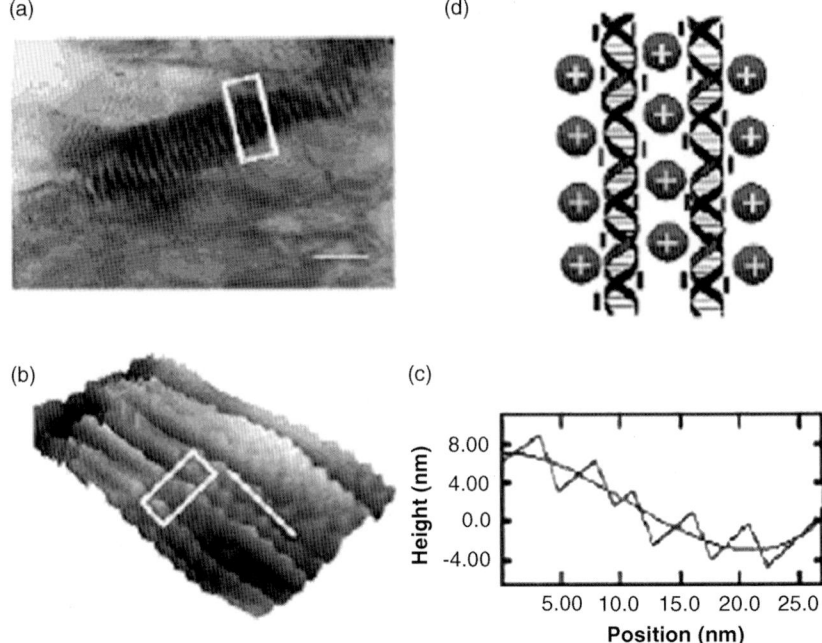

(a)

(b)

(c)

(d)

Figure 1.69 Formation of Au superclusters on a band-like template of oligonucleotides. (a) TEM picture of a Au supercluster mediated by a template of 15-mer oligonucleotides. (b) Scanning tunneling microscopy (STM) image of a Au supercluster grown on a template of 30-mer oligonucleotides. (c) Height change profile along the line in the STM image. (d) Proposed structure model for the band-like Au supercluster. (Reproduced with permission from Kumar et al. (2001). Copyright 2001, Wiley-VCH Verlag GmbH.)

molecules causing a positive surface charge at pH 7. Fifteen- and 30-mer oligonucleotides were hybridized with complementary oligomers. Films of the short double-stranded DNA templates were cast on a Si wafer or quartz substrate. After drying the film, the Au particle solution was dropped onto the film. As shown in the TEM pictures of Figure 1.69, bands of closely packed Au superclusters were formed where the width of the bands is in the range of the length of the DNA duplexes. The positive charges of the lysine molecules bound to the gold particles and the negative charges of the phosphate backbones of the short DNA duplexes cause the assembly to form a "line-by-line" structure in the film. This is similar to the "layer-by-layer" structure of anionic and cationic polyelectrolytes organized onto a planar substrate (see also Section 4.3). Obviously, the initial DNA structures were still sufficiently mobile to relax into the energetically favored band-like line-by-line structures.

Heterogeneous Nucleation and Growth of Metal Clusters The various functional groups of a DNA can also serve as nucleation sites for heterogeneous nucleation of metal clusters starting with aqueous solution of metal salts (Mertig et al., 2002; Seidel, Mertig,

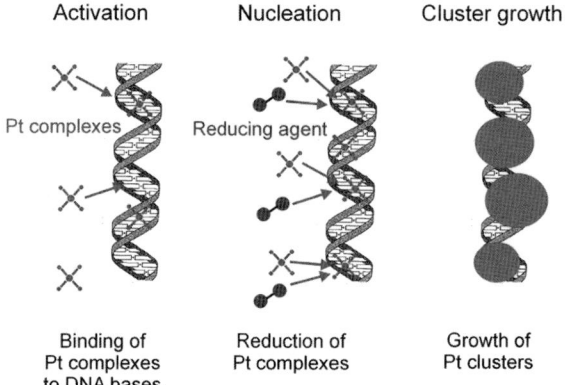

Activation	Nucleation	Cluster growth

Pt complexes   Reducing agent

Binding of Pt complexes to DNA bases	Reduction of Pt complexes	Growth of Pt clusters

Figure 1.70 DNA metallization by heterogeneous nucleation as a two-step process: activation by site-specific binding of metal complexes; formation of seeds by adding reducing agent; and stable cluster growth from seeds. (Reproduced with permission from Mertig and Pompe (2004). Copyright 2004, Wiley-VCH Verlag GmbH.)

and Pompe, 2002; Seidel *et al.*, 2004). By stable growth of these clusters and their aggregation, continuous metallic wires can be produced on the DNA template. This idea is based on the work by Coffer, Bigham, and Li (1996) and Braun *et al.* (1998). Coffer and coworkers had deposited Cd^{2+} ions on plasmid DNA. By reducing them with H_2S, chains of 5 nm CdS clusters at the ring-shaped templates were produced. Braun *et al.* adapted this concept for metallization of λ-DNA. For this purpose, λ-DNA was stretched between electrodes and activated with silver salt solution. Afterwards, the immobilized silver complexes acted as nucleation sites for the growth of silver clusters in the presence of hydroquinone as a reducing agent. Finally, a granular conducting wire of 100 nm in diameter was obtained.

The principle of heterogeneous nucleation and growth of metal clusters on DNA is shown in Figure 1.70. The *electroless metallization of DNA* involves two steps. First, metal ions or complexes are immobilized at specific binding sites. Platinum, and palladium complexes are preferentially bound at the bases. *The DNA bases act as electron donors.* It has been shown that the N7 position of guanine bases is a preferred site for covalent binding of hydrolyzed metal complexes (Lippert, 1999; Colombi Ciacchi, 2002). Other metal ions such as cadmium and copper bind preferentially to the phosphates at the backbone. Typically, the reaction time for sufficient activation of the DNA is in the range of a few hours. For example, the activation with platinum complexes starting with K_2PtCl_4 is saturated after ~10 h. In the following process step, a *reducing agent* is added in order to provide electrons for the further reduction of the bound metal complexes. *Suitable reducing agents* are dimethylaminoborane (DMAB, $(CH_3)_2NH \cdot BH_3$), hydroquinone (HQ, $C_6H_4(OH)_2$), sodium borohydride ($NaBH_4$), or hydrogen. Instead of applying a reducing agent, photoreduction is another option. For example, silver (Berti *et al.*, 2005) and platinum (Erler, Guenther, and Mertig, 2009) particles have been grown by exposing the DNA with metal complexes to 254 nm UV irradiation. It is assumed that the DNA acts as a light harvester that favors the localized metallization. The bound metal complexes are reduced to seeds for the *stable growth of clusters* along the DNA. A typical example of a metallized λ-DNA is shown in Figure 1.71 (Mertig *et al.*, 2002). High-resolution transmission electron microscopy reveals

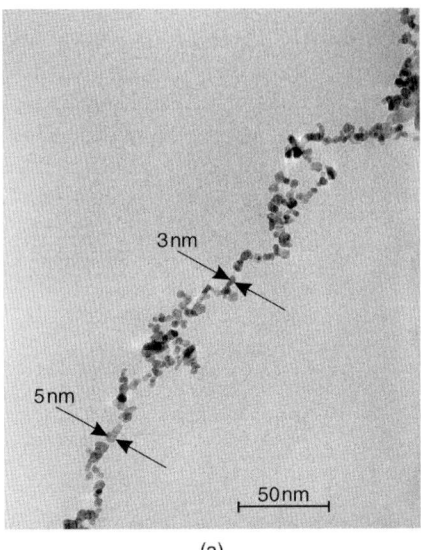

(a)

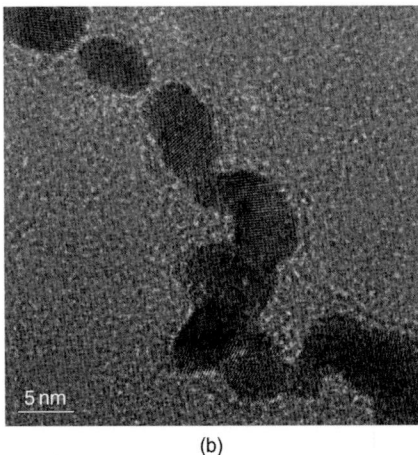

(b)

Figure 1.71 Pt cluster chain grown on λ-DNA. The DNA was incubated in a K_2PtCl_4 solution for 10 h, keeping the ratio of hydrolyzed metal complexes to nucleotides at 65 : 1. The reducing agent DMAB has been applied for the initiation of cluster growth. (a) TEM image of a continuous chain of platinum clusters (diameter: 4 ± 1 nm). (b) High-resolution TEM image of the platinum clusters grown on the DNA. The lattice planes of the single crystals can be seen. Several clusters are fused, thus forming a short conducting path along the DNA. (Reproduced with permission from Mertig and Pompe (2004). Copyright 2004, Wiley-VCH Verlag GmbH.)

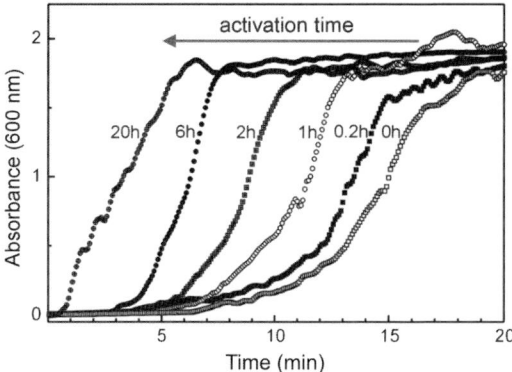

Figure 1.72 Optical absorbance versus time of a hydrolyzed K_2PtCl_4 solution at 600 nm in the presence of λ-DNA. (Reproduced with permission from Mertig and Pompe (2004). Copyright 2004, Wiley-VCH Verlag GmbH.)

that the clusters are single crystals, where the (111) lattice plane distance of $d = 0.227$ nm corresponds to that of bulk platinum. As a characteristic feature of the experiment, there is no indication of homogeneous nucleation of platinum clusters in the solution. This is unexpected because only about 3% of the available Pt complexes are bound to the DNA during the activation. The heterogeneous nucleation sites are highly effective. They act as autocatalytic reaction sites in the subsequent growth process. The influence of the activation on the rate of heterogeneous nucleation is also evident from optical absorption measurements during the metallization reaction. The change of the absorbance indicates the formation of colloidal platinum at the DNA. The reduction time needed for complete metallization of the DNA decreases with longer activation time (Figure 1.72). The role of the nucleotides in the metallization process becomes obvious from the plots of reduction time versus activation time for several ratios of G–C and A–T pairs (Figure 1.73). The reduction time is defined here as the time required to reach half saturation of the absorbance. The lower reduction time for

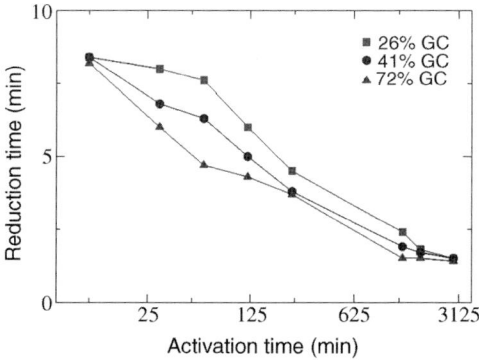

Figure 1.73 Influence of the GC content of the DNA on the metallization kinetics. (Reproduced with permission from Mertig and Pompe (2004). Copyright 2004, Wiley-VCH Verlag GmbH.)

higher G–C content indicates higher binding efficiency of the G bases compared to the A bases for platinum complexes.

The stochastic nature of the growth process and the interface stresses of touching clusters with different crystallographic orientations are the reasons for a fluctuating shape of the ultrathin cluster chains. Therefore, metallized DNA suitable as nanowire for bridging gaps of a few micrometers has to be thicker, typically ≥ 50 nm. In the temperature range from 300 down to 4.2 K, the relation between current and voltage is linear for wires of such type (Richter *et al.*, 2002), indicating metallic conductivity. Interestingly, here the conductivity of the nanowires is only a factor of about 7 lower than that of the bulk material. The conductivity of 2×10^4 S/cm corresponds to a mean free electron path of 2 nm, which is about the same size as the grown clusters (Figure 1.74). Hence, the observation is compatible with the concept of conductivity being governed by the scattering of the electrons at the grain boundaries. The presence of metallic conductivity is confirmed by the linear dependence on temperature in a wide range. The conspicuous deviation at about 30 K and below is due to the *quantum nature of the electron scattering process.* The

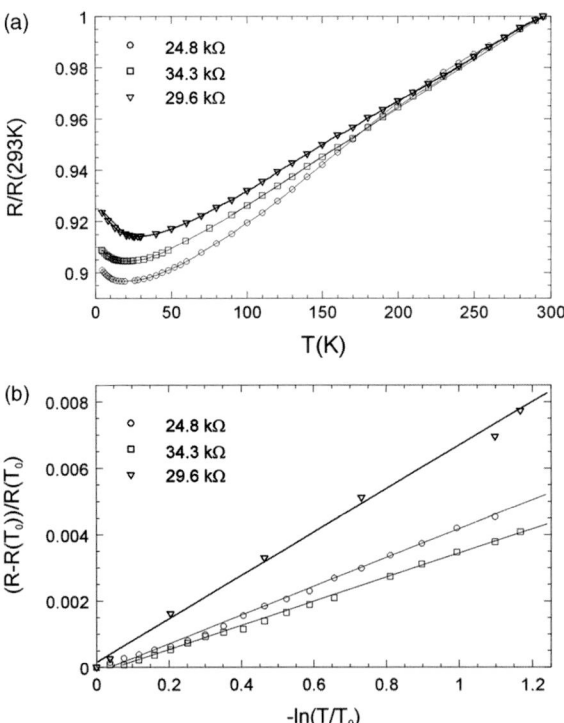

Figure 1.74 (a) Temperature dependence of the resistance of three palladium wires, normalized to their room-temperature resistances of 24.8, 29.6, and 34.3 kΩ. (b) Low-temperature resistance versus $-\ln(T/T_0)$, with $T_0 = 13.5$ K. (Reproduced with permission from Richter *et al.* (2002). Copyright 2002, Springer + Business Media.)

logarithmic dependence of the increase below 30 K indicates that the electron transport in the nanowires corresponds rather to *scattering processes in a thin film* (2D) than in a 1D structure. Annealing of the Pd nanowires for 2 h at 200 °C caused grain growth resulting in lower resistance at room temperature, higher temperature coefficient of the resistance, and disappearance of the quantum behavior at low temperature.

The advantage of DNA for heterogeneous nucleation and for growth of metallic nanowires is still more obvious when additionally the molecular recognition properties are exploited. This opens the way to *molecular lithography*. Braun and coworkers have given an impressive example for such an option by patterning a gold nanowire at molecular resolution (Keren *et al.*, 2002), as shown in Figure 1.75a. The basic idea is a sequence-specific covering of short areas along the DNA *with enzymes*. In the example, RecA was chosen, which facilitates the so-called homologous recombination, a protein-mediated reaction by which two DNA molecules with homologous sequences are linked. RecA monomers can be polymerized on a selected DNA sequence to form a *nucleoprotein filament*. Such filaments are then mixed with the target, a solution of double-stranded DNA. It leads to *site-specific covering of the target DNA due to binding at the sites of the homologous recombination*. This routine process from molecular biology can be implemented in the metallization process. In step (i), RecA monomers polymerize on a single-stranded DNA molecule to form the nucleoprotein filament. In step (ii), the nucleoprotein filament binds to an aldehyde-derivatized double-stranded target DNA at a homologous sequence. The bound aldehyde groups act as localized reducing agents in the following nucleation of metal clusters along the derivatized DNA. In step (iii), incubation in $AgNO_3$ solution results in Ag nuclei arranged along the target DNA at regions unprotected by RecA. In step (iv), the Ag nuclei serve as catalysts for electroless gold deposition, which converts the unprotected regions into conductive gold wires. In this step, hydroquinone is applied as a reducing agent. Thus, the site-specific deposition of RecA on aldehyde-derivatized DNA enables the manufacturing of patterned nanowires with well-defined isolating gaps in between. The efficiency of this process has been demonstrated by producing gold nanowires (diameter 50 nm) with an insulating gap region covered with a 2027-base RecA nucleoprotein filament. The approach described has been refined by changing the sequence of process steps by marking already the natural DNA with a RecA pattern before aldehyde derivatization is performed, as schematically shown in Figure 1.75b (Keren, Berman, and Braun, 2004). The derivatization of the DNA with glutaraldehyde occurs only at regions that are not covered with RecA protein. After removal of excess glutaraldehyde, the RecA protein can be disassembled so that underivatized regions along the DNA are left over. Subsequent metallization occurs only at the aldehyde-derivatized regions whereas the underivatized DNA segments are available for biomolecular manipulation based on the recognition properties of DNA. One interesting option is the immobilization of metallic clusters or semiconducting quantum dots at those sites (as schematically shown in Figure 1.75b), or biotinylated DNA strands to create junctions and networks in nanoelectronic devices.

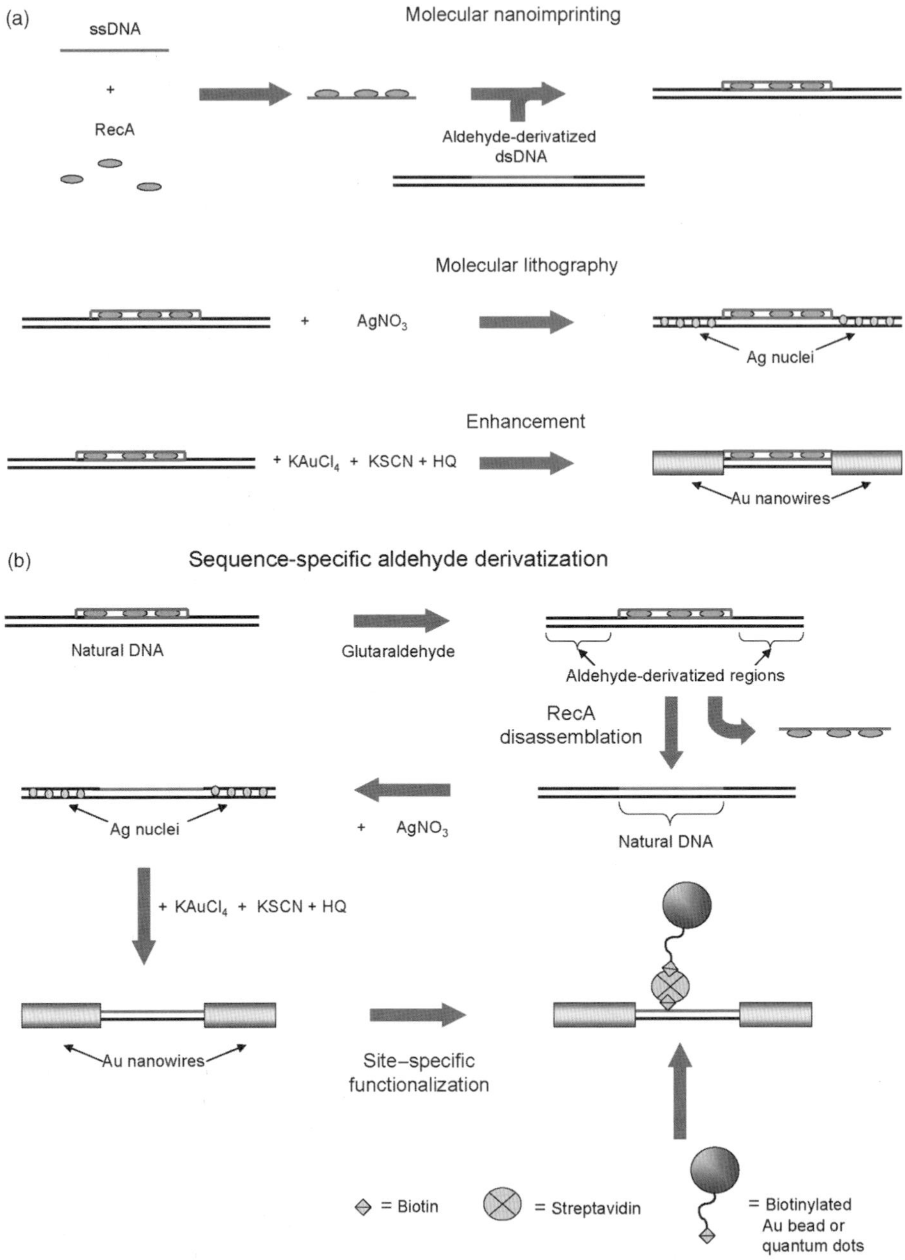

(a)

ssDNA   Molecular nanoimprinting

+

RecA   Aldehyde-derivatized
                                              dsDNA

Molecular lithography

+   AgNO₃

Ag nuclei

Enhancement

+ KAuCl₄ + KSCN + HQ

Au nanowires

(b)   Sequence-specific aldehyde derivatization

Natural DNA   Glutaraldehyde

                                              Aldehyde-derivatized regions

                                              RecA
                                              disassemblation

Ag nuclei   +   AgNO₃

                                              Natural DNA

+ KAuCl₄ + KSCN + HQ

Au nanowires   Site–specific
                         functionalization

◇ = Biotin   ⊗ = Streptavidin   = Biotinylated
                                            Au bead or
                                            quantum dots

Figure 1.75 Schematics of the homologous recombination reaction and molecular lithography. (a) The site-specific recombination of the structured ssDNA with the aldehyde-derivatized dsDNA leads to a patterned template for the following metallization of the unprotected regions of the DNA template. HQ = hydroquinone. (b) Localization of a reducing agent, glutaraldehyde, on selected DNA sequences directs their metallization. DNA retains its recognition capability and other biological functionalities at the nonderivatized region. (Adapted from Keren, Berman, and Braun (2004).)

1.3.2.2 DNA-Based Nanoprobes

The unique properties of metallic and semiconducting nanoparticles (NPs), related to fluorescence, magnetism, and electronic behavior, make them interesting for a broad range of applications including materials technology, biotechnology, and medicine. Mostly applied are NPs of gold, magnetic alloys, ZnS, CdS, CdSe/ZnS, silica, diamond, and nanorods (NRs), such as carbon nanotubes. Typical sizes of the NPs range from 2 to 50 nm. Advanced preparation technologies guarantee well-defined particle sizes and shapes. Variation of sizes and shapes as well as alloying offers additional degrees of freedom for tuning the physical properties, as shown by the examples depicted in Figures 1.76 and 1.77 for the optical properties (extinction and plasmon resonance) of

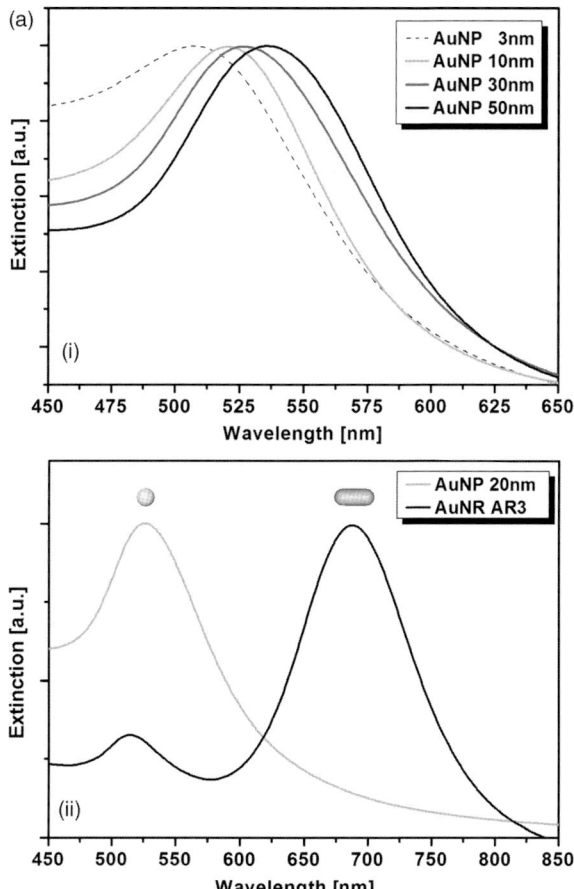

Figure 1.76 (a) UV–Vis spectra of spherical gold nanoparticles and gold nanorods. The peak positions of the plasmon resonance due to collective oscillation of the valence electrons depend on the particle diameter (i), and the aspect ratio (AR) (here AR = 3 with diameter of 20 nm) (ii). (b) Variation of size and shape of Au nanoparticles depending on growth conditions: (i) Au nanospheres; (ii) spherical Au cores with a magnetite shell; (iii, iv) Au nanorods; and (v, vi) prismatic Au nanoparticles. (Reprinted with permission from Lakatos (2013).)

(b)

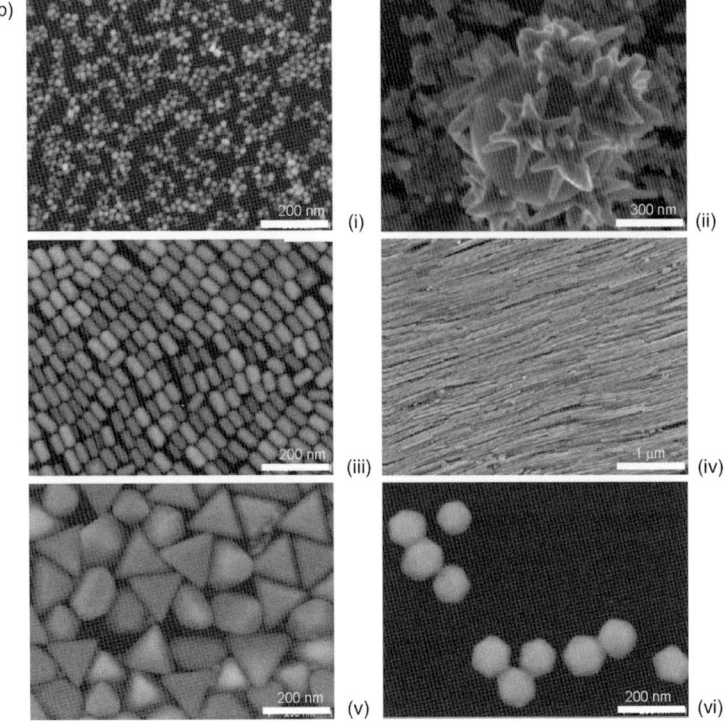

Figure 1.76 (*Continued*)

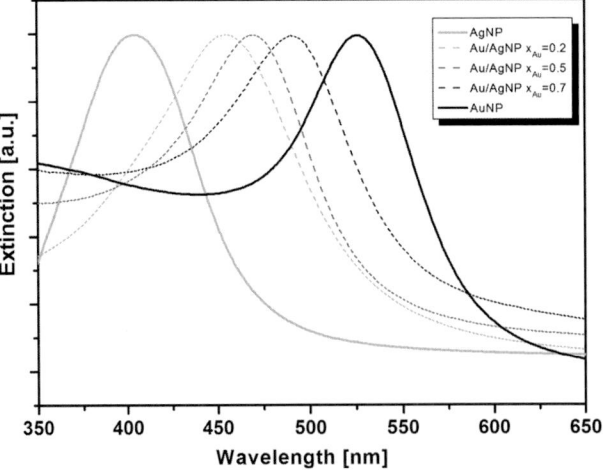

Figure 1.77 Change of the UV–Vis spectra of alloyed Au/Ag nanoparticles. With increasing Au concentration (x_{Au}), the plasmon resonance shifts from yellow to red. (Reprinted with permission from Lakatos (2013).)

gold nanoparticles (AuNPs), gold nanorods (AuNRs), and alloyed Au/AgNPs. Hence, the physical and chemical properties of nanomaterials are tunable. As another extension of versatility, nanomaterials can be functionalized with biomolecules. In this way, molecular recognition patterns can be utilized for biologically directed self-assembly of biofunctionalized nanoparticles into higher order structures. These topics will be discussed in detail in Chapter 7. Respective developments were initiated with DNA-conjugated nanoparticles. Therefore, they are considered here as an introduction into the broader field.

It is not surprising that the breakthrough in the application of DNA-conjugated nanoparticles was realized in biodiagnostics. Nanoprobes capable of reading biomolecular signals and translating them into physical or chemical signals are highly desirable in this field (Rosi and Mirkin, 2005). Diagnostic assays for DNA, proteins, small molecules, or metal ions in life sciences as well as in environmental technology are examples for this development. Research activities are aiming at inexpensive and disposable systems that allow rapid and user-friendly data acquisition. At the same time, the novel systems should exhibit a sensitivity meeting or even exceeding the current standard. In the area of DNA diagnostics, the standard is represented by the polymerase chain reaction (PCR) in combination with molecular fluorophore technology. In PCR, the theoretically achievable sensitivity is nearly unlimited. However, complexity, cost, and restrictions in multiplexing (detection of multiple targets in one assay) set serious bounds in praxis with a sensitivity of about 100 aM. For protein diagnostics, the new techniques have to compete with the sensitivity realized with enzyme-linked immunosorbent assays (ELISA) (~pM limit).

In 1996, the Alivisatos group and the Mirkin group published independently two basic papers on "nanocrystal molecules" combining the recognition properties of DNA with the physical properties of nanobeads (Alivisatos *et al.*, 1996; Mirkin *et al.*, 1996). These structures have been produced by attaching single-stranded DNA oligonucleotides of defined lengths and sequences to individual AuNPs. The thiol-terminated oligonucleotides have been covalently linked to the gold particles. These AuNPs can be assembled into complex structures by adding additional single-stranded DNAs that are partially complementary to every one of the immobilized oligonucleotides (Figure 1.78a). Densely packed lattices of gold clusters are produced by self-assembly. In the case of short oligonucleotides and short target DNA, the small distance between the gold beads leads to mutual electronic interaction. This interaction affects the plasmon resonance spectrum of the particle in the form of a red shift of the plasmon peak. The change of light absorption turns the color of the colloidal gold solution from red to blue. Furthermore, the interaction between the gold beads affects the melting transition of the DNA. The melting transition of the DNA in assemblies with metallic nanoparticles is significantly sharper than that of unmodified DNA. Apparently, the melting of DNA in such aggregates is a highly cooperative process (see also Section 1.2.4), and the presence of the gold nanoparticles obviously enhances the cooperativity (Jin *et al.*, 2003). The transition profile is affected by various factors, such as the DNA surface density, the nanoparticle size, the interparticle distance, and salt concentration in the solvent. An increasing oligonucleotide density on the nanoparticle surface leads to an increase of the

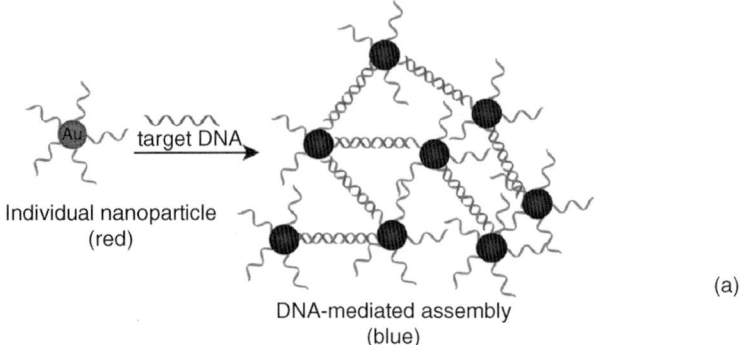

(a)

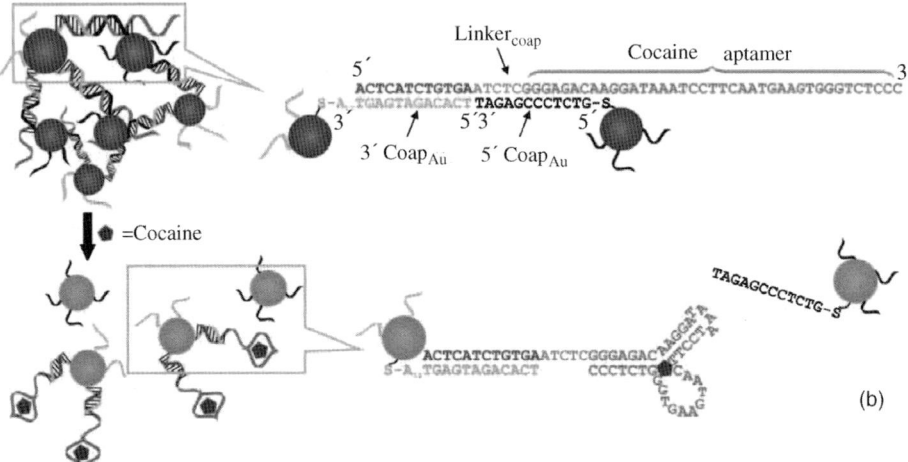

(b)

Figure 1.78 Au nanoparticle-based colorimetric sensors. (a) In the presence of complementary target DNA, oligonucleotide-functionalized AuNPs aggregate, turning the color of the solution from red to blue. (Reproduced with permission from De, Ghosh, and Rotello (2008). Copyright 2008, Wiley-VCH Verlag GmbH.) (b) Aptamer-based colorimetric detection of cocaine by disassembling of the nanoparticles. (Reproduced with permission from Liu and Lu (2006). Copyright 2006, Wiley-VCH Verlag GmbH.)

melting temperature and a sharper profile. Larger particle size is connected with a higher number of links. Increasing the salt concentration causes an enhanced charge screening of the negatively charged DNA, and hence an increase of the melting temperature. Below a critical concentration, melting at room temperature is observed. Increasing the interparticle distance by adding poly$(A)_n$ spacer segments at each end of the DNA duplex increase the melting temperature as the electrostatic repulsion of the nanoparticles is diminished with the larger distance.

In another approach, aptamers, which are single-stranded oligonucleic acid-based binding molecules with the ability to bind a wide range of targets with high affinity and specificity, have been incorporated into such colorimetric sensor. An example has been worked out for cocaine sensing, using a cocaine-specific aptamer by Liu and Lu (2006). For this purpose, the Au nanoparticles were functionalized with two different sequences of single-stranded DNA, one for conjugation and another one with the cocaine aptamer as shown in Figure 1.78b. The sensor assay was made of two kinds of Au nanoparticles, one kind functionalized with a 3'-thiol-modified DNA (3'Coap$_{Au}$) and another one functionalized with a 5'-thiol-modified DNA (5'Coap$_{Au}$). To create aggregates, linker DNA (Linker$_{Coap}$) molecules have been used. The linker was composed of three segments. The first segment hybridized with a 3'Coap$_{Au}$ nanoparticle. The second segment hybridized with the last five nucleotides of a 5'Coap$_{Au}$ nanoparticle. The third segment was the aptamer sequence for cocaine. By adding the Linker$_{Coap}$ to a solution of the two kinds of Au particles, purple-colored aggregates were formed. In the presence of cocaine, the aptamer changed its structure to bind cocaine. As a result, only five base pairs were left to hybridize with 5'Coap$_{Au}$. This link was unstable at room temperature. The 5'Coap$_{Au}$ particles dissociated from the 3'Coap$_{Au}$ particles. The disassembly of the aggregates caused a color change of the solution from purple to red.

The sharp melting transition is one of the most important advantages for use of metallic nanoparticles as colorimetric sensors in chip-based systems. It is the basic concept of the so-called *scanometric assay* (Taton et al., 2000). Different capture strand DNAs are immobilized on a glass chip. The target DNA of interest is recognized by a particular capture strand. A sequence of the captured target DNA is then labeled with a specific oligonucleotide–nanoparticle conjugate. Nonspecifically bound targets are washed away. In the final step, the bound gold particles can be enlarged by catalytic reduction of silver onto the surface. The bound structures are evaluated by light scattering. The scattered light is much more intense than the fluorescent light. For example, the intensity scattered by one 80 nm gold particle is comparable to that emitted from 10^6 fluorescein molecules. As another advantage, the signal is resistant to quenching. As target concentrations as low as 50 fM can be detected with this approach, this technique possesses a sufficiently high sensitivity to possibly replace the costly PCR in medical praxis.

The catalytic silver enlargement of immobilized AuNPs can also be applied for the *electrical detection* of the target molecule (Park et al., 2002). In such a setup, oligonucleotide capture strands are immobilized in the gap between two electrodes (Figure 1.79). Analogous to the previous example, target DNAs together with appropriately functionalized AuNPs form sandwich complexes that can be enlarged by the catalytic silver deposition. The target DNA is detected by the change of the electrical resistance between the two electrodes. For a nonoptimized version, a detection limit of 500 fM has been reported. There is another outstanding feature of the experiment. It allows a stringent differentiation between perfectly complementary DNA sequences and those with a single mismatch. As discussed above, the sandwich structures are characterized by a sharp melting transition. Therefore, when the chip is washed at 50 °C before treatment with silver enhancer solution, the complexes with a single-base mismatch are denatured, whereas the complexes with perfectly complementary sequences remain stable. After the subsequent silver enhancement for 3–20 min, the resistance of gaps spotted with perfectly complementary strands is typically 500 Ω, whereas the gaps with single-base mismatches

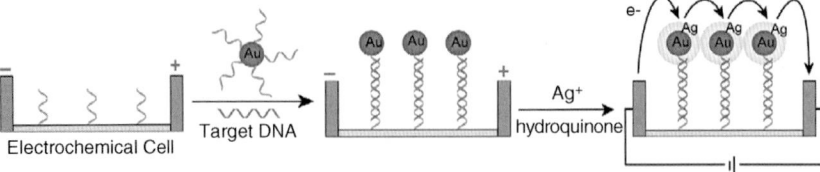

Figure 1.79 Electrical detection of DNA. Capture oligonucleotides whose one end is fixed to a patterned area in the gap between two electrodes bind to a complementary part of the target DNA in the solution. Oligonucleotides whose one end is fixed to an AuNP bind to another complementary part of the target DNA so that the AuNP becomes tethered to the substrate. Catalytic deposition of silver onto the sandwiched structure leads to a composite structure that can be detected electrically by the change of the electrical resistance. (Reproduced with permission from De, Ghosh, and Rotello (2008). Copyright 2008, Wiley-VCH Verlag GmbH.)

behave as insulators, with resistances exceeding $200\,M\Omega$. Differentiation of perfectly and nonperfectly hybridized complexes can also be realized by exploiting the influence of salt solutions on the stability of hybridized structures. Washing the chip with solutions with suitable concentrations of cations (e.g., Na^+) at room temperature before silver amplification is another possibility to denature the single-base-mismatch complexes. The electrical detection approach can be realized in conventional microelectrode arrays. Thus, it is well suited for massive multiplexing, when a larger array of electrode pairs is spotted with different capture oligonucleotides.

The sensitivity of the assay can be increased about 100-fold by exploiting the interaction of metal particles and electromagnetic waves. Under laser irradiation, strong electric fields arise near the particles. Depending on the particle size and shape, the local power density can exceed that of the laser beam by 10^6–10^8. This phenomenon is utilized for surface- or particle-enhanced Raman spectroscopy (SERS). It is one of the most sensitive spectroscopic techniques in chemistry. Multiplexed spectrometric detection has been realized with conjugates of gold nanoparticles and *Raman dye-labeled oligonucleotides* (Cao et al., 2002) (Figure 1.80). Compared to conventional fluorescence assays with multiple fluorophores as labels, Raman spectroscopy is favorable for several reasons: narrow spectroscopic lines, broader spectral window, and single-wavelength laser sufficient for excitation. By the additional silver enhancement, a sensitivity of $\sim$20 fM was reached in a nonoptimized experiment.

The signal amplification with silver-coated AuNPs leads to a maximum sensitivity of $\sim$100 aM in scanometric detection of target DNA, provided that a preceding target amplification by PCR is applied. However, PCR is a time-consuming process, which is a disadvantage for practical use. There is another alternative approach with less expensive and faster indirect target amplification, called the *bio-barcode amplification* (BCA) (Nam et al., 2002, 2004). The basic idea of the BCA approach is summarized in Figure 1.81. Two different types of biofunctionalized probes are applied in the procedure. The first one is a magnetic iron oxide microparticle (MMP) deposited with single-stranded capture oligonucleotides that are complementary to one half of the sequence of the target oligonucleotide (analyte). The second probe is an AuNP (about 30 nm size) functionalized with two different types of oligonucleotides. One

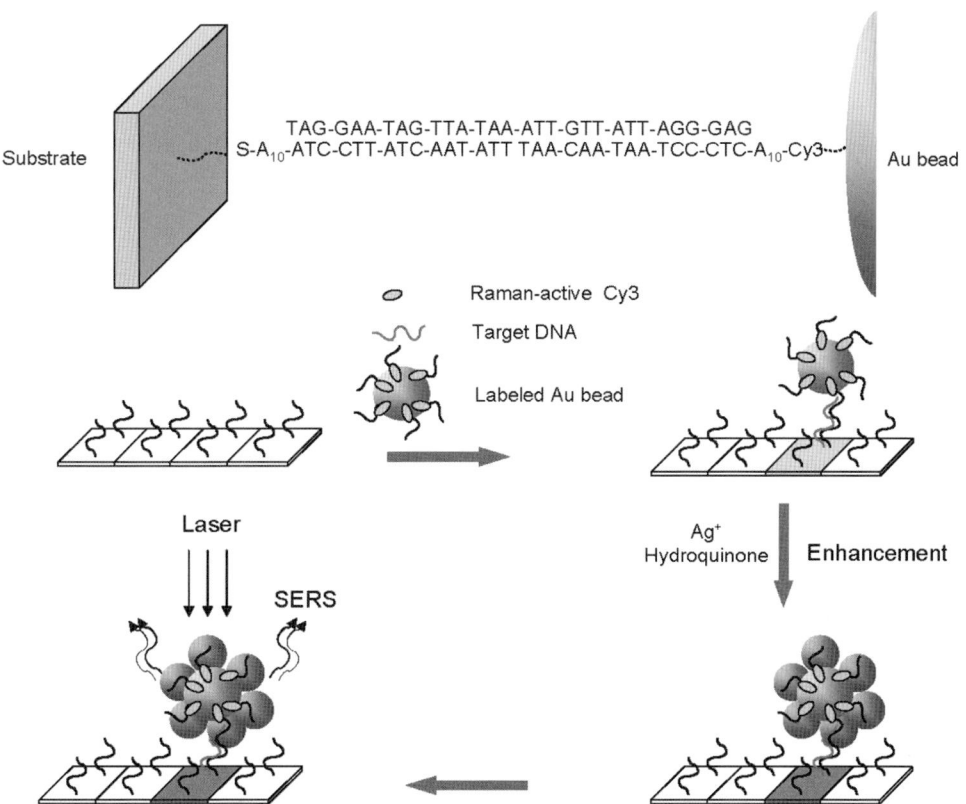

TAG-GAA-TAG-TTA-TAA-ATT-GTT-ATT-AGG-GAG
Substrate   S-A$_{10}$-ATC-CTT-ATC-AAT-ATT TAA-CAA-TAA-TCC-CTC-A$_{10}$-Cy3····   Au bead

Raman-active Cy3

Target DNA

Labeled Au bead

Laser

SERS

Ag$^+$
Hydroquinone   Enhancement

Figure 1.80 Scheme for SERS detection of DNA targets. In a sandwich assay similar to those above, DNA nanoparticles encoded with Raman-active dyes (e.g., Cy3) are hybridized to the surface-immobilized capture/target hybrid and silver enhancement is performed. Upon single-wavelength laser excitation, the particles emit a strong and reproducible Raman spectrum specific to the Raman-active dye chosen. (Reproduced with permission from Thaxton and Mirkin (2004). Copyright 2004, Wiley-VCH Verlag GmbH.)

is complementary to the second half of the target sequence (analyte), and the other one is complementary to the half of a barcode sequence (amplification signal). This second oligonucleotide acts as a unique identification tag of the particular target sequence. In order to realize high amplification, the ratio of barcode binding DNA and target binding DNA has to be large (about 70 : 1), and the total number of bound oligonucleotides per AuNP should be about 300–400. The whole detection process consists of three steps. In step 1, sandwiched structures of MMPs, AuNPs, and target DNA are formed by hybridization in a mixture of the functionalized MMPs and AuNPs together with the target DNA. After hybridization, in step 2, the magnetic sandwiched structures together with unreacted MMPs are separated from the solution by a magnetic field. By repeated washing, the unreacted DNA and AuNPs are removed. Afterward, the magnetic field is switched off. In pure aqueous solution, the particles are heated to 55 °C for 3 min to release the barcode

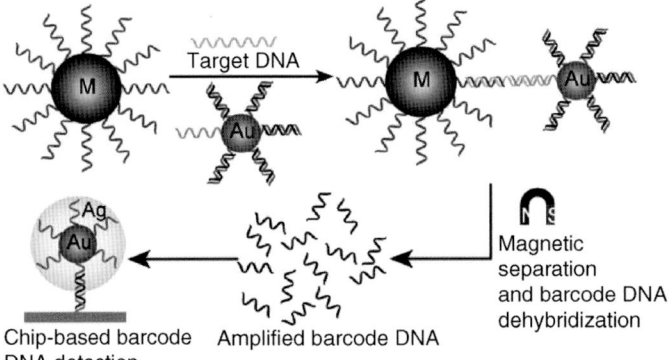

Chip-based barcode   Amplified barcode DNA
DNA detection

Figure 1.81 Schematics of the DNA BCA assay. The two types of biofunctionalized particles used in the BCA: (Au) AuNPs functionalized with two different oligonucleotides. One is complementary to half of the sequence of the target DNA, whereas the second one is complementary to half of a specific barcode sequence. (M) Magnetic iron oxide microparticles functionalized with one type of oligonucleotides that are complementary to the second half of the sequence of the target DNA. After magnetic separation, chip-based DNA detection can be applied for identification of the amplified barcode DNA. (Reproduced with permission from De, Ghosh, and Rotello (2008). Copyright 2008, Wiley-VCH Verlag GmbH.)

DNA by dehybridization. Also in step 2, the MMPs are removed from the solution by reapplying the magnetic field. In step 3, the solution with barcode DNA can be analyzed with the above-mentioned scanometric method. The high ratio of barcode binding DNA to target binding DNA deposited onto the AuNPs leads to very high amplification of the initial signal. Nam *et al.* (2004) have shown that 500 zM $(zM = 10^{-21} M)$ concentration of target DNA ($\sim$10 copies in a 30 µl sample) can be differentiated from a control solution with only noncomplementary DNA.

DNA-Directed Assembly of Single-Walled Carbon Nanotubes   Single-walled carbon nanotubes are one of the most promising artificial nanomaterials for application in electronics, optics, plasmonics, mechanics, thermal transport, biosensing, and drug delivery. The outstanding physical properties, such as electrical and thermal conductivities, mechanical strength, and stiffness, make them a preferred material for the manufacturing of nanocomposites. Several difficulties have to be overcome on the way from preparation to application: separation of bundles of the hydrophobic carbon tubes in aqueous solution, assembling of the SWCNTs into large architectures, and the formation of binding sites for a subsequent functionalization with other nanostructures. This can be done by linking SWCNTs with single-stranded DNA, based on the observation that single-stranded DNA molecules wrap around the hydrophobic SWCNTs. This mechanism has been exploited for dispersion of SWCNTs in aqueous solutions, and sorting them with respect to metallic and semiconducting tubes (Zheng *et al.*, 2003). By the conjugation of SWCNTs with single-stranded DNA, their binding sites can be used for a subsequent self-assembly process (Chen *et al.*, 2007). The basic idea behind this technique is compatible with the above-discussed experiments made with DNA-functionalized AuNPs. Two basic experiments have been performed: (i) By wrapping oligonucleotides around SWCNTs, hybridization reactions with complementary oligonucleotides can be used for the reversible formation of SWCNT aggregates. Mixing of one type of conjugated SWCNTs with another type, conjugated with the complementary oligonucleotides, leads to multicomponent nanocomposites. (ii) In the first process step, thiolated DNA single strands were linked to AuNPs. Subsequently, these AuNPs were mixed with a solution of SWCNTs conjugated with the complementary DNA, which resulted in the formation of SWCNT–AuNP complexes by hybridization.

DNA-Based Protein Assemblies   Covalent or noncovalent coupling facilitates the generation of synthetic DNA–protein conjugates (Niemeyer, 2002). This opens a wide range of opportunities for generating versatile molecular constructs for applications in life sciences and nanotechnology. The addressable assembly of DNA can be combined with the functionality of proteins. The addressability of DNA can be used in such constructs for controlled site-specific deposition of peptides or proteins on material surfaces. Another option is the use of conjugated double-stranded DNA for sensitivity enhancement of protein-based sensors, as explained with an example below. Several successful applications have been reported: self-assembly of high-affinity reagents for immunoassays, fabrication of laterally microstructured protein chips, design of bienzymatic complexes and biometallic aggregates, and DNA-directed immobilization (DDI) of peptides or proteins at implants for controlled cell adhesion (Niemeyer, 2002). Various advanced techniques are available for *covalent conjugation of oligodeoxyribonucleotides* (ODNs) to proteins or peptides. Amino groups or thiol groups can be used for covalent coupling of the ODNs with amino acids. In Figure 1.82, an example is given for the conjugation of an *aminopropylated ODN with a short hexapeptide* GRGDSP mediated by disuccinimidyl tartrate (DST) (for application see the following example for improvement of bone cell interaction with titanium implants in Figure 1.85).

Similarly, *thiol-modified ODNs* can be bound with the *heterospecific cross-linker sulfosuccinimidyl 4-[p-maleimidophenyl]butyrate* (sSMPB) to the protein (Niemeyer and Mirkin, 2004). Another approach is based on the conjugation of a recombinant protein containing a *C-terminal thioester* that is selectively bound to *cysteine conjugates of nucleic acids* (Tomkins *et al.*, 2001). In Figure 1.83, two examples are given for immunoassay applications. An antibody is coupled to DNA oligomers. Covalent conjugates of double-stranded DNA fragments and immunoglobulin (IgG) molecules are used for the detection of antigens. The detection signal of the antigen–antibody interaction can be enhanced by subsequent polymerase chain reaction multiplying the coupled DNA oligomer. Thus, the immuno-PCR (IPCR) increases the sensitivity about 1000-fold compared to the conventional enzyme-linked immunosorbent assay (Hendrickson *et al.*, 1995). The detection sensitivity for small-weight molecules, too, can be enhanced. Fluorescein has been chosen as a model analyte. The free analyte competes with a fluorescein–streptavidin–DNA conjugate. Short biotinylated DNA forms nanocircles bound to streptavidin. The nanocircles

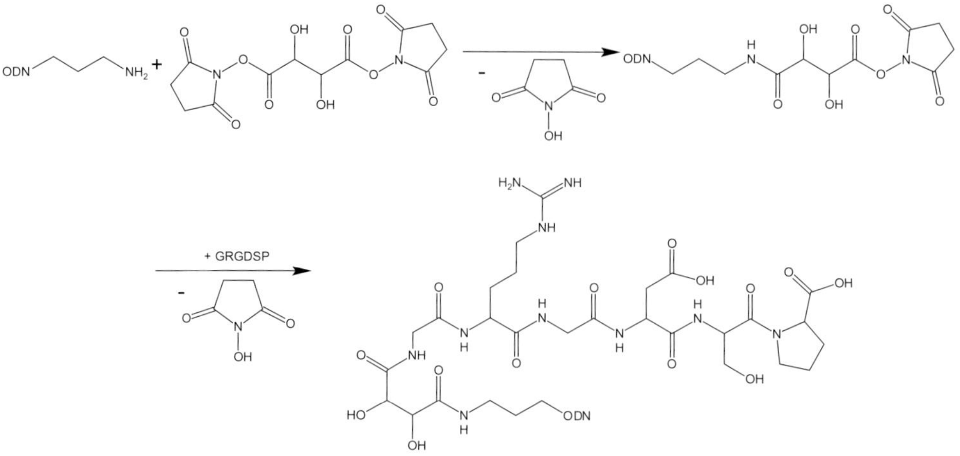

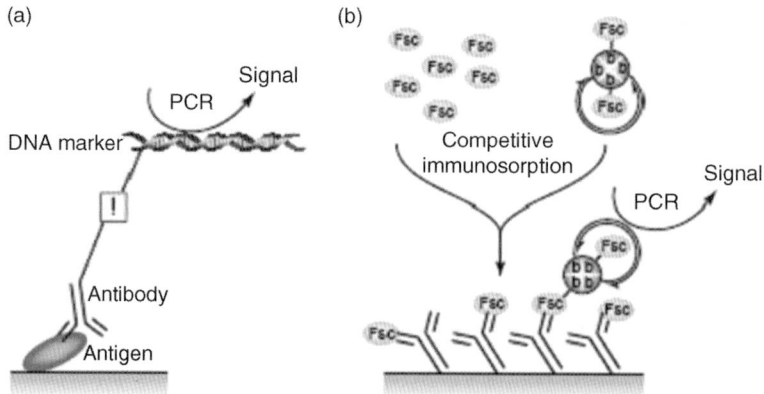

Figure 1.82 Conjugation reaction of aminopropylated ODN and GRGDSP mediated by disuccinimidyl tartrate.

can be functionalized by the coupling of biotinylated fluorescein. After competitive binding, the immobilized conjugate has been detected with PCR.

DNA–streptavidin conjugates have proven to be versatile constructs for the design of innovative nanostructures (Niemeyer, 2002), as indicated in Figure 1.84. Streptavidin covalently bound to *5′-thiol (SH)-modified oligonucleotide* serves as a unique building block. A single-stranded carrier nucleic acid with complementary sequences can act as a template for the formation of well-defined streptavidin aggregates. The biotin-binding sites of streptavidin allow the construction of various

Figure 1.83 (a) Application of DNA–protein conjugates in immuno-PCR. The antigen is detected with a specific antibody that is coupled to a DNA fragment. PCR amplification leads to a high sensitivity of the antigen detection. The key of the experiment is a reliable chemical conjugation indicated by the square box.

(b) Detection of small-weight molecules by competitive IPCR. Fluorescein (Fsc) is chosen as a model analyte. It competes with Fsc–streptavidin–DNA conjugates. The binding of the conjugates can be detected via PCR amplification. (Reprinted with permission from Niemeyer (2002). Copyright 2002, Elsevier.)

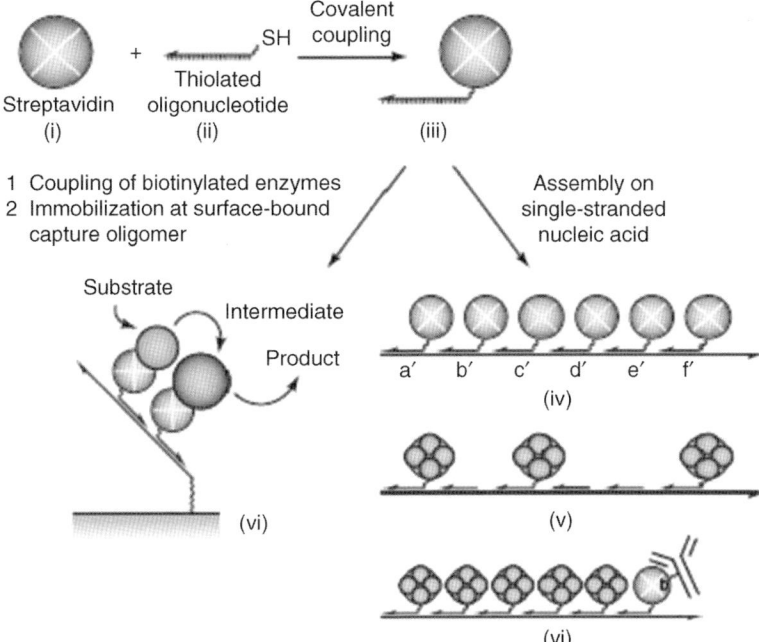

Figure 1.84 Synthesis of covalent DNA–streptavidin conjugates (iii) by covalent coupling of 5′-thiol (SH)-modified oligonucleotides (ii) and streptavidin (i). A set of conjugates (iii) self-assembles in the presence of a single-stranded nucleic acid with complementary sequence stretches (iv). Two examples are shown: fabrication of an antibody-containing metallic cluster chain (vi), and a bienzymatic functional complex (vi). (Reprinted with permission from Niemeyer (2002). Copyright 2002, Elsevier.)

supramolecular functionalities (Niemeyer *et al.*, 2002a). One example has been realized with the fabrication of a chain of gold particles grown by labeling the streptavidin with biotinylated 1.4 nm gold clusters. With a biotinylated immuno-globulin, coupled to the DNA–streptavidin conjugate, additionally incorporated into the chain, the whole construct can be considered as a promising nanoprobe for antigen detection. Another option is the close coupling of two enzymes enabling bienzymatic reactions (Niemeyer *et al.*, 2002b).

DNA–protein conjugates open also interesting options for the design of material interfaces capable of a controlled interaction with living cells, which is of high interest for implantology. Among the favored metallic implant materials in surgery are titanium and its alloys that exhibit good biocompatibility. High strength, good corrosion resistance *in vivo*, and toxicological harmlessness make them widely applied materials. Similar to other implant materials, however, difficulties may arise in cases of systemic diseases, such as diabetes and osteoporosis, or after radiation therapy. A material is needed that promotes the growth of new bone tissue near the implant interface. In such cases, bone tissue growth can be furthered by specific peptides promoting cell adhesion or directing cell proliferation or differentiation. Also, the immobilization of drug delivery units at the implant interface can be beneficial. When

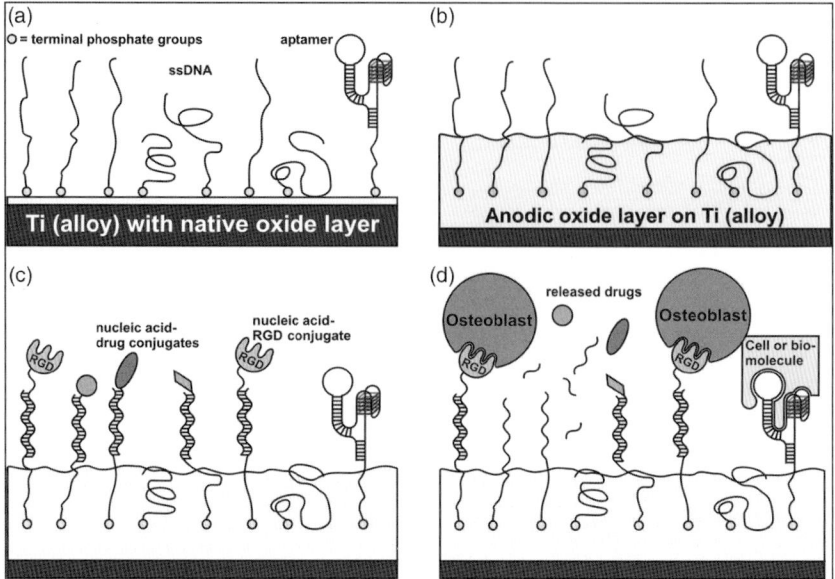

Figure 1.85 Four process steps of functionalization of titanium interfaces for improvement of cell interaction by means of oligonucleotide–protein conjugates. (a) Adsorption of single-stranded anchors of functional nucleic acids. (b) Anodic growth of a TiO_2 layer for partial entrapment of the biomolecules. (c) Hybridization of the nucleic acid conjugates to the anchor strands. (d) Specific binding of osteoblasts to RGD peptides. (Reprinted with permission from Michael *et al.* (2009). Copyright 2009, the American Chemical Society.)

such structures are to be integrated into the interface, care has to be taken that their biological functionalities are not destroyed by interaction with the metal. Degradation can be avoided by applying a coupling layer between peptide and metal surface. Oligonucleotide–peptide/protein conjugates are suitable for this purpose. In particular, this concept has been successfully applied for titanium alloys (Michael *et al.*, 2009). The whole process is subdivided into four steps (see also Figure 1.85):

a) Nucleic acid anchor strands (AS) or functional nucleic acids such as aptamers are adsorbed on the native oxide layer of the titanium implant.
b) The adsorbed nucleic acids are partially entrapped during subsequent anodic growth of a TiO_2 layer.
c) Fixed AS hybridize either with complementary strands (CS) conjugated to biomolecules (e.g., RGD peptides and growth factors) or drugs (e.g., antibiotics and antiphlogistics) or with functional nucleic acids. As an advantage, the technique can be adapted to the specific needs of the patient.
d) After implantation, specific tissue reactions are induced by the biomolecules (immobilized and/or released in a defined way).

The feasibility of this process has been demonstrated by adhesion experiments with bone-forming osteoblasts. As mentioned above, the hexapeptide GRGDSP was chosen as the biologically active model substance because the sequence RGD is an adhesive motif in extracellular matrix proteins of bone cells. The adhesion of the osteoblasts on the conjugate-modified titanium surface is found to be stronger than that on the plain surface. Interestingly, this effect is observed already in the presence of the anchor oligonucleotides alone.

1.3.3
Protein-Based Nanotechnology

The design of hybrid nanomaterials based on proteins is even more challenging than the DNA technologies described above. This is due to the higher complexity of proteins at the level of the primary structure (20 amino acids versus 4 nucleotides) particularly at the level of secondary and tertiary structures that determine the protein function. Therefore, the immobilization strategy has to be selected very carefully, taking into account the given surface chemistry, the particular protein structure, and the intended application. Important issues are the *maintenance of the structural integrity to avoid denaturation*, the stable transfer of the *natural protein configuration*, and an *optimized spacing between the proteins*. Very often a *defined orientation at the interface* is required to ensure that the reactive sites needed for the intended function are exposed and hence accessible to external signals. Known immobilization routes can be classified into those for manufacturing two-dimensional or three-dimensional interfaces, and also into physical, chemical, and bioaffinity-mediated ones.

Physical Immobilization Several adhesive interactions are suitable for conjugation of proteins with a nonbiological material: electrostatic, hydrophobic/hydrophilic, and various kinds of polar interactions (Table 1.10). Usually they produce a random pattern of weakly bound proteins.

The simplest way of controlling the protein adsorption is by tuning of the electrostatic interaction. Plasma treatment of the solid substrate, polyelectrolyte coatings, adsorption of ions (e.g., Mg^{2+}, NH_2^+, and $(C_3H_5O(COO)_3^{3-})$), and pH control are potential options. Figure 1.86 shows an example for patterned

Table 1.10 Mechanisms of protein adsorption.

Interaction	Surface modification	Advantages	Drawbacks
Ionic bonds	Plasma activation, polyelectrolyte coating, tuned ion adsorption	Widespread applicability	Randomly oriented, heterogeneous, weak adsorption, nonspecific bonding, high background signals
Hydrophobic/ hydrophilic interaction	Self-assembled monolayers, lipid bilayers, hydrogels, BSA adsorption	Molecular designed tail groups	
Polar interaction	SAM		

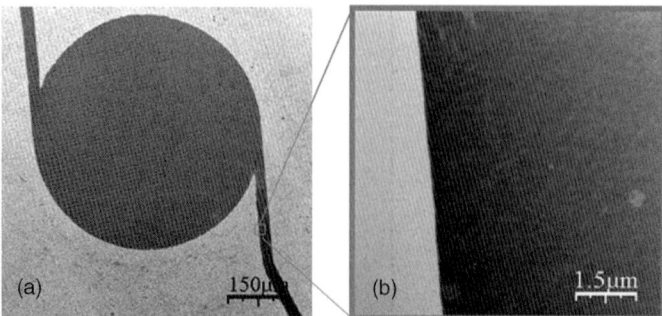

Figure 1.86 Scanning electron micrograph of a recrystallized protein layer (dark structure – bacterial surface protein of *Bacillus sphaericus* NCTC 9602) on a Si wafer. By using a PDMS stamp with an appropriate profile, the protein monomers can be physisorbed in the restricted reaction volume after preadsorption of Mg^{2+} ions. (a) Active probe region of a piezoelectric sensor; (b) the marked area shows the sharp boundary of the protein layer. (Reprinted with permission from Blüher (2008).)

functionalization of sensor surfaces by physisorption of proteins. Bacterial surface layer proteins served as templates for local metallization of probe regions. The adsorption on a silica surface was controlled by addition of Mg^{2+} ions. In a subsequent step, the protein layer acts as a template for the deposition of nanosized noble metal clusters (Pt or Pd) suitable for catalytic gas sensors (for more details see Section 7.3).

Layer-by-layer deposition of differentially charged polyelectrolytes allows the electrostatic interaction to be controlled. Hydrogels with charged surface groups, such as sulfate-modified dextrans or carboxymethyl cellulose, are suitable for adsorbing proteins. Self-assembled monolayers (SAMs) are known for their ability to control electrostatic, hydrophobic, and polar interactions (Mrksich and Whitesides, 1996). SAMs are formed spontaneously by exothermic interaction of the surface-active group with the substrate. Thiols and disulfides are commonly used as head groups on noble metal substrates such as gold, silver, and platinum, whereas silanes form the head groups on nonmetallic oxidic surfaces such as silica and titanium oxide. The side chains are usually alkyl chains or derivatized alkyl chains because they can form densely packed monolayers. The side chains are usually tilted with respect to the surface normal to maximize the van der Waals interactions between the molecules. The interaction of the SAMs with proteins can be tailored by a suitable choice of the tail group. Polar functional groups such as carboxylic acids and hydroxyls form hydrophilic interfaces, whereas nonpolar, organic groups such as methyl or trifluoromethyl form hydrophobic interfaces. The adsorption of proteins and the degree of denaturation correlate with the hydrophobicity of the surface of the SAM. Polyethylene glycol (PEG) has proven to be most suitable for avoiding protein adsorption. Therefore, biosensor surfaces that are to be protected against protein adsorption are often coated with alkanethiols terminated in short oligomers of ethylene glycol (e.g., $HS(CH_2)_{11}(OCH_2CH_2)_nOH$, $n = 2–7$). Another option is the preadsorption of a protein, for example, bovine serum albumin (BSA),

Table 1.11 Non-site-specific immobilization of proteins via covalent binding.

Interaction	Surface modification	Amino acids	Remarks
Amine chemistry (—NH₂)	Carboxylic acid, N-hydroxysuccinimide, aldehyde groups, epoxy	Lysine, hydroxylysine	Multipoint attachment, restricted conformational flexibility, heterogeneous, generally above pH 9
Thiol chemistry (—SH)	Maleimide, pyridyl disulfide, vinyl sulfone	Cysteine	Better orientation, pH 7–9.5
Carboxyl chemistry (—COOH)	Aminated SAM, carbodiimide	Aspartic acid, glutamic acid	Usually major fraction of surface groups
Epoxy chemistry (—OH)	Amino supported	Serine, threonine	

which precludes the adsorption of other proteins. Alternatively, there are also methods for covalent coupling of proteins to the SAM surface. Active tail groups that form amide or disulfide bonds are most suitable for this purpose. Application of mixed SAMs (protein adsorptive and adsorption resistant) offers the option of adsorbing proteins in a patterned manner. The weak binding and the random distribution of proteins are the major drawbacks of physical immobilization mechanisms for the formation of stable biointerfaces.

Chemical Immobilization Functional groups of amino acid side chains are possible binding sites for the chemical immobilization of proteins (see Tables 1.11 and 1.12).

Table 1.12 Site-specific immobilization of proteins via covalent binding.

Interaction	Surface modification	Protein	Remarks
Diels–Alder cycloaddition	Maleimide modified	Cyclopentadiene ligated	Reaction in water at room temperature, high rate
"Click" chemistry, 1,3–dipolar cycloaddition of azide and alkyne	Alkyne modified	Multipoint azide functionalized	High-density immobilization, simple process
α-Oxo semicarbazone ligation	Aldehyde groups	Semicarbazone ligated	
Peptide ligation	Ester glycoaldehyde, thioester	N-terminal cysteine, serine, histidine, threonine containing weak base nucleophiles	
Staudinger ligation	Phosphinothioester	Azido modified	pH 7.4–7.6 without noticeable side chain products

Very often the binding pattern is random as the binding can occur through many side chains of amino acids exposed on the exterior of the protein. In order to realize site-specific immobilization, a particular chemical nanostructuring of the substrate or additional functionalization of the protein is necessary. Site-specific immobilization is favored when the abundance of the functional amino acids exposed by the protein is very small (e.g., cysteine); ideally, the protein should exhibit only one reactive amino acid. Genetic modification of proteins can be a means for designing a suitable side group without affecting the protein function. Under such conditions, the protein can be deposited at a well-defined site with predictable orientation without affecting conformational stability. Various chemical methods can be used for immobilization, depending on the exposed amino acids (Rusmini *et al.*, 2007; Jonkheijm *et al.*, 2008; Goddard and Erickson, 2009). For non-site-specific immobilization, the available functional side groups are given in Table 1.11, and for site-specific covalent immobilization in Table 1.12.

Bioaffinity Immobilization   The immobilization of proteins by means of biochemical affinity reactions is the favored method when the functionality of the protein is to be preserved. It is also advantageous when gentle detachment of a protein or regeneration of a protein coating under near-line processing conditions is intended. As these methods make use of molecular recognition structures, we will discuss the various options later in Chapter 2.

Biofunctionalized Nanoparticles: Artificial Proteins   The large interest in medical and biological application of nanoparticles is partially due to the fact that the tunable size can be well adapted to the sizes of the various molecular and subcellular targets. There are two major areas of nanoparticle application in biology: bioimaging and nanotherapy. Bioimaging concerns targeted optical contrast enhancement, fluorescence imaging, nanosized magnetic resonance imaging (MRI), and locally enhanced spectroscopy (e.g., particle-enhanced Raman spectroscopy). The development of nanotherapy is aimed at administering the therapeutic formulations with high precision to the malignant cell. This is important in view of the fact that in current cancer therapy only 10–100 ppm of intravenously administered monoclonal antibodies reach their targets *in vivo*. This may be improved by a therapeutic means that overcomes the biological barriers on the path to the target. Moreover, the delivery has to be highly selective in order to minimize collateral damage to the tissues. Obviously, multifunctionality of nanoparticles would be highly desirable in nanotherapy. An "ideal" nanoparticle should carry one or more therapeutic agents, permeate biological barriers, avoid uptake by macrophages, target selectively subcellular objects, and deliver the therapeutics in a controlled way after reaching the target. This means that multifunctional nanoparticles should be carrier, sensor, and actor in one. A related design concept, intended for advanced cancer therapy, has been proposed by Ferrari (2005).

The manufacturing of multifunctional nanoparticles ought to be done in a stepwise process starting from a colloidal solution, as proposed by Moyano and Rotello (2011). The nanoparticle is made to surround itself with a polymeric or biochemical shell resulting in a noninteracting core–shell structure. The shell can be functionalized with specific functional groups adapted to a particular biosystem under consideration. Typically, the core nanoparticles are stabilized by surface charges (electrostatic stabilization) or by short polymers (sterical stabilization) (Hunter, 1992). This initial stabilization can interfere with the subsequent step of the formation of the shell. A sufficiently stable and reproducible shell may be obtained by stable chemisorption of linker molecules. Favored molecules are structures that allow the presentation of addressable functional groups in high regularity. The shell is usually formed with short alkyl-based monolayers with specific functional tail groups. However, as the hydrophobic interior of alkyl-based monolayers can result in protein denaturation, oligo- or polyethylene glycol layers (OEG or PEG, respectively) are often applied. Moyano and Rotello (2011) have proven that an optimized shell structure can be obtained by combining an inner shell of alkyl chains with a thin outer shell of OEG. The inner hydrophobic shell serves as a stable linkage to the core whereas the outer shell provides a noninteracting surface. An outer layer consisting of four ethylene glycol repeats has turned out to be optimal. Structures of this type have been successfully prepared with CdSe quantum dots (3.2 nm core diameter) as well as Au core particles (2 nm diameter) (Figure 1.87). The hydrophobic part of the shell was formed by an alkanethiol-based monolayer.

Core–shell constructs have been tested as carriers for a model enzyme, the protease chymotrypsin (ChT). The positively charged enzyme denatures if the core is

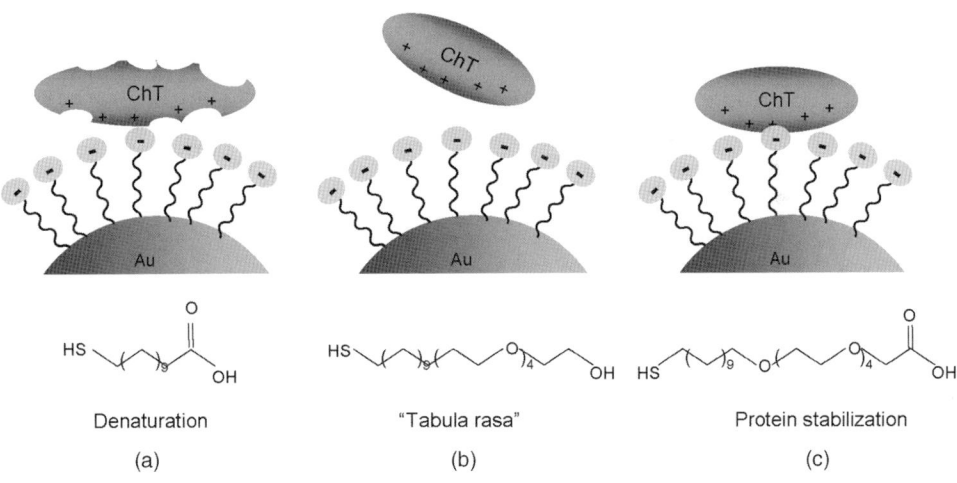

Denaturation   "Tabula rasa"   Protein stabilization

(a)   (b)   (c)

Figure 1.87 Generic approach to a hybrid core–shell structure of biofunctionalized nanoparticles with tailored interaction with proteins. In this example, a positively charged protein (ChT) is considered. (a) Simple alkanethiol-based monolayer results in protein denaturation. (b) TEG-functionalized particles are noninteracting. (c) Termination of the TEG layer with carboxylate groups results in reversible binding of the protein. (Adapted with permission from Moyano and Rotello (2011). Copyright 2011, the American Chemical Society.)

functionalized only with the alkanethiol monolayer (e.g., mercaptoundecanoic acid). By adding an additional outer shell of tetraethylene glycol (TEG) segments, a noninteracting "tabula rasa" structure is formed. The termination of the TEG layer with carboxylate groups leads to reversible binding of the ChT to the nanoparticles functionalized with the carboxylated TEG layer (NP–TCOOH). The enzyme is stabilized with full functionality. The stabilization of ChT at the NP–TCOOH structure leads to an essential increase of the substrate selectivity of the enzyme. Free ChT is a promiscuous enzyme hydrolyzing anionic, neutral, and cationic proteins. Owing to the negatively charged surface of the carrier nanoparticle NP–TCOOH, anionic substrates are repelled, whereas cationic substrates are attracted. This leads to an accelerated hydrolysis of the cationic substrates and a retarded reaction with the anionic ones (Moyano and Rotello, 2011). As mentioned above, the established platform allows systematic variation of the functional groups of the outer shell of the NP–TCOOH complex. For the majority of applications, the conjugation of the core–shell NP with a protein is mediated by noncovalent linkage in order to avoid any irreversible structure changes of the protein. This is usually realized with a tailored content of hydrophobic/hydrophilic and electrostatic interactions between the shell and the targeted protein. By adding specific amino acids to the terminal group in a controlled way, the degree of hydrophobicity can be varied. Alternatively, also bioaffine linkers can be incorporated into the shell. With the tailored linkage of amino acids to terminal groups of the shell structure, protein–protein interactions can be mimicked in a systematic way. The core–shell nanoparticle can be regarded as an "artificial protein" with tunable functional properties.

Core particles consisting of gold provide interesting options for the design of optical biosensors. The concept of a fluorescence assay suitable as a "chemical nose" is pictured in Figure 1.88. In this example, the sensor is based on the competitive binding of a fluorescent anionic polymer (e.g., poly(*p*-phenylene ethynylene) (PPE)) or the anionic green fluorescent protein (GFP with pI 5.92) and four analytes (A, B, C, and D) to cationic gold NP–TNH$_2^+$R complexes. The cationic tail groups are additionally functionalized with different short terminal groups making a "chemical nose" sensor array. (The six different R_1, . . . ,R_6 result in the six sensing systems S_1, . . . ,S_6.) The fluorescence of PPE (or GFP) is quenched when PPE (or GFP) is bound at the gold complex and is switched on when PPE (or GFP) is released. Depending on the size of terminal group R_i, the extent of quenching for a given bound fluorescent anion varies among the sensing systems. The evaluation of the patterns of fluorescence response of the sensing systems S_1, . . . ,S_6 for known concentrations of the individual proteins (A, B, C, and D) provides the information for identification of proteins in mixtures with unknown composition.

Tasks T1

Task 1.1: Calculate the strain u of a semiflexible elastic polymer chain loaded with a tensile force f in thermal equilibrium. Which condition has to be fulfilled in order that the polymer can be described as inextensible rod? Consider a polymer chain

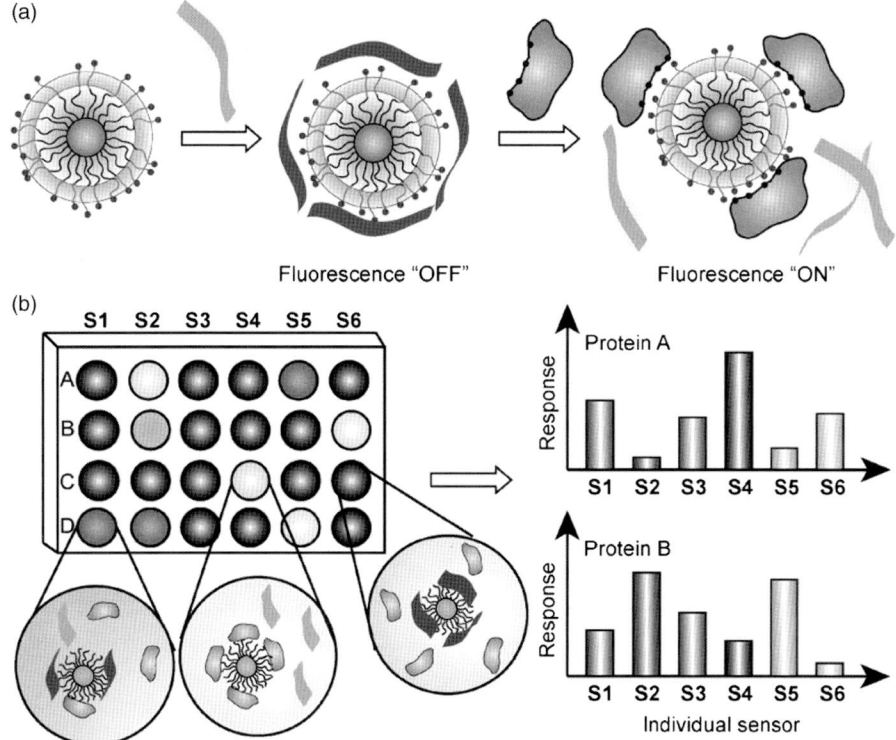

(a)

Fluorescence "OFF"   Fluorescence "ON"

(b)

S1 S2 S3 S4 S5 S6

A
B
C
D

Protein A

Response

S1 S2 S3 S4 S5 S6

Protein B

Response

S1 S2 S3 S4 S5 S6

Individual sensor

Figure 1.88 Schematic drawing of a "chemical nose" sensor array based on nanoparticles and fluorescence assay. (a) The competitive binding between protein and quenched PPE or GFP leads to the fluorescence light-up. (b) The combination of an sensor array generates fingerprint response patterns for individual proteins. (Reproduced with permission from De, Ghosh, and Rotello (2008). Copyright 2008, Wiley-VCH Verlag GmbH.)

with Young's modulus of 1 GPa and circular cross section. Estimate its minimum diameter so that this condition is fulfilled at tensile loads of up to $f = 10$ pN under the assumption that the coupling between strain u and twist density ω can be neglected in the balance of the total elastic energy.

Task 1.2: Calculate the elastic energy stored in a short bent DNA segment of length l_0 when the radius of curvature equals a constant value R along the segment.

Task 1.3: Calculate the Young's moduli of the four biopolymers listed in Table 1.6 by using the experimentally determined values of the bend persistence length A and a continuum approximation for the second moment of inertia I of the polymer chains. From Eq. (1.5) second moments of inertia are obtained for the different filament geometries as summarized in Figure 1.89. Note that the actin filament forms a helical fiber with an elliptical cross section.

Task 1.4: Calculate the coil diameter of λ-DNA in thermal equilibrium. λ-DNA consists of 48 502 base pairs.

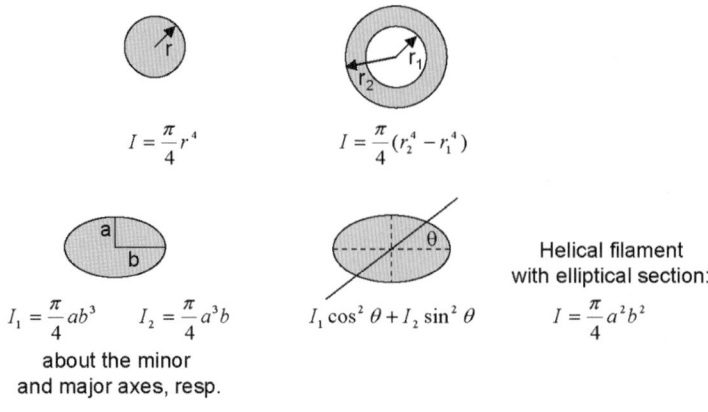

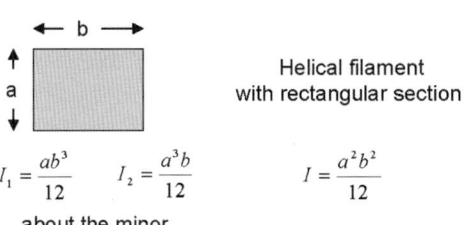

$$I = \frac{\pi}{4} r^4$$

$$I = \frac{\pi}{4}(r_2^4 - r_1^4)$$

$$I_1 = \frac{\pi}{4} ab^3 \qquad I_2 = \frac{\pi}{4} a^3 b$$

about the minor
and major axes, resp.

$$I_1 \cos^2 \theta + I_2 \sin^2 \theta$$

Helical filament
with elliptical section:

$$I = \frac{\pi}{4} a^2 b^2$$

Helical filament
with rectangular section:

$$I_1 = \frac{ab^3}{12} \qquad I_2 = \frac{a^3 b}{12}$$

$$I = \frac{a^2 b^2}{12}$$

about the minor
and major axes, resp.

Figure 1.89 Second moments of inertia *I* of filaments. (Adapted from Howard (2001).)

Task 1.5: How long has an uncatalyzed first-order reaction to run in order that 1% of a substrate is transformed? For the activation energies of the uncatalyzed reaction ΔG_{uncat}, we assume values between 80 and 100 kJ/mol.

References

Aich, P., Labiuk, S.L. *et al.* (1999) M-DNA: a complex between divalent metal ions and DNA which behaves as a molecular wire. *Journal of Molecular Biology*, **294** (2), 477–485.

Aldaye, F.A. and Sleiman, H.F. (2007) Modular access to structurally switchable 3D discrete DNA assemblies. *Journal of the American Chemical Society*, **129** (44), 13376–13377.

Alivisatos, A.P., Johnsson, K.P. *et al.* (1996) Organization of 'nanocrystal molecules' using DNA. *Nature*, **382** (6592), 609–611.

Banavar, J.R., Hoang, T.X. *et al.* (2004) Unified perspective on proteins: a physics approach. *Physical Review E: Statistical, Nonlinear, and Soft Matter Physics*, **70** (4 Pt 1), 041905.

Benke, A. (2007) Aufbau nanoskopischer Netzwerke aus DNA und Bindeproteinen. Ph.D. thesis, Technische Universität Dresden.

Benke, A., Mertig, M. *et al.* (2011) pH- and salt-dependent molecular combing of DNA: experiments and phenomenological model. *Nanotechnology*, **22** (3), 035304.

Berti, L., Alessandrini, A. *et al.* (2005) DNA-templated photoinduced silver deposition. *Journal of the American Chemical Society*, **127** (32), 11216–11217.

Blüher, A. (2008) S-Schichtproteine als molekulare Bausteine zur Funktionalisierung mikroelektronischer Sensorstrukturen. Ph.D. thesis, Technische Universität Dresden.

Bockelmann, U., Thomen, P. *et al.* (2002) Unzipping DNA with optical tweezers: high sequence sensitivity and force flips. *Biophysical Journal*, **82** (3), 1537–1553.

Bosaeus, N., El-Sagheer, A.H. *et al.* (2012) Tension induces a base-paired overstretched DNA conformation. *Proceedings of the National Academy of Sciences of the United States of America*, **109** (38), 15179–15184.

Braun, E., Eichen, Y. *et al.* (1998) DNA-templated assembly and electrode attachment of a conducting silver wire. *Nature*, **391** (6669), 775–778.

Brochard-Wyart, F. (1995) Polymer chains under strong flows: Stem and flowers. *Europhysics Letters*, **30** (7), 387.

Bustamante, C., Marko, J.F. *et al.* (1994) Entropic elasticity of lambda-phage DNA. *Science*, **265** (5178), 1599–1600.

Cao, Y.C., Jin, R. *et al.* (2002) Nanoparticles with Raman spectroscopic fingerprints for DNA and RNA detection. *Science*, **297** (5586), 1536–1540.

Chen, J.H. and Seeman, N.C. (1991) Synthesis from DNA of a molecule with the connectivity of a cube. *Nature*, **350** (6319), 631–633.

Chen, Y., Liu, H., Ye, T., Kim, J., and Mao, C. (2007) DNA-directed assembly of single-wall carbon nanotubes. *Journal of the American Chemical Society*, **129** (28), 8696–8697.

Clausen-Schaumann, H., Rief, M. *et al.* (2000) Mechanical stability of single DNA molecules. *Biophysical Journal*, **78** (4), 1997–2007.

Cluzel, P., Lebrun, A. *et al.* (1996) DNA: an extensible molecule. *Science*, **271** (5250), 792–794.

Coffer, L., Bigham, S., and Li, X. (1996) Dictation of the shape of mesoscale semiconductor nanoparticle assemblies by plasmid DNA. *Applied Physics Letters*, **69** (25), 3851.

Colombi Ciacchi, L.C. (2002) Growth of platinum clusters in solution and on biopolymers: the microscopic mechanisms. Ph.D. thesis, Technische Universität Dresden.

De, M., Ghosh, P.S., and Rotello, V.M. (2008) Applications of nanoparticles in biology. *Advanced Materials*, **20**, 4225–4241.

Douglas, S.M., Bachelet, I., and Church, G.M. (2012) A logic-gated nanorobot for targeted transport of molecular payloads. *Science*, **335**, 831–834.

Douglas, T. and Young, M. (1999) Virus particles as templates for materials synthesis. *Advanced Materials*, **11** (8), 679–681.

Erler, C. (2010) Synthesis of metallic nanowires using integrated DNA molecules as templates. Ph.D. thesis, Technische Universität Dresden.

Erler, C., Guenther, K., and Mertig, M. (2009) Photo-induced synthesis of DNA-templated metallic nanowires and their integration into micro-fabricated contact arrays. *Applied Surface Science*, **255**, 9647–9651.

Ferrari, M. (2005) Cancer nanotechnology: opportunities and challenges. *Nature Reviews. Cancer*, **5**, 161–171.

Festag, G., Klenz, U., Henkel, T., Csáki, A., and Fritzsche, W. (2005) Biofunctionalization of metallic nanoparticles and microarrays for biomolecular detection, in *Biofunctionalization of Nanomaterials* (ed. C. Kumar), Wiley-VCH Verlag GmbH, Weinheim.

Freeman, R., Finder, T. *et al.* (2009) Multiplexed analysis of Hg^{2+} and Ag^{+} ions by nucleic acid functionalized CdSe/ZnS quantum dots and their use for logic gate operations. *Angewandte Chemie – International Edition*, **48** (42), 7818–7821.

Goddard, J.M. and Erickson, D. (2009) Bioconjugation techniques for microfluidic biosensors. *Analytical and Bioanalytical Chemistry*, **394** (2), 469–479.

Günther, K., Mertig, M. *et al.* (2010) Mechanical and structural properties of YOYO-1 complexed DNA. *Nucleic Acids Research*, **38** (19), 6526–6532.

Guo, X., Gorodetsky, A.A. *et al.* (2008) Conductivity of a single DNA duplex bridging a carbon nanotube gap. *Nature Nanotechnology*, **3** (3), 163–167.

Haldane, J.B.S. (1930) *Enzymes*, MIT Press edition 1965; ISBN 0-262-58003-9.

Harnack, O., Ford, W.E., Yasuda, A., and Wessels, J.M. (2002) Tris(hydroxymethyl) phosphine-capped gold particles templated by DNA as nanowire precursors. *Nano Letters*, **2** (9), 919.

He, Y., Ye, T. *et al.* (2008) Hierarchical self-assembly of DNA into symmetric supramolecular polyhedra. *Nature*, **452** (7184), 198–201.

Hendrickson, E.R., Truby, T.M. *et al.* (1995) High sensitivity multianalyte immunoassay using covalent DNA-labeled antibodies and polymerase chain reaction. *Nucleic Acids Research*, **23** (3), 522–529.

Hoang, T., Trovato, A., Seno, F., Banavar, J.R., and Maritan, A. (2006) Marginal compactness of protein native structures. *Journal of Physics: Condensed Matter*, **18**, S297–S306.

Howard, J. (2001) *Mechanics of Motor Proteins and the Cytoskeleton*, Sinauer Associates, Inc. Publishers, Sunderland, MA.

Hunter, R.J. (1992) *Foundations of Colloidal Science*, vol. **1**, Clarendon Press, Oxford.

Jin, R.C., Wu, G.S. *et al.* (2003) What controls the melting properties of DNA-linked gold nanoparticle assemblies? *Journal of the American Chemical Society*, **125** (6), 1643–1654.

Jonkheijm, P., Weinrich, D. *et al.* (2008) Chemical strategies for generating protein biochips. *Angewandte Chemie – International Edition*, **47** (50), 9618–9647.

Kauert, D.J., Kurth, T. *et al.* (2011) Direct mechanical measurements reveal the material properties of three-dimensional DNA origami. *Nano Letters*, **11** (12), 5558–5563.

Keren, K., Berman, R.S., and Braun, E. (2004) Patterned DNA metallization by sequence-specific localization of a reducing agent. *Nano Letters*, **4** (2), 323–326.

Keren, K., Krueger, M. *et al.* (2002) Sequence-specific molecular lithography on single DNA molecules. *Science*, **297** (5578), 72–75.

Knez, M., Sumser, M., Bittner, A.M., Wege, Ch., Jeske, H., Martin, T.P., and Kern, K.

(2004) Spatially selective nucleation of metal clusters on the tobacco mosaic virus. *Advanced Functional Materials*, **14**, 116–124.

Kuklenyik, Z. and Marzilli, L.G. (1996) Mercury(II) site-selective binding to a DNA hairpin. Relationship of sequence-dependent intra- and interstrand cross-linking to the hairpin–duplex conformational transition. *Inorganic Chemistry*, **35** (19), 5654–5662.

Kumar, A., Pattarkine, M., Bhadbhade, M., Mandale, A.B., Ganesh, K.N., Datar, S.S., Dharmadhikari, C.V., and Sastry, M. (2001) Linear superclusters of colloidal gold particles by electrostatic assembly on DNA templates. *Advanced Materials*, **13** (5), 341–344.

Lakatos, M. (2013) Aufbau colorimetrischer Sensoren mit Goldnanopartikeln für unterschiedliche Analytgrößen. Ph.D. thesis, Technische Universität Dresden.

Lee, J.S., Latimer, L.J. *et al.* (1993) A cooperative conformational change in duplex DNA induced by Zn^{2+} and other divalent metal ions. *Biochemistry and Cell Biology*, **71** (3–4), 162–168.

Lemanov, V.V. (2000) Piezoelectric and pyroelectric properties of protein amino acids as basic materials of soft state physics. *Ferroelectrics*, **238** (1), 211–218.

Lezon, T.R., Banavar, J.R., and Maritan, A. (2006) The origami of life. *Journal of Physics: Condensed Matter*, **18**, 847–888.

Lippert, B. (ed.) (1999) *Cisplatin: Chemistry and Biochemistry of a Leading Anticancer Drug*, Wiley-VCH Verlag GmbH, Weinheim.

Liu, D., Park, S.H. *et al.* (2004) DNA nanotubes self-assembled from triple-crossover tiles as templates for conductive nanowires. *Proceedings of the National Academy of Sciences of the United States of America*, **101** (3), 717–722.

Liu, H., Chen, Y., He, Y., Ribbe, A.E., and Mao, Ch. (2006) Approaching the limit: can one DNA oligonucleotide assemble into large nanostructures? *Angewandte Chemie – International Edition*, **45**, 1942–1945.

Liu, J. and Lu, Y. (2006) Fast colorimetric sensing of adenosine and cocaine based on a general sensor design involving aptamers and nanoparticles. *Angewandte Chemie – International Edition*, **45**, 90–94.

Liu, Y., Lin, C., Li, H., and Yan, H. (2005) Aptamer-directed self-assembly of protein arrays on a DNA nanostructure. *Angewandte Chemie – International Edition*, **44** (28), 4333.

Marko, J.F. and Siggia, E.D. (1995) Stretching DNA. *Macromolecules*, **28**, 8759–8770.

Mertig, M., Ciacchi, L.C., Seidel, R., Pompe, W., and De Vita, A. (2002) DNA as a selective metallization template. *Nano Letters*, **2**, 841–844.

Mertig, M. and Pompe, W. (2004) Biomimetic fabrication of DNA-based metallic nanowires and networks, in *Nanobiotechnology* (eds C. Niemeyer and Ch. Mirkin), Wiley-VCH Verlag GmbH, Weinheim.

Michael, J., Schönzart, L. *et al.* (2009) Oligonucleotide–RGD peptide conjugates for surface modification of titanium implants and improvement of osteoblast adhesion. *Bioconjugate Chemistry*, **20** (4), 710–718.

Mirkin, C.A., Letsinger, R.L. *et al.* (1996) A DNA-based method for rationally assembling nanoparticles into macroscopic materials. *Nature*, **382** (6592), 607–609.

Moyano, D.F. and Rotello, V.M. (2011) Nano meets biology: structure and function at the nanoparticle interface. *Langmuir*, **27** (17), 10376–10385.

Mrksich, M. and Whitesides, G.M. (1996) Using self-assembled monolayers to understand the interactions of man-made surfaces with proteins and cells. *Annual Review of Biophysics and Biomolecular Structure*, **25**, 55–78.

Nam, J.M., Park, S.J. *et al.* (2002) Bio-barcodes based on oligonucleotide-modified nanoparticles. *Journal of the American Chemical Society*, **124** (15), 3820–3821.

Nam, J.M., Stoeva, S.I. *et al.* (2004) Bio-bar-code-based DNA detection with PCR-like sensitivity. *Journal of the American Chemical Society*, **126** (19), 5932–5933.

Nelson, D.L. and Cox, M.M. (2008) *Lehninger – Principles of Biochemistry*, W. H. Freeman and Company, New York.

Nelson, P. (ed.) (2008) *Biological Physics – Energy, Information, Life*, W.H. Freeman and Company, New York.

Niemeyer, C.M. (2002) The development of semisynthetic DNA–protein conjugates. *Trends in Biotechnology*, **20** (9), 395–401.

Niemeyer, C.M., Adler, M. *et al.* (2002a) Supramolecular DNA nanocircles with a covalently attached oligonucleotide moiety. *Journal of Biomolecular Structure and Dynamics*, **20** (2), 223–230.

Niemeyer, C.M., Koehler, J. *et al.* (2002b) DNA-directed assembly of bienzymic complexes from *in vivo* biotinylated NAD (P)H:FMN oxidoreductase and luciferase. *ChemBioChem: A European Journal of Chemical Biology*, **3** (2–3), 242–245.

Niemeyer, C.M. and Mirkin, C.A. (eds) (2004) *Nanobiotechnology*, Wiley-VCH Verlag GmbH, Weinheim.

Odijk, T. (1995) Stiff chains and filaments under tension. *Macromolecules*, **28**, 7016–7018.

Park, S.J., Taton, T.A. *et al.* (2002) Array-based electrical detection of DNA with nanoparticle probes. *Science*, **295** (5559), 1503–1506.

Patolsky, F., Weizmann, Y. *et al.* (2002) Au-nanoparticle nanowires based on DNA and polylysine templates. *Angewandte Chemie – International Edition*, **41** (13), 2323–2327.

Pauling, L. and Corey, R.B. (1951) Configurations of polypeptide chains with favored orientations around single bonds: two new pleated sheets. *Proceedings of the National Academy of Sciences of the United States of America*, **37** (11), 729–740.

Rakitin, A., Aich, P. *et al.* (2001) Metallic conduction through engineered DNA: DNA nanoelectronic building blocks. *Physical Review Letters*, **86** (16), 3670–3673.

Richter, J., Mertig, M., Pompe, W., and Vinzelberg, H. (2002) Low-temperature resistance of DNA-templated nanowires. *Applied Physics A*, **74** (6), 725.

Rosi, N.L. and Mirkin, C.A. (2005) Nanostructures in biodiagnostics. *Chemical Reviews*, **105** (4), 1547–1562.

Rothemund, P.W. (2006) Folding DNA to create nanoscale shapes and patterns. *Nature*, **440** (7082), 297–302.

Roy, S., Vedala, H. *et al.* (2008) Direct electrical measurements on single-molecule genomic DNA using single-walled carbon nanotubes. *Nano Letters*, **8** (1), 26–30.

Rusmini, F., Zhong, Z. *et al.* (2007) Protein immobilization strategies for protein biochips. *Biomacromolecules*, **8** (6), 1775–1789.

Sastry, M., Kumar, A., Datar, S., Dharmadhikari, C.V., and Ganesh, K.N. (2001) DNA-mediated electrostatic assembly of gold nanoparticles into linear arrays by simple drop-coating procedure. *Applied Physics Letters*, **78** (19), 2943.

Scholtz, J.M., Qian, H. *et al.* (1991) Parameters of helix–coil transition theory for alanine-based peptides of varying chain lengths in water. *Biopolymers*, **31** (13), 1463–1470.

Seeman, N.C. (1982) Nucleic acid junctions and lattices. *Journal of Theoretical Biology*, **99** (2), 237–247.

Seeman, N.C. (2003) DNA in a material world. *Nature*, **421**, 427–431.

Seeman, N.C. and Belcher, A.M. (2002) Emulating biology: building nanostructures from the bottom up. *Proceedings of the National Academy of Sciences of the United States of America*, **99** (Suppl. 2), 6451–6455.

Seidel, R., Ciacchi, L.C. *et al.* (2004) Synthesis of platinum cluster chains on DNA templates: conditions for a template-controlled cluster growth. *Journal of Physical Chemistry B*, **108** (30), 10801–10811.

Seidel, R., Mertig, M., and Pompe, W. (2002) Scanning force microscopy of DNA metallization. *Surface and Interface Analysis*, **33**, 151–154.

Shih, W.M., Quispe, J.D. *et al.* (2004) A 1.7-kilobase single-stranded DNA that folds into a nanoscale octahedron. *Nature*, **427** (6975), 618–621.

Smith, S.B., Cui, Y. *et al.* (1996) Overstretching B-DNA: the elastic response of individual double-stranded and single-stranded DNA molecules. *Science*, **271** (5250), 795–799.

Smith, S.B., Finzi, L. *et al.* (1992) Direct mechanical measurements of the elasticity of single DNA molecules by using magnetic beads. *Science*, **258** (5085), 1122–1126.

Tanaka, K., Clever, G.H. *et al.* (2006) Programmable self-assembly of metal ions inside artificial DNA duplexes. *Nature Nanotechnology*, **1** (3), 190–194.

Taton, T.A., Mirkin, C.A. *et al.* (2000) Scanometric DNA array detection with nanoparticle probes. *Science*, **289** (5485), 1757–1760.

Thaxton, C.S. and Mirkin, C.A. (2004) DNA–gold nanoparticle conjugates, in *Nanobiotechnology* (eds C.M. Niemeyer and C.A. Mirkin), Wiley-VCH Verlag GmbH, Weinheim.

Tomkins, J.M., Nabbs, B.K. *et al.* (2001) Preparation of symmetrical and unsymmetrical DNA–protein conjugates with DNA as a molecular scaffold. *ChemBioChem: A European Journal of Chemical Biology*, **2** (5), 375–378.

Wang, M.D., Yin, H. *et al.* (1997) Stretching DNA with optical tweezers. *Biophysical Journal*, **72** (3), 1335–1346.

Watson, J.D. and Crick, F.H. (1953) Molecular structure of nucleic acids: a structure for deoxyribose nucleic acid. *Nature*, **171** (4356), 737–738.

Wirtz, D. (1995) Direct measurement of the transport properties of a single DNA molecule. *Physical Review Letters*, **75** (12), 2436–2439.

Yoo, K.H., Ha, D.H. *et al.* (2001) Electrical conduction through poly (dA)–poly(dT) and poly(dG)–poly(dC) DNA molecules. *Physical Review Letters*, **87** (19), 198102.

Zhang, Y. and Seeman, N.C. (1994) Construction of a DNA-truncated octahedron. *Journal of the American Chemical Society*, **116** (5), 1661.

Zheng, M., Jagota, A. *et al.* (2003) Structure-based carbon nanotube sorting by sequence-dependent DNA assembly. *Science*, **302** (5650), 1545–1548.

Zimmermann, R., Freudenberg, U., Schweiß, R., Küttner, D., and Werner, C. (2010) Hydroxide and hydronium ion adsorption – a survey. *Current Opinion in Colloid & Interface Science*, **15**, 196–202.

2
Molecular Recognition

2.1
Case Study

Antibody–Antigen Interaction   In evolution, the immune system has been developed as a cellular recognition system, based on antibody proteins. The immune system is able to distinguish between "own" and "foreign" on the molecular level and initiate responses neutralizing the foreign substances. Essential components of the recognition system are the *antibodies*, which interact specifically with the foreign ligands, the *antigens*. One particular antibody specifically binds one structural unit (epitope) of the antigen. Owing to the presence of multiple epitopes, an antigen is usually bound by numerous different antibodies. The immune system is made up of a humoral and a cellular component. B lymphocytes (B cells), the key cells of the *humoral immune system* in the body fluids, produce soluble proteins called antibodies or *immunoglobulins* (Igs) that make up about 20% of all blood proteins. After binding the ligand (a bacterium, a virus, or large molecules) and its identification as a foreign component to be destroyed, the antibody activates leukocytes such as macrophages, which destroy the antibody-bound bacterium or virus by phagocytosis. *Polyclonal antibodies*, which are produced by many different B cells, bind a particular epitope of the same antigen. In contrast, *monoclonal antibodies* are produced by identical B cells and bind the same epitope.

The *cellular immune system* consists of T lymphocytes (T cells), which are equipped with antibody-like receptor molecules (T-cell receptors) on their surface. Cytotoxic T cells (T_C cells) or killer cells are activated by the simultaneous interaction between surface molecules and surface-exposed molecules of antigen-presenting cells, which are eventually destroyed. A second group of T cells, the helper T cells (T_H cells), initiate the destruction of the bound ligand by emitting signal molecules (so-called cytokines) that activate macrophages. The role of T_H cells is dramatically obvious in HIV infection: First the human immunodeficiency virus infects all the T_H cells, which finally leads to a total breakdown of the whole immune system. About 10^{18} different antibodies in the human body provide a giant pool of recognition units, with an enormous potential for use in bionanotechnology. *Immunoglobulin G* (IgG) constitutes the largest group of antibodies in the blood serum. It consists of four

Bio-Nanomaterials: Designing Materials Inspired by Nature, First Edition. Wolfgang Pompe, Gerhard Rödel, Hans-Jürgen Weiss, and Michael Mertig.
© 2013 Wiley-VCH Verlag GmbH & Co. KGaA. Published 2013 by Wiley-VCH Verlag GmbH & Co. KGaA.

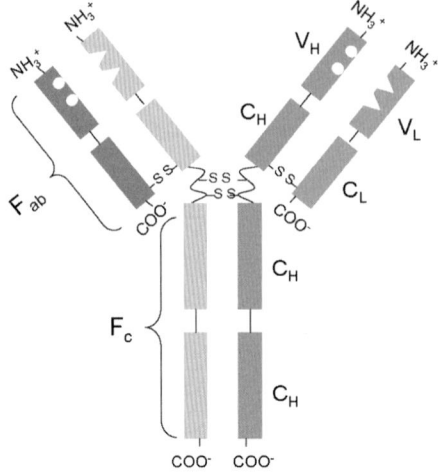

C = constant domain
V = variable domain
H,L= heavy, light chains

Figure 2.1 Immunoglobulin G. Pairs of heavy and light chains form a Y-shaped structure. The two antigen binding sites are formed by the amino-terminal domains of light and heavy chains. Hinges separate the antigen binding fragments F_{ab} and the basal fragment F_c.

polypeptide chains: two heavy and two light chains. As shown in Figure 2.1, the molecule forms a Y-shape. Each heavy chain is linked by two and one disulfide bridge(s), respectively, with the other heavy chain and one of the two light chains (see Figure 2.1). The amino-terminal domains exhibit a high variability in their amino acid sequence and form the antigen binding site, while the residual domains (one in the case of light chains and three in the case of heavy chains) have a constant structural composition.

Besides the IgG antibodies, there are four other classes of antibodies in vertebrates (IgA, IgD, IgE, and IgM). While IgD and IgE antibodies possess a similar structure as IgGs, the IgM group is characterized by a pentamer of Y-shaped immunoglobulin units cross-linked with disulfide bridges, and monomers, dimers, and trimers of Y-shaped molecules are present in the IgA group. The chemical complementarity between the antibody and the specific epitope is realized in several ways: fitting shapes capable of conformational changes for optimized fitting, charge distributions forming electrostatic bonds, and nonpolar and hydrogen binding groups. The dissociation constant is typically in the range of $K_d \approx 10^{-10}$ M. With more than one binding site per antibody, the affinity is higher because the complete dissociation would require two or more bond breaks. As the dissociation events are independent of each other, the probability for a simultaneous release of more than one bond is smaller than that of a single bond. It is the product of the independent single site probabilities.

Aptamers Aptamers are biomolecules that bind to specific targets; hence, they are potentially useful in biotechnology in similar ways as antibodies are. Two groups, aptamers based on nucleic acids and on peptides, are distinguished. Nucleic acid aptamers are short sequences of nucleic acids that are obtained by *in vitro* selection from a large random pool of sequences via repeated cycles of binding to a certain target. Peptide aptamers consist of a variable peptide loop of 10–20 amino acids attached at both ends to a protein scaffold. These constraints increase the selectivity for target molecules, eventually resulting in a high binding affinity to specific targets, comparable to antibodies.

Haptens Big foreign structures such as viruses, bacteria, or macromolecules with $M_r > 5000$ are typical antigens, whereas small molecules usually have no antigenic potential. The immune system does not respond to such molecules that are products of cellular metabolism. However, such small molecules or *haptens* can elicit an immune response if covalently attached to bulky carriers such as proteins. Antibodies produced in response to protein-linked haptens are able to bind the small molecule even in its free form. Exemplary haptens of interest for biotechnology are biotin, fluorescein, digoxigenin, and dinitrophenol.

Cellular Signal Transduction and Cell Adhesion The cooperative function of cells in any organism is based on highly efficient information transfer systems. Cells communicate via a well-tuned exchange of information on the chemical composition of the extracellular matrices, the metabolic activities at different places within the whole organisms, and changes of the environment. Often the signals are detected by specific receptors at the plasma membrane. Upon binding the signal is converted to cellular responses, for example, specific biochemical reactions. This means that such receptors are part of *signal transduction systems*. Commonly, the following six basic receptor types are classified (Nelson and Cox, 2008):

1) *G protein-coupled receptors*: Binding of a ligand to the receptor activates an intracellular guanosine triphosphate (GTP) binding protein (G protein), which controls enzymes that generate intracellular second messengers.
2) *Receptor tyrosine kinases*: These plasma membrane receptors are enzymes. Binding of ligands activates the phosphorylation of cytosolic plasma proteins.
3) *Receptor guanylyl cyclases*: These plasma membrane receptors stimulate the enzymatic formation of cyclic guanosine monophosphate (cGMP), which acts as a second messenger for activation of cytosolic kinases.
4) *Gated ion channels*: These membrane channels respond to the binding of the ligand or to changes of the membrane potential with opening or closing of the channel, respectively, thereby controlling the influx and efflux of ions.
5) *Adhesion receptors*: These receptors interact with macromolecular components of the extracellular matrix and upon binding transfer the information to the cytoskeleton.
6) *Nuclear receptors*: Binding of specific ligands, for example, steroids such as the hormone estrogen, results in the translocation of the receptor proteins to the nucleus, where they act as transcriptional modulators of a specific subset of genes.

The high selectivity of each transduction system is due to the molecular complementarity of signal and receptor molecules. The transduction is realized by noncovalent interactions, facilitating fast reactions similar to the antibody–antigen interactions. The specialization of the different transduction systems to specific cells implies an additional specificity of the signal transfer. The high affinity of receptors to the signal molecules with dissociation constants (K_d) typically $<10^{-10}$ M allows a highly sensitive signal transduction. Often the ligand–receptor interactions exhibit cooperativity allowing small changes in the ligand concentration to cause major changes in receptor activation (see also Section 1.2.4). The sensitivity can be further increased by a cascade of internal enzymatic processes, where the chemical signal is transformed by an enzyme that activates other enzymes. Thus, the number of involved molecules increases with the rate of an induced enzyme cascade. Among the six types of signaling mechanisms, the group of adhesion receptors is of specific interest for the design of biomaterials. The adhesion receptors, also called integrins, mediate the adhesion of cells to each other and to the extracellular matrix. The integrins transfer signals in both directions between the extracellular matrix and the cytoskeleton. Integrin receptors are composed of α and β subunits that are characterized by a large extracellular extension and a short cytoplasmic domain. 18 different α and 8 different β subunits are known allowing multiple combinations that facilitate the formation of different receptors interacting specifically with different proteins of the extracellular matrices in the various tissues. For example, specific integrins are known to bind collagen ($\alpha_{1,2}\beta_1$), fibronectin ($\alpha_5\beta_1$), or blood platelets ($\alpha_{IIb}\beta_3$). The ligand proteins possess characteristic amino acid sequences (e.g., Arg–Gly–Asp (RGD)) that are recognized by integrins. Inside the cell, the integrins interact with cytoskeletal proteins next to the plasma membrane, such as talin, α-actin, vinculin, or paxillin. In "outside-in" signaling, the extracellular domain of the integrin undergoes major conformational changes caused by load-induced alterations of the extracellular matrix. This conformational change is transferred via the cytoplasmic tails of the α and β subunits to intracellular proteins linked with the cytoskeleton. The mechanical load transfer causes a reorganization of the cytoskeleton finally leading to an intracellular signal cascade. Alternatively, signals from the *inside* can cause essential conformation changes of the extracellular domains of the integrins that modify the adhesion to the extracellular matrix or to adjoining cells (for more details see Chapter 3).

2.2
Basic Principles

The Crane Principle for Molecular Recognition Molecular recognition is one of the fascinating characteristics of interacting biomolecules. The remarkable ability of certain biomolecules to engage in highly selective bonding with other biomolecules is essential for various processes such as the regulation of gene expression and metabolic pathways, as well as the control of enzymatic reactions.

Outstanding examples are the hybridization of single-stranded DNA and RNA, the self-assembly of higher order supramolecular protein structures, and bio-mineralization, The selective bonding is mediated by molecular recognition, which is a short term for the complex phenomenon that some regions of a (bio)molecule are destined to fit and bond exclusively to a certain region of another (bio)molecule. This is favored if the following conditions known as *Crane principle* are met (Crane, 1950):

1) multiple and weak bonds;
2) form fit of contact regions.

2.2.1
Complementary Interaction between Proteins and Ligands

Proteins can be subdivided into fibrous proteins and others. Fibrous proteins serve as structural elements of cells and tissues. *Long-term stable quaternary interactions* between identical polypeptide chains provide mechanical strength, as realized, for example, with the various forms of collagen (see also Section 6.2.8.3). Another large group of proteins is characterized by "fleeting" interactions that make them suitable actors in a variety of physiological processes, for example, in the catalytic support of metabolic processes sustaining the life functions of cells, transport of oxygen between tissues, immune response, and contraction and relaxation of muscles. All these processes require molecular recognition. The essential element of protein-based molecular recognition is the reversible binding of proteins to other molecules. The bound molecules, called ligands, vary in size from small to big, and include proteins. In agreement with the Crane principle, the transient interaction between protein and ligand that facilitates fast recognition and exchange of information is characterized by one or more of the following features:

i) Binding sites on protein and ligand match in size, shape, charge, and hydro-philicity or hydrophobicity.
ii) Conformational changes induced by the interaction may result in an even better fit between the protein and the ligand.
iii) Several binding sites with specific affinities.
iv) Chemical reaction of the ligand triggered by contact with the protein.

The latter feature is characteristic for the transient binding between an enzyme (protein) and a substrate (ligand). In this case, the binding site is called catalytic or active site. The time dependence of the molar concentrations for the reversible binding between protein and ligand, $P + L \leftrightarrow PL$, follows the rate equation

$$\frac{d[PL]}{dt} = k_a[P][L] - k_d[PL], \tag{2.1}$$

where k_a and k_d are the rate constants for association and dissociation, respectively. In equilibrium, the concentration ratio of bound and unbound complexes is called

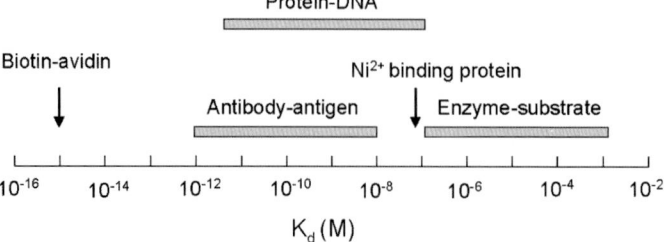

Figure 2.2 Typical examples of protein dissociation constants. (Adapted from Nelson and Cox (2008).)

the *association constant* K_a:

$$\frac{[PL]}{[P][L]} = \frac{k_a}{k_d} \equiv K_a. \tag{2.2}$$

It is a measure for the *affinity* of the ligand to the protein. Often the reciprocal of K_a, called the *dissociation constant* K_d, is used for the description of the equilibrium. The fraction θ of ligand binding sites on the proteins occupied by the ligands in equilibrium can be expressed by K_d and [L]:

$$\theta = \frac{[PL]}{[PL] + [P]} = \frac{K_a[P][L]}{K_a[P][L] + [P]} = \frac{[L]}{[L] + K_d}. \tag{2.3}$$

This reveals an interpretation of K_d: It is the molar concentration of ligands when half of the binding sites ($\theta = 1/2$) are occupied in equilibrium. A high binding affinity of the protein corresponds to a small value of the dissociation constant. In biological systems, the dissociation constants cover a very wide range. The very tight, almost irreversible binding between avidin and biotin is characterized by $K_d = 1 \times 10^{-15}$ M. The fast reaction rate of enzymes is related to much higher K_d values: 10^{-7} M $< K_d < 10^{-3}$ M. An overview of K_d ranges typical for protein–ligand reactions is given in Figure 2.2.

2.2.2
Cooperative Protein–Ligand Interaction

The model described by Eq. (2.1) is the simplest case for a reaction between a protein and a ligand, with every binding reaction being a statistically independent event. However, this is not always the case. For example, hemoglobin in erythrocytes (the red blood cells) which is responsible for the oxygen transport in blood, exhibits four oxygen binding sites, one per subunit of the tetrameric protein (see Figure 1.17). The concentration of oxygen is high in the lungs (13.3 kPa partial pressure) but low in the tissue (~4 kPa). Any flowing liquid which is able to dissolve oxygen would transport it from high to low concentrations but blood is much more efficient owing to a peculiar property of hemoglobin. Hemoglobin undergoes conformational changes depending on the oxygen concentration. If the oxygen concentration is high as, for example, in the lungs, it is in a state of high affinity to oxygen (R state); hence it takes up oxygen and binds it tightly. The low oxygen concentration in the tissue triggers a transition to a state of low affinity (T state); hence, the oxygen is readily released. The change of oxygen affinity coupled to conformation change is a cooperative effect: Binding of one O_2 to one subunit of hemoglobin in the T state changes the conformation such that the oxygen affinity of the other subunits is increased, eventually leading to the R state. Proteins with the property that binding of a ligand to one site affects the binding at other sites are called *allosteric proteins*. Allosteric proteins play an important role as enzymes for regulating the efficiency of the reaction with the substrate.

The cooperative binding of hemoglobin has been first analyzed by Hill in 1910. Following the reaction rate model explained above, the *Hill model* can be expressed in a modified reaction equation

$$P + nL \leftrightarrow PL_n \qquad (2.4)$$

and the corresponding rate equation is

$$\frac{d[PL_n]}{dt} = k_a[P][L]^n - k_d[PL_n]. \qquad (2.5)$$

The association constant for the high-affinity state is obtained as

$$\frac{[PL_n]}{[P][L]^n} = \frac{k_a}{k_d} \equiv K_a. \qquad (2.6)$$

The fraction θ of occupied ligand binding sites follows a sigmoid dependence on the ligand concentration:

$$\theta = \frac{[PL_n]}{[PL_n] + [P]} = \frac{K_a[P][L]^n}{K_a[P][L]^n + [P]} = \frac{[L]^n}{[L]^n + K_d}. \qquad (2.7)$$

With decreasing dissociation constant (high-affinity state), the $\theta = f([L])$ relation approaches a step-like shape. It is not obvious that all four subunits of hemoglobin are necessarily involved in the cooperative binding mechanism. In order to evaluate the extent of cooperativity in a given experimental situation, the data can be analyzed by starting from the rearranged Eq. (2.7):

$$\log\left(\frac{\theta}{1-\theta}\right) = n\log[L] - \log K_d. \qquad (2.8)$$

The slope in a $\log[\theta/(1-\theta)]$ versus $\log[L]$ plot yields the Hill coefficient n_H, which represents the effective number of interacting subunits. In the case of hemoglobin, $n_H = 1$ is obtained in the low-affinity state (T state), $n_H = 3$ in the transition between the low- and high-affinity states, and $n_H = 1$ again in the high-affinity state (R state). This means that cooperative binding prevails only for intermediate oxygen pressure, where almost three empty binding sites with high affinity are available. At high oxygen pressure, only one binding site is involved in the equilibrium reaction.

2.2.3
The Enzyme-Linked Immunosorbent Assay

The importance of antibodies in the regulation of the immune system makes them key molecules in medical diagnostics and drug development. Therefore, versatile analytical techniques have been developed for their detection, such as the enzyme-linked immunosorbent assay (ELISA). Such assays have found widespread applications in molecular biology, mainly owing to their robustness. Four variants derived from the basic principle are briefly characterized in the following. They lead to the detection of an antibody or an antigen by exploiting an enzymatic reaction for signal amplification (see also Figures 2.3–2.6). Traditionally, the tests have been designed with chromogenic reporters and substrates that produce color change when the analyte is present. Novel ELISA-like assays also use fluorogenic, electrochemiluminescent, or real-time PCR reporters to generate the signal.

The Indirect ELISA The test is aimed at measuring the concentration of an antibody of interest in a sample. On a microtiter plate, usually a 96-well polystyrene plate, a

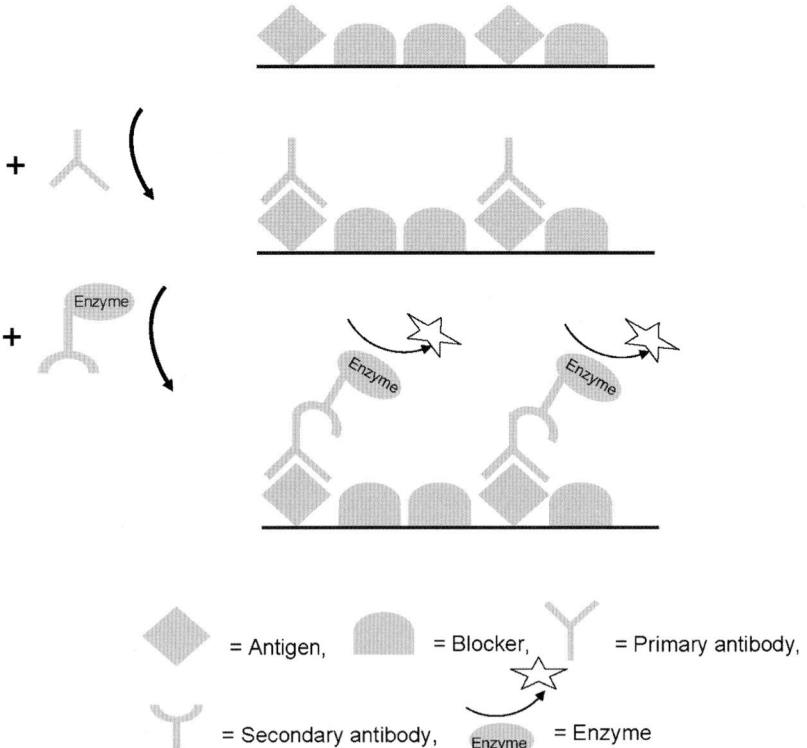

Figure 2.3 The indirect ELISA.

buffered solution of a corresponding antigen is placed. After adhesion of the antigen, a nonreacting protein (e.g., bovine serum albumin (BSA or casein) is added to block the uncoated areas of the plastic surface (Figure 2.3). Now the sample solution containing, among other proteins, the primary antibody (analyte) of interest, which binds specifically to the immobilized antigen, is added. In the final step, a solution with a secondary antibody with an attached enzyme is added, which binds the constant domain F_c of the primary antibody. After carefully washing the unbound antibodies away, a substrate for this enzyme is added to the wells, typically causing a color change upon reaction. By spectrometry, the color change can be quantified as a measure for the concentration of the primary antibody in the sample. Alternatively, the signal can be detected by means of fluorescence or electrochemical reaction.

The "Sandwich" ELISA With this assay, the concentration of a particular antigen in a sample can be determined. The basic layer of the assay is formed by a defined amount of the capture antibody directed against the antigen of interest (Figure 2.4). Free areas of the polymer surface are blocked to avoid nonspecific binding of proteins. In the next step, the antigen-containing sample is added. After careful washing to remove unbound antigen, a solution with the primary antibodies is

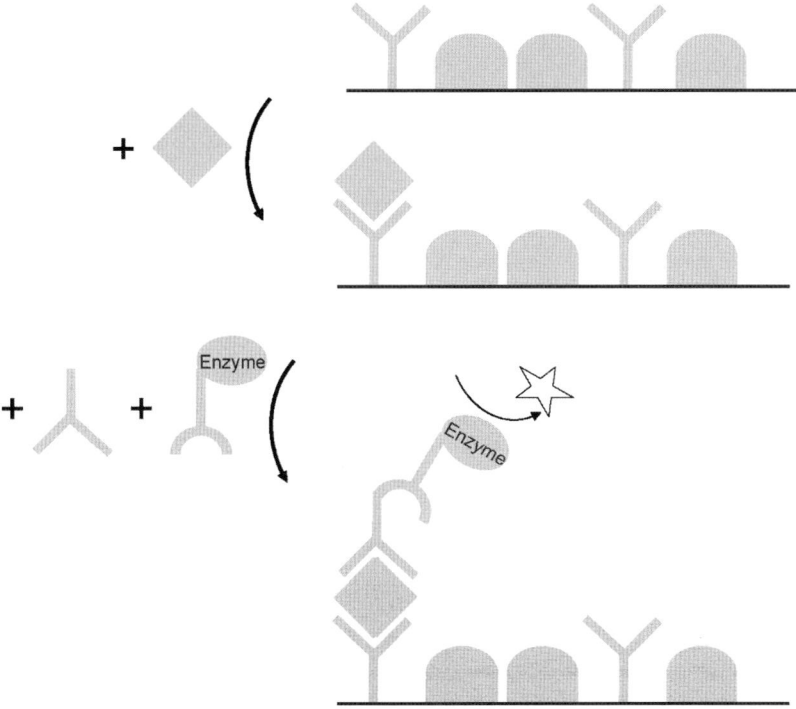

Figure 2.4 The "sandwich" ELISA. (Symbols see Fig. 2.3.)

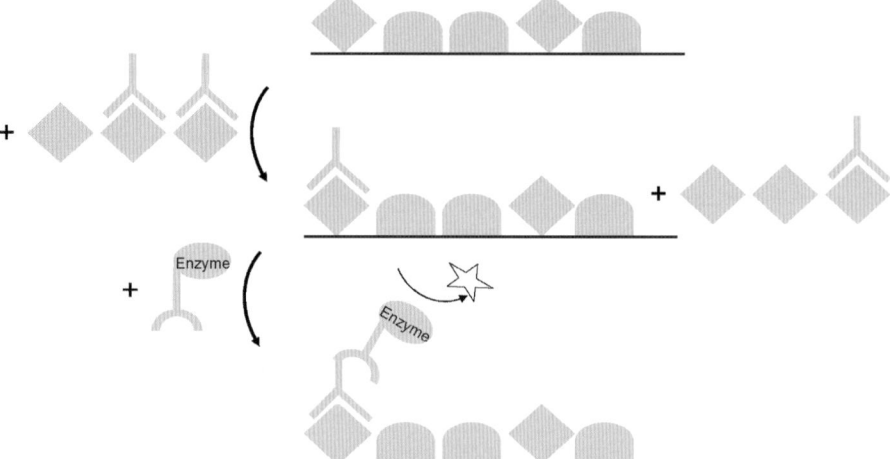

Figure 2.5 The "competitive" ELISA. (Symbols see Fig. 2.3.)

added that also bind to the antigens. Finally, the constant domain F_c of these antibodies is marked with a secondary enzyme-linked antibody. Again an enzymatic reaction of an appropriately chosen substrate is used for colorimetric, fluorescent, or electrochemical signal detection.

The "Competitive" ELISA In this case, the surface is coated with antigens and blocked to prevent nonspecific protein binding (Figure 2.5). A known amount of unlabeled antibodies is added to the sample containing the antigen of interest, resulting in the formation of antibody–antigen complexes. Upon addition of this solution to the reaction well, exchange reactions between the complexes and the surface-bound antigens are initiated. After washing, enzyme-linked secondary antibodies label the surface-bound primary antibodies. As the antigens in the sample compete with the surface-bound antigens for primary antibody binding, the enzymatic activity inversely correlates with the antigen concentration in the sample.

The Lateral Flow Test With the *lateral flow test*, a simple "point of care" option of an immunosorbent assay has been developed that finds already wide application in medical diagnostics, including tests for pregnancy, HIV, malaria, and drugs. The main advantage of the test is the minimal effort for sample or reagent preparation. The basic principle of the test is shown in Figure 2.6. The test sample flows along a porous substrate (nitrocellulose membrane) driven by capillary forces. Most tests give only qualitative information about the presence of a certain analyte. After placing the sample on a test strip, a colored reagent is added to the sample. Suitable reagents are nanometer-sized gold particles (red color) or latex beads (blue color) that are functionalized with an antibody directed to the target analyte. The sample solution with the colored reagents moves along the test strip by diffusional flow and

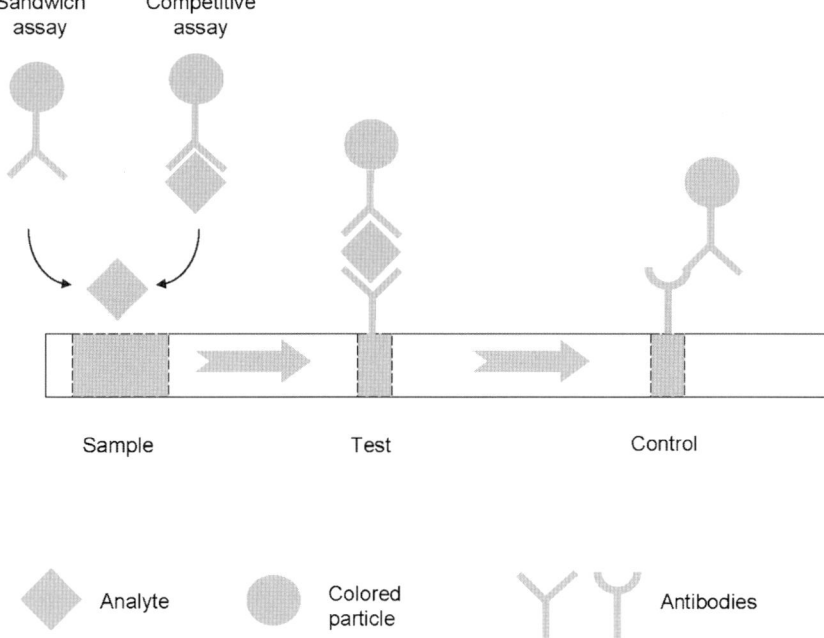

Figure 2.6 Lateral flow test: sandwich assay.

encounters at a certain distance two deposited lines. For a *sandwich assay*, the sample is mixed with colored particles that are labeled with antibodies directed to the target analyte. The first encountering line is soaked with the same antibodies. When the analyte is present in the sample, it can be bound at the first encountering line. Then an intensive coloring will be visible there. In a *competitive assay*, the colored particles to be mixed with the sample are already labeled with the target analyte. The first encountering line is soaked with antibodies to the target, leading to a competition of the unlabeled analyte in the sample with the particle-linked analyte. The test line will show a strong coloring when no unbound target analyte is present in the sample. The correct functioning of the test strip is monitored by a control line soaked with antibodies that bind the free colored particles.

2.3
Engineering of Biomolecular Recognition Systems

2.3.1
Engineering of Protein-Based Bioaffine Materials

2.3.1.1 Interfacing Mechanisms of Proteins via Bioaffinity
The reversible binding between proteins and ligands, called bioaffine interaction, enables the organisms to respond rapidly to changing situations. This interaction

Table 2.1 Bioaffine immobilization of proteins.

Interaction	Surface modification	Protein	Remarks
Avidin–biotin supramolecular complex	Streptavidin, double layer of biotin/streptavidin	Biotinylated	One of the strongest noncovalent bonds, good control of binding site density, with NeutrAvidin pH-dependent detachment possible
His tag coordination complex	Nickel chelate complex like nitriloacetic acid, or multichelator beads	$(His)_n$-tagged	Highly specific binding, entirely reversible upon addition of a competitive ligand
Affinity capture ligand system	Target protein	Fusion protein ligated with capture tag	Good control of orientation and surface density
Protein A/ protein G-mediated immobilization	Layer-by-layer architecture of protein A (or G) and antibodies		Highly oriented antibodies
DNA-directed immobilization	DNA oligomers	DNA–protein conjugates with complementary oligomers	High site specificity, high stability, regeneration possible, immobilization of multiple compounds in one step possible
Targeting carbohydrate moiety	SAM of conjugates of dithioaliphatic and aminophenylboronic acids	Glycoproteins	

has proven a useful tool in the development of bio-inspired materials. It offers advantages over other techniques of protein immobilization on nonbiological surfaces: preferred orientation of the reaction partners, optional detachment of the bound protein without denaturation, and easy regeneration or cleaning of substrates. Well-proven techniques in protein analytics and purification based on bioaffine interactions can serve as a guide for engineering applications. There are several approaches to the site-specific deposition of proteins on a substrate (Table 2.1). As the dissociation constants of the various protein–ligand complexes cover a wide range between 10^{-16} and 10^{-3} M (Figure 2.2), the bond strength can be tailored to suit the market need. The interaction between avidin and biotin provides one of the most favored options for bioaffine protein immobilization. Avidin from chicken egg and *Streptomyces*-derived streptavidin bind biotin with extreme affinities, which are among the strongest known. Functionalization of surfaces with avidin or streptavidin is therefore an excellent means for the immobilization of biotinylated molecules. Similarly, biotinylated surfaces capture and efficiently bind avidin or streptavidin, either in their isolated form or fused to a target protein.

Genetic engineering of target proteins with suitable tag sequences is a useful approach that, however, always requires (i) that the protein's function is not affected by the tagging and (ii) that the size of the tag is small enough to meet requirements for transduction of the biological signal to the technical sensor device (see Section 2.3.2). A small tag of consecutive histidine residues (polyhistidine tag) is a highly useful, common modification. This tag binds with relatively low affinity (dissociation constant in the micromolar range) to media with bound bivalent metal ions (nickel or cobalt). Bound proteins can be eluted with high concentrations of imidazole. Other genetic modifications that allow protein immobilization comprise the cellulose binding domain and the chitin binding domain. These tags allow efficient binding of target proteins to polyglycan-functionalized surfaces and lectins such as concanavalin A, which bind multiple carbohydrates (Nelson and Cox, 2008). The protein A- or protein G-mediated immobilization has been developed for the preparation of sensor surfaces functionalized with antibodies. It relies on specific interaction domains of protein A or protein G with the constant F_c region of antibodies. Antibodies bound via protein A or protein G thus expose their variable domain, which is responsible for the detection of the antigen, in an accessible position. By coupling the protein with a single-stranded DNA, the whole spectrum of DNA-directed immobilization techniques (see Section 1.3.2) is also suitable for the immobilization of target proteins. Possible options for conjugation are covalent attachment, streptavidin bridges, and bifunctional linkers (e.g., bis(sulfosuccinimidyl)suberate (BS)).

Engineering of Peptides with Specific Binding to Inorganic Materials During the past 15 years, learning from nature has initiated many activities in materials science laboratories with the aim to exploit biomolecular mechanisms for the manufacturing of inorganic nanomaterials by "soft" processing techniques. In particular, mimicking the processes of biomineralization has attracted the attention of materials scientists in search for novel functional materials (see Chapter 6). The use of peptides as structure-controlling motifs is one option for realization. In view of the fact that peptides can be composed of 20 natural amino acids, it is obviously a challenge to find compositions with structures optimally adapted to particular purposes. Various combinatorial techniques are available for the exploration of peptide structures. The basic step of such techniques is always to test the binding of a large group of peptides to the ligand of interest. It can be realized with large peptide "libraries." Either the surface of host cells (yeast cells or bacteria) or phages can be used for displaying the peptides. Combinatorial binding studies in connection with genetic engineering of the chosen library facilitate the selection and evolution of optimized binding peptides for a particular inorganic material.

Combinatorial Host Cell Display Peptide libraries can be set up by means of yeast cells and bacteria and making use of the proven techniques of genetically controlled display of peptides on the cell surface. Surface display using the yeast *Saccharomyces cerevisiae* has been proven successful owing to the well-established protein expression system and the rigid cell wall (Peelle *et al.*, 2005a). As an additional advantage, the nearly spherical cells of about 4 μm can be easily observed with conventional microscopy. The combinatorial selection of the peptides binding to the solid material surface is based on the so-called panning technique, which is schematically explained in Figure 2.7.

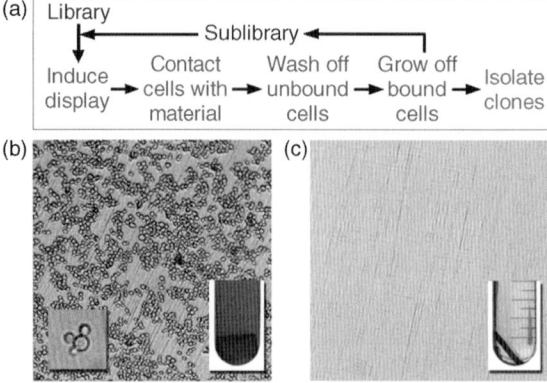

Figure 2.7 Cell display panning method. (a) Identification of material-specific yeast display library members that bind to a solid surface. (b) Optical microscopy. Yeast library expressing surface-displayed scFv polypeptides bound to single-crystal CdS during panning. (c) No background binding is seen with the same yeast library grown under conditions that repress the expression of the surface-displayed polypeptides. Insets show budding yeast cells bound to CdS (b, left) and yeast cultures (b and c, right) 24 h after the corresponding CdS crystals were incubated in glucose-based media under "growth off" conditions that repress expression of library clones. (Reprinted with permission from Peelle *et al.* (2005a). Copyright 2005, Elsevier.)

In the example, a library of *S. cerevisiae*, displaying about 10^9 different human single-chain antibodies (scFvs) and shorter polypeptides, has been used as an initial cell pool. ScFvs are a minimal form of functional recombinant antibodies (Hagemeyer and von Zur Muhlen, 2009), which consist only of the variable regions of the heavy and light chains of a full IgG antibody (Figure 2.1), connected by a small linker peptide. The scFvs can be presented on the cell surface. The library clones were induced by shifting the cells to galactose-based growth media. A material of choice (CdS in the example) was exposed repeatedly to the yeast library for 1–24 h, washed with media, and visualized by light microscopy (Figure 2.7b). In the first panning round, the material was exposed for 24 h to the cell suspension. In the following step, the bound yeast cells were stimulated to "grow off" the surface by placing the substrate in glucose-based media. Under these conditions, the expression of the library clones is repressed (Figure 2.7b, inset). The "grow off" allows rescue of all formerly bound cells. This sublibrary was subsequently used for the second screening round. The repeated cycling is also called "biopanning," derived from the common term "panning" for gold washing. After four screening rounds, shorter growth periods were used to increase the stringency of selection. After seven screening rounds, the panning was finished. Plasmid DNAs of yeast clones from the selected binding cell population were isolated and sequenced. Bioinformatic analysis revealed three plasmids that encoded fragments of 47, 70, and 70 amino acids, derived from full-length scFv genes, exhibiting 100, 91, and 90% homology to the variable heavy chain domains of the IgG antibody.

The semiquantitative yeast cell display also allows a deeper analysis of specific peptide–material interaction as demonstrated by Belcher and coworkers (Peelle *et al.*, 2005b). In order to reveal general design rules for engineering inorganic material-specific peptides, the four II–VI semiconductors CdS, CdSe, ZnS, and ZnSe and Au were chosen. The authors addressed the questions: (1) Which amino acid functional groups are sufficient for binding short peptides to these five inorganic material surfaces? (2) How do neighboring amino acid functional groups and their spatial arrangement in a peptide sequence modulate the binding strength? To answer these questions experimentally, genetically engineered peptides were displayed as fusions to the C-terminus of Aga2, a native surface-exposed protein of yeast cells. First, homohexapeptides X_6 of all 20 naturally occurring amino acids were engineered. The material samples were contacted with induced yeast cells in wells under fluidic shear provided by rocking at a steady rate of 2 cycles/s for 1 h and then washed for 30 min. The surface area fraction of the samples covered by bound yeast cells was taken as a measure of the binding strength. The results showed that homohexamers of histidine (H) and tryptophan (W) with an aromatic secondary nitrogen in the side chain functional group as well as cysteine (C) and methionine (M) with a thioether group in the side chain are sufficient for binding (Figure 2.8). The amino acids glutamate (E), aspartate (D), tyrosine (Y), serine (S), threonine (T), lysine (K), and arginine (R), which exhibit metal coordination properties, show no significant binding when displayed as homohexamers. The outstanding role of histidine H_6 is visible as it mediates binding of yeast to all five materials, whereas the other three amino acids show a material-specific interaction.

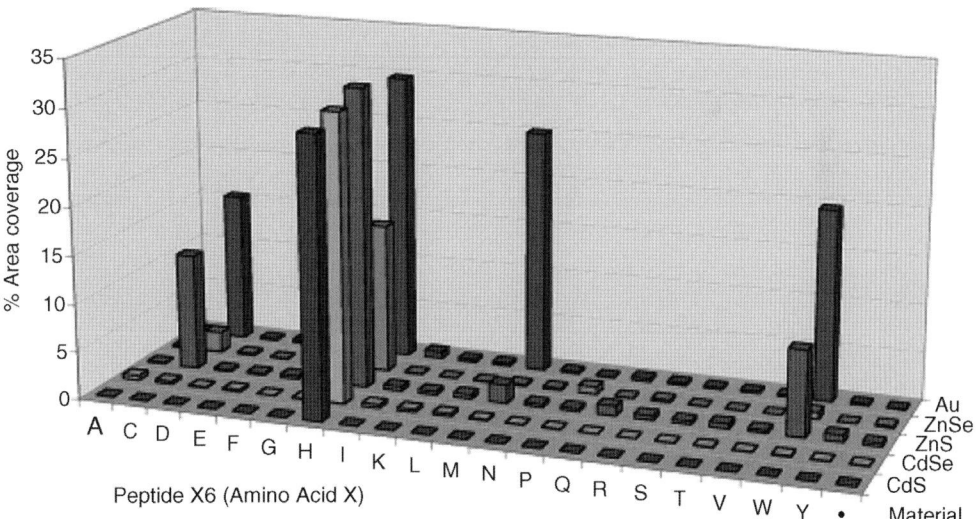

Figure 2.8 Binding efficiency of homohexamers of various amino acids displayed on the surface of yeast and contacted with every one of the five solid materials. (Reprinted with permission from Peelle *et al.* (2005b). Copyright 2005, the American Chemical Society.)

The second question was addressed by investigating the binding behavior of interdigitated peptides of the general form XHXHXHX, where X was one of the 19 other natural amino acids. The percentage area coverages caused by the various peptides were normalized to the alanine (A)-modulated construct AHAHAHA, which was intermediate in comparison to the other 18 amino acids. Alanine has only a methyl side chain. Binding experiments on CdS, ZnS, and Au showed that in general K and W, as well as glycine and basic amino acids, upmodulate the binding relative to A, whereas acidic, polar, and hydrophobic residues without a nitrogen downmodulate the binding.

Besides the identification of polypeptides that bind to specific materials, yeast cells displaying scFvs and fragments are candidates for further potential applications of cell material-specific interaction. For example, Peelle *et al.* (2005a) demonstrated the selective binding of yeast cells on a metal–insulator heterostructure. It has been shown that an isolated FePt clone binds selectively to FePt in the presence of silicon nitride and silica. As proposed by Peelle *et al.* (2005a), by transposing the selected peptides to other cell types such as *Pseudomonas*, the directed materials interaction might be used for self-healing biofilms for corrosion prevention. Engineered cells could also be used for controlled attachment to electrodes, a relevant issue, for example, for neurons, which are to be directly attached to Au electrodes (Straub *et al.*, 2001; Chiappalone *et al.*, 2003).

Combinatorial Phage Display Libraries   The phage display technique is based on the interaction of genetically modified coat proteins of viruses with a target structure. The idea of phage displays has been established by Smith (1985) using the minor coat protein, p3, of the filamentous phages (e.g., M13, fd, and f1) for the design of the fusion construct. The M13 phage is mostly used in phage display assays (Nam *et al.*, 2006). In comparison to the yeast display, phage displays deal with much smaller particles. The use of bacteriophages for the display of the fusion constructs is supported by well-elaborated genetic methods for the engineering of large libraries of fusion constructs of proteins or peptides. Furthermore, bacteriophages are monodisperse with characteristic shapes. This regularity makes them suitable as templates for the preparation of artificial nanomaterials. They can form self-organized liquid crystalline structures (for details see Chapter 7). Hence, they not only are intermediate carriers of targeted polypeptides, but can also be used as building units for the manufacturing of hybrid inorganic–organic nanostructures with higher complexity. The general scheme of a combinatorial phage display process is considered in Figure 2.9, which concerns the search for a peptide with high binding affinity to a certain inorganic material (see also Section 1.1.2). For example, various peptide encoding DNA sequences can be ligated into the p3 gene encoding the minor coat protein at the rounded tip of the M13 phage. The recombinant phages are used to infect *Escherichia coli* bacteria, in which new phages exhibiting the genetically modified p3 coat protein are generated and continuously released from the host cells. Within a short time, a large library of different genetically modified phages (typically from 10^8 to 10^9 viruses) can be produced. In the following step, the whole library of phages is added to a dish with the substrate of interest to allow the eventual

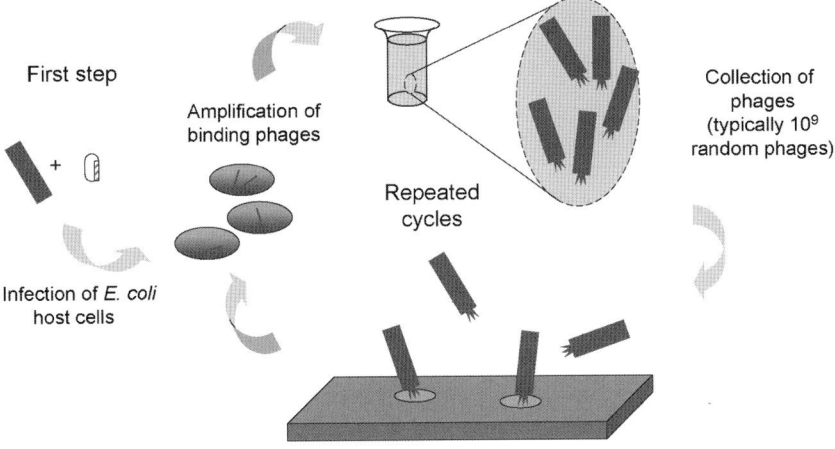

Figure 2.9 Selection of specific binding peptides by the phage display technique. *First step*: Infection of the host cells with recombinant M13 phages bearing fusion constructs of a gene encoding a surface protein and random peptide sequences. *Subsequent steps*: Repeated cycling of phages with genetically modified coat proteins, surface binding selection. (Adapted from Gazit (2007).)

binding of suitable phages. Upon washing, only phages displaying fusion proteins with an affinity to the target material remain attached. By applying a buffer with lower pH value, the bound phages can be eluted from the inorganic substrate and added to the bacterial host for another cycle of amplification. The phage DNA (or RNA in the case of RNA phages) can be mutated, for example, in a random or directed manner via the polymerase chain reaction (Nelson and Cox, 2008), giving rise to a new family of phages with slightly changed surface proteins. The new library in the second cycle contains significantly more phages with a high binding affinity. After several cycles, almost all phages in the library show optimized binding strength. The DNA of these phages is now sequenced to identify the peptides with high binding efficiency. The same procedure can be applied to generate phages which bind with high affinity to a certain target material via fusions of the major coat protein p8. As discussed in Chapters 6 and 7, phages with such properties are, for example, suitable as templates for the production of various nanowire structures.

Whaley *et al.* (2000) reported on such an example with a phage display library designed for specific interaction of the M13 phage with different single-crystal semiconductors (GaAs(100), GaAs(111), InP(100), and Si (100)). The minor coat protein, p3, was selected for the fusion construct. Five copies of the p3 protein are located on one end of the phage particle contributing only 10–16 nm to the total of 900 nm of the phage length (see Figure 1.19). By selective binding of the recombinant p3 protein, the M13 phage can be applied as a marker for a specific material, for instance, on a structured nanomaterial. Starting with a combinatorial library of random 12 amino acid peptides fused to the p3 protein, phages could be selected that bound to a specific crystal. Upon five reamplification and subsequent selection steps under more stringent conditions, phages were identified that could discriminate not only the chemical composition but also the lattice planes of crystals of the same compound, such as GaAs(100) and GaAs(111). The phages were additionally labeled with 20 nm gold beads. As a result of the studies, peptide sequences have been identified that exhibit a pronounced binding affinity to semiconductors, pure metals, metal oxides, and intermetallic compounds. In the peptide sequences binding to GaAs, preferentially amino acids with *uncharged polar groups or Lewis bases* were found (histidine, tryptophan, cysteine, methionine, serine, asparagine (N), and glutamine (Q)). As a possible binding mechanism, the donation of electrons to the semiconductor surface has been proposed by the authors.

In the following example, the presence of different proteins in the protein coat has been exploited for the engineering of *multifunctional nanostructures*. The background of the study was the development of a new type of electrodes for a lithium ion battery with higher capacity (Nam *et al.*, 2006). Cobalt oxide (Co_3O_4) nanowires show excellent properties as electrodes for lithium ion batteries. Anodes composed of Co_3O_4 nanowires have been produced with an extremely large reversible storage capacity for Li ions, approximately three times that of the capacity of carbon-based anodes used in commercial batteries. A further increase of the storage capacity by about 30% can be realized by doping the Co_3O_4 nanowires with Au nanoparticles. The Au nanoparticles probably improve the electronic conductivity and can catalyze the electrochemical reaction on the nanoscale. A bifunctional virus can be used for the synthesis of hybrid gold–cobalt nanowires of such kind if it provides appropriate fusion constructs that allow the site-specific deposition of the two materials (Figure 2.10). Again the M13 virus has been used as the basic structure. The *bifunctional virus* template was designed to simultaneously express the *tetraglutamate* Co_3O_4 *nucleation motif* and a *gold binding motif*. Tetraglutamate (EEEE) was fused to the N-terminus of the major coat p8 protein by directed polymerase chain reaction. This clone, called E4 virus, can fulfill three tasks: (i) The tetraglutamate motif serves as a template for the deposition of positive metal ions. (ii) The negatively charged tetraglutamate acts as a blocking motif for negatively charged colloidal gold particles. (iii) The negatively charged E4 virus can be ideally assembled at positively charged electrolyte polymers acting as a substrate for the electrode formation. For increasing the capacity of the electrodes by doping with Au nanoparticles, additionally a material-specific peptide motif was engineered into the major coat p8 protein. By phage display, the gold binding peptide motif (LKAHLPPSRLPS) was identified. This allows bifunctional AuE4 viruses to be designed that express both Au- and Co_3O_4-specific peptides with the virus coat. First, host bacteria were transformed with a plasmid encoding the gold binding version of p8. Afterward, these plasmid-bearing host cells were infected with the E4 viruses. As a result, a small percentage of the newly formed phages displayed both the E4p8 and the gold binding version of p8. Upon incubating the hybrid AuE4 virus with a 5 nm gold colloid suspension, Au nanoparticles bound to the gold binding motif were randomly distributed among the E4p8 proteins. After removing unbound gold nanoparticles by centrifugation, Co_3O_4 was nucleated and grown via the E4 functionality. Finally, a Co_3O_4 wire

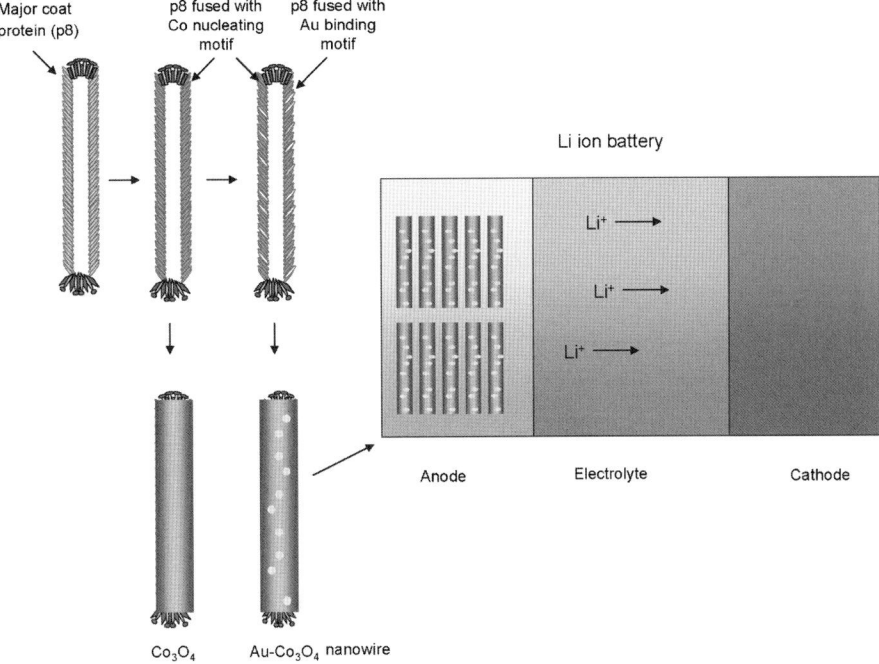

Major coat protein (p8)

p8 fused with Co nucleating motif

p8 fused with Au binding motif

Li ion battery

Li+ ——→

Li+ ——→

Li+ ——→

Anode   Electrolyte   Cathode

Co₃O₄   Au-Co₃O₄ nanowire

Figure 2.10 Schematic drawing of the virus-enabled synthesis and assembly of nanowires as electrode materials for Li ion batteries. (Adapted from Nam *et al.* (2006).)

with interdispersed gold nanoparticles was obtained. By self-assembly, the metallized viruses were deposited as a thin film of highly ordered nanowires at the anode of a Li ion battery (see also Section 7.2.6 on liquid crystals and Section 7.3.3 on bioengineering of functional materials by self-assembly of viruses). The specific capacity of the hybrid nanowires was estimated to be at least 30% higher than that of plain Co_3O_4 nanowires. The higher lithium storage capacity probably results from the catalytic effects of Au nanoparticles on the reaction of Li with Co_3O_4.

2.3.2 Engineering of Sensing Biofunctionalized Materials

Medical diagnostics is presently the main field of application for materials with protein-based bioaffine interfaces. There are already several successful examples, including the specific detection of bacteria or viruses, the use of immunoassays for antibody and antigen determination, the monitoring of metabolic products of the cells, and the intracellular targeting. Furthermore, in biotechnology there is growing interest in improving the process control by extending the measurable process parameters to biomolecular information, which characterize the activity of microorganisms in a bioreactor. Similarly, the increasing importance of environmental issues in our life requires the development of sensor systems that allow to quantify the impact of pollutions on living organisms. Progress in this field could make an essential contribution to improvements of the worldwide problems of limited resources, for instance, of drinking water. All these problems have led to an accelerated development of biosensors.

2.3.2.1 **Design Principles of Biosensors** A biosensor detects an analyte by combining a biological sensing element with a physical or chemical detector unit. It is composed of three parts:

- A sensing biological element, which produces the primary signal upon contact with the analyte, consists of individual biomolecules (antibodies, antigens, receptors, enzymes, nucleic acids, etc.), complex biological structures (e.g., ion channels), tissues, whole cells, or biomimetic components.
- A transducer element that transforms the primary signal into a physical or chemical signal. In the case of whole-cell sensors, the transducer could comprise cells that produce secondary signals (see Chapter 4).
- A reader unit that facilitates the further processing of the transducer signal and its display.

Numerous biosensor designs that are based on this modular structure led to a wide variety of commercially available systems. Usually the transducer element determines the technical complexity, the reliability, and the suitability for use in a clinical or industrial environment. Techniques transforming biological signals into physical or chemical signals are compiled in Table 2.2.

Optical techniques are most suitable for flexible integration into complex systems, whereas electrochemical and piezoelectric methods are favored for lab-on-chip designs.

Table 2.2 The main physical and chemical transducer variants applied in biosensors.

Transduction principle	Examples
Optical	ELISA
	Colorimetric sensor
	Fluorescence assay
	Chemiluminescence assay
	Surface and particle plasmon spectroscopy
	Surface-enhanced Raman spectroscopy
	Interference spectroscopy
Electrochemical	Amperometric detection
	Potentiometric detection
	Impedimetric detection
	Ion-selective field effect transistor (ISFET)
Mechanical	Quartz microbalance
	Surface acoustic waves
	Piezoelectric sensor
	Multicantilever force spectroscopy
Calorimetric	Microcalorimetry
Magnetic	Magnetic beads

2.3.2.2 **Integration of Sensing Biological Elements and Transducer Units**   The key problem in biosensor design is the suitable combination of sensing element and transducer. In this regard, the following conditions should be met as far as possible:

- The arrangement of the sensing biological element must allow undisturbed interaction with the analyte.
- Nonspecific interaction of the analyte with the transducer has to be excluded.
- The sensing element must be stable under normal storage conditions.
- The assay should be stable in repeated measurements.
- The response should be reproducible and change monotonically with the analyte concentration.
- Ideally, regeneration of the sensing biological element should be possible under near-line conditions.

According to the arrangement of their components, heterogeneous and homogeneous biosensors are discriminated. Either the sensing biological element is arranged on a separate macroscopic carrier (*heterogeneous biosensors*), or the carrier size is in the same range as the sensing element (*homogeneous biosensors*). As these geometric differences lead to different strategies in handling the biosensor in a typical process environment (for instance, coupling of the sensing element with the transducer unit, and sensor regeneration under conditions of an online process), specific aspects of both concepts will be discussed separately in the following.

Heterogeneous Biosensors   The immobilization of the sensing biological element on the carrier usually follows a generic approach as shown in Figure 2.11.

In the initial process step, the substrate surface is functionalized with binding groups whose density determines the final density of the sensing sites. In the second step, a coupling layer is deposited, which facilitates the site-specific and oriented binding of the sensing molecules in the third step. The coupling layer also has to preclude adsorption of other biomolecules. Moreover, thickness and chemical composition of the coupling layer must be chosen such that denaturation of the sensing molecules by interaction with the substrate or coupling layer is avoided. Occasionally, specific conditions have to be met. For example, if the transducer unit is integrated in the substrate as in the case of surface plasmon resonance (SPR) spectroscopy, the coupling layer must be thin enough to allow signals from the sensing element to reach the active zone of the transducer (in the case of SPR, the interaction zone of the surface plasmons excited in a thin gold metal film on the substrate surface; for more details see below). The final step, the stable and oriented deposition of the sensing biological element, is usually realized with covalent or strong bioaffine bonds. For the reason of cost efficiency, the design should also include an option for the (near-line) regeneration of the sensing elements, for example, by the implementation of distinct reaction sites which reproducibly allow a complete dissociation and rebuilding of the damaged sensing element or coupling layer. Dissociation reactions can be initiated by temperature change (e.g., dissociation of oligonucleotide pairs), pH shift (e.g., dissociation of the NeutrAvidin–biotin complex), or electrochemical dissolution (e.g., of thin metal

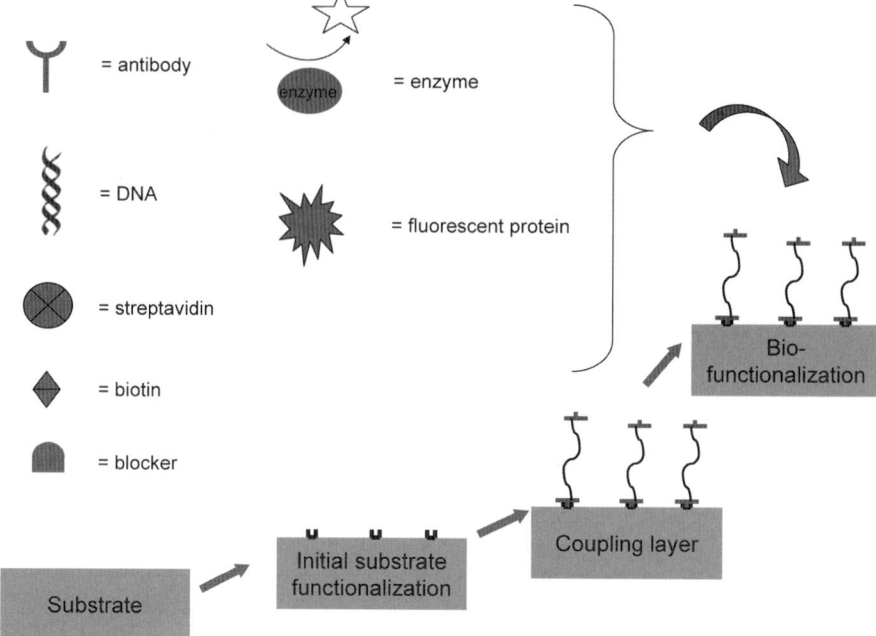

Figure 2.11 Successive formation of the sensing biological element. (Adapted from Goddard and Hotchkiss (2007).)

films on the carrier surface). Three typical examples for the formation of the sensing biological elements and their coupling with a transducer unit are described in the following. They demonstrate the efficiency of the modular design approach.

Modular Solid Sandwich Immunoassay   There are several realizations of biosensors based on the connection of the primary biological sensing event with the subsequent formation of a physical or chemical signal in the transducer by means of an enzymatic reaction. This approach is derived from the basic ideas of the enzyme-linked immunosorbent assay. One favored enzyme for such design is the horseradish peroxidase (HPR), a glycoprotein with four lysine residues that allow labeling with molecules via conjugation. HPR catalyzes the formation of colored, fluorescent, or luminescent products when incubated with a suitable substrate. An example for a solid sandwich immunoassay for application in bioreactors in a near-line processing regime (Eubisch, 2012) is described in the following. The enhanced chemiluminescence of luminol reacting with hydrogen peroxide in the presence of HPR and an additional "enhancer" molecule has been exploited for transduction (Figure 2.12).

Based on this reaction, a modular solid sandwich immunoassay was developed by applying the generic process described above, for manufacturing the sensing

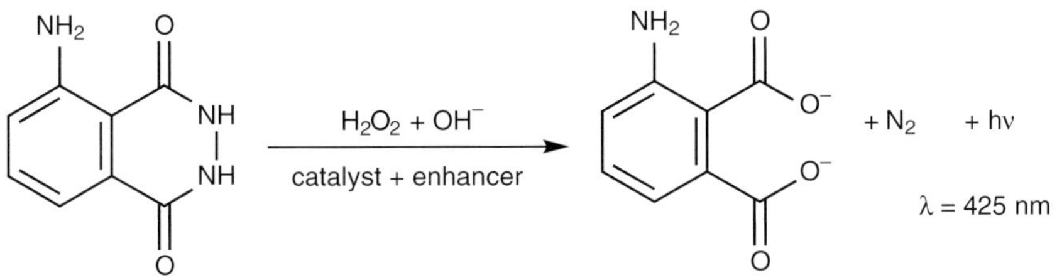

Figure 2.12 Oxidation reaction of luminol with hydrogen peroxide catalyzed by horseradish peroxidase in the presence of an additional enhancer that leads to light emission.

biological element. The complete signal detection and transduction process is performed in a microfluidic system. Two options provide high flexibility of the near-line regeneration of the sensing biological element:

1) Coupling layer formed with biotinylated photolinker molecules (benzophenone–polyethylene glycol (PEG)$_3$–biotin). The planar substrate is coated with a thin polymer film (cyclic olefin copolymer). The benzophenone residue of the photolinker molecules photolyzes upon exposure to UV irradiation and is inserted in C—H bonds of the copolymer substrate (Figure 2.13). The biotin group of the photolinker allows the subsequent coupling with biological structures via the bioaffine avidin system by biotin–NeutrAvidin–biotin linkages. The advantage of the process is the high flexibility for manufacturing patterned coupling layers. However, it has been proven that the biotin–NeutrAvidin–biotin linkages are not optimal for repeated regeneration of the sensing layer under near-line conditions.

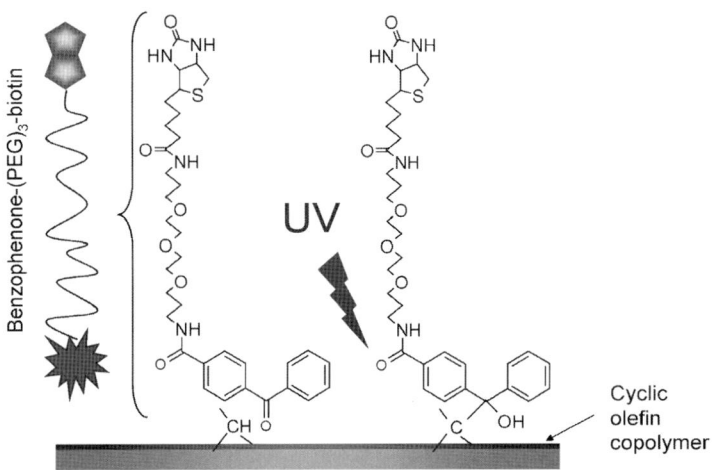

Figure 2.13 Immobilization of biotin by a UV-induced photolysis of benzophenone molecules. (Reprinted with permission from Eubisch (2012).)

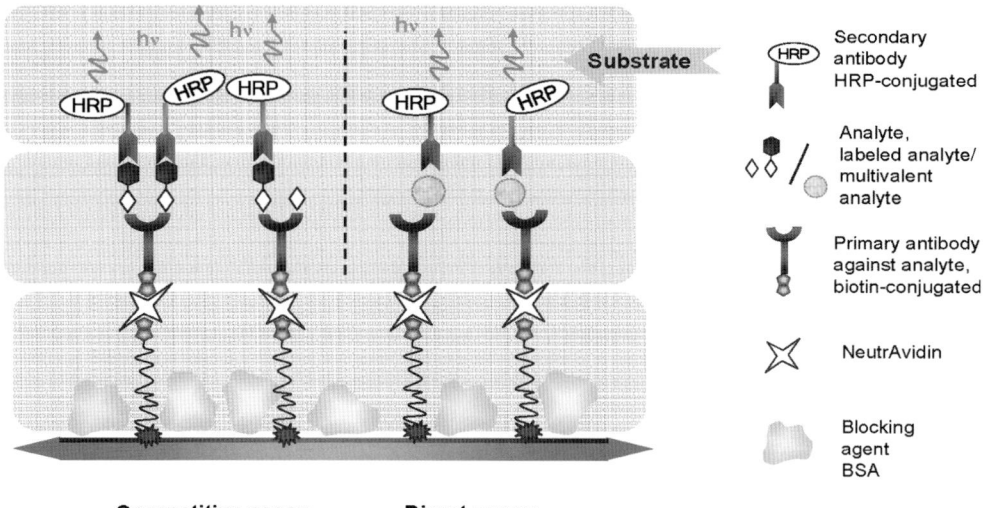

Competitive assay   **Direct assay**

Figure 2.14 Design of the competitive and the direct immunoassay. (Reprinted with permission from Eubisch (2012).)

2) Coupling layer consisting of thiol–(PEG)$_3$–biotin deposited on a FePt interlayer. The coupling layer is formed on a thin FePt film deposited by thiol chemistry on the substrate. The thiol groups form covalent bonds with the metal film, and the biotin end groups allow a reliable linkage of the biomolecular sensing element via the biotin–streptavidin–biotin coupling. With this setup, two kinds of sandwich immunoassays were realized (Figure 2.14): a competitive and a direct one. The components of the immunoassays were successively attached *in situ* by affinity reactions under flow conditions. In order to avoid nonspecific binding of any component, possible additional binding sites were blocked on the substrate with bovine serum albumin. The amount of binding analyte was characterized with an enzyme-conjugated secondary antibody. The intensity of chemiluminescence after substrate (luminol + H$_2$O$_2$ + enhancer) incubation was measured with a photo-diode. One advantage of the competitive assay is the large range of analyte concentration for which it can be applied, as in the example shown in Figure 2.15 with hemagglutinin (HA) peptide as a model analyte. HA-tagged enhanced green fluorescent protein (HA-EGFP) was chosen as competitor. For a fixed competitor concentration, the chemiluminescence intensity versus concentration makes a sigmoidal curve in a semilogarithmic plot. By changing the competitor concentration, the detection range can be adjusted over five orders of magnitude.

A direct immunoassay for detection of the human influenza A virus (subtype H3N2) is shown in Figure 2.16. As the virus exhibits the HA peptide on its surface, biotin-conjugated and HPR-conjugated HA antibodies were used as primary and secondary antibodies, respectively. The measurement was performed in a micro-fluidic system with constant flow rate. This explains the shift of the signal curve with

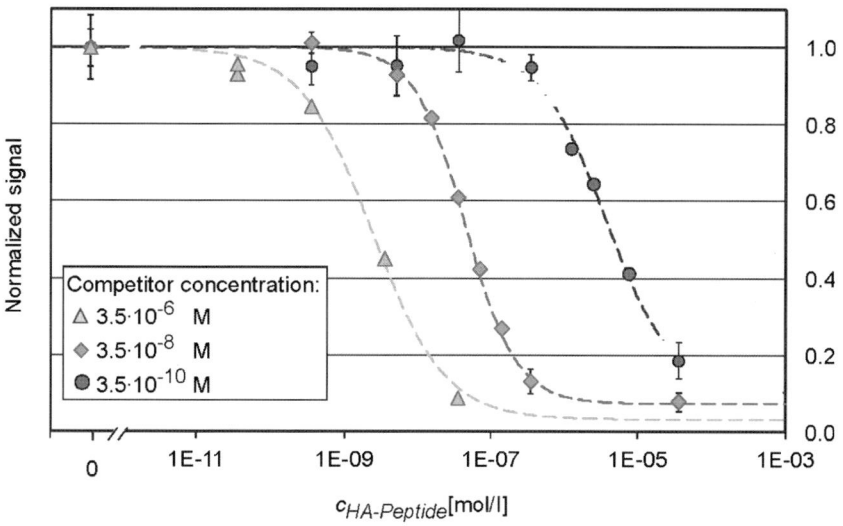

Figure 2.15 Competitive chemiluminescence immunoassay for the measurement of the HA peptide concentration. The detection range can be adjusted by a suitable choice of the competitor concentration. (Reprinted with permission from Eubisch (2012).)

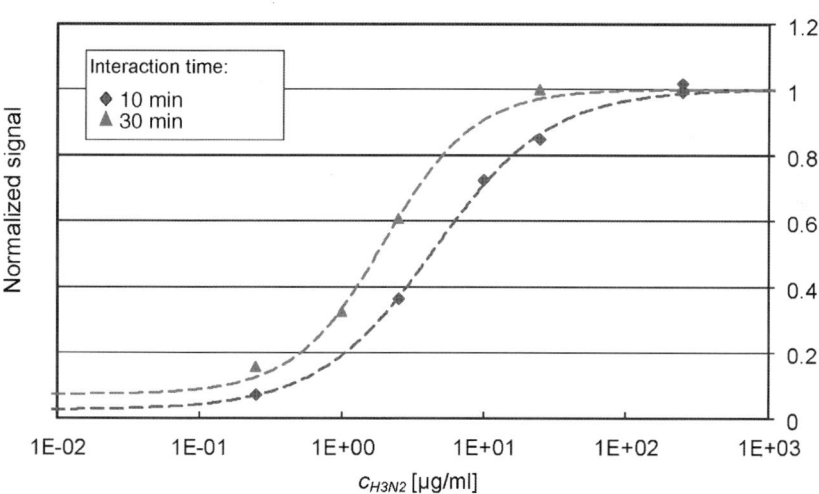

Figure 2.16 Direct chemiluminescence immunoassay of human influenza A virus (subtype H3N2) using primary HA antibodies and secondary HPR-conjugated HA antibodies. The luminescence intensity versus virus concentration in the fluid was measured at constant flow rate for two interaction times of the sensor with the fluid. (Reprinted with permission from Eubisch (2012).)

increasing interaction time caused by the larger amount of presented analyte. In analogy to the competitive assay, the detection range can be extended to smaller analyte concentrations by increasing the reaction time. The sensitivity can be further enhanced by increasing the "active volume" of the fluid interacting with a sensing layer. This can be realized with a lower height of the microfluidic channel and lower flow rate.

Surface Plasmon Resonance Measurement Surface plasmon resonance measurements are based on the light reflection at the interface between transparent media, coupled to plasmon excitation in a metal film. An incident beam is split at the interface into a transmitted and a reflected beam, except for the case of total reflection without a transmitted beam (Figure 2.17). This case is realized under two conditions: The incident beam must be in the medium with the higher refractive index, and the angle of incidence must exceed a critical value depending on the refractive indices of the two media. As a peculiarity of total reflection, the electromagnetic field of the reflected beam transcends across the interface for a short distance, a phenomenon known as evanescent field, which decays exponentially with a decay length δ comparable to the wavelength. The evanescent field "feels" the presence of thin layers at the interface, which can be exploited for the characterization of layer properties. For reasons explained below, the glass substrate used in SPR measurements (see Figure 2.17a) is coated with a metal film (typically

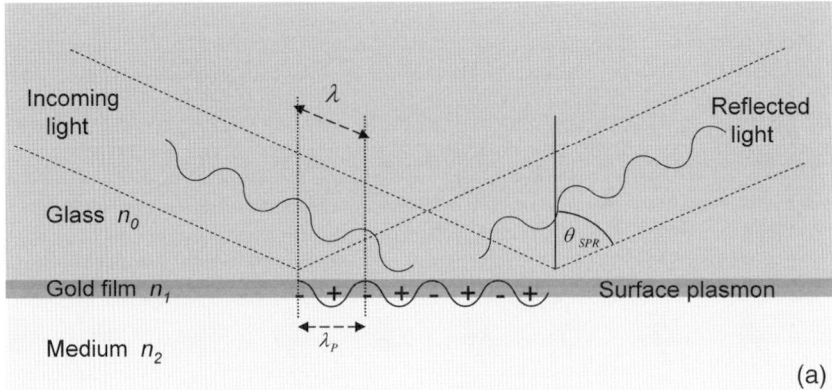

(a)

Figure 2.17 Surface plasmon resonance measurement. (a) Kretschmann configuration of glass substrate and gold film. n_i with $i = 0, 1, 2$ are the refractive indices of the substrate, the gold film, and the surrounding medium, respectively. ε_i are the permittivities of the three materials. λ is the wavelength of the incoming light. *Note:* The ratio of thickness of the gold layer and the glass substrate in the scheme does not correspond to the real experimental situation with $d_{gold}/d_{glass} \ll 1$. (b) Reflectivity versus angle of incidence. The reflectivity values of four different substrates are compared: a bare glass substrate, a glass substrate deposited with a thin gold film, a glass substrate deposited with a gold film and a biopolymer film of thickness d, and the same layered coating with the biopolymer film of thickness $2d$. The biopolymer layer causes a shift of the angle of the plasmon resonance θ_{SPR} to higher values with increasing film thickness.

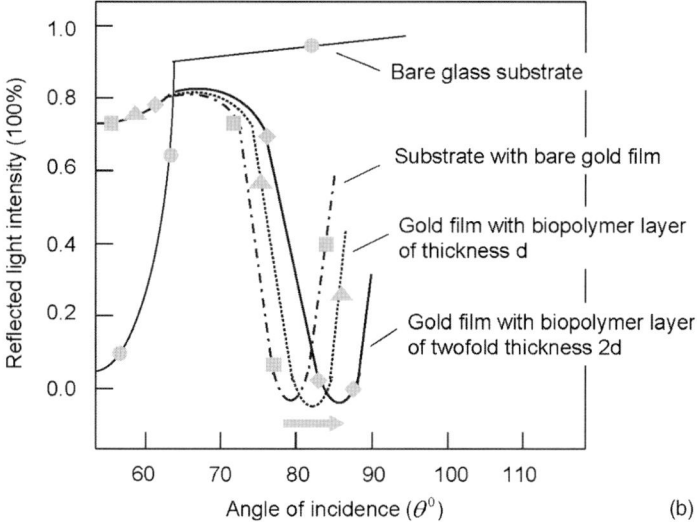

Figure 2.17 (*Continued*)

gold). It is so thin that it practically does not absorb the evanescent wave, except at a certain angle of incidence of the light beam, where the conducting electrons in the metal film are excited into collective oscillations called plasmons. The effect is largest with monochromatic p-polarized light. The resonance condition is fulfilled when the wavelength of the plasmon, propagating parallel to the surface of the gold film, matches the projection of the wavelength of the incoming light. This condition is met for a particular value θ_{SPR} of the angle of incidence. Since the absorbed power is missing in the reflected beam, the angle of plasmon resonance θ_{SPR} can be found by measurement of the reflected beam intensity depending on the angle of incidence. The angle of plasmon resonance is affected by the presence of additional layers on the metal layer, which make it suitable for the characterization of biological thin layers. For example, association and dissociation of analytes at a sensing biological layer placed on the gold film can be detected in this way, provided that the coupling layer is sufficiently thin so that the evanescent field extends into the sensing layer.

A typical layer design to be used for the detection of small proteins by receptor–ligand interaction is explained in the following example with estrogen receptors (ERs) to be detected in a test solution by means of a DNA–protein recognition mechanism. The specific binding of the ER to the so-called estrogen-responsive element (ERE) harboring DNA was exploited for this purpose (D. Gelinsky-Wersing, private communication, Technische Universität Dresden, 2011). The ER protein is composed of six domains, with the C-terminal domain exhibiting two Zn^{2+} finger motifs that specifically interact with the ERE sequence of the DNA. The coupling layer was deposited on the gold film by thiol chemistry (Figure 2.18). Here the sulfur atoms coordinate to the gold atoms. As the interface has to be "inert" with respect to

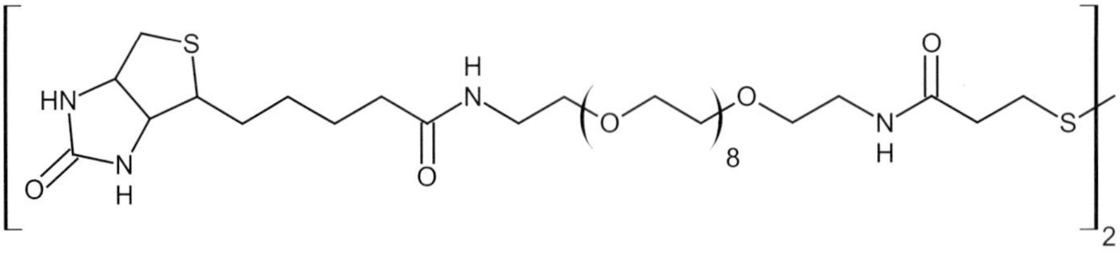

Figure 2.18 Structure of biotin–PEG–disulfide used as coupling molecule.

an unspecific adsorption of biomolecules, a self-assembling monolayer (SAM) of polyethylene glycol was selected for the coupling layer polymer. Furthermore, the interface should offer terminal groups that allow a site-specific binding of the recognition unit. In the present example, the coupling layer was made up by self-assembling biotin–PEG–disulfide molecules. By means of the biotin tail groups, a second layer of NeutrAvidin molecules was assembled via the bioaffinity mechanism (Figure 2.19). This layer controls the oriented deposition of biotinylated DNA molecules harboring EREs. The stepwise changes of the angle of incidence θ_{SPR} in the SPR diagram reflect the coupling of the ERE with the NeutrAvidin layer and the following binding of the ER target molecules.

One-Dimensional Biosensor Arrays With the manufacturing of metallic and semiconducting nanowires such as carbon nanotubes or silicon nanowires, the way has been opened for the development of a new generation of biosensors. There is one effect that makes them very interesting for biosensing. The nanowires are so thin that a single biomolecule attached to the surface can cause a measurable change in the conductance. Thus, nanowires are well suited for single-molecule detection. Furthermore, the progress in nanotechnology enabling large-scale integration of nanowires in microelectronic circuitries provides options for the simultaneous detection of a large number of different biomolecules on one chip. The development of nanowire-based biosensors is closely connected with the work of the group of Charles Lieber (Patolsky *et al.*, 2004). The potential of such an approach has been impressively demonstrated with the electrical detection of single viruses. The measurement was based on a simple experimental setup. The electric current was conducted from the source electrode to the drain electrode. The conductance of the wire or tube can be controlled by a third electrode, the gate electrode, between the source and drain electrodes. In the experiment, the gate electrode has been realized by antibodies immobilized on the nanowires (see also Figure 2.20). In this so-called field effect transistor (FET) configuration, a positive charge of the gate decreases the conductance of p-type nanowires (holes as positive charge carriers) and increases the conductance of n-type nanowires (electrons as negative charge carriers). The antibodies can be used for binding specifically charged macromolecules to be detected. The resulting conductance change can be measured with extreme sensitivity. In the particular experiment, the biomolecular recognition facilitated the detection of

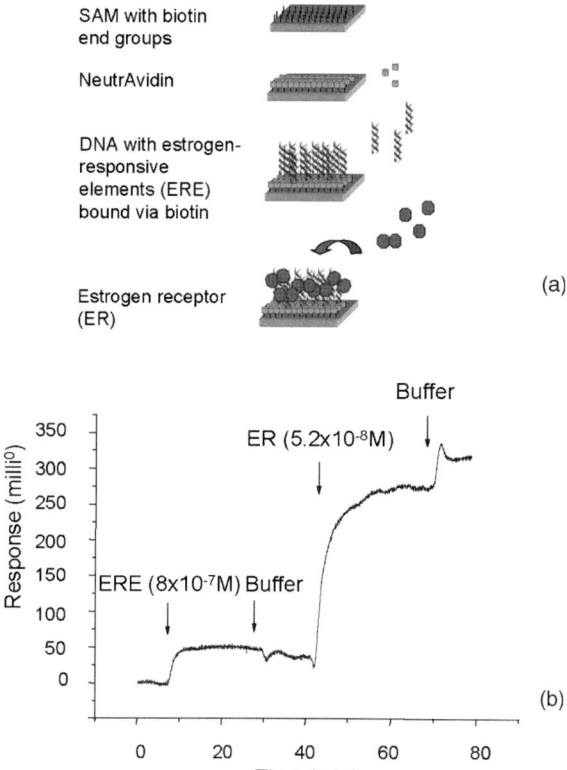

SAM with biotin
end groups

NeutrAvidin

DNA with estrogen-
responsive
elements (ERE)
bound via biotin

Estrogen receptor
(ER)

(a)

(b)

Figure 2.19 Design of the interface of an SPR sensor. (a) Schematic of the three-layer interface composed of biotinylated SAM, NeutrAvidin, and estrogen-responsive elements. (b) Stepwise changes of SPR signals due to coupling with the EREs and the ERs. The change of shift was measured in comparison to a control sample without ERE coating. The last increase caused by rinsing with a buffer indicates some change of the control layer. (Courtesy of D. Gelinsky-Wersing.)

influenza A virus, as explained in Figure 2.20. The measurement was performed in a microfluidic system (flow rate of 0.15 ml/h). The frequency of single binding events scales linearly with the virus concentration in the solution (see also inset in Figure 2.21). The high time resolution of the measurement (on/off times) provides information about the binding kinetics of the virus at the receptor. In the example shown, monoclonal anti-hemagglutinin antibodies were coupled to aldehyde-terminated silicon nanowires functionalized with 3-(trimethoxysilyl)propyl aldehyde. Receptors were coupled in phosphate buffer containing 4 mM sodium cyanoborohydride. Unreacted aldehyde surface groups were subsequently passivated with ethanolamine. A reference wire not functionalized with the receptors did not show a change of conductance in the presence of viruses. In addition to the detection of

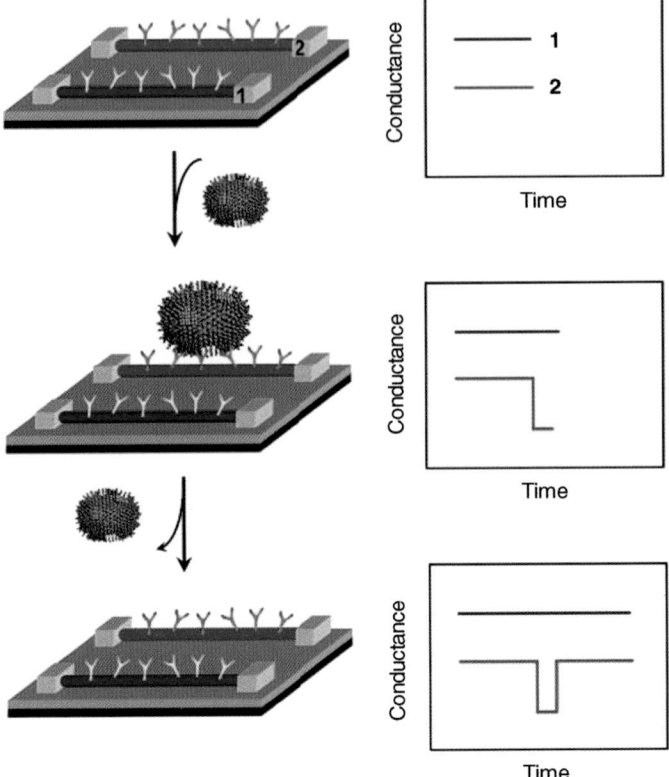

Figure 2.20 Nanowire-based detection of single viruses. Only one of the two nanowires modified with different antibody receptors responds to binding and debinding of viruses by conductance changes. (Reprinted with permission from Patolsky *et al.* (2004). Copyright 2004, National Academy of Sciences, USA.)

molecules and viruses, the influence of the pH on the binding kinetics can be investigated, and the isoelectric point of viruses or molecules can be determined by means of this concept.

Whereas the thermodynamics of affinity reactions limits the sensitivity of conventional transduction techniques based on macroscopic devices, bionanoelectronics with one-dimensional charge transport allows to scale down the sensitivity to the level of single molecules (Noy, Artyukhin, and Misra, 2009). For example, in the case of antibody–antigen reactions, the sensitivity of macroscopic biosensors is one or two orders of magnitude below the dissociation constants K_d, which are typically in the range from 10^{-8} to 10^{-12} M. In the case of antibodies immobilized on a nanowire, binding of a single antigen would result in a measurable response.

Homogeneous Biosensors The functionalization of nanoparticles with antibodies facilitates the design of various kinds of nanosized homogeneous biosensors. One

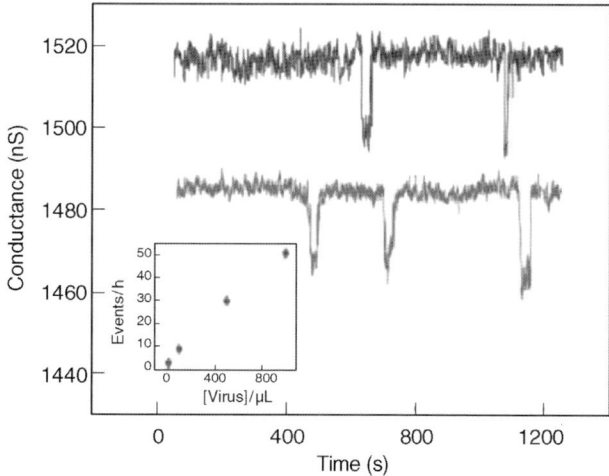

Figure 2.21 Selective detection of influenza A viruses. *Inset*: The frequency of the conductance change scales linearly with the virus concentration. (Reprinted with permission from Patolsky *et al.* (2004). Copyright 2004, National Academy of Sciences, USA.)

advantage of this class of sensors consists in the large variety of physical and chemical properties of nanoparticles usable for the signal transduction, their small size that allows the application for intracellular measurements, and a cost-efficient manufacturing in many cases. The following examples will give an impression of the potential of nanomaterials in biosensor development.

Bioconjugated Fluorescent Silica Particles for Biomarking of Single Bacteria The rapid and reliable detection of single bacteria is a major challenge in medicine, biotechnology, and more recently in the fight against bioterrorism. Traditional detection of small traces of bacteria is a time-consuming procedure based on amplification and enrichment of the target bacteria. Typical process times are in the range of several hours, but 1 h or less would be highly desirable. A possible solution was presented by Zhao *et al.* (2004) with the development of a rapid bioassay based on bioconjugated nanoparticles. A bacterium exhibits numerous antigens at its surface, which can be recognized by specific antibodies. Therefore, binding of monoclonal antibodies that are specific for the targeted bacterium can be used for the detection process. The antibody can be linked with a fluorescent marker for the design of the bioassay. In order to realize an ultrasensitive assay, fluorescent silica nanoparticles (diameter $\approx 60\,$nm) conjugated with antibodies have been applied. Every silica nanoparticle contains thousands of fluorescent dye molecules (tris(2,2'-bipyridyl)dichlororuthenium(II) hexahydrate (RuBpy)). The amine-functionalized silica nanoparticles are made to react with succinic anhydride in *N,N*-dimethylformamide solution under N_2, resulting in the formation of carboxyl groups on the silica surface for conjugation of the antibodies. The advantage of the bioconjugated silica beads with encapsulated fluorescent dye molecules consists in the more than 1000 times greater fluorescence signal compared to that of the antibodies labeled with dye molecules. Thus, a highly amplified and reproducible signal is obtained from every antigen–antibody binding event. Scanning electron microscopy reveals that a single bacterium binds hundreds of antibody–nanoparticle conjugates. Thus, spectrofluorimetry provides strong signals from single bacteria. By using a flow cytometer with a micrometer-sized capillary, a highly efficient detection system has been developed that allows probing of sample volumes of a few picoliters in about 60 s for single bacteria. With the same method, the option of multiple pathogen quantification has also been realized by simultaneous detection of three pathogenic species (*E. coli* O157, *Salmonella typhimurium*, and *Bacillus cereus* spores) with three kinds of specific antibody–nanoparticle conjugates.

Nanoparticle-Based Bio-Barcode for Ultrasensitive Detection of Proteins   The idea to increase the sensitivity of a homogeneous sensor assay by an additional internal amplification mechanism has also been realized in the bio-barcode concept, which was introduced in Section 1.3 in connection with indirect target amplification for nanogold particle-based DNA sensors. A modified version of this concept for

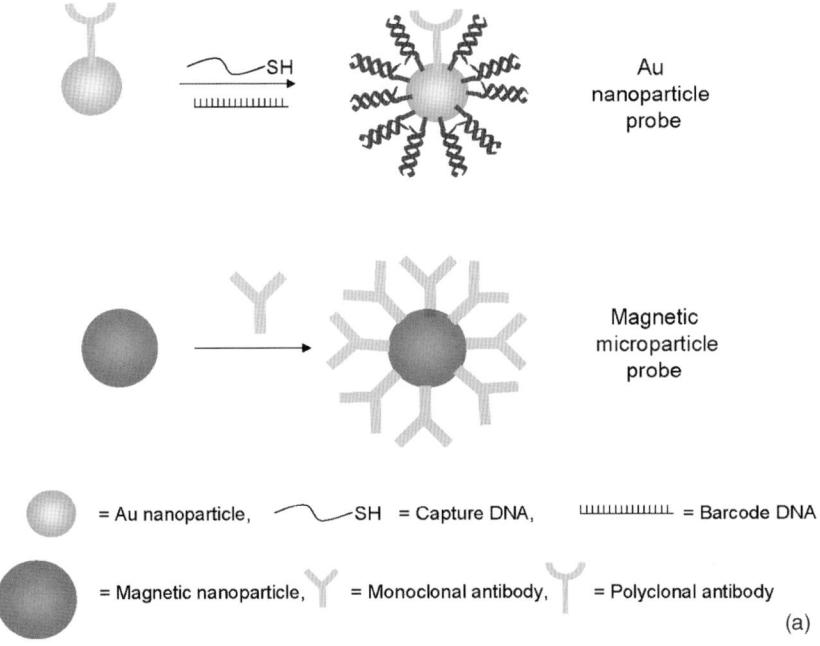

(a)

Figure 2.22 Schematic of the barcode assay method for protein detection. (a) Magnetic microparticles functionalized with monoclonal antibodies. Gold nanoparticles functionalized with polyclonal antibody and a large number of barcode DNA. (b) Target protein capture with MMP probes. (c) Amplification by magnetic separation, dehybridization of barcode DNA, and DNA detection options. (Adapted with permission from Nam *et al.* (2003). Copyright 2003, Wiley-VCH Verlag GmbH.)

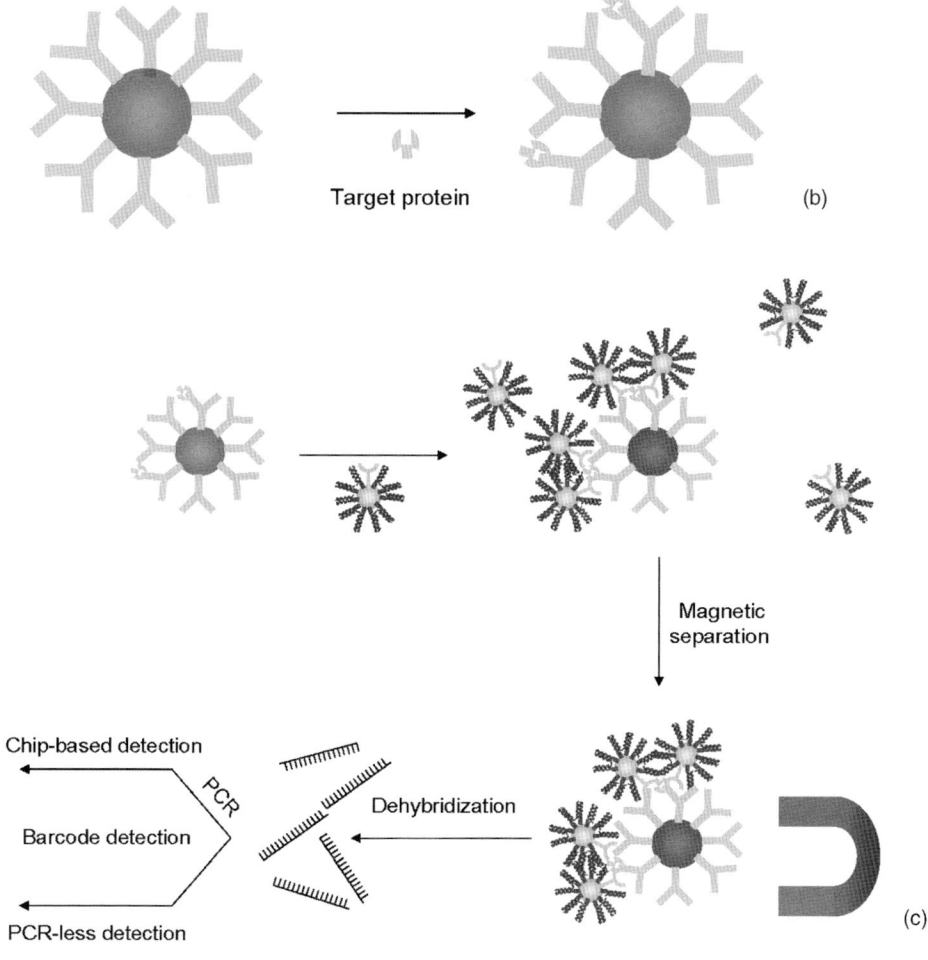

Target protein

(b)

Magnetic
separation

Chip-based detection

PCR

Dehybridization

Barcode detection

PCR-less detection

(c)

Figure 2.22 *(Continued)*

antigen detection that facilitates a sensitivity in the attomolar range was proposed by Nam *et al.* (2003). For example, the prostate-specific antigen can be detected with a concentration of $30\,aM = 30 \times 10^{-18}\,M$. The basic steps of the measurement are schematically shown in Figure 2.22. Similar to the DNA bio-barcode test, two different probes are used:

i) Magnetic iron oxide microparticles (MMPs) functionalized with a small number of monoclonal antibodies that specifically bind the target protein. To avoid unspecific protein adsorption, the free surface areas are blocked with bovine serum albumin.

ii) Gold nanoparticles (AuNPs) functionalized with a polyclonal antibody that specifically binds to other sites of the target protein. In addition, a large number of hybridized barcode oligonucleotides (about 30 nm long) are attached to the gold nanoparticles.

In the detection process, the MMPs and the NPs first form sandwich complexes together with the target proteins by interaction of the monoclonal and polyclonal antibodies with the target. After separation of the magnetic sandwiches and the nonreacted MMPs from the solution by applying a magnetic field and repeated washing, the magnetic particles are again suspended in pure water. The barcode sequences are stripped off by heating. The large number of barcode DNA can now be identified by a PCR-less detection, for example, with the scanometric detection after enhancement of the gold nanoparticles with catalytic silver deposition. An alternative detection can be realized with the more time-consuming PCR. It is the main advantage of the barcode concept that in the final step only the specific DNA identifying the target protein has to be detected. The intended amplification of the primary signal can be realized by either the PCR-less or the PCR-based process leading to the remarkably high sensitivity. Any disturbing background signals from the protein have been excluded in that last process step. It is also obvious that the method enables multiplexed detection of different proteins without serious limitations by applying different pairs of MMP and NP probes.

Gold Nanoparticles as Plasmonic Nanosensors for Imaging Intracellular Biomarkers Small size and biocompatibility of gold NPs are the reasons for their successful application as biomarkers. As an additional advantage, the change of their plasmon resonances as a result of close linking can be exploited for the development of DNA bioassays, as already outlined in Section 1.3. In the mentioned example of DNA bioassays, the gold NPs (size typically between 10 and 50 nm) were linked by hybridizing complementary oligonucleotides. The same principle can be used for the detection of specific proteins by immobilization of a corresponding antibody on the gold nanoparticle. By combining this concept with the possible transfection of cells with gold nanoparticles, plasmonic nanosensors for intracellular imaging of specific protein structures have been developed. Two problems have to be solved in order to achieve intracellular targeted imaging of specific proteins. The gold NPs have to be conjugated with the appropriate antibody and have to be delivered, combined with a transfection reagent (Figure 2.23). Transfection describes the process of introducing NPs inside the cells. It is a stepwise procedure consisting of four major steps: (1) interaction of the transfection complex with the extracellular matrix; (2) translocation through the cellular membrane connected with endosome formation; (3) intracellular trafficking; and (4) endosomal escape. In order to overcome the cell membrane barrier, translocation of NPs into cells is facilitated either physically or chemically. For NPs, the most widespread method is the use of chemical carriers also known as transfection reagents. The commonly used transfection reagents are cationic lipid formulations, cell-penetrating peptides, or dendrimers (Zuhorn and Hoekstra, 2002; Pujals *et al.*, 2006). It is a challenge of a targeted transfection process to make sure that, after the endosomal escape in the

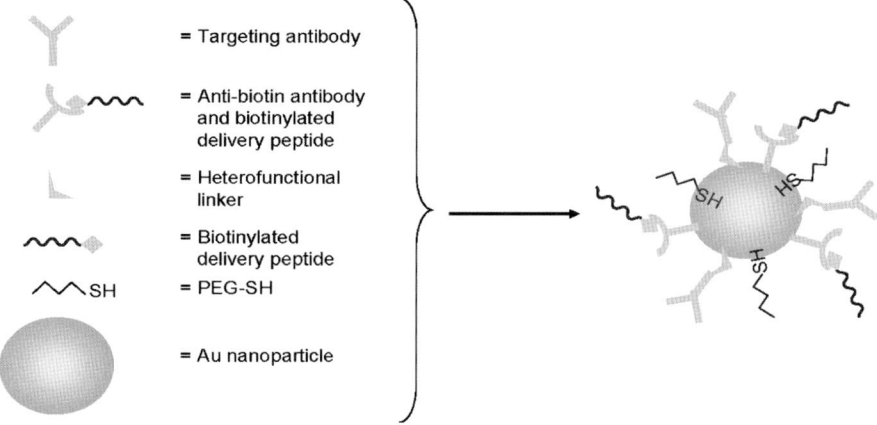

Figure 2.23 Gold nanoparticle-based contrast agent with both targeting and delivery components.

cytosol, the antibody–NP complexes are still fully functional for binding the target unit in the cell.

Kumar *et al.* (2007) have reported on the targeting of actin filaments, one of the major proteins of the cytoskeleton, with 20 nm gold NPs. The active polymerization of actin monomers forming a network of actin filaments plays an essential role in many cellular processes (for details see Sections 3.2 and 7.2). Plasmonic nano-particles can be used as biomarkers for investigating the kinetics of such network formation, as they provide a contrast enhancement. In the experiments, the gold NPs were conjugated with two kinds of antibodies: anti-actin antibodies and anti-biotin antibodies. For this purpose, the constant domains F_c of the antibodies were connected with heterofunctional linkers consisting of an alkane terminating in a dithiol tether and an amide-bonded adipic hydrazide. After oxidizing the carbohy-drate on the constant domain F_c of the antibody to an aldehyde, it was made to react with the hydrazide group of the heterofunctional linker. In the subsequent step, the antibody–linker complexes were conjugated to the gold NPs via the thiol linker. The remaining free areas of the gold surface were blocked with thiolated poly-ethylene glycol (PEG-SH) in order to avoid unspecific protein adsorption. Finally, the transfection reagent, a biotinylated peptide (TAT-HA2), was connected with the anti-biotin antibody. Here the HA2 domain is the influenza virus hemagglutinin protein that disrupts the integrity of the endosomal lipid membrane releasing the endo-somal content to the cytosol. Thus, all functions needed for a successful transfection and targeting were realized by a stepwise attachment of the various building blocks at the gold NP surface. The functionality of the nanoprobes was demonstrated. After binding of the gold NP conjugates via the anti-actin antibodies at the actin filaments, the filopodia of transfected fibroblasts were marked with strong red signals imaged with a fluorescence microscope using a 633 nm HeNe laser line. The bright red color was caused by the plasmon resonances of aggregates of the 20 nm gold NPs along the actin filaments.

Similarly, antibody-conjugated gold NPs have been used for targeting mitochondria with the intention to develop new physically based therapies (Mkandawire, 2010). Mitochondria are membrane-enclosed organelles (0.5–10 μm in diameter) found in most eukaryotic cells. They are responsible for energy production of the cell, and they are also involved in cellular differentiation and cell death. The envelope of the mitochondria consists of the outer mitochondrial membrane (OMM) and the inner mitochondrial membrane (IMM) with the intermembrane space (IMS) in between. Both membranes are composed of various lipids and proteins. A damage-free structure of the membranes is a precondition for the functionality of the mitochondria. Thus, apoptosis (cell death) can be induced by damaging the membranes. This makes mitochondria a promising target for cancer therapy. Cancer cells can be destroyed by chemotherapy that induces apoptosis either extrinsically by targeting cell surface receptors or intrinsically by direct interaction with the mitochondria. Promising studies on the use of AuNPs for cancer therapies on the cellular level have become known recently. AuNPs strongly absorb light of a certain wavelength owing to surface plasmon resonance. This mechanism can be exploited for the localized overheating of biological targets, which offers options for photothermal therapies. It has been proven that site-specific aggregation of AuNPs at the membrane of cancer cells or alternatively inside cancer cells induces cell death. By appropriate positioning of the AuNPs, preferably immediately on mitochondria, the laser beam power required for treatment can be minimized. The main steps of targeted transport of AuNPs to mitochondria are indicated in Figure 2.24.

As seen in the above example, the AuNPs have to be combined with a transfection agent and a targeting protein in order to be suitable for a targeted cell transfection. Citrate-stabilized AuNPs of 20 nm (for easy visualization by electron microscopy) were used in the study. For targeting, mitoTGFP, a mitochondria localizing green fluorescent protein, was chosen. This protein contains a mitochondrial targeting sequence derived from the protein cytochrome *c* oxidase subunit COX8 at its N-terminus, a protein present in the inner mitochondrial membrane. The conjugation was performed by incubation of the gold NPs in solutions of the antibodies with an appropriate concentration ratio. Electrostatic interactions caused a stable random binding of the protein at the negatively charged gold surfaces.

As transfection reagent, poly(propylene imine) (PPI) dendrimers were used. Dendrimers are highly ordered, branched macromolecules that are synthesized by multistep reactions via convergent or divergent synthetic approaches (Bosman *et al.*, 1999). Dendrimers are classified by generation, which refers to the number of these reactions. The surface of the dendrimer can be functionalized with specific surface groups. The interior core tends to be shielded off from the solvent, and it can be used to entrap small molecules through hydrophobic interactions. The core (generation G0) of PPI dendrimers is made up of 1,4-diaminobutane. In order to achieve a good biocompatibility, oligosaccharide-modified PPI dendrimers were

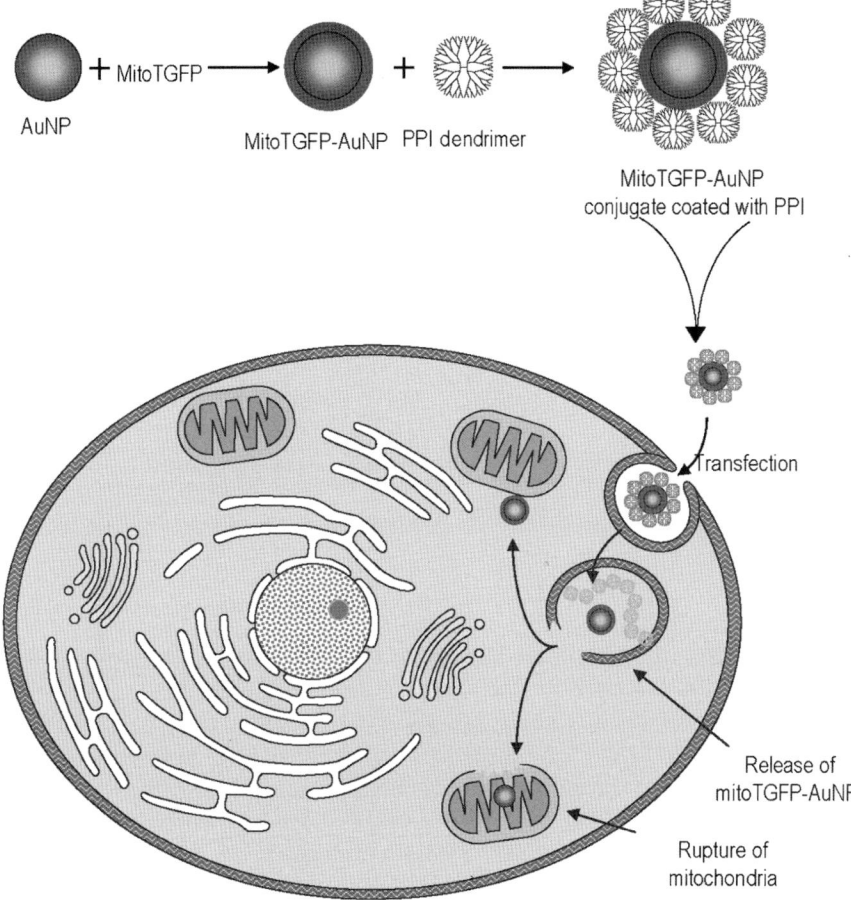

Figure 2.24 Proposed multistep process of the destruction of cancer cells by delivering AuNPs to the mitochondria causing rupture. (Reprinted with permission from Mkandawire (2010).)

used, whose surface amino groups were substituted with maltotriose. Figure 2.25 shows a fourth-generation maltotriose-modified PPI dendrimer used in this study.

For the successful transfection of human carcinoma cells (HeLa cell line), mixtures of mitoTGFP-conjugated AuNPs and PPI dendrimers were prepared. As can be seen in Figure 2.26, transfection complexes were transferred into cells. For comparison, solutions of mitoTGFP-conjugated AuNPs without dendrimers were applied. AuNPs were exclusively observed by transmission electron microscopy (TEM) inside the cancer cells in the presence of the dendrimers. Obviously, the targeting protein was still functional after escape from the endosome. A possible intracellular trafficking pathway of mitoTGFP–AuNPs inside the cancer cell may be deduced from Figure 2.27. As seen in the TEM images, the outer mitochondrial membrane was seriously damaged. Probably proteins located in the intermembrane

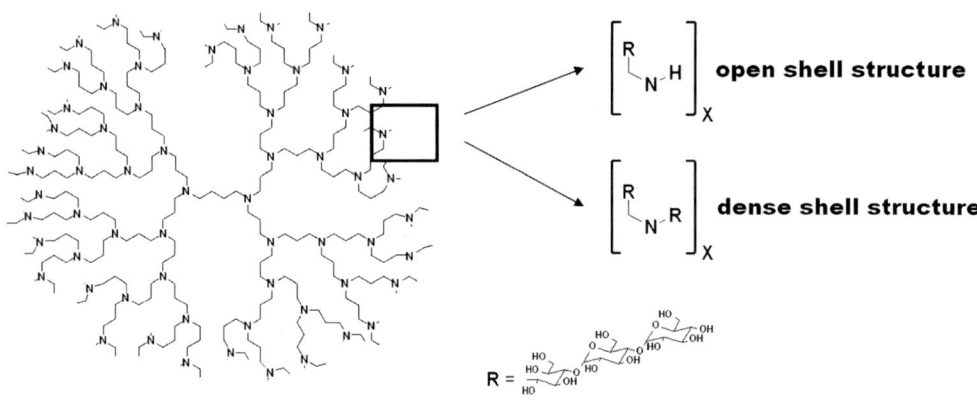

R = Maltotriose

PPI 4th generation (G4, x=32)

open shell structure

dense shell structure

Figure 2.25 Maltotriose-modified fourth-generation PPI dendrimers. The dendritic structure is characterized by "layers" between each focal point also called generations. The dendrimer generation is defined as the number of focal points when going from the core to the surface. For instance, a generation 4 (G4) dendrimer has four cascade points between the core and the surface. The core is sometimes denoted generation "zero" (G0), as no focal points are present. Depending on the number of maltotriose units at the surface, so-called open shell structures or dense shell structures can be produced. In the transfection experiments, open shell structures were used.

space such as cytochrome c were released into the cytosol, thereby activating a signaling cascade, which later led to programmed cell death as observed in the experiment. In conclusion, AuNP conjugates can induce apoptosis by disruption of the outer mitochondrial membrane, which makes them a potential means for the development of localized nonthermal physical therapies.

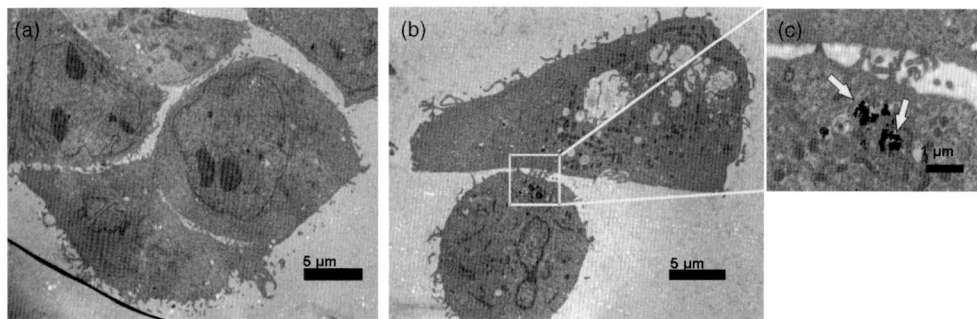

Figure 2.26 TEM micrographs of breast cancer cells incubated with mitoTGFP–AuNPs, fixed 1 h after stopping transfection: (a) in the absence of PPI–G4 dendrimer; (b) same in the presence of glycodendrimer; (c) detail from (b) showing AuNPs as black spots or clusters (see arrows). Transfection was effected in the absence of serum in all cases. Scale bar: (a and b) 5 μm; (c) 1 μm. (Reprinted with permission from Mkandawire (2010).)

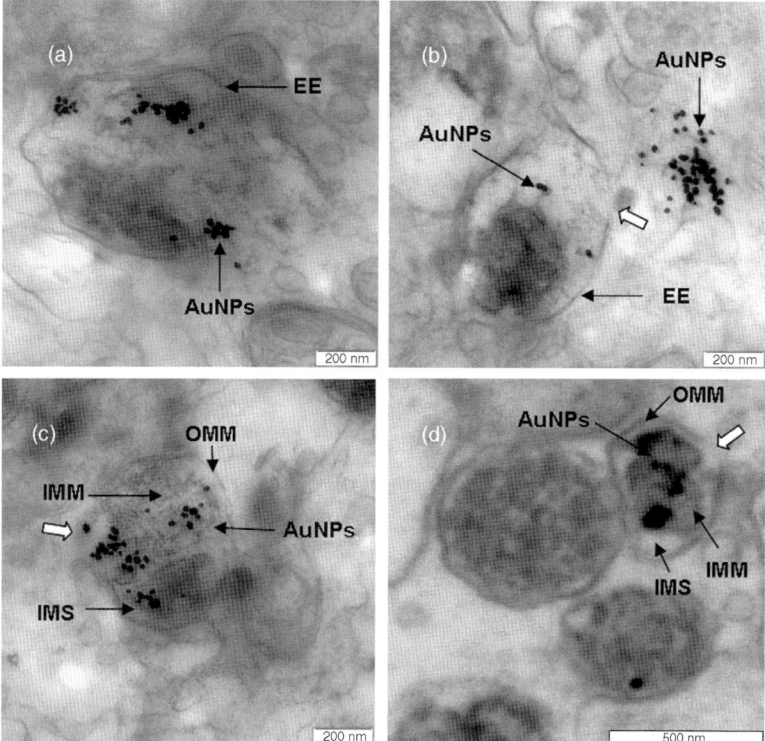

Figure 2.27 TEM micrographs of breast cancer cells transfected with mitoTGFP–AuNPs in the presence of maltotriose-modified PPI dendrimers, fixed 24 h after stopping transfection. (a) MitoTGFP–AuNPs contained in an early endosome (EE). (b) Ruptured early endosome with a few AuNPs inside and some AuNPs released in the cytosol in close proximity to the endosome; possible position of membrane rupture is indicated by white arrow. (c) MitoTGFP-conjugated AuNPs, seen as black spots, making their way into mitochondria via rupture of the outer mitochondrial membrane and getting associated with the inner mitochondrial membrane. (d) AuNPs localized at the IMM, membrane rupture (white arrow) as a possible entry site of the mitoTGFP–AuNPs, and intermembrane space. Scale bar: (a, b, c) 200 nm; (d) 500 nm. (Reprinted with permission from Mkandawire (2010).)

References

Bosman, A.W., Janssen, H.M. *et al.* (1999) About dendrimers: structure, physical properties, and applications. *Chemical Reviews*, **99** (7), 1665–1688.

Chiappalone, M., Vato, A. *et al.* (2003) Networks of neurons coupled to microelectrode arrays: a neuronal sensory system for pharmacological applications. *Biosensors & Bioelectronics*, **18** (5–6), 627–634.

Crane, H.R. (1950) Principles and problems of biological growth. *The Scientific Monthly*, **70**, 376–389.

Eubisch, A. (2012) *Modulare Anbindung von Biomolekülen an Oberflächen zur Anwendung in prozesstauglichen Biosensoren*. Ph.D. thesis, Technische Universität Dresden.

Gazit, E. (2007) *Plenty of Room for Biology at the Bottom: An Introduction to*

Bionanotechnology, Imperial College Press, London.

Goddard, J.M. and Hotchkiss, J.H. (2007) Polymer surface modification for attachment of bioactive compounds. *Progress in Polymer Science*, **32** (7), 698–725.

Hagemeyer, C.E., von Zur Muhlen, C. *et al.* (2009) Single-chain antibodies as diagnostic tools and therapeutic agents. *Thrombosis and Haemostasis*, **101** (6), 1012–1019.

Kumar, S., Harrison, N. *et al.* (2007) Plasmonic nanosensors for imaging intracellular biomarkers in live cells. *Nano Letters*, **7** (5), 1338–1343.

Mkandawire, M. (2010) In vitro *interaction of nanoparticles with mitochondria for surface enhanced Raman spectroscopy and cell imaging*. Ph.D. thesis, Technische Universität Dresden.

Nam, J.M., Thaxton, C.S. *et al.* (2003) Nanoparticle-based bio-bar codes for the ultrasensitive detection of proteins. *Science*, **301** (5641), 1884–1886.

Nam, K.T., Kim, D.W. *et al.* (2006) Virus-enabled synthesis and assembly of nanowires for lithium ion battery electrodes. *Science*, **312** (5775), 885–888.

Nelson, D.L. and Cox, M.M. (2008) *Lehninger – Principles of Biochemistry*, W.H. Freeman and Company, New York.

Noy, A., Artyukhin, A., and Misra, N. (2009) Bionanoelectronics with 1D materials. *Materials Today*, **12** (9), 22–31.

Patolsky, F., Zheng, G.F. *et al.* (2004) Electrical detection of single viruses. *Proceedings of the National Academy of Sciences of the United States of America*, **101** (39), 14017–14022.

Peelle, B.R., Krauland, E.M. *et al.* (2005a) Probing the interface between biomolecules and inorganic materials using yeast surface display and genetic engineering. *Acta Biomaterialia*, **1** (2), 145–154.

Peelle, B.R., Krauland, E.M. *et al.* (2005b) Design criteria for engineering inorganic material-specific peptides. *Langmuir*, **21** (15), 6929–6933.

Pujals, S., Fernández-Carneado, J. *et al.* (2006) Mechanistic aspects of CPP-mediated intracellular drug delivery: relevance of CPP self-assembly. *Biochimica et Biophysica Acta*, **1758**, 264–279.

Smith, G.P. (1985) Filamentous fusion phage: novel expression vectors that display cloned antigens on the virion surface. *Science*, **228** (4705), 1315–1317.

Straub, B., Meyer, E. *et al.* (2001) Recombinant maxi-K channels on transistor, a prototype of iono-electronic interfacing. *Nature Biotechnology*, **19** (2), 121–124.

Whaley, S.R., English, D.S. *et al.* (2000) Selection of peptides with semiconductor binding specificity for directed nanocrystal assembly. *Nature*, **405** (6787), 665–668.

Zhao, X., Hilliard, L.R., Mechery, S.J., Wang, Y., Bagwe, R.P., and Jin, S. (2004) A rapid bioassay for single bacterial cell quantitation using bioconjugated nanoparticles. *Proceedings of the National Academy of Sciences of the United States of America*, **101** (42), 15027–15032.

Zuhorn, I.S. and Hoekstra, D. (2002) On the mechanism of cationic amphiphile-mediated transfection. To fuse or not to fuse: is that the question? *Journal of Membrane Biology*, **189**, 167–179.

3
Cell Adhesion

In bionanotechnology, living cells are often regarded as highly organized entities equipped with a manifold of unique machineries that are organized in sophisticated assembly lines. In short, the cell is compared to a complex factory. A factory needs reliable supply lines for energy and raw materials or pre-products, paths for the emission of products and waste, and last but not least communication with the environment. All this is realized via the cell wall. Therefore, the cell interface is more than a mere boundary forming the containment for the cell. It is a highly structured functional layer involving biomolecular sensors, switching gates, and actuators. In the biological evolution, nature has created specific structures to realize all the mentioned functions in an optimized way. In a mechanistic view, we can understand the cell interface as an active structure equipped with various sophisticated biomolecular "relay stations". Living cells use all kinds of chemical, physical, and mechanical processes for their communication with the environment. Chemical reactions often supported by enzymes, coupled mass and charge transport, passive or active optical information exchange such as bioluminescence, and mechanical interaction forces are examples for the wide range of possibilities. Coupling processes between the cell interface and the internal biomolecular machineries of the cell add to this variety. Cascades of internal processes such as phosphorylation or conformational changes of specific proteins combined with diffusive or directed transport of small ions and proteins act as transducers of the incoming signals into outgoing cell responses.

Deeper investigation of the various biophysical and biochemical mechanisms realized with these interface structures is useful for purposes such as controlling the interaction of specific cells with an artificial environment in medical applications, cell-based regenerative therapy of human tissues, and optimizing the cell activity in a bioreactor for applications in pharmacy, food industry, and materials engineering. One can expect that the detailed study of particular interface structures will provide inspiration for innovative artificial nanostructures to be used as nanosensors, nanoactuators, nanofilters, and similar tools. For this purpose, mechanical interactions of the cell with the environment seem to offer feasible options for biologically motivated materials engineering. Considering this, it makes sense to look at the basic problem of cell adhesion first. Adhesion controls the function of cells fundamentally. By attachment to neighboring cells or to the *extracellular matrix*

Bio-Nanomaterials: Designing Materials Inspired by Nature, First Edition. Wolfgang Pompe, Gerhard Rödel, Hans-Jürgen Weiss, and Michael Mertig.
© 2013 Wiley-VCH Verlag GmbH & Co. KGaA. Published 2013 by Wiley-VCH Verlag GmbH & Co. KGaA.

(ECM), the cell gets an essential part of the information needed to control its morphology, dynamics, metabolic processes, and fate. The ECM is the immediate vicinity of the cell with its insoluble and soluble biomolecules. The cells and the matrix together form soft or hard tissues. These matrices are composed of so-called *structural macromolecules* and *functional macromolecules*. The structural macromolecules form the framework and biomechanical components of the matrix whereas the functional macromolecules realize specific properties of the matrix (e.g., acidic macromolecules for control of biomineralization; see also Section 6.2). The structure-forming macromolecules are mainly proteins (e.g., collagen in bone, cartilage, blood vessels, silicatein in sponges, and silk-like proteins in mollusk shells) and polysaccharides (e.g., hyaluronic acid in cartilage, chitin in mollusk shells, and cellulose in plant cells). Other macromolecules such as fibronectin, laminin, and vitronectin are of major relevance for the biomechanical properties of the ECM. They act as ligands that are involved in the mechanical interaction of the cell with the ECM. The cell surface is equipped with *specific adhesive protein complexes*, the so-called *adhesomes*, forming a sophisticated mechanosensitive transducer and actuator complex. The adhesome can be considered as one particular cellular "relay station" that establishes a mechanical link to the internal metabolic reaction cascades. The whole complex of interacting units covers a hierarchy of structures from the mesoscopic scale of the ECM down to the nanometer size of specific adhesive epitopes of few amino acids. This structure, which will be discussed in the following, is shown schematically in Figure 3.1.

Basic units of the adhesome are the *integrin-based adhesion complexes* that are responsible for the mechanical coupling between the ECM and the *cytoskeleton*, in particular with the actin filaments. Thus, a discrete external mechanical force can be transferred from the ECM to the cell (in the sensor mode). The response to such external signals can be realized either locally, affecting the adhesion sites, or globally by activating internal biochemical signal pathways. The adhesome can also act the other way around (in the actuator mode) by transferring energy released in *intracellular biochemical processes* to the ECM. This can facilitate cell motility or restructuring of the ECM, for example.

3.1
Case Study

In the process of tissue development, mammalian cells control the functional optimization of micro- and macrostructure of the ECM. This is done by means of combined biochemical and mechanical processes. The following example demonstrates how the mechanotransduction mechanism of the cell governs the structural optimization of the ECM. In a model experiment, the reorganization of fibronectin (FN) ligands of the ECM by endothelial cells has been studied (Pompe *et al.*, 2005). FN is a large fibrous glycoprotein (MW = 440 000) that has separate domains that bind ECM components such as fibrin, heparin, and collagen as well as adhesion receptors such as integrins. The binding energy with integrins is relatively high,

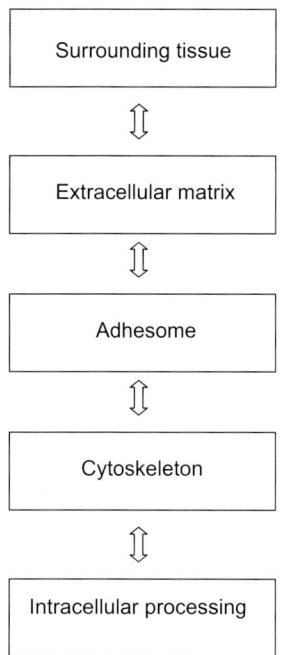

Figure 3.1 Functional units controlling the cell adhesion.

about $20k_B T$ for FN with $\alpha_5\beta_1$, for example. Varying strength of the linkage of FN molecules with the ECM scaffold has been simulated with maleic anhydride copolymers. The kind of interaction between the binding copolymer and the FN molecules has been changed by the choice of the comonomer as well as by hydrolyzation of the anhydride groups into carboxyl groups of the copolymer. The bonding of FN molecules is strong and covalent if mediated by the anhydride but weak and physisorbed if mediated by its hydrolyzed form. The strength of physisorption can be tuned by the degree of hydrophobicity of the comonomer. It ranges from the hydrophobic poly(octadecene-*alt*-maleic anhydride) (POMA) with $n = 16$ alkyl groups in the side chain of the comonomer over poly(propene-*alt*-maleic anhydride) (PPMA) with $n = 1$ alkyl group in the side chain to the most hydrophilic poly(ethene-*alt*-maleic anhydride) (PEMA) with no side chain (for more details of the materials chemistry see Section 3.3). In the experiment, the three copolymers have been deposited as thin films onto glass substrates. After coating the copolymer films with a monolayer of randomly distributed FN molecules, endothelial cells have been seeded. Subsequently, the reorganization of the FN molecules has been studied by fluorescence microscopy and scanning force microscopy (SFM). FN molecules bound covalently and those bound by physisorption to the copolymer film have been compared with respect to the degree of reorganization (see Figure 3.2). The fluorescence images of the FN fibrils (red) and the linked focal adhesion

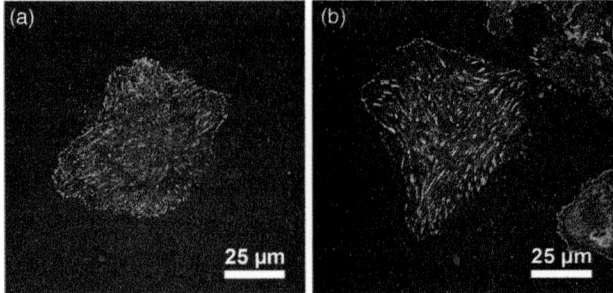

Figure 3.2 Fluorescence images illustrating the adhesion of endothelial cells 60 min after seeding on fibronectin-coated poly(propene-*alt*-maleic anhydride). Fibronectin (red) and phosphotyrosine (green) are visualized. (a) Homogeneous distribution, covalently bound. (b) Textured fibril distribution of physisorbed fibronectin.

(phosphotyrosine molecules, green) of the endothelial cells show that in the case of physisorption the initially homogeneously distributed FN molecules have assembled into fibrils. Obviously, there is a direct relation between the fibril distribution and the intracellular distribution of the focal adhesion.

A higher resolution of the structure of aligned FN fibrils allows an even deeper insight into the mechanical interaction of the cells with the ECM. SFM (Figure 3.3) reveals that the fibrils are composed of paired nanofibrils with a spacing ranging from 70 to 400 nm. Each pair starts from a separate focal adhesion. The distance between the two paired nanofibrils is governed by the strength of the interaction of the fibronectin with the underlying substrate. It has been quantified by using the three different comonomers for the maleic anhydride copolymer film. As seen in Figure 3.4, weaker binding to the copolymer film in the order POMA, PPMA, and PEMA leads to closer spacing of the nanofibrils arranged in pairs (Pompe *et al.*, 2005). These spacings are multiples of some unit of about 71 nm. The observation that the spacings of the fibrils come in multiples of some unit provides insight into

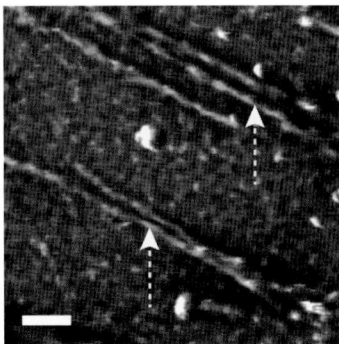

Figure 3.3 Scanning force microscopy image of paired fibronectin nanofibrils on a PPMA film. Scale bar: 500 nm.

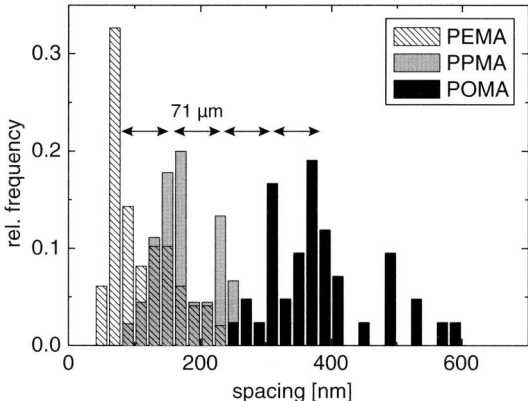

Figure 3.4 Histogram of spacings of paired FN fibrils on three different polymer coatings with 45 measurements for each substrate.

the underlying cellular process governing the reorganization of the ECM. It can be understood by considering the interaction of the FN with the cell structure in more detail. The FN molecules are linked to a component of the cytoskeleton, the actin stress fibers. The links are provided by the focal adhesion (see Figure 3.5). They are formed by aggregation of integrins diffusing within the cell membrane. During clustering of integrins in the cell membrane, actin filaments congruently assemble into actin stress fibers, which terminate at the focal adhesion. Each stress fiber is composed of numerous actin filaments. Actin filaments are cross-linked by α-actinin molecules. The spacing of the cross-linked actin filaments is about 34–39 nm. The α-actinin-linked molecules are thought to be arranged in a square lattice. In order to bear tensile forces exerted by myosin motors inside the actin stress fibers, the actin filaments have to be arranged in an antiparallel way, which would provide a repeating unit with twice the spacing of the α-actinin cross-linked actin filaments. FN molecules bound to integrins are assembled along the border of the focal adhesion. Therefore, the repeating unit of the spacing of the FN nanofibrils of 71 nm coincides with twice the spacing of the α-actinin cross-linked actin filaments. The process is supported by a directed transport of the FN–integrin complexes with myosin motor proteins along the actin stress fibers (Pompe *et al.*, 2006).

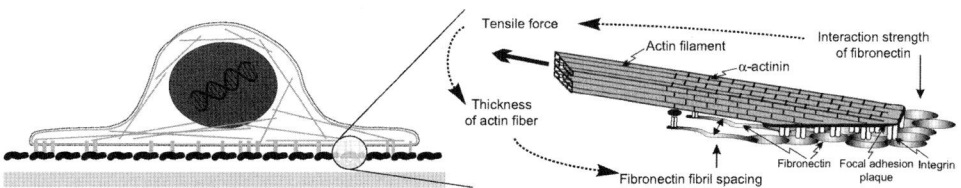

Figure 3.5 Proposed model for the FN fibrillogenesis explaining the typical spacing of nanofibrils. (Adapted from Pompe *et al.* 2005.)

3.2
Basic Principles

The cell interface is equipped with a large set of biochemical, physical, and mechanical transduction systems for information exchange with the environment. The related biomolecular realizations of biochemical recognition processes and their application for biosensor development have been considered in Chapter 2. This chapter covers mechanotransduction mechanisms and their possible mimicking in materials engineering. A few basic principles inherent in mechanotransduction systems at the cell interface are outlined.

3.2.1
The Cellular Mechanotransduction System

While a progenitor cell turns into a mature cell, which involves growth, proliferation, and differentiation, it receives plenty of relevant physical and chemical information from the surrounding tissue. Among the possible external signals, mechanical information plays an exceptional role as the tissues differ much in the geometry of their microstructures and their mechanical properties as quantified by the parameters of elasticity, nonlinear stress–strain behavior, viscosity, and adhesion strength. Therefore, biological evolution has produced very efficient mechanical regulatory systems on the cellular level. For example, animal cells from solid tissues such as neurons from brain, myocytes from muscle, or osteoblasts and osteoclasts from bone need permanent mechanical information input from their vicinity to be viable. They are "anchorage dependent". Mechanical signals can be sent from adjacent cells or from the ECM. Typical forces of such interactions are in the range of a few nanonewtons (nN) acting on areas of a few μm^2. The mechanical behavior can be described quantitatively by a stress–strain relation including strain rate dependences in a more general case. Characteristic parameters of such relationships are the elastic constants (often only reduced to Young's modulus E of the uniaxial stress–strain test), yield stress, friction coefficients, viscosities, and the maximum adhesive stress. Furthermore, the coupling of the mechanical properties with chemical reactions or mass transport (e.g., diffusion constants) has to be considered. As the behavior of each biosystem is essentially governed by the living cells, it is obvious that finally the kinetics of the developing cell system coupled with the mechanical properties of the surrounding ECM comes into play. This can give rise to feedback processes in the mechanical behavior. A striking example is the development of bone tissue, discussed in Chapter 6.

The complexity of the mechanical properties of a living cell and the surrounding ECM needs a rigorous reduction. In order to decouple the very complex situation, we consider the mechanical interaction of an individual cell with its surrounding ECM, with the aim to include interactions with surrounding cells later. (See the example of bone remodeling in Chapter 6.) Let us begin with the macroscopic aspect. The various tissues differ much in their elastic properties (Figure 3.6). Neurons develop their shape in a very soft matrix with an elastic modulus in the range of about

E (kPa) of cellular microenvironments

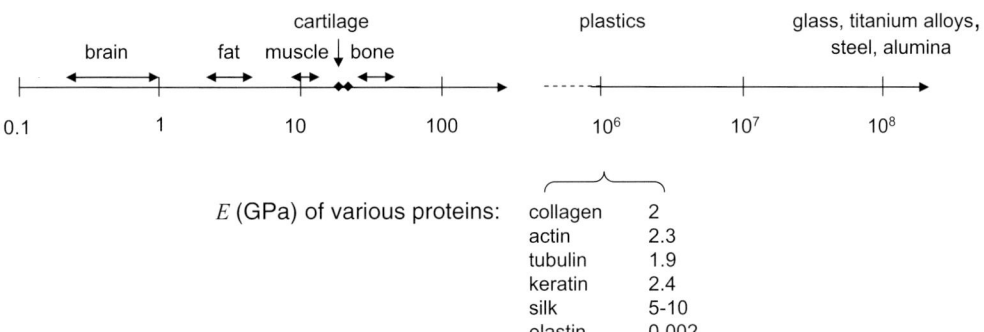

E (GPa) of various proteins:

collagen	2
actin	2.3
tubulin	1.9
keratin	2.4
silk	5-10
elastin	0.002

Figure 3.6 Characteristic ranges of the stiffness of various tissues.

$E_{brain} \sim 0.2$–1 kPa, whereas bone progenitor cells grow and differentiate in a stiff collagen network with $E_{coll} \sim 25$–40 kPa. The values for fat, muscle, and cartilage tissues are in between. Note that the stiffness of all tissues is much lower than that of traditional man-made biocompatible materials such as tissue culture plastics (3–5 GPa), glass (50–90 GPa), titanium alloys (105–120 GPa), steel (200 GPa), and alumina (435 GPa).

The cells are sensitive coupling structures that allow a fine-tuned detection of the mechanical stiffness of the surrounding tissue. These signals induce internal biochemical reaction cascades of the cells that finally govern the cell fate. Under cell fate, we understand the proliferation, differentiation, and maybe also the programmed cell death, the apoptosis. We see that it is necessary to get a deeper understanding of this coupled mechanical sensor and actuator system at the cell interface. This so-called mechanotransduction system plays a decisive role for any molecular-based biomaterials development. The mechanical link between the extracellular region and the intracellular space is realized by assemblies of complex protein structures, the *focal adhesion* arranged at the cell membrane. The focal adhesions form discrete bonds to the ECM via transmembrane proteins (e.g., integrins). Alternatively, there can be discrete links to neighboring cells formed by other transmembrane arrangements. Calcium ion-controlled cell contacts, the so-called *cadherins* (for "calcium-dependent adhesion"), can be mentioned as another important example of such transmembrane proteins.

Connected via several other proteins (e.g., vinculin, talin, and α-actinin) inside the focal adhesion, the cytoskeleton inside the cell generates a contraction force by means of *motor proteins* such as myosin (Howard, 2001). They generate displacements of the cytoskeleton filaments by using internal energy resources of the cell (ATP). The cytoskeleton is also connected to the cell nucleus and other cellular structures. It performs three fundamental functions in the living cell. First, it creates and stabilizes the external shape of the cell, and facilitates motility and mitosis.

Second, it acts as an important carrier for signal transfer between various cell regions. Third, it can guide intracellular mass transport along individual filaments. The first task means that it should be composed of different protein filaments (e.g., actin filaments, microtubules, and intermediate filaments; see also Chapter 7) that can generate tensile stress or bear compression. The active reorganization of the cytoskeleton facilitates the control of the cell shape. The cytoskeleton can be modeled by a prestressed network with a biochemically governed structure. The second task of signal transduction can be realized by variation the mechanical stresses in the cytoskeleton without large-scale restructuring. Changing contraction of the cytoskeleton, for example, can generate biochemical signals in downstream signaling pathways. Finally, components of the cytoskeleton (e.g., the microtubules) can also act as pathways for a directed motion of motor proteins (e.g., myosin motors). Such mechanisms are used in the cell for physical transport, which is realized with vesicles attached to the motor proteins (Howard, 2001).

In the natural evolution process, various highly sensitive mechanotransduction systems have been developed in cell membranes. About 160 proteins have been found to be involved in the force transfer between cell and ECM. Most of them are intrinsic constituents of the adhesion sites. Others are transiently linked with the adhesion sites and act as signaling molecules. There are various types of integrin-mediated adhesion types: *focal adhesions*, *podosomes*, and *invadopodia*. Among them, focal adhesions are the structures studied in most detail. A focal adhesion can be considered as a complex composed of four interacting modules regulating the mechanical interaction between the ECM and the cytoskeleton: the actin linking module, the actin polymerizing module, the ECM binding module, and the adhesion signaling module (Figure 3.7).

The Actin Linking Module Actin filaments of the cytoskeleton are linked to the cytoplasmic domains of integrins through various anchoring proteins such as talin, integrin-linked kinase (ILK), tensin, and Wech. It is assumed that two molecules of talin are necessary to connect α-integrin/β-integrin dimers with actin filaments. Furthermore, there are proteins such as vinculin that trigger the clustering of integrin complexes. Their association strengthens the actin–integrin link.

The Actin Polymerizing Module The focal complexes or nascent focal adhesions are spots of about 70–100 nm in diameter, which are composed of several hundred protein molecules. The formation of the focal adhesion is mediated by the actin-related protein 2/3 (Arp2/3) molecules. They are found underneath the lamellipodia, which are thin, flat cellular extensions. About 2–4 μm from the lamellipodia tip in the lamella, the composition of the actin network is changed. Here the Arp2/3 proteins disappear and motor proteins myosin II are found. They are responsible for the contractility of the actin network. Their activity contributes to the assembling of individual focal complexes into focal adhesions. Mature focal adhesions are elongated and localized at bundles of actin filaments, the so-called stress fibers. They also contain the cross-linking and structure-organizing α-actinin (see also above). The polymerization of the actin filaments contributes to the stress generation at the focal

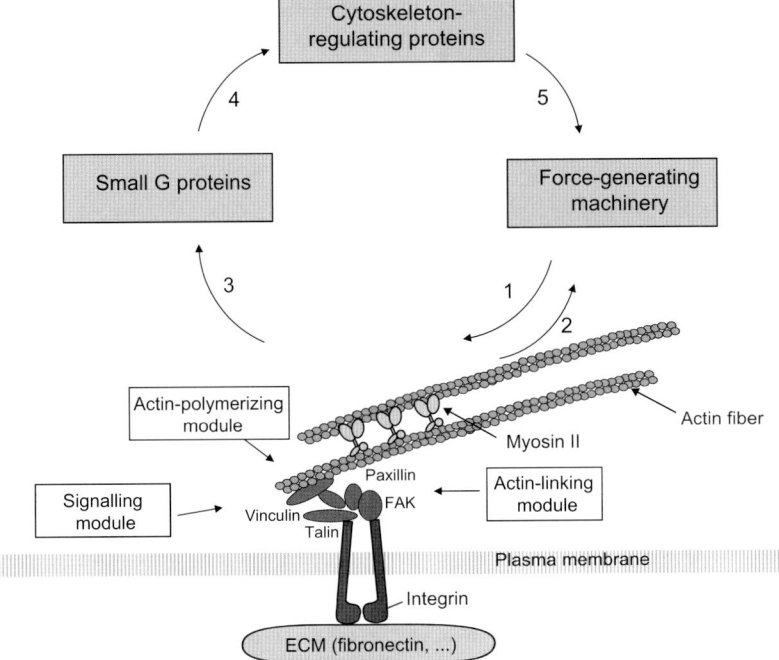

Figure 3.7 Regulating principles and feedback loops at focal adhesions. (1) Actin polymerization and myosin II-dependent contractility exert forces in concert with external load transmitted by the ECM, which affect mechanosensitive proteins of the focal adhesion complex, consisting of an actin linking module such as talin and vinculin, an ECM binding module (integrins), an associated actin polymerizing module, and a signaling module (e.g., FAK and p130CAS). (2) The force generating actin cytoskeleton reacts in an integrated manner to the ECM characteristics and applied mechanical forces. (3) Such signals activate guanine nucleotide exchange factor and GTPase activating proteins, providing downstream regulation signals of small G proteins (e.g., Rho and Rac). (4) G protein-triggered depolymerization/polymerization of actin filaments in turn affects the force generating machinery of the complex cytoskeleton network (5) (see also Section 7.2). (Adapted from Geiger *et al.*, 2009.)

adhesion. The actin filament (F-actin) is a filamentous assembly of G-actin monomers. The monomers polymerize two by two forming a right-handed spiral of two filaments with a pitch of 36 nm. In addition, each monomer binds ATP and hydrolyzes it to ADP. So every actin molecule in the filament is complexed to ADP (see also Section 7.2).

The ECM Binding Module The connection of the focal adhesion to the extracellular microenvironment is facilitated via the binding of integrins to the ligands of the ECM. Integrins are heterodimeric transmembrane receptors that facilitate force and signal transduction by clustering, ligation, and conformational changes. Up to now,

18 α and 8 β subunits are known, which can assemble 24 different integrins. They are able to specifically bind to ECM proteins such as collagen type I and collagen type IV, laminin, fibronectin, vitronectin, fibrin, thrombospondin, and osteopontin. Bond energies lie in the range of $10k_B T$; however, the bond can become much stronger under tensile cell forces, which is known as "bond strengthening" or "clutch bond." For example, an $\alpha_5\beta_1$-integrin binds a fibronectin ligand under a load of 9 pN with a lifetime of almost 100 s.

The Adhesion Signaling Module The focal adhesion mechanosensor is a complex network of interconnected molecular mechanosensing units. They react cooperatively to forces generated within or without the cell. Although the complete transduction mechanism is not known until now, key elements have already been identified. As schematically shown in Figure 3.7, the signaling module connects the ECM–integrin links with the force generating motor proteins localized at the stress fibers. A force signal generated at the focal adhesion activates proteins (e.g., paxillin, p130CAS, and focal adhesion kinase (FAK)). They transfer that signal to guanine nucleotide exchange factors (GEFs) that catalyze the exchange of GDP for GTP. This process leads to the activation of the GTPases such as Rho and Rac. These G proteins activate other signaling cascades that finally cause motion of motor proteins and stress response of the cytoskeleton. Among the proteins that transfer integrin signals to GEFs and GAPs, the focal adhesion kinase has been studied in detail. Its structural change (e.g., phosphorylation of particular amino acids) can be used for detection of the activation stage of cells under the influence of a particular external signal.

Coming back to the whole signaling chain connected with the integrin–cytoskeleton network, we see that it includes a forward signal transfer from the ECM to the intracellular space. At the same time, a feedback process exists. Force signals caused by actin polymerization and the motor protein (myosin II) activity are transferred from the cell to the ECM. This feedback process finally results in a characteristic reorganization of the ECM as we have seen in the case study above.

3.2.2
Mechanical Impact of the ECM on Cell Development

As already mentioned at the beginning of the previous section, the fate of a cell is often governed by the mechanical signals exerted by the environment. From the mechanical point of view, we can classify such signals into those caused by external forces, such as the stresses in the bone skeleton of a runner, and self-stresses caused by internal sources, such as growth stresses due to mineralization of a collagen matrix in growing bone or stresses caused by changes of the cytoskeleton of a living cell resulting from biochemical processes in the cell. It makes sense to distinguish between the influence of local point-like forces and the average mechanical behavior of the surrounding matrix. Whereas the description of local interaction needs the consideration of molecular details of the mechanotransduction system, the average interaction can be described by a continuum model of the surrounding tissue. These two structural levels provide also the main approaches for materials scientists for controlling the cell fate in tissue engineering. The following examples will show how the interplay of the global interaction characterized by the elastic stiffness of the ECM and the local interaction expressed by the adhesion strength governs the cell fate.

Influence of the Stiffness of the ECM In Figure 3.8, results of a model study are shown that demonstrate the influence of the stiffness of a surrounding medium on the cell differentiation (Engler *et al.*, 2006). Various surrounding media with different stiffness have been modeled with polyacrylamide (PAAm) gels cross-linked with bisacrylamide. The Young's modulus of the gel has been tailored by varying the concentration of the cross-linker. Mesenchymal stem cells (MSCs) have been cultured on these substrates of different stiffness. Young's moduli were chosen to mimic brain tissue, muscle tissue, and collagenous bone tissue. During 1 week in culture, a significant reprogramming of the cells occurred. When growing on substrates with stiffness varying in the order brain-like, muscle-like, and collagenous bone-like, the stem cells developed branching, spindle-like, or polygonal shapes, respectively. Also, the transcriptional profiles of neurogenic, myogenic,

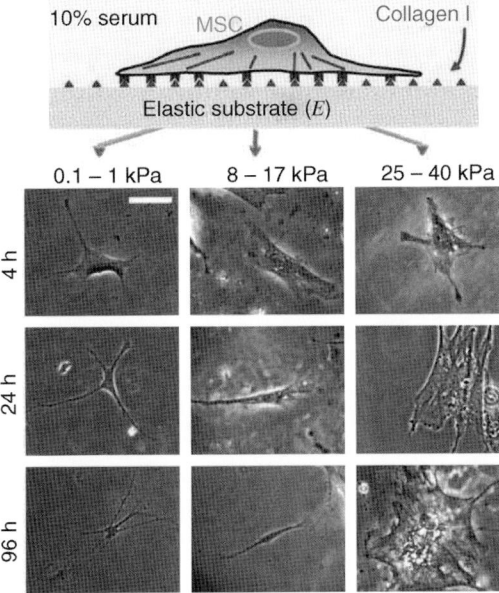

Figure 3.8 Influence of substrate elasticity on differentiation of mesenchymal stem cells. Chosen substrate moduli correspond to characteristic values of different tissues: $E_{brain} \sim 0.1{-}1$ kPa; $E_{muscle} \sim 8{-}17$ kPa; $E_{collagen} \sim 25{-}40$ kPa. The development of cell shape with time (after seeding) has been imaged by fluorescence microscopy. Scale bar: 20 μm. (Reprinted with permission from Engler *et al.* (2006). Copyright 2006, Elsevier.)

and osteogenic markers showed the same effect of a specific differentiation depending on the substrate stiffness as indicated by the cell shapes.

Influence of the Adhesion Strength on the Cell Fate   In order to find out the effect of the anchorage forces on the cell development and compare it with the effect of stiffness, model experiments are helpful again, as demonstrated with the following example. Human endothelial cells were cultured on a PAAm hydrogel (Pompe *et al.*, 2009). The stiffness of the hydrogel was varied in the range $E = 2.6$–12.6 kPa by cross-linking with bisacrylamide at different concentrations. A thin layer of a maleic anhydride copolymer was used to tune the anchorage of fibronectin acting as adhesion ligand at the PAAm substrate. Similar to the experiment discussed in the case study, two comonomers differing in the amount of hydrophobic alkyl groups were applied. Strong or weak anchorage of cells was realized with poly(styrene-*alt*-maleic anhydride) (PSMA) or PEMA films, respectively. The various combinations of PAAm substrates with different stiffness and maleic anhydride copolymer films with different anchorage strength for the adhesion ligand fibronectin allowed the contributions of the mechanical processes regulating cellular adhesion to be separated. This means a clear distinction between the influences of matrix stiffness and adhesion strength on the cell activity. Figure 3.9 shows that the cell spreading is governed by the elastic modulus of the PAAm hydrogel. The endothelial cells have the same shapes on hydrogels with equal Young's modulus independent of the adhesive maleic anhydride copolymer layer. Higher substrate stiffness causes larger spreading of the cells.

   The analysis of the local stresses showed that the work transferred by the cell into hydrogel substrates with equal stiffness depended on the anchorage layer. Using traction force measurements, the maximum stress T_{max} caused by each cell in the hydrogel and the stored elastic energy per cell in the hydrogel were evaluated. Mean

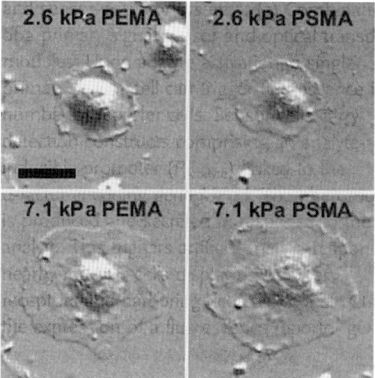

Figure 3.9 Phase contrast images of adherent endothelial cells after 1 h of cell culturing on PAAm hydrogels of different Young's modulus (2.6 and 7.1 kPa) coated with two different maleic anhydride copolymer films (PEMA and PSMA). Scale bar: 30 μm. (Reprinted with permission from Pompe *et al.* (2009). Copyright 2009, Elsevier.)

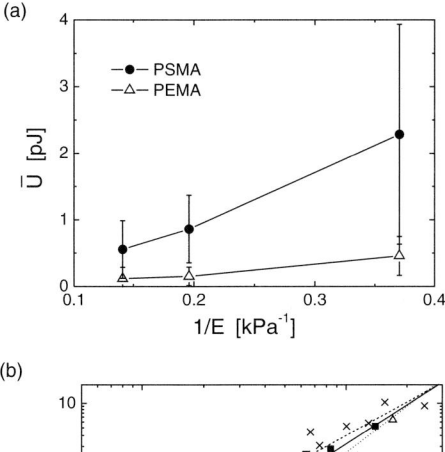

(a)

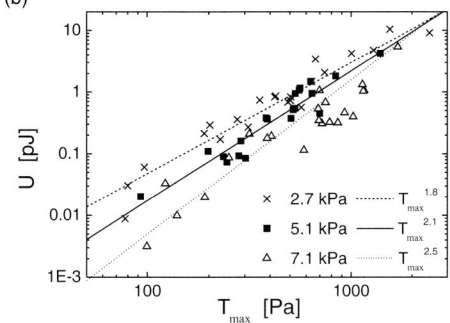

(b)

Figure 3.10 Strain energy U built-up in PAAm hydrogel substrates coated with maleic anhydride copolymer anchorage layers (PSMA or PEMA). (a) Average $\bar{U}$ of the two possible copolymer layers plotted versus inverse Young's modulus of the hydrogel. (b) U plotted versus T_{max}^2 for each cell irrespective of the type of the maleic anhydride copolymer coating. (Reprinted with permission from Pompe *et al.* (2009). Copyright 2009, Elsevier.)

maximum stresses $\bar{T}_{max} = 700$ and $300\,Pa$ had been obtained for the PSMA and PEMA coatings, respectively. Also, the average stored elastic energy $\bar{U}$ is significantly larger in hydrogels coated with PSMA than in those coated with PEMA copolymers. As seen in Figure 3.10, the stored elastic energy scales with inverse Young's modulus $1/E$ of the hydrogel. For intermediate hydrogel stiffness, a nearly ideal T_{max}^2 scaling of U is observed, whereas lower or higher stiffness results in slightly lower or higher powers of T_{max}, respectively. The evaluation of the experimental data shows that deformations are concentrated in a thin layer (about 5–10 μm) of the PAAm hydrogel coatings with an overall thickness of about 100 μm (see also Task 3.2).

As mentioned above, the changing traction forces in the cytoskeleton network generate specific biochemical responses along the downstream signaling pathways. One of the relevant signal proteins involved in such a pathway is the FAK, which localizes specifically to the focal adhesions. Its phosphorylation at specific amino acids activates binding sites for a number of other regulatory signaling molecules. Thus, it has an important function in cell adhesion, migration, and spreading, and it couples actively into signaling cascades in the cytoskeleton assembly, force generation, and even cell proliferation. Interestingly, it was shown in the discussed experiments that variation of hydrogel stiffness and anchorage strength causes phosphorylation of tyrosine residues at different sites of the amino acid chain (phosphorylation of Tyr^{397} is controlled by hydrogel stiffness, whereas that of Tyr^{861} is governed by stress signals related to changes of the adhesion force). Such signaling mechanisms are quite important as it is known that FAK phosphorylation at the specific sites regulates downstream activation of RhoA and ROCK and thereby of cytoskeletal tension and cell growth. Therefore, the elastic response of the ECM and the adhesion forces have to be considered as two independent exogenous cues that regulate the fate of cells in tissues via the intracellular mechanotransduction mechanism. At the same time, the activated downstream signaling pathways generate feedback mechanisms that can give rise to compositional and morphological changes of the ECM.

3.2.3
Influence of the Microenvironment Topology on the Cell Spreading and Development

The mechanotransduction system can be used by the cell to explore the topography of the surrounding medium with nanoscale resolution. Such topology information is essential for the motility, and also for proliferation, differentiation, and apoptosis. In surgery, for example, it is well known that the surface roughness of implants essentially influences the successful integration of the implant into the newly forming tissue. Usually, structured surfaces lead to a better integration. In cell culturing, the size of the microcontainment of the growing cells has a significant influence on cell proliferation and differentiation. As we will see in the following section, the fate of stem cells can be controlled by such processes. Furthermore, textured patterning of the surface morphology of plane substrates can be used for directing the growth of cells.

An interesting phenomenon of geometric influences on the cell behavior is related to the nanoscale topography. In view of our studies in the previous sections, it is not surprising that there is a relation between nanoscale topography of surfaces and the fate of adhering cells. As we have seen, the key elements being responsible for information transfer across the cell membrane, the focal adhesions, are assembled from building units with the characteristic size of the diameters of integrin heads of about 10–15 nm or two cross-linked actin filaments in the range of about 34–39 nm. There are already some very informative experimental results that clearly show that this size range coincides indeed with the characteristic length scale that governs the topography-directed cell fate. An informative model experiment has been worked out by Spatz and coworkers (Cavalcanti-Adam *et al.*, 2007) to get deeper insight into the geometrically controlled cellular process. The starting point was a simplifying experimental situation where the adhesive interaction of the cell was realized via amino acid epitopes (Arg–Gly–Asp (RGD)) that bind to integrins of fibroblasts. In order to realize a tunable assembly of binding sites, first a substrate was covered with a quasi-hexagonal lattice of 6 nm gold nanoparticles. The interparticle spacing was varied in a controlled way in the range from about 60 to 110 nm. The gold nanoparticles were imbedded in micelles of diblock copolymers. A closely packed monolayer of such micelles was deposited on glass cover slips. The interparticle distance was determined by the length of the copolymer. By treatment with hydrogen plasma, the polymer was removed, leaving a hexagonal array of gold nanoparticles. Precisely structured arrays of binding sites were prepared, which allowed a detailed study of the influence of the density of binding sites on the cell spreading and motility. The gold nanoparticles were functionalized with the integrin $\alpha_{IIb}\beta_3$-specific cyclic peptide (cRGDfK) containing the RGD sequence (Geiger *et al.*, 2009). In the study, rat fibroblasts were plated on the substrates with different binding site spacing. As a reference system, a substrate homogeneously deposited with densely packed RGD motifs ($\sim$40 000 RGD molecules/μm^2) was also studied. Characteristic features of the cell spreading for three different epitope patterns (homogeneous, interparticle spacing of 58 nm, and interparticle spacing of 108 nm) are shown in Figure 3.11.

Whereas the cell shape looked similar for homogeneous epitope distribution and 58 nm spacing, it was significantly different for 108 nm spacing. The projected cell surface was much smaller, and ruffling of the cell membrane was observed. The experiments have shown that the lateral spacing between the integrin ligands regulates the cell spreading and adhesion stability. For a lateral spacing of binding sites >73 nm, the formation of stable focal adhesion and stress fibers was not observed. Then only a delayed and transient spreading was found, and the cell dynamics was found to be different. Cells cultured on substrates with a small distance of binding sites ($\leq$58 nm) behaved fairly immobile, whereas on substrates with the large binding site distance ($\sim$108 nm) the cells changed their positions rapidly, combined with an erratic formation of protrusions and retractions. In conclusion, we see that the mechanotransduction system represents a

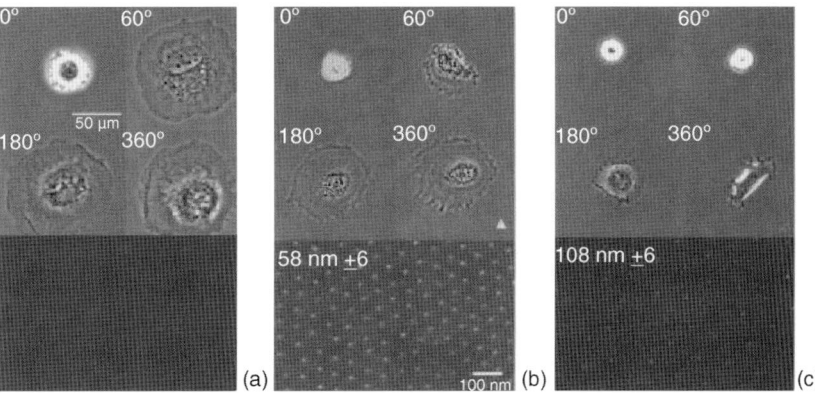

Figure 3.11 Phase contrast images of cell spreading on surfaces: (a) homogeneously coated with RGD peptide, (b) 58 nm spacing RGD nanopattern, and (c) 108 nm spacing RGD nanopattern. (Reprinted with permission from Cavalcanti-Adam *et al.* (2007). Copyright 2007, Elsevier.)

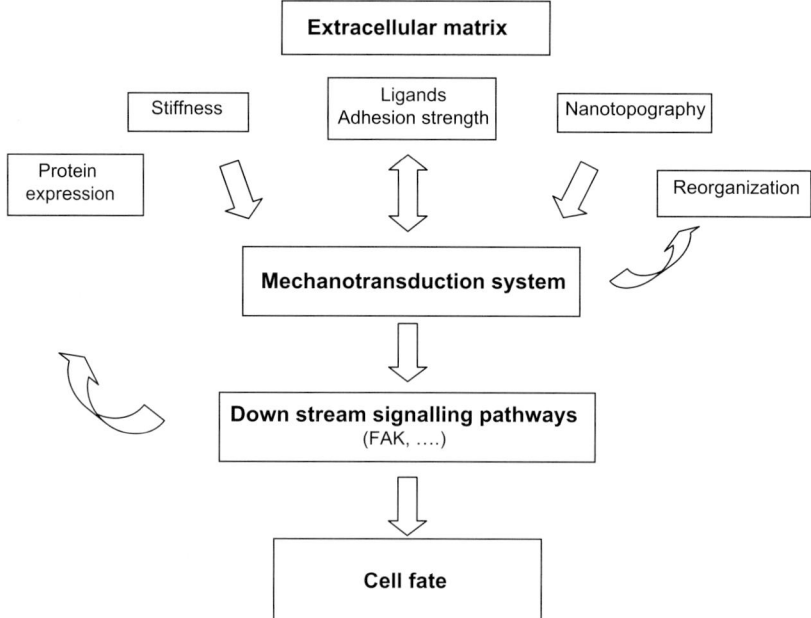

Figure 3.12 Mechanically regulated cellular signal and feedback mechanisms.

crucial link between the single cell and the developing tissue (Figure 3.12). It is used by the cell to generate information about the ECM. This concerns the local mechanical behavior as well as the global response of the surrounding medium. The medium can be regarded as an "effective" ECM including the presence of other cells. As we have seen, the mechanotransduction acts in both directions transferring signals from the outside to the internal processing units of the cell, as well as in the opposite direction. The force-driven morphological reorganization of the neighboring ECM and the compositional change of the ECM by expression of proteins are important feedback mechanisms. All of these processes are promising keys for a materials engineering-based control of the cell fate.

3.3
Bioengineering

3.3.1
The Basic Approach and Goals

The interaction of living cells with an artificial material is a topic of rapidly growing interest. There are two major motivations for exploring this challenging field of research (Smith, 2005). First, in medicine and biotechnology there is the hope that direct coupling of optical, microelectronic, or micromechanical devices with living tissue may provide novel conceptions for measuring and controlling particular functions of the living organism. It would be a great benefit if the biological response of a single living cell to external signals could be directly detected by coupling with physical systems. The idea of a "lab on chip" based on whole-cell sensors is regarded as an essential contribution to realize the transition from the macroscopic diagnostics of complex reactions of biological organisms to the molecular analytics of the relevant basic processes. In order to realize this idea, living cells have to be *immobilized on the chip surface* by providing a quasi-three-dimensional "natural" microenvironment so that they can be observed under conditions as close as possible to those in the living organism. Mainly driven by regenerative medicine there is another field of research that deals with materials development for mimicking the biological cell interaction. Here it is intended to control the cell development with the aim to produce specific constituents to be assembled into artificially generated tissues, starting from appropriate progenitor cells. For this purpose, the materials engineer has to find new solutions that facilitate the smooth transition from culturing cells under *in vitro* conditions to the *in vivo* evolution of new tissue. This task can be tackled from very different starting points concerning the cells to be used. Along the evolution chain of cells from embryonic stem cells via adult stem cells to mature cells of the tissue to be regenerated, various approaches are possible. For a solution, several aspects have to be considered: the already existing knowledge about the available resources of cell cultures, the elaborated methods for *in vitro* proliferation and differentiation of progenitor cells, the response of the tissue to be regenerated to the foreign cells, and ethical issues if embryonic stem cells are involved. As we will see in the following, materials engineering can be helpful here in various ways.

As an introduction to the problems involved in materials development for application in regenerative medicine, let us first have a look at the various existing tasks (Figure 3.13). Any therapy starts with the selection of stem cells or progenitor cells from an individual. As a next step, the cells have to be multiplied. The exceptional ability of stem cells to self-renew and differentiate into various mature cell types makes them favorite objects of research. Stem cells can be divided into four major classes: (1) embryonic stem (ES) cells, (2) tissue-specific fetal stem cells, (3) adult stem cells, and (4) induced pluripotent stem (iPS) cells. ES cells from the so-called blastocyst of the very early embryo are referred to be pluripotent as they are the origin of all kinds of tissues of the embryo. Fetal stem cells and adult stem cells are also multipotent but less compared to ES cells. Among the adult stem cells, hematopoietic blood stem cells, epithelial stem cells, and mesenchymal stem cells play a major role in regenerative therapy. A hematopoietic stem cell, for example, can lead to at least nine different cell fates. Mesenchymal stem cells are widely used for research approaches to engineer mesenchymal tissue such as bone, cartilage, skin, and tendon (for examples see Section 6.3). The recent discovery that mature cells can be redifferentiated into stem cells (induced pluripotent stem cells) appears to be promising with respect to future developments. It could be the favored way to avoid ethic problems related to working with embryonic stem cells. The fate of the stem cell is controlled by its microenvironment involving size and morphology of the containment, neighboring cells, local biochemical composition, compositional

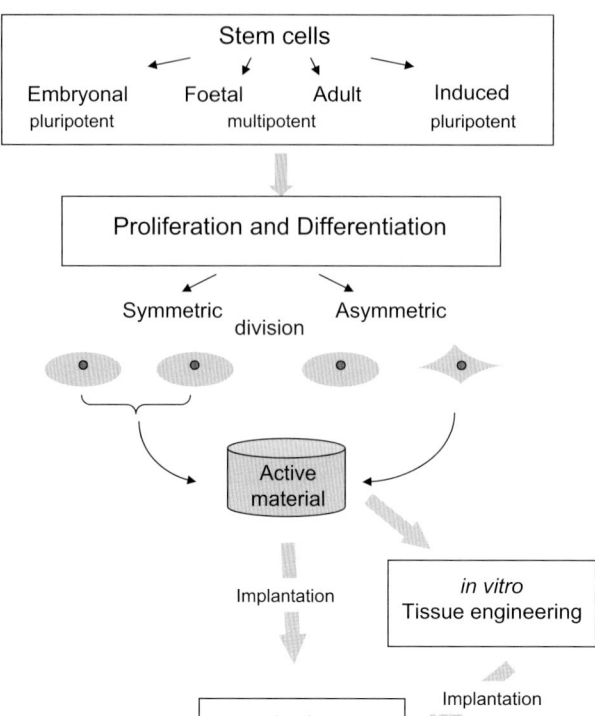

Figure 3.13 Various process routes for regenerative therapy. Stem cells or progenitor cells are isolated from an individual. An active material is used to create an appropriate microenvironment (niche) for the cell development into a tissue. This process can occur *in vitro* by using a bioreactor or *in vivo* after implantation of the active material loaded with the selected cells.

gradients, and other local factors. Often all of these parameters characterizing the microenvironment are also summarized under the term "niche." The fate of a particular stem cell is always related to its existence in a specific niche. The physical and biochemical characteristics of the niche govern the processes of self-renewal and differentiation. At this point, materials engineering comes into play. It can provide substrates that offer suitable niches for several purposes: cell culturing, cell association, or cell delivery (Figure 3.14). The material can be used in a bioreactor with the aim to *proliferate and differentiate (or redifferentiate) the cells in vitro* to form a specific tissue prior to transplantation. It can also serve as a *carrier of stem cells or progenitor cells as well as of signal molecules for chemotactic cell interaction*. Such carrier can be directly implanted into the patient to facilitate *in vivo* autoregeneration of damaged tissue. In transplantation biology, the cell type is differentiated between allogenic and autologous cells. Allogenic cells are of the same species but antigenically distinct, whereas autologous cells are derived or transferred from the same individual's body. Per definition, the application of allogenic cells typically includes

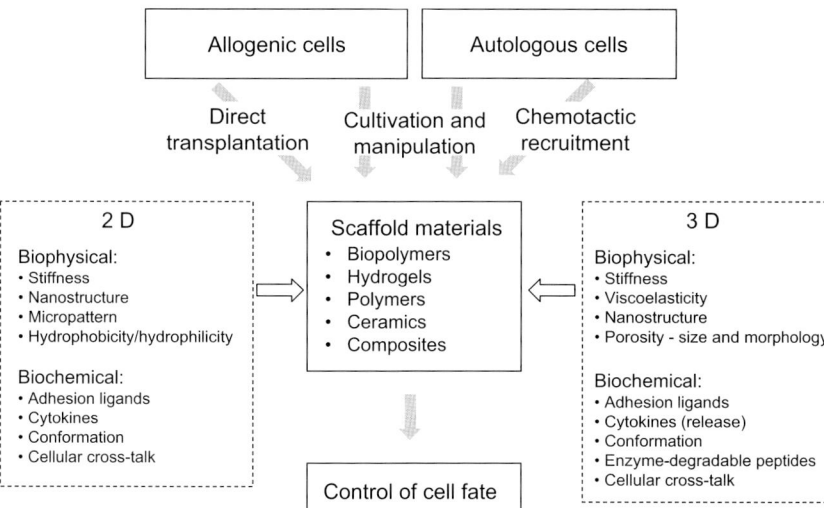

Figure 3.14 Biomaterials used to recruit and differentiate host cells. The incorporation of the cells into the selected scaffold occurs *ex situ* for allogenic cells or *in situ* for autologous cells. There is a rich toolbox for functionalization of the scaffold in order to control the cell fate.

the *ex situ* incorporation of the living cells into the intended implant material. When autologous (healthy) cells are used for the tissue regeneration, the active biomaterial can be formed *in situ* and can induce a chemotactic recruitment of the cells by an appropriate functionalization.

For the engineering of artificial tissue, nowadays the three basic cues structure, mechanical properties, and biomolecular mediators are used for controlling the cell fate. The large variety of available biocompatible materials allows the mechanical parameters to be varied in a wide range, from very soft hydrogels to rigid ceramics or composites. Variation of porosity is a convenient way of fine-tuning the stiffness (Figure 3.14). Variation of macroporosity is suitable for controlling the invasion of cells into the scaffold and the supply with nutrients. A variety of techniques have become available for tailoring the topography of 2D scaffolds: nanostructuring of surfaces by deposition of nanobeads, nanoimprinting, electrolytic etching, self-assembling monolayers, or optical lithography. The functionalization of the surface of the artificial 2D and 3D scaffolds with components of the ECM, synthesized peptides, or compounds that act as delivery systems for cellular signal molecules (e. g., cytokines, antibiotics, inhibitors, or small molecules) is offering a nearly infinite richness for biochemical control of the cell fate.

As outlined above, it is useful to classify the possible approaches into three major groups: (i) utilization of biomaterials (scaffolds) and cells for the *in vitro* generation of tissue (tissue engineering), (ii) combination of cells and scaffolds for immediate implantation in terms of a matrix-assisted cell therapy, and (iii) the approach of *in situ* tissue engineering, in which cells are directed toward a concentration gradient of chemoattractive agents (chemokines). For (ii), sometimes the misleading term

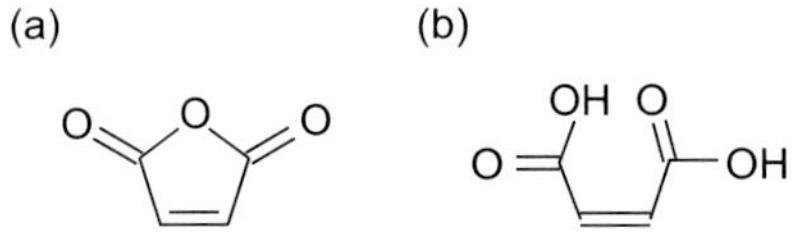

(a)   (b)

Figure 3.15   Chemical structures of (a) maleic anhydride and (b) maleic acid.

"*in vivo* tissue engineering" is also used. In the following, we will consider a number of typical examples. It appears that common concepts are underlying all of them, independent of the particular applications.

3.3.2
Tailored Surfaces for In Vitro Culturing of Cells

3.3.2.1 A Modular Polymer Platform for Mechanically Regulated Cell Culturing at Interfaces

Reactive copolymer films can be used as a versatile *in vitro* platform for simultaneous control of elasticity, adhesion, and biochemistry at an interface between a solid substrate and a cell. As an interesting example, maleic anhydride copolymers films have been developed for this purpose (Pompe *et al.*, 2003). Maleic anhydride, $C_2H_2(CO)_2O$, is highly reactive against primary amines and alcohols, a property that is widely used in biotechnology. Also, it easily hydrolyzes in an aqueous environment into maleic acid, *cis*-HOOC—CH=CH—COOH, leaving a rather nonreactive moiety (Figure 3.15).

The reactivity of the functional maleic anhydride can be combined with various comonomers (e.g., ethene, propene, butene, styrene, octadecene, vinyl methyl ether, etc.) varying the overall polymer characteristics in hydrophobicity and polarity (Figure 3.16). Attachment of adhesion ligands can be controlled by conversion of the anhydride by hydrolysis, reaction with functional amines, or conversions to other chemical moieties. In addition, important parameters such as the accessibility

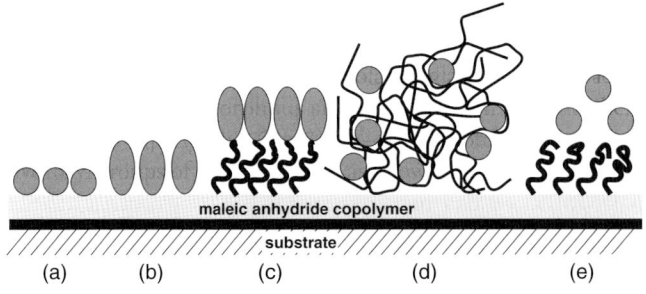

(a)   (b)   (c)   (d)   (e)

Figure 3.16   Schematic of the options for interface engineering with maleic anhydride copolymer films. (a) Binding of biomolecules (e.g., proteins) by physisorption, (b) simple covalent binding, (c) covalent binding via spacers, (d) entrapment in hydrogels, and (e) repulsion. (Reprinted with permission from Pompe *et al.* (2003). Copyright 2003, the American Chemical Society.)

(a)

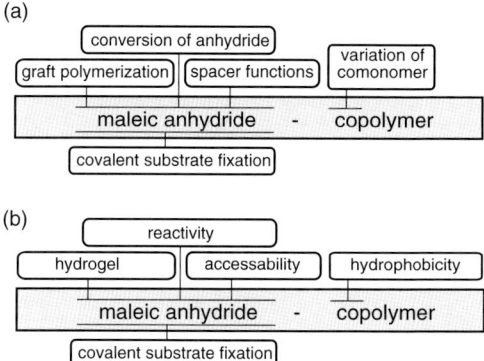

(b)

Figure 3.17 The maleic anhydride copolymer platform: (a) chemical functionalities and (b) their influence on the structure development of the biointerface. (Reprinted with permission from Pompe *et al.* (2003). Copyright 2003, the American Chemical Society.)

of ligands and the flexibility of coupled ligands can be changed by introducing appropriate spacers (polymer chains) (Figure 3.17). Thus, it is possible to create an interface that allows a fine-tuned interaction with cell receptors.

As explained above, the extensive use of stem cells for regenerative therapy is one of the ambitious visions of advanced medicine. For realization, the two problems of maintenance and maturation of stem cells under *in vitro* conditions have to be solved. This requires mimicking of the natural microenvironment by means of an artificial material providing niches for the stem cells to interact with the ECM, with other cells via mutual contact, and with freely diffusible or immobilized cytokines. The presence of oxygen and various metabolic products in suitable concentrations is essential for such interactions. All of them together control the balance between cycling and quiescence of the cell. It finally leads to proliferation or apoptosis, self-renewal, or differentiation of the cell. As already pointed out in this chapter, there are a number of means available for the design of artificial niches directing the stem cells along a certain path of development. The relevant material parameters in this connection are stiffness, geometric constraints (2D or 3D niches), micro- and nanotopography, and adhesive ligands as well as specific cytokines and signaling molecules. The following examples demonstrate possible design strategies for tailored polymer surfaces for the controlled evolution of stem cells and blood progenitor cells.

Adhesion-Controlled Stabilization of Stem Cells in a Quiescent State   The first example concerns the culturing of hematopoietic stem cells (HSCs). HSCs are already of clinical relevance for treating cancer and hematological disorders. For *in vitro* tissue engineering, it is often necessary to keep the cells in a quiescent state as long as they are not yet implanted. The experiment described here is to find out suitable conditions for keeping the cells quiescent (Kurth *et al.*, 2009). This is achieved by controlled functionalization of the surface of a 2D carrier material. A silicone substrate with micrometer-sized cavities, 10 μm deep and 15–80 μm in diameter,

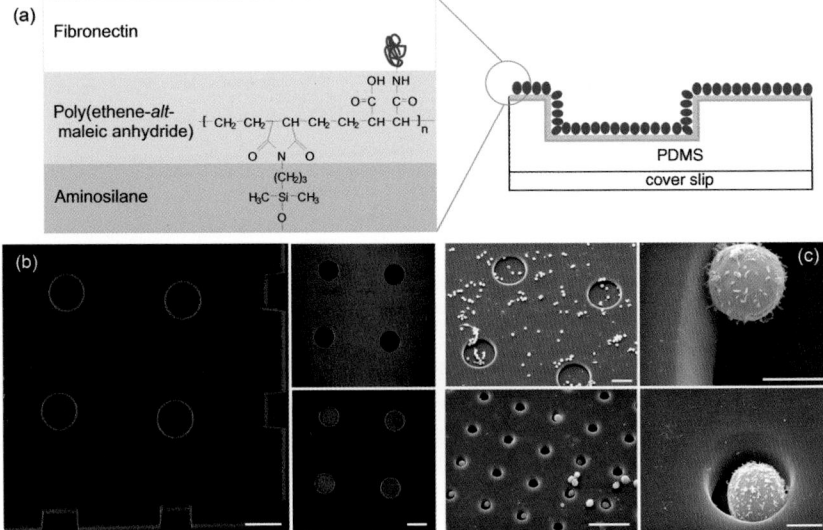

Figure 3.18 Protein immobilization on a microstructured surface. (a) After coupling of poly(ethene-*alt*-maleic anhydride) to the silicone surface fibronectin was covalently attached via its lysine side chain to the anhydride moieties. The surface quality was assayed by 3D imaging of fluorescently labeled fibronectin by confocal laser scanning microscopy (b) and scanning electron microscopy (c). HSC adhesion in the microcavities has been visualized in (c). Scale bar: (b) 40 μm; (c) 50 μm (left) and 5 μm (right). (Reproduced with permission from Kurth *et al.* (2009). Copyright 2003, the Royal Society of Chemistry.)

is chosen for mimicking the geometric constraints of an *in vivo* bone marrow HSC niche with its highly porous structure. Fibronectin is presented as adhesive ligand by applying the maleic anhydride copolymer platform (Figure 3.18). The cells are grown by adding a solution of cytokines (stem cell factor (SCF), thrombopoietin (TPO), and FMS-like tyrosine kinase 3 ligand (FL3)) to the growth medium. By varying the external cues (geometry and cytokine concentration), three main observations have been made in the experiments: (i) HSCs growing in the small 3D cavities preserve their immature state *in vitro* but those adhering to the ECM-coated planar surface lose stem cell surface markers and differentiate, thus losing their multipotent state. (ii) HSCs deposited in the smallest cavities (15 μm) show a much decreased expansion rate, which means they have been more or less in a quiescent state. (iii) The adhesion-related quiescence and maintenance of stem cells can be suppressed by nonphysio-logically high levels of growth factors. The experiment shows that proliferation signals driven by cytokines compete with counteracting signals from adhesive receptors and geometric constraints. The balance between cytokine level and adhesion capability is pictured schematically in Figure 3.19. Experimental results have led to the suggestion that HSC exposure to adhesive ligands and spatial constraints at moderate cytokine levels may preserve the stem cell character, decrease proliferation, and provide better engraftment in the tissue.

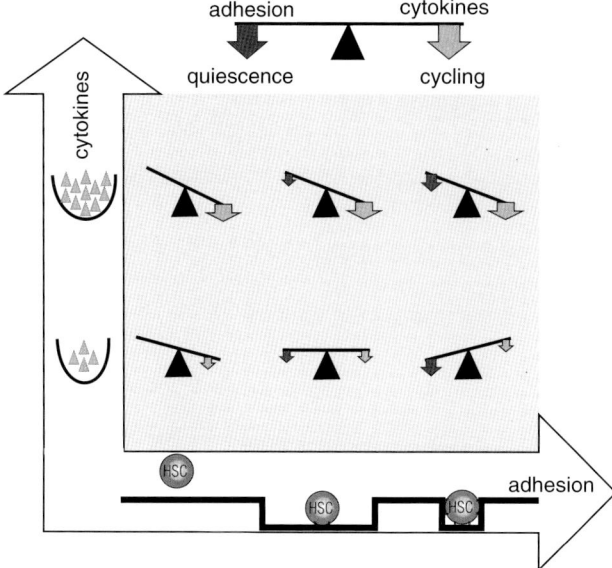

Figure 3.19 Balance of cytokine level and adhesion in cell cycling. Adhesion-related signaling is overbalanced by an excess of cytokine signaling, which results in an enhanced stimulation of cell cycling. With lower cytokine levels, increasing adhesion causes decreasing cell cycling. (Reproduced with permission from Kurth *et al.* (2009). Copyright 2009, the Royal Society of Chemistry.)

Immobilization of Cytokines at Surfaces Another example concerns the effect of cytokines immobilized on the substrate (Alberti *et al.*, 2008). In living tissue, cytokines exist mostly in a diffusible form. Therefore, the question arises whether it would be possible to present such cytokines immobilized at the surface of a biomaterial to be implanted without significant loss of efficacy. Again the maleic anhydride copolymer platform has been used for the immobilization of the cytokines. In the particular case, octadecene was the comonomer of choice. Three options of cytokine immobilization have been tested: covalent attachment to POMA, covalent attachment to flexible polyethylene glycol (PEG) spacer arms (PEG7) tethered to POMA (POMA–PEG7), and noncovalent binding to an ECM deposited on top of hydrolyzed POMA (see Figure 3.20). As ECM collagen I and fibronectin was alternatively used.

Two cytokines have been tested. First, the question has been answered whether the immobilization of the leukemia inhibitory factor (LIF) would change its functionality in comparison to that of the diffusible form. *In vivo*, the diffusible form of LIF inhibits the differentiation of mouse embryonic stem cells (mESCs). This means it sustains their pluripotency. The experiments have shown that the immobilized LIFs also facilitate a stable self-renewal of mESCs without loss of pluripotency. Among the various immobilization modes, the covalent attachment to POMA and POMA–PEG7 worked better than the physisorption at the ECM.

(a)   (b)   (c)

POMA   POMA - PEG7   POMA - matrix

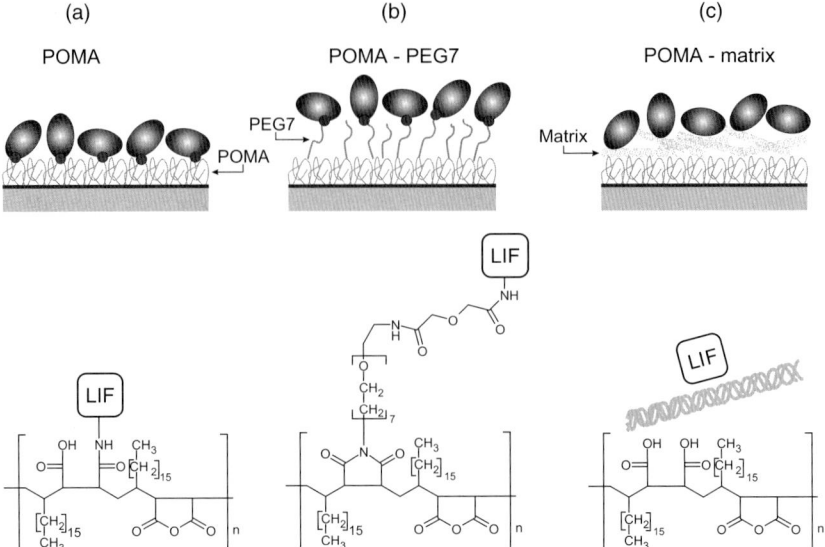

Figure 3.20 Different strategies for immobilization of the LIF by using a POMA interlayer. (a) Covalent attachment to POMA, (b) covalent attachment to flexible POMA–PEG7, and (c) noncovalent binding to an ECM deposited on top of hydrolyzed POMA. (Reproduced with permission from Alberti et al. (2008). Copyright 2008, the Nature Publishing Group.)

Another cytokine, the stem cell factor (SCF), has been immobilized by applying the same immobilization techniques. SCF induces the expansion of an erythroleukemia cell line Mo7E, a model for blood progenitor cells. In this case, the immobilized cytokines proved highly active. They efficiently enhance the cell proliferation with increasing SCF concentration at the substrate interface. In conclusion, the immobilization of cytokines on the surface of a substrate with the maleic anhydride copolymer platform is a very promising technique for cell culturing in regenerative therapy.

3.3.2.2 Regulation of Cell Fate by Nanostructured Surfaces
Judging from the basic experiments by Spatz et al. discussed above, nanoscale structuring of surfaces offers additional options for influencing the cell fate. The key idea of such approach – the regulation of the cell fate by the density of integrins inside the focal adhesion adhering to the surface of a substrate – has been applied by Park et al. (2007) with an innovative processing of nanostructured titanium surfaces. Ti-based alloys are of outstanding importance as implant materials owing to their good biocompatibility. They find broad application in orthopedics for load-bearing implants such as hip prostheses or in dentistry for the reconstruction of tooth roots. Surfaces of Ti can be nanostructured by controlled anodization. By treatment of the Ti surfaces in a phosphate–fluoride electrolyte at voltages ranging from 1 to 20 V, vertically aligned and densely packed TiO_2 nanotubes are produced with well-controlled diameters in the range between 15 and 100 nm (Figure 3.21).

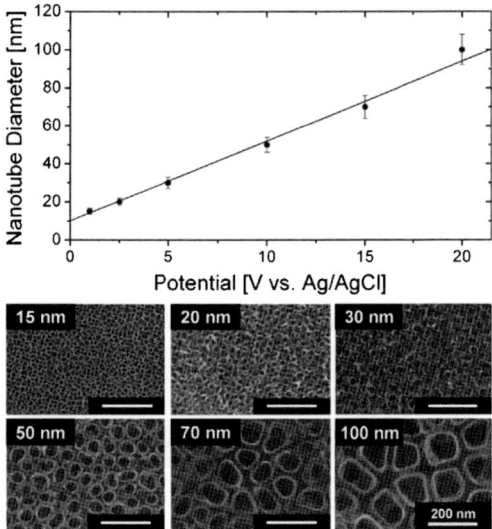

Figure 3.21 Titanium surfaces covered with self-aligned TiO$_2$ nanotubes. The different structures have been generated by anodizing in a phosphate–fluoride electrolyte with several voltages between 1 and 20 V. (Adapted with permission from Park *et al.* (2007). Copyright 2007, the American Chemical Society.)

Mesenchymal stem cells have been plated on the nanostructured surfaces in culture medium in order to study their behavior. As a reference, a polished TiO$_2$ surface has been included in the experiment. Tube diameters between 15 and 30 nm provided optimum results concerning adhesion, spreading, and growth, well above the performance of the smooth surface. Larger tubes provided poor results. Most remarkable is the high apoptosis rate of the cells stuck to 100 nm tubes. In agreement with these observations, the osteogenic differentiation of the stem cells is most pronounced for the surfaces with the smallest nanotubes. After 2 weeks of incubation in an osteogenic differentiation medium, enhanced deposition of calcium phosphate has been detected for the 15 nm structured surfaces by colorimetric analysis of the surfaces stained with a dye. All these observations can be explained by considering the fine structure of the adhesive contact for the different surfaces. Cells grown on 30 nm structures show extensive formation of focal adhesion contacts, whereas those grown on 100 nm structures are less spread and develop long protrusions with only a few focal adhesions. The high density of focal adhesions correlates with the density of binding sites for integrin clusters offered by the TiO$_2$ ridges at the nanostructured surfaces. As discussed above, signaling cascades propagating across these binding sites can induce cell proliferation and differentiation. Signals triggering apoptosis are emitted from regions of missing adhesive contacts. In agreement with the study using gold nanoparticles for surface structuring, we can state that optimum cell response is obtained with binding site distances equal to a certain minimal distance of integrins related to the internal focal adhesion structure.

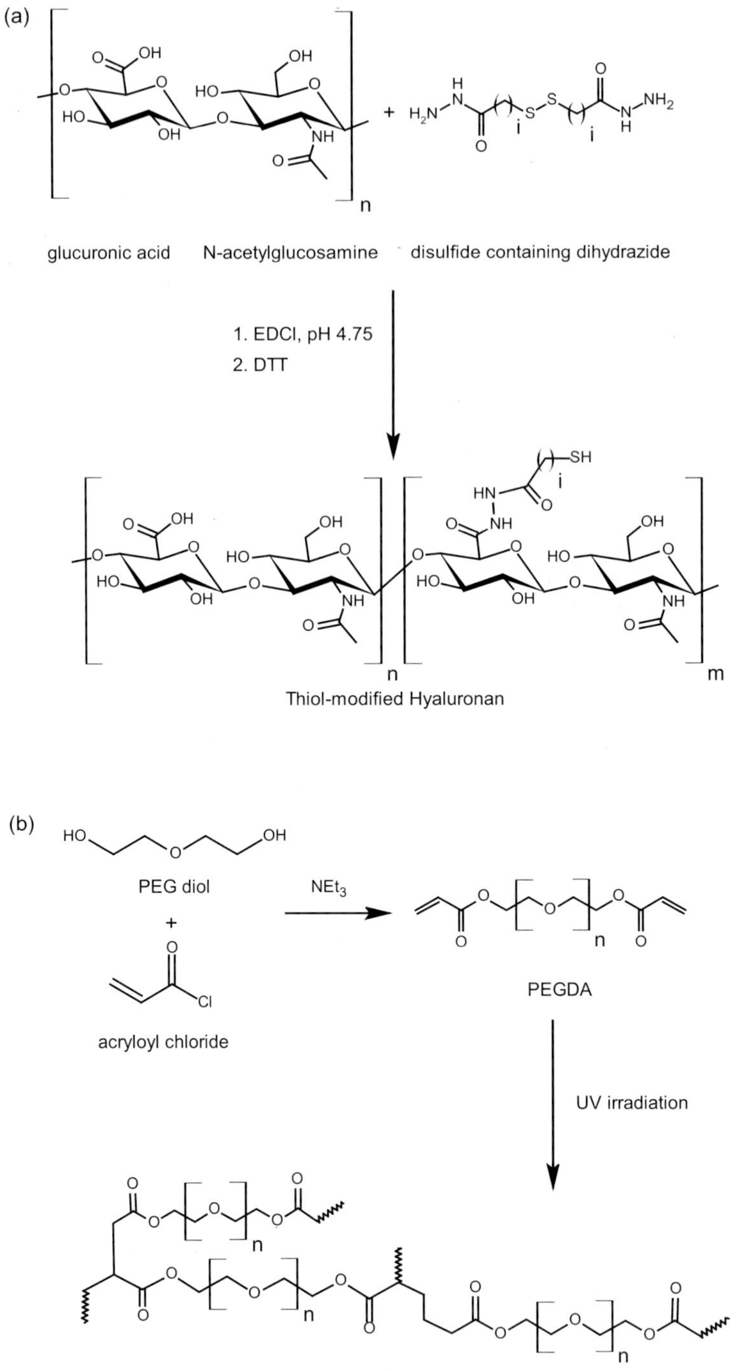

(a)

glucuronic acid   N-acetylglucosamine   disulfide containing dihydrazide

1. EDCl, pH 4.75
2. DTT

Thiol-modified Hyaluronan

(b)

PEG diol

NEt₃

PEGDA

+

acryloyl chloride

UV irradiation

3.3.3
Three-Dimensional Scaffolds for Tissue Engineering

Three-dimensional (3D) scaffolds for tissue engineering act as matrices supporting the cell activity required for regeneration of damaged tissue. From a simplistic
point of view, it serves as a transient or permanent mechanical substitute of a
destroyed ECM without disturbing the surrounding living tissue. However, in
reality it serves as a biomechanical and biochemical signaling system governing
the tissue regeneration including biological degradation and generation of new
tissue. With the selection of suitable materials one has to observe side effects such
as toxic degradation products and inflammatory responses of the tissue. Any
foreign material will cause an inflammatory reaction. Such reactions have to be
minimized in order to avoid the growth of a fibrotic scar between material and
tissue. Under the aspect of mechanical control of the cell activity, synthetic or
natural polymers are the favored scaffold materials for most tissues. Soft tissues,
such as nerves, require a malleable scaffold, whereas for the regeneration of skin a
flexible polymer is favored. For cartilage regeneration, a polymer scaffold is
suitable because of its damping properties. For the design of appropriate polymer
scaffolds, it would be advantageous to find a modular approach that allows a
stepwise engineering of a large class of materials meeting various demands
(Shoichet, 2010). One starting point is a polymer that is characterized by low
protein adsorption and cell adhesion. It can reduce an unfavorable response of the
immune system to the implanted material. Also, such polymer should offer
options for a selective immobilization of specific proteins and/or for inducing the
adherence of particular cells on a 2D scaffold or their in-growth into a 3D scaffold.
With this intention, the polymer can be adapted to a particular tissue by
conjugation with specific components of the natural ECM such as adhesion
proteins, glycosaminoglycans, and cytokines.

PEG is one of the widely used synthetic polymers for such a development. PEG is a
hydrophilic polymer, usually with a layer of ordered water molecules attached to the
chain. If applied to surfaces, it reduces any peptide or protein adsorption and cell
adhesion. Besides surface modification of materials by PEG via grafting of PEG
molecules or formation of self-assembled alkyl monolayers with PEG end groups,
PEG is widely applied as cross-linked 3D hydrogel as thin layer or bulk material. In
this connection, various techniques are available: photo-cross-linking with UV
irradiation of PEG diacrylate (Figure 3.22), cross-linking via vinyl sulfone end
groups, and cross-linking of activated acid groups of heparin molecules by means

◀━━━

Figure 3.22 Two examples for formation of 3D
polymer scaffolds used to limit protein and cell
adhesion. (a) Hyaluronan can be modified
using a disulfide-containing dihydrazide that
produces a thiol-modified hyaluronan that is
then cross-linked through disulfide bond
formation. (Adapted from Shu *et al.* (2002).)

(b) PEG diol is reacted with acryloyl chloride to
give poly(ethylene glycol) diacrylate PEGDA,
which is cross-linked upon UV irradiation.
(Adapted with permission from Cruise *et al.*
(1998) and Shoichet (2010). Copyright 2010,
the American Chemical Society.)

of amine-terminated PEG. These options may also allow enzyme-degradable peptides to be incorporated into the hydrogel network, which facilitates the design of biodegradable scaffolds.

Similar scaffolds with low protein adsorption and cell adhesion can be derived from polysaccharides, such as hyaluronan (also called hyaluronic acid), which is also present as a native polysaccharide in the ECM (see also Section 1.1.3). It can be cross-linked via modification with disulfide-containing dihydrazide (Figure 3.22). Disulfide bonds formed between the polysaccharide chains adjust stiffness and strength. As a further advantage, hyaluronan can be enzymatically degraded with hyaluronidase into nontoxic remains. Whereas hyaluronan does not adhere to the majority of cells, there are a few cell types, including tumor cells, with specific hyaluronan receptors (e.g., CD44). This can be exploited for the development of materials with mixed adhesive and anti-adhesive behavior with respect to different cell types.

Another important platform for scaffold manufacturing is based on the use of collagen, the natural biopolymer. There are about 20 different collagens. They facilitate the different properties of the various tissues (for more details see Chapter 6). Similar to PEG and hyaluronan, collagen can be cross-linked. Thus, the mechanical and physical properties of collagen can be tailored in a wide range. The potential of collagen as an engineering material goes far beyond medical applications. Known as tanning of leather, the cross-linking of skin collagen has been developed to high perfection from ancient handcraft to advanced chemical processing (Reich, 2007). Today more than 1 million ton of leather is manufactured per year worldwide. Its remarkable mechanical and thermal properties as well as the controllable interaction with water (from hydrophilic to highly hydrophobic) explain the wide range of applications from shoes, gloves, and clothes to furniture and components for automotive industry.

3.3.4
Switchable Substrates and Matrices

In bioengineering, there is a growing interest in influencing the cell behavior by using substrates or matrices that facilitate a controlled variation of cell adhesion and permittivity for nutrients and metabolic products. In order to prepare confluent cell films for tissue engineering, for example, it would be useful to be able to switch from an adhering to a repelling substrate after the first growth period. As a common problem in biotechnology, a particular cell strain is to be included into the reaction at a certain stage of the process. In this connection, it would be convenient to be able to switch that strain in a controlled way from quiescence to an active stage. This could possibly be realized by switching the permittivity of the coating of encapsulated cells from impermeable to permeable. There are so-called *stimuli-responsive polymers* (SRPs) that offer promising options for the design of scaffolds with externally regulated cell interaction. The SRPs undergo a phase transition from a soluble to an insoluble state upon physical or chemical changes in the environmental media, such as temperature, pH, magnetic or electric fields, light,

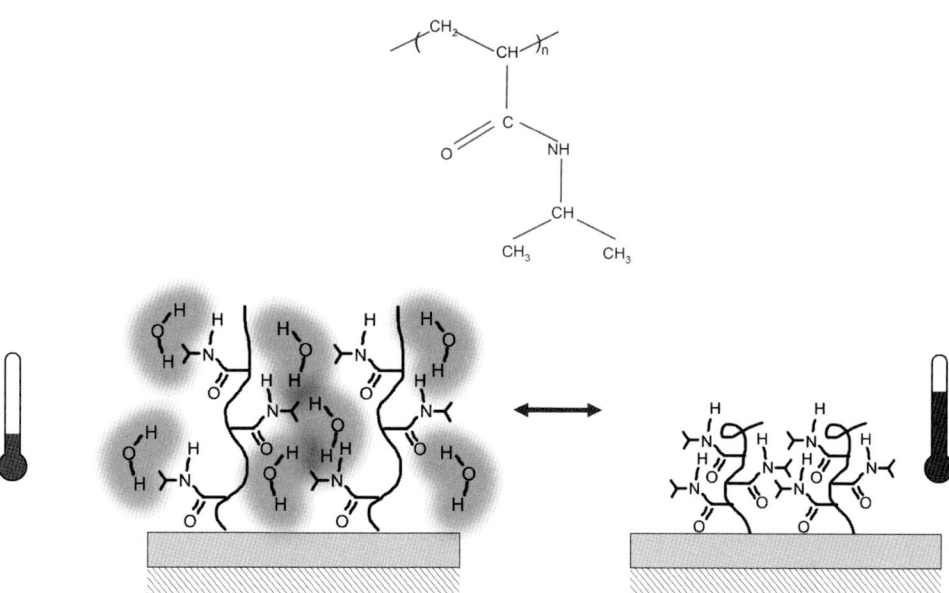

Figure 3.23 Structure of PNIPAAm and schematic of the temperature-controlled phase transition of PNIPAAm at $T = T_p$. For $T > T_p$, the collapsed hydrophobic structure is stabilized by hydrogen bonds between the polymer chains (left). For $T < T_p$, hydrogen bonds of individual PNIPAAm chains with water molecules lead to an extended hydrophilic polymer layer (right). (Courtesy of Kaufmann (2013).)

ion strength, redox potential, or addition of specific molecules. In an aqueous medium, the phase transition from the soluble to the insoluble state is immediately connected with shrinkage of the polymer. Among these SRPs, poly(N-isopropylacrylamide) (PNIPAAm) plays an exceptional role for the mechanical control of cell interaction. PNIPAAm shows a phase transition at a temperature T_p of around 30 °C in water (Figure 3.23). For $T > T_p$ the polymer collapses, and it is insoluble in water, whereas for $T < T_p$ it is soluble. Such behavior is also known as LCST (lower critical solution temperature) behavior. The stimuli-responsive behavior of PNIPAAm originates from hydrogen bonds formed between the polar groups of the PNIPAAm chains and water molecules.

The entropy-driven transition fits in the large group of conformation transitions of polymers interacting with a solvent. The driving force for the transition can be derived from the change of the free enthalpy between the collapsed and extended conformations ($\Delta G = G_{\text{col}} - G_{\text{ext}} = \Delta H_{\text{bond}} - T\Delta S_{\text{tot}}$) of the total system consisting of the polymer and the solvent. Below T_p, the conformation is governed by the hydrogen bonds between the polymer chains and a layer of water molecules. Above T_p, these bonds are disfavored as the total entropy of the system can be increased by liberation of water molecules. The change of the hydrogen bonds is connected with an increase of the enthalpy ($\Delta H_{\text{bond}} > 0$). The total change of the entropy is caused by a decrease of the conformational entropy

($\Delta S_{con} < 0$), when the polymer chains switch into the more constrained collapsed conformation. However, this part is overcompensated by the liberation of the water molecules, which causes the dominating entropy increase ($\Delta S_{bond} > 0$). This leads to an increase of the total entropy ($\Delta S_{total} = \Delta S_{con} + \Delta S_{bond} > 0$). In conclusion, this entropy gain leads to a negative change of the free enthalpy for $T > T_p = \Delta H_p / \Delta S_{tot}$, which means a collapsed PNIPAAm conformation. There is an obvious parallel to the known cold denaturation of proteins at low temperature (typically for $T < 20\,°C$). Usually the configurations of proteins are stabilized by hydrophobic interactions that correspond to the collapsed configuration of PNIPAAm. The cold denaturation of the proteins is also entropy driven. At low temperature, the gain of enthalpy related to hydrogen bonds of the protein with water molecules leads to the destabilization of the hydrophobic domains observed as cold denaturation. The phase transition is connected with a significant volume change.

The thickness of a PNIPAAm layer deposited on a plane substrate increases typically by a factor of about 2. The transition between a soluble (hydrophilic) and an insoluble (hydrophobic) conformation leads to a distinct change of the interaction with cells. Above T_p cells adhere to PNIPAAm layers, whereas below T_p the cells detach from the swollen polymer. The transition of the polymer to the swollen hydrophilic stage leads to a detachment of the ECM proteins secreted from the cell. Therefore, also the receptors of the mechanotransduction system lose their contact to the polymer substrate. The formerly spread cell is detached and takes a round shape.

The absolute value of T_p as well as the sharpness of the phase transition depends on the particular presentation of the polymer. The transition behavior can be drastically altered if the SRPs are attached to solid supports (Kaufmann *et al.*, 2010). Grafting, cross-linking density, and intermolecular forces from the supporting substrate influence the transition. Additional constraints for the hydrophobic and hydrophilic domains of the SRP are decisive for the instability of a given initial conformation. Therefore, the processing of the polymer can be used to tune the transition temperature T_p. There are various preparation routes of PNIPAAm layers, such as polymer brush synthesis, layer-by-layer deposition, end-group grafting, and cross-linking by UV, plasma, electron beam, and chemical induction of cross-linking. Copolymerization of PNIPAAm is another powerful means for adjusting T_p to a favorable temperature range, for improving the swelling degree, or for accelerating the phase transition. Possible choices are the copolymerization of PNIPAAm with monomers exhibiting charged groups such as carboxyl groups (e.g., acrylic acid, methacrylic acid, and carboxyalkylacrylamide (CAAm)) (Kaufmann *et al.*, 2010) or other hydrophilic groups (e.g., PEG) (Schmaljohann *et al.*, 2003). Thus, the responsiveness can be tuned by balancing the ratio of hydrophilic and hydrophobic components. The transition point can also be controlled via pH value by introducing additional charged groups. In Figure 3.24, an example is given for the influence of the composition of a PNIPAAm copolymer on the phase transition, with CAAm used as a comonomer. The swelling onset transition has been determined from the distinct onset of dissipation in the copolymer films subjected to oscillation

(a)

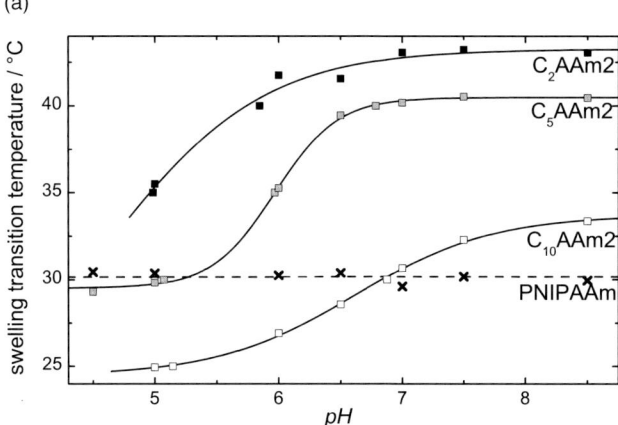

(b)

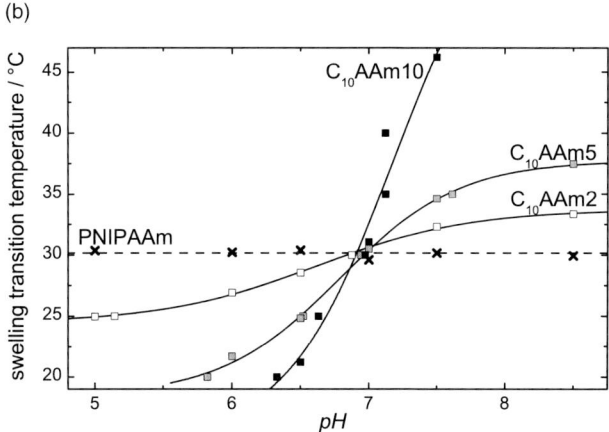

Figure 3.24 Swelling onset transition temperature of PNIPAAm–CAAm copolymers versus pH value. (a) Influence of the alkyl spacer length for given comonomer content (2 mol%). (b) Influence of the comonomer content for given alkyl spacer length ($n = 10$). (Reproduced with permission from Kaufmann *et al.* (2010). Copyright 2010, the Royal Society of Chemistry.)

on a quartz crystal microbalance (QCM). Figure 3.25 shows the structure of the PNIPAAm–C_nAAmy copolymer. Alkyl spacer chains allow additional hydrophobic domains to be incorporated. At a given pH, an additional hydrophobic domains lower T_p (Figure 3.24a) and a higher comonomer content with fixed alkyl spacer length increases T_p (Figure 3.24b). A higher concentration of carboxyl groups leads to a higher hydrophilicity of the copolymer. This also explains the increase of T_p with pH for a given comonomer content caused by a larger deprotonation of the carboxyl groups. The general transition behavior of SRPs, as governed by the interaction of hydrophilic and hydrophobic domains, strongly resembles the protein folding behavior as discussed in Chapter 1.

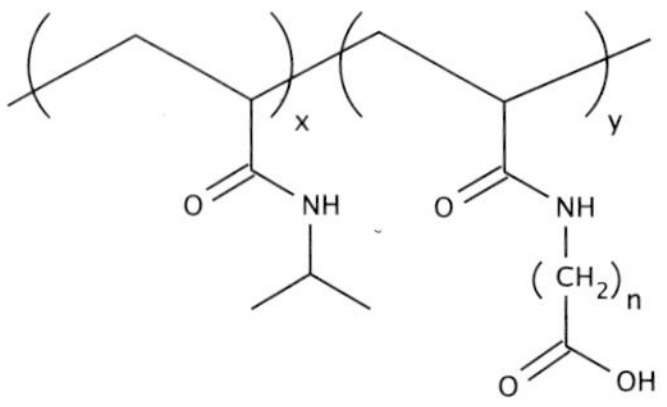

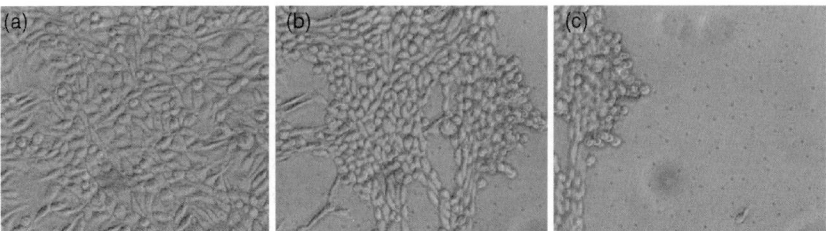

Figure 3.25 Chemical structure of the PNIPAAm–C_nAAmy copolymers. The abbreviation C_nAAmy in Figure 3.24 denotes the length of the alkyl spacer ($n = 2$, 5, and 10) and the content of comonomer ($y = 2$, 5, and 10 mol%). (Reproduced with permission from Kaufmann *et al.* (2010). Copyright 2010, the Royal Society of Chemistry.)

The control of adhesion and detachment of cells on thermoresponsive polymers offers a large variety of applications for engineering cell sheets. In regenerative therapy, there is an obvious demand for growth of confluent cell sheets. To provide an example, we refer to a study by Nitschke *et al.* (2007) for harvesting human corneal endothelial sheets. The corneal endothelium is critical for corneal transparency. It forms a boundary between the corneal stroma and the anterior chamber and regulates the stromal hydration. Impaired vision or even blindness may result from disruption of the cell layer due to accidental or surgical trauma. Corneal transplantation can be done with high reliability. In view of the shortage of donor corneas, confluent cell sheets grown on and harvested from a thermoresponsive PNIPAAm–copolymer substrate are highly welcome. Diethylene glycol methacrylate (DEGMA) has been used as comonomer for this purpose. Human cornea endothelial cells have been cultivated on the collapsed (hydrophobic) PNIPAAm–copolymer substrate at 37 °C. After 24 h, a confluent sheet was reached. After cooling to 30 °C, such cell sheets detached without any enzymatic or mechanical treatment from the swollen (hydrophilic) substrate (Figure 3.26).

Together with the cells, a thin layer of ECM is detached. Such stabilizing ECM layer is essential for the subsequent successful transplantation of the confluent cell sheet. The stability of the detached confluent cell sheets enables them to be formed into cell assemblies of various shapes. One example is the formation of multicellular spheroids. Such shape is of interest for toxicological tests, for developing hybrid

Figure 3.26 Cell detachment of mouse fibroblast from a PNIPAAm–DEGMA substrate. Micrographs of cells at standard cultivation at 37 °C (a) and in the time course of cooling for 30 min to 30 °C (b and c) showing the detachment of the confluent cell sheet (image size: $436 \times 346\ \mu m^2$).

artificial organs, for the study of tumor environments, and for evaluation of the effects of chemotherapy or radiation therapy on tumors. Shaping detached cell sheets into spheroids is surprisingly easy: The sheet is transferred into a hydrophobic dish where it spontaneously rearranges itself into a spheroid. Hence, the size of the sphere is determined by the area of the sheet. The area depends on the particular patterning of the SRP substrate. An exciting example for the versatility of such shape engineering has been given by Matsuda (2004). He has manufactured self-supporting tubular endothelial cell sheets. A capillary tube was coated with an aqueous solution of PNIPAAm-grafted gelatin and air-dried. Then endothelial cells were seeded in the tube and cultured for 4 days at 37 °C. The following infusion of the culture medium into the tube at 20 °C resulted in the spontaneous detachment and removal of a tubular vascular tissue composed of endothelial cells and supra-molecularly organized ECM produced by endothelial cells on the basal side of the tissue. It is obvious that techniques of such kind open a wide field for the manufacture of shape-engineered tissues to be used in transplantation, for modeling organs for *in vitro* studies, and for preparing structures composed of different cell types that could serve as models for cellular communication.

Task T3

Task 3.1: In order to control the cell behavior by the mechanical properties of the PAAm hydrogel, any interaction of the cells with the underlying glass substrate has to be avoided. This condition is fulfilled when the thickness of the hydrogel is larger than a critical value given by the depth of the deformed layer caused by the work of the adherent endothelial cells. Estimate this value by using the experimental data given in Figure 3.10 for the two polymer coatings, PSMA and PEMA. For the estimate you can assume mean maximum stresses $\overline{T}_{max} = 700$ and $300\,\text{Pa}$ for the PSMA and PEMA coatings, respectively. In agreement with the experimental data, it can be assumed that the averaged 2D stress field in the PAAm surface layer is nearly isotropic.

References

Alberti, K., Davey, R.E., Onishi, K., George, S., Salchert, K., Seib, F.P., Bornhäuser, M., Pompe, T., Nagy, A., Werner, C., and Zandstra, P.W. (2008) Functional immobilization of signaling proteins enables control of stem cell fate. *Nature Methods*, **5** (7), 645–650.

Cavalcanti-Adam, E.A., Volberg, T., Micoulet, A., Kessler, H., Geiger, B., and Spatz, J.P. (2007) Cell spreading and focal adhesion dynamics are regulated by spacing of integrin ligands, *Biophys. J.* **92**, 2964–2974.

Cruise, G.M., Scharp, D.S., and Hubbel, J.A. (1998) Characterization of permeability and network structure of interfacially photopolymerized poly(ethylene glycol) diacrylate hydrogels. *Biomaterials*, **19** (14), 1287–1294.

Engler, A.J., Sen, S., Sweeney, H.L., and Discher, D.E. (2006) Matrix elasticity directs stem cell lineage specification, *Cell* **126**, 677–689.

Geiger, B., Spatz, J.P., and Bershadsky, A.D. (2009) Environmental sensing through focal adhesions, *Nat. Rev. Mol. Cell Biol.* **10**, 21–33.

Howard, J. Mechanics of Motor Proteins and the Cytoskeleton (2001), Sinauer Associates.

Kaufmann, M., Jia, Y., Renner, L., Gupta, S., Kuckling, D., Werner, C., and Pompe, T. (2010) Tuneable swelling of thermo- and pH-responsive copolymer films. *Soft Matter*, **6**, 937–944.

Kaufmann, M. (2013) Lipid Bilayer Supported by Multi-Stimuli Responsive Polymers. Ph.D. thesis, Technische Universität Dresden.

Kurth, I., Franke, K., Pompe, T., Bornhäuser, M., and Werner, C. (2009) Hematopoietic stem and progenitor cells in adhesive microcavities. *Integrative Biology*, **1**, 427–434.

Matsuda, T. (2004) Poly(*N*-isopropylacrylamide)-grafted gelatin as a thermoresponsive cell-adhesive, mold-releasable material for shape-engineered tissues. *Journal of Biomaterials Science, Polymer Edition*, **15**, 947–955.

Nitschke, M., Gramm, S., Götze, T., Valtink, M., Drichel, J., Voit, B., Engelmann, K., and Werner, C. (2007) Thermo-responsive poly(NiPAAm-*co*-DEGMA) substrates for gentle harvest of human corneal endothelial cell sheets. *Journal of Biomedical Materials Research Part A*, **80** (4), 1003–1010.

Park, J., Bauer, S., von der Mark, K., and Schmuki, P. (2007) Nanosize and vitality: TiO$_2$ nanotube diameter directs cell fate. *Nano Letters*, **7** (6), 1686–1691.

Pompe, T., Zschoche, S., Herold, N., Herold, N., Salchert, K., Gouzy, M.F., Sperling, C., and Werner, C. (2003) Maleic anhydride copolymers: a versatile platform for molecular biosurface engineering. *Biomacromolecules*, **4** (4), 1072–1079.

Pompe, T., Renner, L., and Werner, C. (2005) Nanoscale Features of Fibronectin Fibrillogenesis Depend on Protein-Substrate Interaction and Cytoskeleton Structure, *Biophys. J.* **88**, 527–534.

Pompe, T., Starruss, J., Bobeth, M., and Pompe, W. (2006) Modeling of pattern development during fibronectin nanofibril formation, *Biointerphases* **1**, 93–97.

Pompe, T., Glorius, S., Bischoff, T., Uhlmann, I., Kaufmann, M., Brenner, S., and Werner, C. (2009) Dissecting the impact of matrix anchorage and elasticity in cell adhesion, *Biophys. J.* **97**, 2154–2163.

Reich, G. (2007) *From Collagen to Leather*, BASF Service Center, Media and Communication, Ludwigshafen, Germany.

Renner, L., Pompe, T., Salchert, K., and Werner, C. (2004) Dynamic Alterations of Fibronectin Layers on Copolymer Substrates with Graded Physicochemical Characteristics, *Langmuir* **20**, 2928–2933.

Schmaljohann, D., Oswald, J., Joergensen, B., Nitschke, M., Beyerlein, D., and Werner, C. (2003) Thermo-responsive PNiPAAm-*g*-PEG films for controlled cell detachment. *Biomacromolecules*, **4** (6), 1733–1739.

Shoichet, M.S. (2010) Polymer scaffolds for biomaterials applications. *Macromolecules*, **43**, 581–591.

Shu, X.Z., Liu, Y. Luo, Y., Roberts, M.C., and Prestwich, G.D. (2002) Disulfide cross-linked hyaluronan hydrogels. *Biomacromolecules*, **3** (6), 1304–1311.

Smith, J.P. (2005) Medical and biological sensors: a technical and commercial review. *Sensor Review*, **25** (4), 241–245.

4
Whole-Cell Sensor Structures

The progress in recombinant DNA technology has provided novel options in bio-sensor development. Alternative to the common biosensors based on the reactions of specific biomolecules and the subsequent transformation of the biochemical signal into a physical or chemical signal, the new approach is based on the response of reporting cells to their environment. Such analysis is often necessary when the impact of a polluted environment on the physiological status of living organisms is to be predicted. The limited information obtained with the restricted number of biomolecules selected for the common biosensors would not be adequate to the complexity of the problem. In order to make use of test cells in biosensors, measurable physical or chemical signals indicating the physiological status have to be identified.

4.1
Case Studies

There are conspicuous examples where microorganisms rapidly respond to environmental changes with physical signals. Numerous *marine bacteria* such as *Photobacterium phosphoreum*, *Photobacterium leiognathi*, *Vibrio harveyi*, and *Vibrio fischeri* are distinguished by the ability to *emit luminescent light*. Ocean areas up to $10\,000\,\mathrm{km}^2$ have been seen glowing with the bioluminescence of myriads of such bacteria over several nights. Toxic compounds interfering with the bacterial metabolism cause the bioluminescence to wane or vanish. This makes bio-luminescence a suitable indicator of the toxicity of aquatic media. Relevant data can be obtained typically within less than 24 h, which is remarkably fast in comparison to established methods such as toxicological animal-based tests or *in vitro* tests, commonly used for screening critical chemicals.

The *dinoflagellates* are a second large group of bioluminescent unicellular marine microorganisms constituting a major part of the phytoplankton. Under mechanical stress (hydrodynamic shear flow), they emit blue-green (470–490 nm) light flashes (about 100 ms). The intensity of the emitted light pulses depends on the metabolic state of the cells. Thus, similar to photobacteria, they can indicate toxicity. The dinoflagellates *Pyrocystis lunula*, *Ceratocorys horrida*, and *Lingulodinium polyedrum* have been tried as test organisms.

Bio-Nanomaterials: Designing Materials Inspired by Nature, First Edition. Wolfgang Pompe, Gerhard Rödel, Hans-Jürgen Weiss, and Michael Mertig.
© 2013 Wiley-VCH Verlag GmbH & Co. KGaA. Published 2013 by Wiley-VCH Verlag GmbH & Co. KGaA.

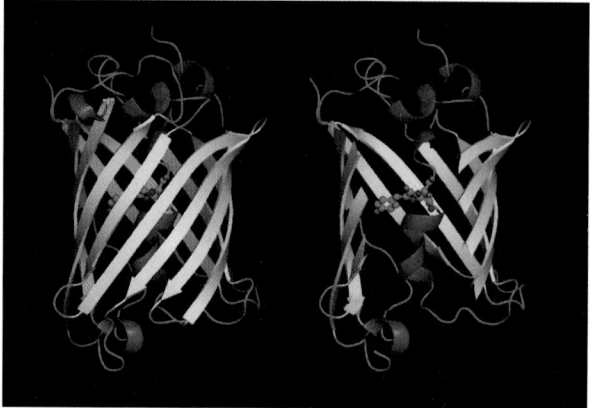

Figure 4.1 GFP with chromophore highlighted as ball-and-stick in the center of one wholly reproduced protein and cutaway version to show the chromophore. Modeled and rendered with PyMOL (Raymond Keller, Wikimedia Commons).

Optical emission can be observed in natural organisms also as a result of excitation by light. Several marine organisms respond to blue light with a strong green fluorescence. The *green fluorescent protein* (GFP) of the jellyfish *Aequorea victoria* is most promising for bionanotechnology. It is composed of 238 amino acids with a molecular mass of 26.9 kDa. It shows two excitation peaks in the blue spectral range, a major one at 395 nm and a minor one at 475 nm. Its emission maximum is at 509 nm in the green range of the visible light. Its 3D structure is an 11-stranded β-barrel with the chromophore in the center (Figure 4.1). The chromophore is formed by a complex of three amino acids Ser^{65}–Tyr^{66}–Gly^{67}. It has the advantage that no other protein or cofactor is necessary for its function, although oxygen is needed for the maturation of the chromophore. Furthermore, it can be genetically engineered and easily be transferred to a variety of other cells. There it can act as a fluorescent marker for specific proteins to which it is fused. There is another advantage of *genetic modification*. The structural robustness of GFP allows a modification with enhanced fluorescence, the so-called enhanced green fluorescent protein (EGFP), to be designed. Furthermore, proteins emitting other wavelengths have been prepared by exchanging the residues of particular amino acids. Variants emitting blue (BFP), cyan (CFP), yellow (YFP), and red (mRFP1) are now available. The high scientific impact of the exploration of the structure and function of the GFP and its genetic modification in life sciences has been honored with the Noble Prize in Chemistry to Martin Chalfie, Osamu Shimomura, and Roger Y. Tsien in 2008.

4.2
Basic Principles

Genetic Design of Reporting Cells The breakthrough in the development of whole-cell biosensors came with the first genetically designed *reporting cells* tailored for specific analytes. Prokaryotes (bacteria), monocellular eukaryotes (e.g., yeast), and even mammalian cells can be modified such that they emit optical or electrical signals or express specific proteins (e.g., enzymes) in response to particular chemical, physical, or biological stimuli. The modification is usually done by means of recombinant DNA technology. Let us first consider the principle of protein expression in a cell (Figure 4.2).

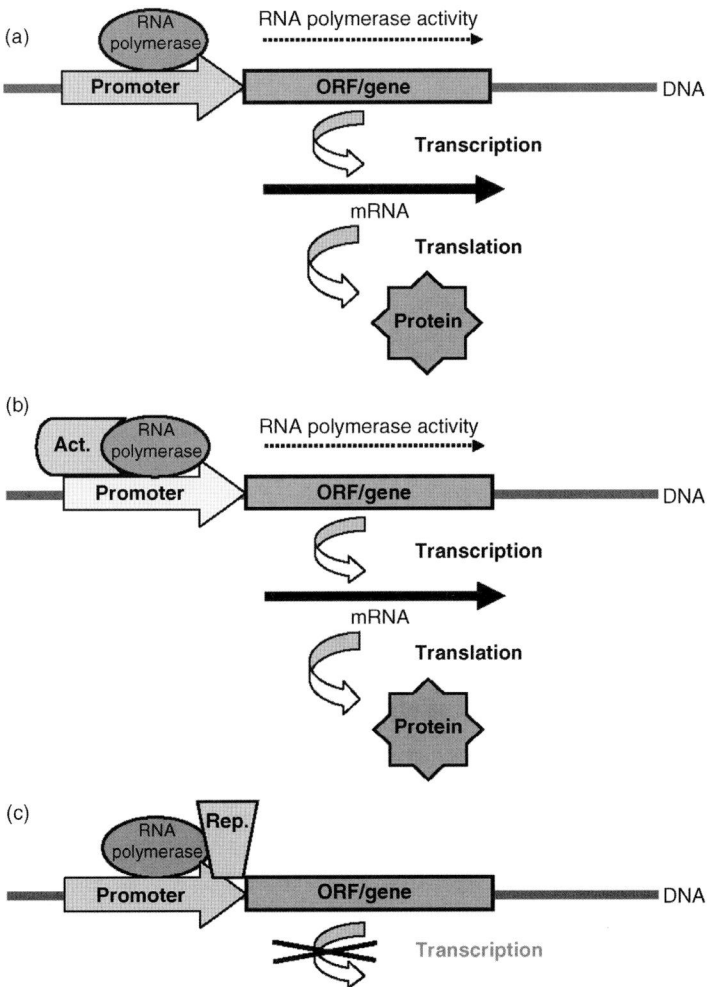

Figure 4.2 Gene expression by controlled transcription. (a) On the DNA level, a promoter adjacent to an ORF of a gene of interest recruits the RNA polymerase, which, in turn, transcribes the information of the DNA into mRNA (transcription). The mRNA is translated (translation) into the amino acid sequence of the coded protein. (b) For a number of promoters, an activator protein (Act.) is required to efficiently recruit the RNA polymerase and transcribe the ORF. (c) Often expression is inhibited by a repressor protein (Rep.) attached to the promoter region, thereby inhibiting RNA polymerase activity.

The information for the assembly of a protein molecule from amino acids is located on the DNA at the gene known as *open reading frame* (ORF) of the protein. The information of the ORF can be translated directly into the primary amino acid sequence of a protein. In front of the ORF, a specific DNA sequence is located, the so-called *promoter*. This sequence acts as a switch on the genetic level. Normally, the promoter sequence recruits the RNA polymerase, which transcribes the ORF/gene into mRNA. This *transcription* is often the target for the control of gene expression. The mRNA as a direct copy of the DNA is *translated* into the respective protein. With a number of genes, an *activator* attached to the promoter is required for an efficient recruitment of the RNA polymerase there. This activator protein interacts with the RNA polymerase, which in turn is able to start transcription, that is, the genetic switch is "on." This leads to gene expression or an "upregulation" of the specific gene and to an increase in the concentration of the gene product (reporter protein). The activity of the activator itself can be modulated, for example, by the presence or absence of specific analytes. On the other hand, gene transcription can be blocked by certain *repressor* proteins. They bind to the promoter, thereby inhibiting RNA polymerase activity and thus prohibiting transcription, which means the genetic switch is "off." The ORF will not be expressed then. This leads to a decrease in the concentration of the respective gene product – the reporter protein. Again, the activity and binding of the repressor to the promoter can be modulated by specific analytes.

If the ORF codes for a marker protein, as a fluorescent protein, for example, the regulation at the transcriptional level can easily be monitored by changing marker protein concentration as revealed by the intensity of fluorescence. The challenge in the design of reporting cells consists in finding promoter sequences acting as

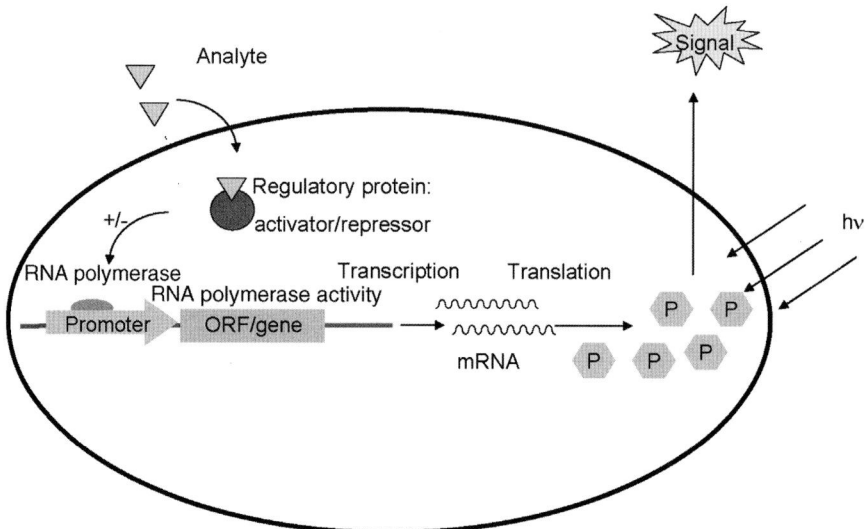

Figure 4.3 Molecular chain of events in a whole-cell biosensor expressing fluorescent proteins induced by a specific analyte. (P=marker protein.) (Adapted with permission from Daunert *et al.* (2000). Copyright 2000 American Chemical Society.)

genetic switches triggered by the presence and absence of specific analytes and in combining these with appropriate marker genes. For the sequence of events along the signal pathway, see Figure 4.3.

In the following section, we will consider examples of regulatory and reporter proteins. They represent a large toolbox available to the designer of whole-cell sensors.

4.3
Bioengineering

The general task for preparation of whole-cell biosensors consists in the genetic design of a sensor cell that responds with specific reporter proteins to a particular external signal and in the integration of a certain number of such viable cells into an appropriate physical or chemical transducer coupled with a signal processing unit (Figure 4.4). As to be shown in the following, for increasing the sensitivity of a whole-cell sensor, possible communication pathways between different cells can be incorporated into the cell assay by using *pheromones* as reporting molecules. A pheromone (from Greek "phero": to bear + "hormone": impetus) is a chemical factor that triggers a (social) response in members of the same species.

The basic principle underlying cellular sensing offers the following two concepts regarding the genetic construction of respective biosensor cells:

i) *Implementing a complete molecular recognition complex from a natural biosensing cell into a host cell*

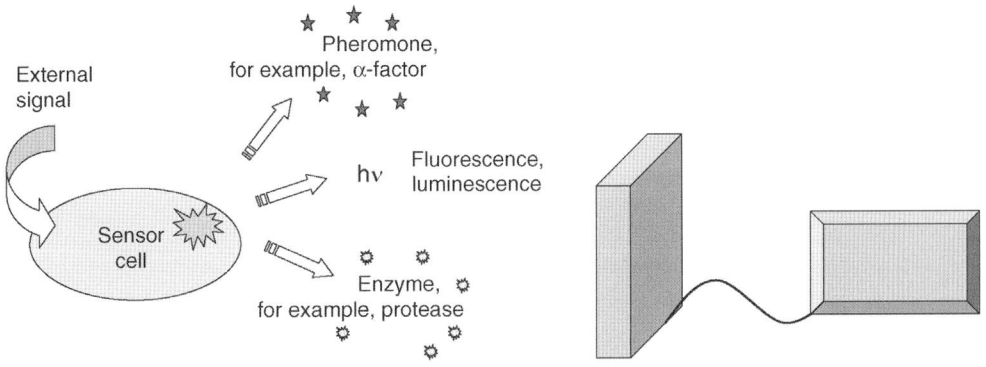

Signaling cells ⟶ Reporting proteins ⟶ Transducer ⟶ Signal processing

Figure 4.4 Main building blocks of a whole-cell sensor. An external signal is transformed by sensor cells into reporter proteins. They are detected by physical or chemical transducers (e.g., optical, electrical, and electrochemical) linked with an integrated signal processing unit.

As an example, the primary components of the mammalian olfactory signaling pathway have been transferred to the wild-type yeast strain *Saccharomyces cerevisiae* YPH501 (Radhika *et al.*, 2007). There the olfactory receptor signaling was coupled to green fluorescence expression. Such "olfactory yeast" can, for instance, indicate the presence of explosive warfare agents polluting the environment.

ii) *Combining a regulatory DNA sequence with a reporter gene in order to create a readable response of the cell to a specific signal*

Very often it is intended to express the GFP (or corresponding modified variants, see above) in response to different signals. This is realized by combining the *EGFP*-encoding DNA sequence (= ORF) with a signal-responsive/specific regulatory element (= the promoter/genetic switch). One example of such combination is given in the following section. For developing engineered whole-cell biosensors, a large toolbox of molecular building blocks can be found. According to the mode of action explained above, the biosensing cells can be regarded as consisting of sensing elements and reporting structures. Whereas the sensing elements control the selectivity of the biosensing cell, the reporting structures determine the sensitivity.

Sensing Elements The term "sensing elements" covers the *regulatory proteins* or their genes and the related promoters. There is a large variety of exogenous and endogenous signals (Table 4.1).

Reporter Proteins The expression of the reporter gene provides a measure for the strength of the analytical signal. The signal has to stand out from the background of other endogenous substances. Recognition can be realized, for instance, by a fluorescent protein, an enzyme, a receptor, or an antibody. Reporter genes and proteins have been thoroughly investigated with the aim to identify cell signaling mechanisms and the possible cross talk between various intracellular information

Table 4.1 Examples of promoters and regulatory genes in bacterial whole-cell-sensing systems.

Target response/analyte	Promoter	Regulatory gene
Oxidative damage (hydrogen peroxide)	*katG*	*oxyR*
Growth limitation	*uspA*	Unknown
Osmotic pressure	*osmY*	*rpoS*
Phosphate starvation	*phoA*	*phoB, phoM, phoR, phoU*
Nitrogen starvation	*glnA*	*glnB, glnD, glnG, glnL*
Antimonite/arsenite	*ars*	*arsR*
Copper	*cupI*	*AceI*
Lead	*pbr*	*pbrR*
Mercury	*mer*	*merR*
Toluene	P_u	*xylR*

Source: Adapted from Daunert *et al.* (2000).

Table 4.2 Examples of reporter proteins, their chemical reactions, and means of detection.

Reporter protein	Reaction	Detection
Firefly luciferase	Luciferin $+ O_2 +$ ATP $\rightarrow$ oxyluciferin $+$ AMP $+ P_i + h\nu$ (550–575 nm)	Bioluminescence
Bacterial luciferase	$FMNH_2 +$ R-CHO $+ O_2 \rightarrow$ FMN $+$ $H_2O +$ RCOOH $+ h\nu$ (490 nm)	Bioluminescence
Green fluorescent protein	Formation of an internal chromophore	Fluorescence
β-Galactosidase	Hydrolysis of β-galactosides	Colorimetric, fluorescence, electrochemical, chemoluminescence

Source: Adapted from Daunert *et al.* (2000).
$FMNH_2$, reduced flavin mononucleotide; R-CHO, long chain fatty aldehyde.

pathways. For instance, by using mammalian cell-based biochips, this has contributed to the exploration of new targets of human diseases and derived therapies and to the development of biological screening techniques for novel drugs. Several reporter proteins relevant for whole-cell biosensing are given in Table 4.2.

Luciferase is a name for any enzyme that catalyzes a light-emitting reaction. Firefly luciferase catalyzes the oxidation of benzothiazolyl-thiazole luciferin to oxyluciferin in the presence of ATP, O_2, and Mg^{2+} producing CO_2, phosphate ions, and light. For bacterial luciferase, the substrate of the catalytic oxidation is a reduced flavin mononucleotide ($FMNH_2$) and a long-chain fatty aldehyde (R-COH). The advantage of the GFP is its autofluorescence, which means it does not need a cofactor or an exogenous substrate to produce light. For β-galactosidase (β-Gal), several detection methods have been developed. The advantage of a colorimetric assay is its simplicity. However, it is not very sensitive. Other methods with higher sensitivity, such as those based on chemoluminescence and electrochemical phenomena, have been applied.

Bioluminescent and Fluorescent Assays The impact of organic and inorganic noxious substances on living cells poses an omnipresent health hazard; hence, it is necessary to evaluate and quantify it. The monitoring of critical changes of our environment (e.g., the risk management of drinking water resources) and the evaluation of the risk potential of novel products (e.g., nanomaterials such as carbon nanotubes, fullerenes, or precious metal nanoparticles) are efforts aimed at meeting this demand. The conventional chemical and physical diagnostics have turned out to be not fully adequate to the problem because of the complexity of biological responses to the exposure to noxious substances. Whole-cell sensors are of interest also for medical diagnostics (point-of-care diagnostics) or for the control of microbiological processes in bioreactors. The diagnostics have to be done with the aid of living organisms, preferably nonvertebrates in order to avoid animal tests, which are cumbersome, expensive, and questionable from an ethic point of view.

Microorganisms equipped with a specific biosensing pathway are well suited for such purposes. Among such test systems, whole-cell sensors based on bioluminescent bacteria have been established as a successful screening method. This development had been initiated by making use of the luminescent strain of the bacterium *V. fischeri* as it has proven to be sensitive across a broad spectrum of chemicals (Girotti *et al.*, 2008). With the progress in genetic engineering of microorganisms, the tools have become available to modify nonlight-emitting organisms living in particular habitats into luminescent cells with a specific response to a contaminant of interest. Today, the dominating reporter genes used in genetically modified microorganisms are the *lux* genes from bioluminescent bacteria or firefly (*Photinus pyralis*) and the *gfp* gene from the fluorescent jellyfish (*A. victoria*). Advantages of *bioluminescent bacterial assays* for toxicity screening are their comparatively rapid response, sensitivity, reproducibility, and cost-efficiency. They can be used for the assessment of toxicity in water, sediments, and soils. Preferably bacterial strains taken from the habitat to be tested are transformed into bioluminescent cells by genetic manipulation, as it can be assumed that the vitality of the modified organisms is optimal in their natural environment.

Regardless of the fact that most whole-cell biosensors reported in literature are based on bacteria, *genetically modified yeast cells* can overcome some disadvantages of bacteria-based biosensors. The following problems can arise with bacterial assays:

i)   Bacteria are often fragile with respect to typical environmental conditions. This limits the available in-use time and storage options.
ii) The tolerable ranges of temperature, pH values, and osmotic pressure are essential limits for the operating parameters.
iii) The physiological behavior of prokaryotic cells such as bacteria differs much from that of eukaryotes such as yeast cells and human/mammalian cells.

Advantages of yeast cells are as follows:

i)   In comparison to bacteria they are more robust. For example, they can be stored after air-drying (at 28–40 °C) longer than 1 year in a nitrogen atmosphere or under vacuum conditions without essential loss of activity.
ii) With their complete genome revealed, the two model organisms *S. cerevisiae* (the so-called baker's yeast) and *Schizosaccharomyces pombe* (fission yeast) offer the best preconditions for a directed genetic engineering of yeast cells with a specific response to external signals. Examples of possible promoters and reporter proteins used are given in Tables 4.1 and 4.2, respectively.

With the following example of the design of yeast cell sensors to detect the biological availability of macronutrients in bioreactors, the general concept of genetic modification of yeast cells can be explained in detail. The availability of the three main macronutrients nitrogen, phosphorus, and sulfur had to be monitored by measuring the intensity of fluorescence signals. Based on studies of (Boer *et al.*, 2003) promoter fragments of "signature" genes are known that indicate limitation of one of these nutrients in *S. cerevisiae*. Examples are given in Table 4.3. Under limitation, the corresponding genes are upregulated or downregulated.

Table 4.3 Examples of *S.cerevisiae* genes responding to nutrient limitation.

	Gene	
Nutrient limitation	Upregulated	Downregulated
Nitrogen limitation	GAP1, DAL5	GNP1
Phosphorus limitation	PHO11, TMA10	RPS22B
Sulfur limitation	PDC6, BDS1	SSU1

These respective regulatory sequences that are activated upon nutrient starvation of cells were linked with the *gfp* reporter gene encoding GFP (Gross, 2012). In Figure 4.5, a proof of concept is given for monitoring the availability of nitrogen and phosphorus. With beginning of starvation the cells start to fluoresce.

A two-color assay may be designed by using a second fluorescent protein with a different color, for example, red. Two different promoters that are responsive to the same signal are used for the expression of the fluorescence reporter genes whose excitation and emission spectra are different. For instance, in case of a nutrient-responsive sensor, upon limitation, the promoter can be specifically upregulated or downregulated causing fluorescence increase or decline, respectively. The assay can be realized in two ways. In a mixture of two single-color cell strains, each sensor strain can harbor one detection construct and one fluorescent reporter protein. Alternatively, the genes of different fluorescent reporters can be expressed in the same cell. The performance of a dual-color *S. cerevisiae* sensor with simultaneous expression of the genes for green and red fluorescent reporter proteins in the same cell is shown in Figure 4.6.

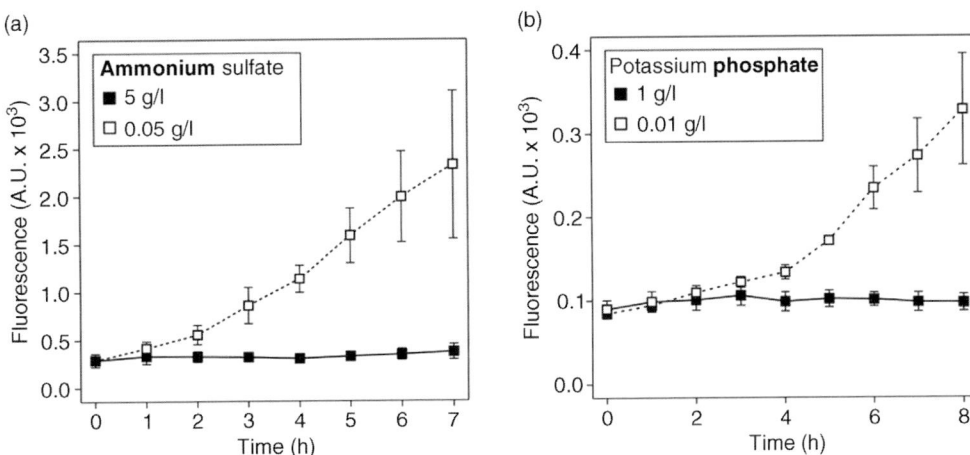

Figure 4.5 Yeast cell sensor for the monitoring of nitrogen and phosphate deficiency. Genetically modified *S. cerevisiae* cells respond to nitrogen (a) or phosphate (b) deficiency with the expression of GFP. The change of fluorescence (A.U.=arbitrary units.) in the limited media (filled icons) is indicated in comparison to the nonlimited medium (empty icons) with increasing time. (Reproduced with permission from Gross (2012).)

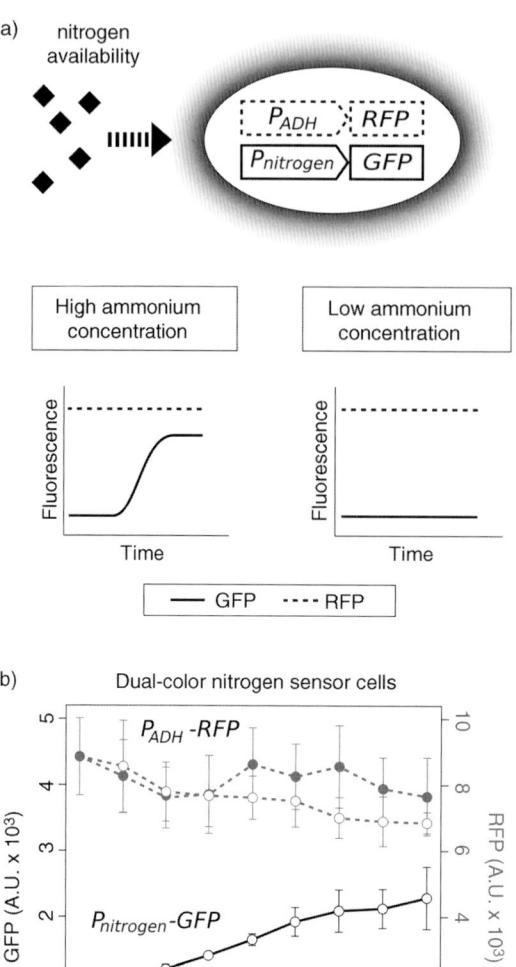

(a) nitrogen availability

P_{ADH} : RFP

$P_{nitrogen}$ GFP

High ammonium concentration

Low ammonium concentration

Fluorescence

Time

Fluorescence

Time

—— GFP ···· RFP

(b) Dual-color nitrogen sensor cells

P_{ADH} -*RFP*

$P_{nitrogen}$-*GFP*

GFP (A.U. × 10³)

RFP (A.U. × 10³)

Time (h)

● high ammonium concentration

○ low ammonium concentration

Figure 4.6 Evaluation of a dual-color nitrogen *S. cerevisiae* sensor. Nitrogen sensor strain BY4741 harboring a detection plasmid of a GFP was transformed with a plasmid of a red fluorescent protein (RFP) yielding dual-color sensor cells. Time trends of green and red fluorescence of the strain in mineral medium (solid lines) and after shift to nitrogen limitation (dashed lines) are shown. (a) Schematic drawing of the different responses of the two fluorescent proteins due to the different promoters. (b) Experimental data showing the fluorescence of about 10^6 cells. (Reproduced with permission from Gross (2012).)

Communicating Sensor–Actor Systems and Signal Amplification Based on the Yeast–Pheromone System   The biology of yeast cells offers another interesting option for the design of a whole-cell biosensor differing from the usual ones by the feature that signaling unit and the reporter unit are not joined in the same cell. It is based on the fact that yeast can produce two types of haploid cells, named α-cells and a-cells, for the purpose of sexual reproduction. They communicate by exchanging two signal molecules (pheromones). Cells of the mating types a and α secrete a dodecapeptide called a-factor and a tridecapeptide called α-factor, respectively. The signals are perceived by cell surface receptors of cells of the opposite mating type. After binding to the receptor, the related pheromone response pathway is induced. This leads to cell cycle arrest, expression of mating genes, and stretching of the cell into a pear-like shape, the so-called shmoo tip. When combining different cell types in a sensor–actor system, the basic idea consists in the realization of primary sensor units, signal amplifiers, signal transducers, and actor units in separate cells that are capable of joint action. Three different modular design principles are schematically represented below. In the three designs, an analyte-inducible promoter ($P_{Analyte}$) always controls the α-factor-encoding gene *MFα1* in the primary sensor cell. By emitting pheromones, the primary sensor cell can activate transducer cells of mating type a (*MATa*), as shown in Figure 4.7a. The transducer cells respond with the production of fluorescent proteins. The modular design allows signal amplification by addition of a separate *amplifier cell type* to the communication chain (Figure 4.7b). This *MATa* cell is genetically modified in a way that it expresses a high number of α-factor in response to a low number of α-factor in the medium. Furthermore, the *MATa* cells can also fulfill *actuator functions* as, for instance, the expression of an enzyme (e.g., a protease that may permeabilize the shell structure of a drug carrier) (Figure 4.7c).

Gross *et al.* (2011) demonstrated the feasibility of the modular yeast-based sensor systems in a model experiment. Three cell types were kept in cocultures: a cell type AM with constitutive (permanent) expression of α-factor, a cell type FM that expresses α-factor after induction by α-factor, and a cell type FE that expresses EGFP induced by α-factor. The cell fractions were mixed in various ratios. In view of the general scheme shown in Figure 4.7, the AM cells can be considered as a model for the primary sensor cells, triggering the subsequent events in the configurations (a), (b), or (c). The FM cells act as amplifiers as shown in the event chain (b), and the FE cells are the transducer cells expressing EGFP. The measured fluorescence of mixtures of two or three of these cell types consisting of about 10^6 cells was compared with the natural fluorescence of a solution containing FE cells only. Figure 4.8a shows a comparison between the coculture of FM and FE cells and the coculture of AM and FE cells 8 h after mixing. The presence of FM cells did not induce any increase of the natural fluorescence of the FE cells since no α-factor has been produced in the coculture. But the presence of AM cells did, which can be ascribed to the constitutive expression of α-factor. A corresponding effect has been observed with the changing cell shape. The AM cells effect pear shape formation and fluorescence of the FE cells as visible in Figure 4.8b. Interestingly, for a ratio of AM : FE cells as low as 1 : 50, no expression of EGFP is observed after 4 h cocultivation (Figure 4.8c and d). However, the addition of low amounts of "amplifier" cells FM to the AM–FE cocultures leads to a significant increase of the fluorescence (Figure 4.8c and d).

(a)

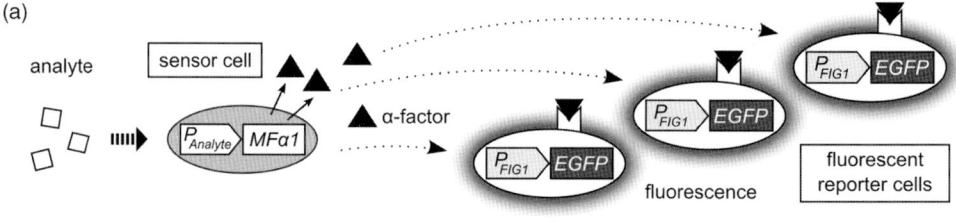

(b)

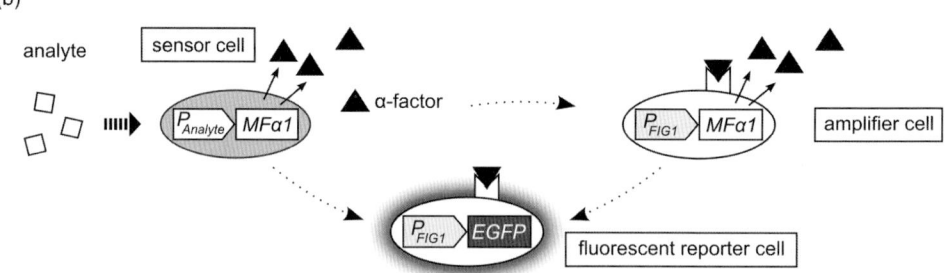

(c)

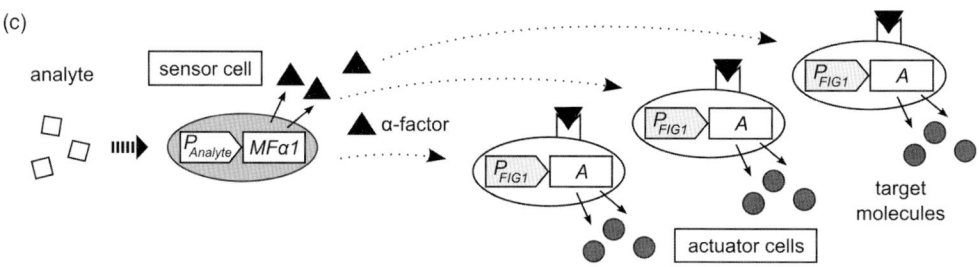

Figure 4.7 Concept of *S. cerevisiae* pheromone-based cell–cell communication for biosensors and sensor–actuator systems. (a) Combination of a primary signal sensor and optical transducer modules: Upon analyte activation, a single primary sensor cell can trigger fluorescence in a number of reporter cells. Sensor cells carry detection constructs comprising an analyte-inducible promoter ($P_{Analyte}$) linked to the α-factor-encoding gene *MFα1*. The pheromone is produced and secreted in response to the analyte. This triggers artificial mating response in nearby reporter cells displaying α-factor receptors and harboring detection constructs for the expression of a fluorescence reporter gene (e.g., *EGFP*) from the pheromone-responsive *FIG1* promoter (P_{FIG1}). (b) Combination of a primary signal sensor, an amplifier, and an optical transducer module: An amplification circuit is generated by additional implementation of cells that link the *FIG1* promoter to the *MFα1* gene. These amplifier cells increase the level of α-factor molecules in response to α-factor. (c) Combination of a primary signal sensor with various actuator modules: The system is highly flexible with respect to the output reporter proteins. In a sensor–actuator setup, actuator cells may produce and release various target molecules (enzymes, metabolites, etc.). (Adapted from Gross *et al.* (2011).)

Patterned Immobilization of Living Cells in Sensor Assays The successful use of whole-cell biosensors depends decisively on a stable immobilization of the cells in an appropriate matrix (Ben-Yoav *et al.*, 2011). It has to facilitate a reliable and fast signal transport and a sufficient exchange of nutrients and metabolic products of the

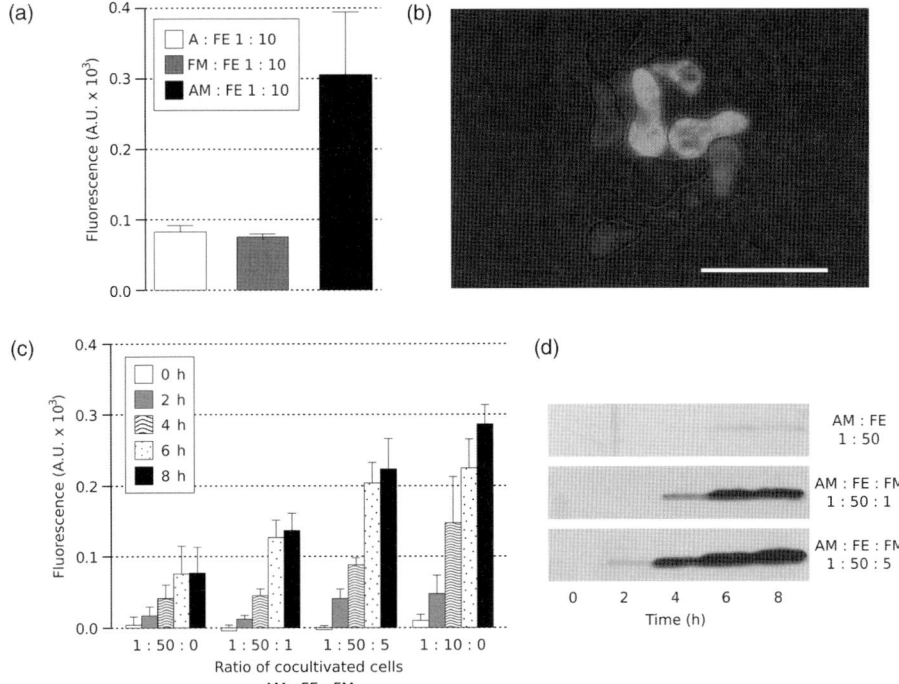

Figure 4.8 Amplification of α-factor signaling with genetically modified *S. cerevisiae* cells. (a) Fluorescence of FE cells (expressing EGFP upon exposure to α-factor) mixed with unmodified a-cells (no α-factor), α-factor-inducible FM cells, or AM cells expressing α-factor constitutively. The signal of 10^6 cells was determined 8 h after mixing. (b) Pear-like shape and fluorescence of FE cells 8 h after mixing with AM cells. Scale bar = 20 μm. (c) Fluorescence of cocultivated cells. AM : FE 1 : 50, AM : FE: FM 1 : 50 : 1, AM : FE: FM 1 : 50 : 5, AM : FE 1 : 10, all normalized against A : FE cells 1 : 10 without α-factor. (d) Western blot analysis of EGFP (approximately 27 kDa) in cocultures. With the western blot, specific proteins can be detected in a given sample of a cell extract. Gel electrophoresis is applied to separate native proteins on the basis of their charge and by 3D structure. The proteins are transferred to a membrane where they are identified using antibodies specific to the target protein. (Reproduced with permission from Gross (2012).)

cells. Cells may be immobilized on a membrane that is attached to the surface of a physical transducer, or they can be brought in between a double-membrane sandwich that is attached to the transducer surface. As another option, embedding the cells in a porous polymer such as agarose, alginate, gelatin, or silica gel yields a more reliable design. Figure 4.9 shows a related example. A glass substrate is covered with a cationic polyelectrolyte (e.g., poly-L-lysine) to get a positively charged base layer for the following deposition steps. First, a droplet of cells solved in an aqueous alginate solution is placed on the substrate. The alginate solution provides best conditions for cell vitality. The outer region of the alginate can be hardened by intrusion of Ca^{2+} ions (for details of that process, see Section 6.3.2.2). The outer

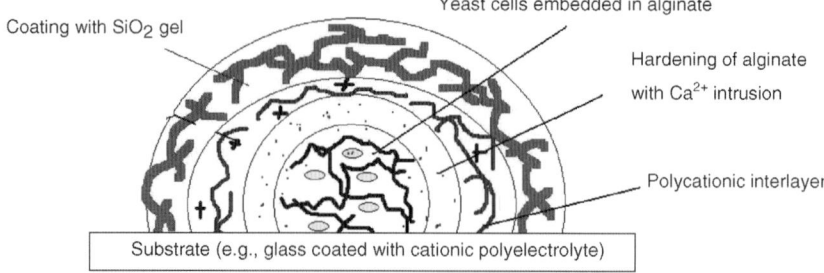

Figure 4.9 Structure of a multishell coating of yeast cells immobilized on a glass substrate. A droplet of yeast cells immersed in an aqueous alginate solution is stabilized by a porous three- layer shell composed of an alginate layer cross- linked by Ca^{2+} ions, a cationic polyelectrolyte coupling layer, and a silica gel layer for mechanical stability.

shell consists of two layers. The silica gel gives the system the necessary mechanical stability without limiting the necessary molecular transport through the nanoporous structure. The intermediate layer of the cationic polyelectrolyte provides sufficiently strong connection of the silica gel to the underlying negatively charged alginate.

As a final step toward application, whole-cell sensors usually have to be integrated into a microfluidic system by deposition on a suitable carrier (usually a plane substrate). Well-proven techniques from microfabrication are available for this purpose (Weibel *et al.*, 2007). Hence, cells and their extracellular environment can be manipulated with high reliability at micrometer and nanometer scale. Plotting and printing are the two options suitable for the deposition of both biomolecules and living microorganisms. Plotting is based on the deposition of droplets of a few nanoliters on a substrate. This facilitates the generation of patterns with characteristic structure lengths of a few hundreds of micrometers (Figure 4.10a). The nanoplotter is enclosed in a chamber in order to maintain sufficient humidity during the process. The preparation of various precursor solutions (e.g., alginate with embedded yeast cells or silica sol for the outer layer) allows the deposition of composite structures. Typically, there are 10^2–10^3 cells per droplet. Embedding the cells in a low-viscous aqueous solution of alginate enables a soft passage of the cells through the capillary of the piezoelectric pipette. Multilayered structures can be produced by a stepwise deposition of the different layers. Figure 4.10b shows a pattern of yeast cells on a glass substrate, with spot size 0.5 mm.

The other option, microcontact printing, can be performed with a resolution of about 1 μm. The key element is a structured stamp usually made of poly(dime- thylsiloxane) (PDMS). The structure details on the stamp surface typically have widths of 1–1000 μm and heights of 0.1–100 μm. Microcontact printing is usually applied for the transfer of molecules; however, it can be applied also for the transfer of cells. The stamp is coated with a solution of the molecules to be transferred. After short drying of the coated stamp, the structure is stamped on a substrate surface, provided that the "ink" tends to stick more tightly to the substrate than to the stamp. The sticking tendency can be modified by coating the substrate with a suitable

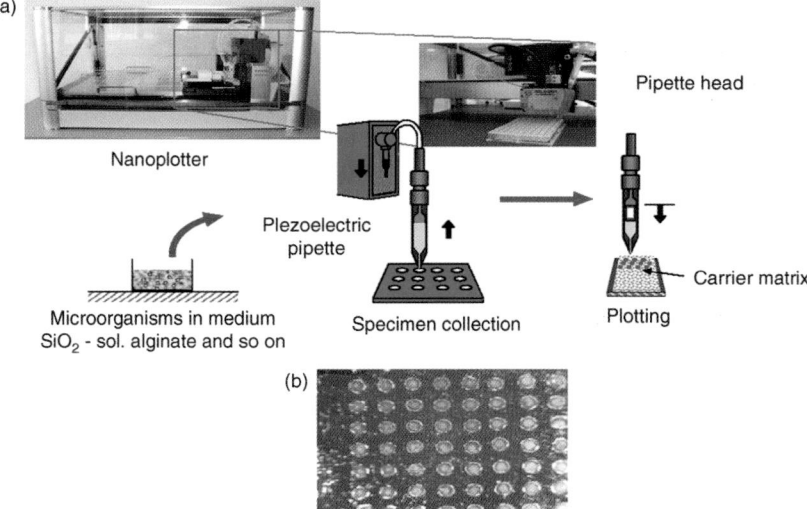

Figure 4.10 Placement of yeast cells with a nanoplotter equipped with a piezoelectric pipette. (a) Setup of the nanoplotter (www .gesim.de/upload/PDFs/NP-En-20). It allows the placement of living cells embedded in a multishell coating on a glass substrate. (b) Droplets harboring about 100 yeast cells each, of size 0.5 mm, on a glass substrate.

substance, a self-assembling monolayer (SAM) with terminal ligands. Such kind of technique, with a hydrogel stamp made from agarose, was used for the patterned immobilization of luminescent *V. fischeri* on an agar surface (Weibel *et al.*, 2005). After "inking" the stamp with an aqueous suspension of the bacteria, the excess liquid was absorbed by the stamp. The hydrogel agarose keeps the cells hydrated during the stamping process.

References

Ben-Yoav, H., Melamed, S. *et al.* (2011) Whole-cell biochips for bio-sensing: integration of live cells and inanimate surfaces. *Critical Reviews in Biotechnology*, **31** (4), 337–353.

Boer, V.M., de Winde, J.H. *et al.* (2003) The genome-wide transcriptional responses of *Saccharomyces cerevisiae* grown on glucose in aerobic chemostat cultures limited for carbon, nitrogen, phosphorus, or sulfur. *The Journal of Biological Chemistry*, **278** (5), 3265–3274.

Daunert, S., Barrett, G. *et al.* (2000) Genetically engineered whole-cell sensing systems: coupling biological recognition with reporter genes. *Chemical Reviews*, **100** (7), 2705–2738.

Girotti, S., Ferri, E.N. *et al.* (2008) Monitoring of environmental pollutants by bioluminescent bacteria. *Analytica Chimica Acta*, **608** (1), 2–29.

Gross, A. (2012) *Genetically tailored yeast strains for cell-based biosensors in white biotechnology.* PhD thesis. Technical University, Dresden.

Gross, A., Rödel, G. *et al.* (2011) Application of the yeast pheromone system for controlled cell–cell communication and signal amplification. *Letters in Applied Microbiology*, **52** (5), 521–526.

Radhika, V., Proikas-Cezanne, T. *et al.* (2007) Chemical sensing of DNT by engineered olfactory yeast strain. *Nature Chemical Biology*, **3** (6), 325–330.

Weibel, D.B., Diluzio, W.R. *et al.* (2007) Microfabrication meets microbiology.

Nature Reviews Microbiology, **5** (3), 209–218.

Weibel, D.B., Lee, A. *et al.* (2005) Bacterial printing press that regenerates its ink: contact-printing bacteria using hydrogel stamps. *Langmuir*, **21** (14), 6436–6442.

5
Biohybrid Silica-Based Materials

5.1
Case Studies

Silica plays an outstanding role in nature as one constituent of structure-forming materials. Amorphous hydrated silica can be found in many organisms such as diatoms, protists, sponges, mollusks, and higher plants. The beautiful architecture of diatoms as well as the remarkable mechanical and physical properties of spicules of marine glass sponges can serve as examples for the richness of biological evolution. In nature, biogenic mechanisms have been developed that are very different from the traditional man-made high-temperature processes of glass manufacturing by melting sand and additives to get silica products of complex shape. The natural source for silica formation is orthosilicic acid, $Si(OH)_4$. Its concentration is typically in the range between 1 and $100\,\mu M$ in the natural environment. The biogenic silica formation in diatoms and marine sponges represents two basic mechanisms that play a role in biomineralization: biotemplating and biocatalysis (Sumper and Brunner, 2008; Ehrlich, 2010; Jeffryes et al., 2011).

The *unicellular diatoms* shape the silica into hierarchical structures. The biosilica wall is composed of two Petri dish-like shells with slightly differing sizes (Figures 5.1 and 5.2). Each shell (theca) consists of a valve and girdle bands (Figure 5.2). The basic silica-forming units are the silicon acid transporters (SITs) and the silica deposition vesicles (SDVs) . Multiple SITs were found in each diatom species. The SITs are transmembrane proteins that interact with silicic acid without catalyzing its polymerization. According to the so-called alternating access model for SIT-mediated silicic acid transport, the uptake of silicic acid occurs by conformation change of the SIT proteins. In an extracellularly facing conformation, silicic acid is bound at a specific motif of amino acids (GXQ: glycine, various amino acids, and glutamine). After a conformational change, there are other GXQ residues adjacent to other transmembrane segments of the SIT that enable the release of the silicic acid into the cell (Hildebrand et al., 2006). By the regulation of expression and activity of SITs during the cell cycle, the uptake of silicic acid is controlled as required for cell wall formation. For cell proliferation, the new cell wall structures are formed inside the SDVs (Figure 5.2). The SDVs are a kind of "reaction vessels" that contain various biomolecules

Bio-Nanomaterials: Designing Materials Inspired by Nature, First Edition. Wolfgang Pompe, Gerhard Rödel, Hans-Jürgen Weiss, and Michael Mertig.
© 2013 Wiley-VCH Verlag GmbH & Co. KGaA. Published 2013 by Wiley-VCH Verlag GmbH & Co. KGaA.

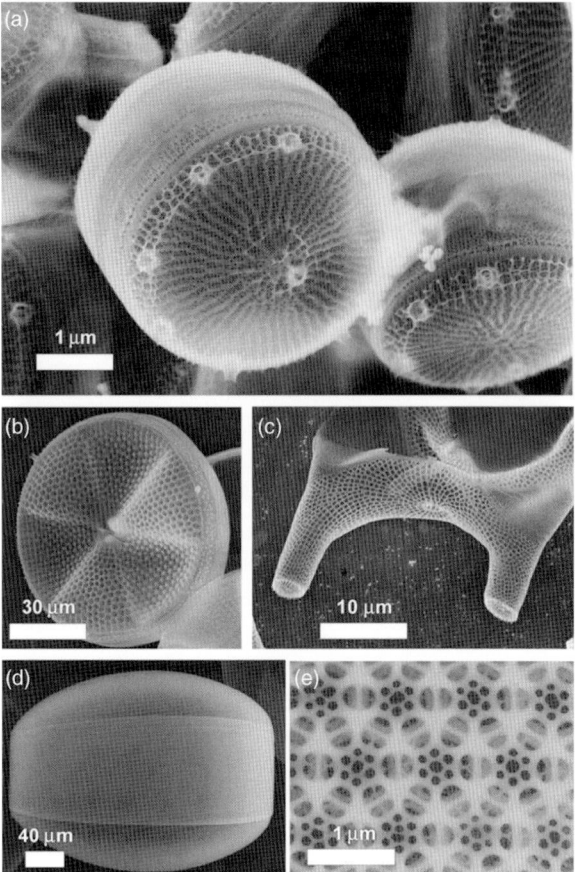

Figure 5.1 SEM images of a few representative diatom genera. (a) *T. pseudonana*; (b) *Actinoptychus* sp.; (c) *Eucampia zodiacus*; (d and e) *Coscinodiscus granii*, complete cell wall and high-resolution image of the valve patterning. (Reproduced with permission from Sumper and Brunner (2008). Copyright 2008, Wiley-VCH Verlag GmbH.)

(polyamines and specific polypeptides (silaffins and silacidins)) that control the formation of silica needed for the new cell wall (for more details of the process of silica formation in the SDV, see Section 6.2.7). In the first step, two valve SDVs produce two new silica valves complementary to the two existing valves. In the second step, the new girdle bands are deposited in the girdle band SDV simultaneously with plasma growth after cell separation.

Recently, Kröger and coworkers (Scheffel *et al.*, 2011) described the discovery and intracellular location of six novel proteins (cingulins) that are integral components of a silica-forming organic matrix (microrings) in the diatom *Thalassiosira pseudonana*. The cingulin-containing microrings are specifically associated with girdle bands, which constitute a substantial part of diatom biosilica. Remarkably, the microrings

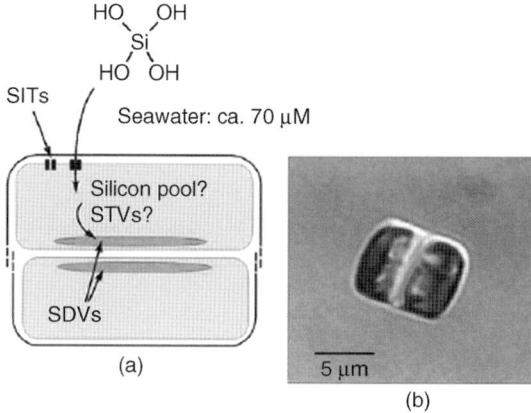

Figure 5.2 Schematic description of silicon uptake and cell wall formation in diatoms during cell division (a: SIT, silicon transporter; STV, silicon transport vesicle; SDV, silica deposition vesicle) and fluorescence microscopic image of a *T. pseudonana* cell during cell division (b). (Reproduced with permission from Sumper and Brunner (2008). Copyright 2008, Wiley-VCH Verlag GmbH.)

exhibit protein-based nanopatterns that closely resemble characteristic features of the girdle band silica nanopatterns. Intriguingly, also structural polysaccharides such as β-chitin in the form of filigree nanofibrillar networks have been shown for the first time to be present within siliceous cell walls of *T. pseudonana* (Brunner *et al.*, 2009). These network-like scaffolds resemble the diatom cell walls in size and shape and consist of cross-linked fibers with an average diameter of approximately 25 nm. The diameter of the fibers varies between approximately 5 and 50 nm. It is tempting to speculate that the chitin-based networks provide the scaffold structure for silica deposition while other biomolecules possess real templating activity and deposit silica on these superstructures in analogy to well-known calcium carbonate bio-mineralization processes. It is also possible that the chitin-based networks are necessary to mechanically stabilize the cell walls.

Diatoms produce a huge amount of organic compounds from CO_2 by photo-synthesis (about 20% of global carbon fixation and 40% of marine primary productivity on Earth). Therefore, they also contribute the main part to the biogenic Si cycle. In this connection, it is interesting that Gautier *et al.* (2006) have shown that some diatom strains are able to dissolve amorphous silica. A surprisingly high initial dissolution velocity of about 1 μm/h was observed after entrapment of the alga *Cylindrotheca fusiformis* in porous silica gel. *Marine siliceous sponges* are another example for producing biogenic silica. They form needle-like spicules of silica that stabilize the organism, are responsible for anchoring of the sponge body into the sandy bottom, and provide defense against predators. The spicules are also pro-duced in membrane-enclosed vesicles. Each spicule contains axially aligned proteins or chitin, which may serve as a mechanical reinforcement. The protein also fulfills an essential task in the formation process of the spicule. First of all, there is the

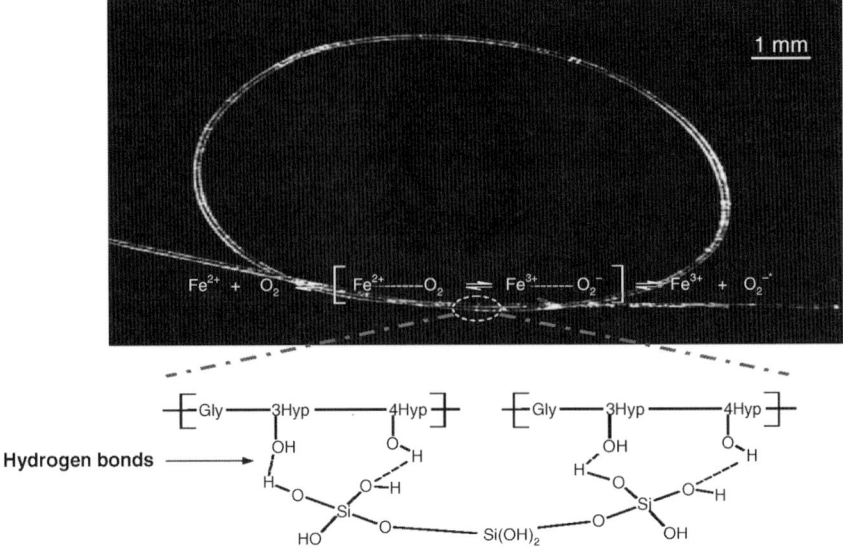

Figure 5.3 Demonstration of the flexibility of the slender stalk by which the glass spong *Hyalonema sieboldii* is attached to the ground. The stalk is a giant spicule whose strength is due to the plywood-type organization of collagen nanofibrils in the silica matrix. (Courtesy by H. Ehrlich.)

possibility that the protein acts as a template for silica deposition similar to the diatoms.

Recently, Ehrlich and coworkers (Ehrlich, 2010; Ehrlich *et al.*, 2010) have shown that the basal spiculae of glass sponge *Hyalonema sieboldi* (length 50 cm, diameter typically 0.5–1.5 mm) contain twisted multifibrillar collagen (overall diameter 1–5 μm, diameter of the microfibrils 10–25 nm). The collagen fibrils are arranged in plywood architecture. Starting from an inner axial channel containing an axial filament, coaxial cylinders of silicified 2D collagen networks form the spicule. The layers are about 1 μm in thickness and are connected to each other by protein fibers. Using slow-alkali etching of the spicules, hydroxylated fibrillar collagen that contains an unusual Gly–3Hyp–4Hyp motif was isolated. Ehrlich and coworkers have shown that under *in vitro* conditions, isolated glass sponge leads to an unusual high deposition rate of amorphous silica in a solution of orthosilicic acid. Obviously, the hydroxyl groups of the Gly–3Hyp–4Hyp motif create hydrogen bonds with the hydroxyl groups of polysilicic acid as shown in Figure 5.3. The hydroxylated collagen fibrils appear to be the basis for the extraordinary mechanical and optical properties of the spicules. Interestingly, this example of collagen-based biosilicification takes place under deep-sea conditions at temperatures between 0 and 4 °C.

Evolution has invented another efficient way of silica deposition, as became apparent by analyzing the spicules of the siliceous sponge *Tethya aurantia* (length about 2 mm, diameter about 30 μm) (Shimizu *et al.*, 1998). Treatment with 2 M HF/8 M NH$_4$F (pH 5) dissolved the silica and bared the axial protein filaments (length 2 mm, diameter 1 μm). Isolated spicules and axial filaments are shown in Figure 5.4.

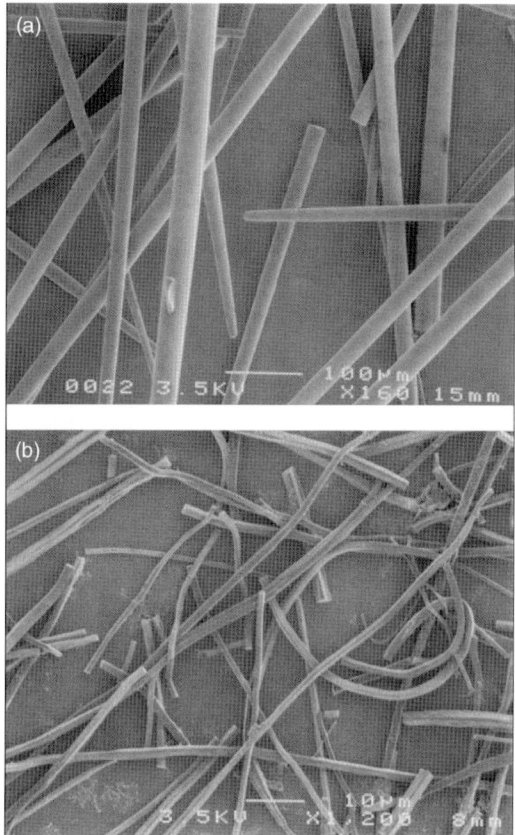

Figure 5.4 Scanning electron micrograph of isolated silica spicules (a) and filaments (b) from *T. aurantia*. (Reproduced with permission from Shimizu *et al.* (1998). Copyright 1998, National Academy of Sciences, USA.)

The protein filament consists of three proteins, designated *silicateins* α, β, and γ, with ratio $\alpha:\beta:\gamma = 12:6:1$. The amino acid compositions of the three proteins are highly similar. They form a repetitive structure within the filament with a periodicity of 17.2 nm. Silicatein α shows a higher similarity to the enzymes of the cathepsin L subfamily of papain-like cysteine proteases. Silicatein α is characterized by clusters of the three hydroxyl-containing amino acids – serine, tyrosine, and threonine. There are strong indications that these highly localized concentrations of hydroxyls act as *catalytic centers* for the condensation reaction of silicic acid. The enzymatic activity has been demonstrated by *in vitro* experiments with natural silicatein as well as with the recombinant protein. However, it seems to be lower than the physiological *in vivo* activity observed in animals. In living animals, a growth rate of spicules up to 5 μm/h has been observed.

Schroeder *et al.* (2006) reported in a study on spicule formation in the marine demosponge *Suberites domuncula* that in addition to silicatein, there is a second

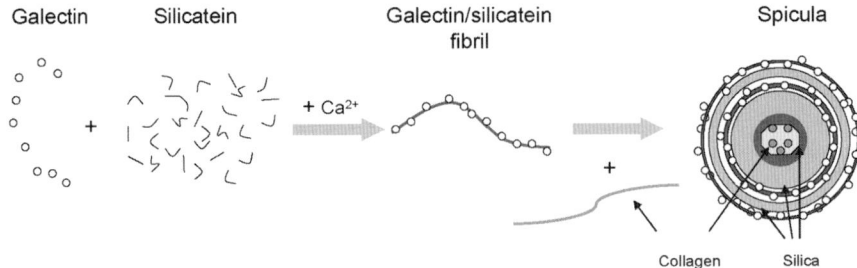

Figure 5.5 Growth of spicules from demosponges. Galectin molecules associate in the presence of Ca^{2+} to fibrils that allow binding of silicatein molecules. Collagen fibers orient the silicatein–galectin fibrils concentrically around the growing spicules. Biosilica deposition is mediated in two directions, originating both from the silicatein–galectin fibrils and from the surface of the core structure consisting of axially aligned collagen fibers. Finally, the third silica lamella is formed that is layered onto the previous two lamellae. (Adapted from Schroeder et al. (2006).)

protein, galectin, that helps catalyzing the spicule formation. Spicules are initially formed within sclerocytes, which are specialized cells for spicule nucleation. When the spicules reach a length of about 10 μm, they are extruded from the cells and completed extracellularly (with a final length of about 200 μm). The spicules have a layered cross section. During the first stage of the intracellular synthesis, the first silica layer is grown with a channel filled with a compact axial filament (thickness about 1–2 μm). The channel also contains very thin protein strings in the space around the filament. During the extracellular completion, two additional silica layers are deposited concentrically around the central fiber. Obviously, the growth of that structure is guided by a complex protein network. By applying a mild procedure for silica dissolution, with a lysis buffer supplemented with 4 M urea, the various protein filaments have been isolated. It has been shown that the network is composed of a very fine string- and net-like structure (fiber diameter between 0.5 and 4 nm) as well as solid rods (diameter between 10 and 14 nm). The fine strings are composed of associates of the two enzymes, *silicatein α* and *galectin 2*. Galectin 2 is a lectin with two galactose binding sites and a hydrophobic region at the N-terminus. It is able to self-associate in the presence of Ca^{2+}. The rods are formed by fibrillar collagen. Based on these observations, the following model for the extracellular growth stage has been proposed (Figure 5.5).

In the extraspicular space, galectin and silicatein are assembled along the spicule nucleus. In the presence of Ca^{2+} ions, galectin molecules form three-dimensional self-associates that allow also the binding of silicatein molecules. The collagen rods oriented parallel to the spicule axis form a frame for two galectin/silicatein networks. These networks are the sites for the catalytic deposition of two concentric silica layers surrounding the primary spicule nucleus. Finally, the third silica film is formed, which is deposited onto the previously formed layers. The model has been additionally supported by *in vitro* experiments where the rate of silica deposition catalyzed by a mixture of the galectin and silicatein exceeded the rate in the absence of one of the two enzymes. Obviously, the silica is the result of the *combined action of catalytic centers and a structure-directing template*. The combination of layered structure with fiber reinforcement explains the remarkable mechanical stability of such spicules.

Recently, Singh *et al.* (2008) have shown that the bacterium *Actinobacter* sp., when stressed by incubating in a 10^{-3} M K_2SiF_6 solution, forms 10 nm sized crystalline Si/SiO_2 nanoparticles in only 48 h in the extracellular space. This behavior differs much from that of the above-mentioned examples. All of them produce amorphous SiO_2 only. The surprising observation of both crystalline Si and SiO_2 leads to the hypothesis that the bacterium expresses nonspecific reductases for the formation of Si as well as hydrolyzing enzymes (oxidases) for the SiO_2 deposition. From earlier reports, it was known that the bacterium synthesizes iron oxide and iron sulfide nanoparticles, which supports the assumption that this bacterium secretes reductases and oxidizing enzymes.

5.2
Basic Principles

5.2.1
Preparation of Silica-Based Xerogels

As outlined above, biogenic silica is formed as nanostructured composites consisting of silica and a biopolymer. Biogenic silica has been grown in aqueous media near neutral pH at room temperature. Therefore, it is not surprising that research aimed at producing man-made silica-based biohybrid materials is intensely pursued. Pioneered by Carturan *et al.* (1989), the great potential of sol–gel techniques for this purpose has been demonstrated in the past 20 years (Böttcher, 2000; Böttcher *et al.*, 2004; Avnir *et al.*, 2006; Ruiz-Hitzky, Ariga, and Lvov, 2008; Depagne *et al.*, 2011). By this type of processes, biomolecules or living cells can be embedded in a nanoporous silica (or other metal oxide) matrix. What are the advantages of such materials? The silica matrix is distinguished by good mechanical, thermal, and photochemical stability, and also high transparency extending into the deep UV region. Its open porosity as well as pore sizes can be controlled in wide ranges. It is nontoxic and biologically inert. Finally, materials based on silica can be produced at room temperature near pH 7, as will be explained in the following.

The Silica Sol–Gel Process The whole process of the formation of silica gel is based on the *condensation reaction* of orthosilicic acid in aqueous solution:

$$(OH)_3Si—OH + HO—Si(OH)_3 \rightarrow (OH)_3Si—O—Si(OH)_3 + H_2O. \qquad (5.1)$$

As a result of this reaction, the *siloxane bond* Si—O—Si is formed. Also, the other three OH groups can take part in the reaction with other silicic acid molecules, thus forming a *three-dimensional network*. However, as silicic acid exists in aqueous solution only in very small concentration, other precursors have to be chosen for

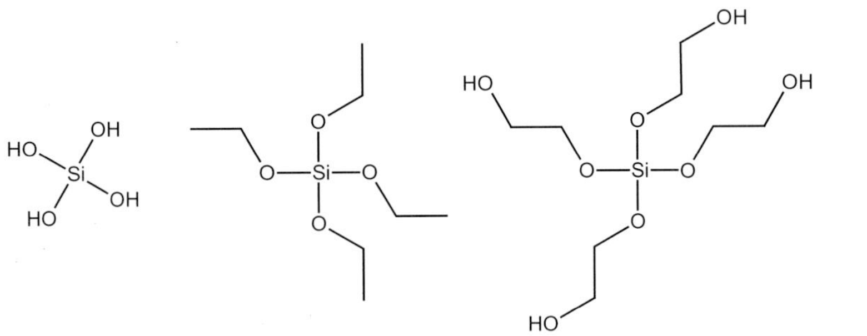

Orthosilicic acid   Tetraethoxysilane   Tetrakis(2-hydroxyethyl) orthosilicate

$Si(OH)_4$ $Si(OC_2H_5)_4$ $Si(OC_2H_4OH)_4$

TEOS   THEOS

Figure 5.6 Structural formulas of silica precursors.

practical use. One option is water glass (Na_2SiO_3), the sodium salt of the silicic acid. For pH <7, gel forms via the reaction

$$Na_2SiO_3 + 2HCl + (x-1)H_2O \rightarrow SiO_2 \cdot xH_2O + 2NaCl. \tag{5.2}$$

Sodium chloride can be a disadvantage in the case of an intended entrapment of biomolecules or microorganisms. Therefore, alkoxides with the general structure $Si(OR)_4$ are the favored precursors for silica gel formation. Here Si is joined via an O-bridge with organic groups R. Examples of often used compounds are given in Figure 5.6.

In contact with water, partial hydrolysis leads to the formation of silanol groups:

$$Si(OR)_4 + nH_2O \rightarrow (HO-)_n Si(-OR)_{4-n} + nR-OH. \tag{5.3}$$

It is a very slow reaction that can be accelerated by shifting pH from neutral to acidic or alkaline because protons or hydroxyl groups act as strong electrophilic or nucleophilic agents, respectively, and catalyze the process. The following *poly-condensation into small clusters* is caused by the silanol groups. There are two possible reaction channels:

$$(RO)_{4-n}Si(OH)_n + (HO-)_n Si(-OR)_{4-n} \rightarrow$$
$$(OH)_{n-1}(RO)_{4-n}Si-O-Si(-OR)_{4-n}(OH)_{n-1} + H_2O, \tag{5.4}$$

$$Si(OR)_4 + (HO-)_n Si(-OR)_{4-n} \rightarrow$$
$$(RO)_3Si-O-Si(-OR)_{4-n}(OH)_{n-1} + HO-R. \tag{5.5}$$

At neutral pH, hydrolysis is slow but condensation is fast. Hence, the size distribution of the nanoparticles in the sol can be controlled via pH. Then the so-called *nanosol* with particle sizes <20 nm is formed. By increasing the

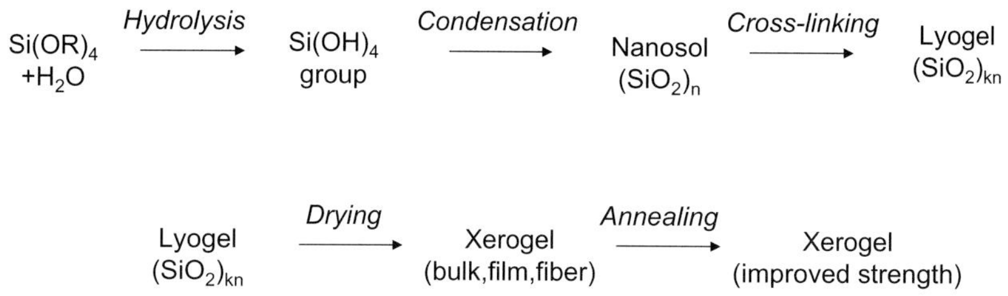

$$\begin{array}{ccccc}
\text{Si(OR)}_4 & \xrightarrow{\text{Hydrolysis}} & \text{Si(OH)}_4 & \xrightarrow{\text{Condensation}} & \text{Nanosol} & \xrightarrow{\text{Cross-linking}} & \text{Lyogel} \\
\text{+H}_2\text{O} & & \text{group} & & \text{(SiO}_2)_n & & \text{(SiO}_2)_{kn}
\end{array}$$

$$\begin{array}{ccccc}
\text{Lyogel} & \xrightarrow{\text{Drying}} & \text{Xerogel} & \xrightarrow{\text{Annealing}} & \text{Xerogel} \\
\text{(SiO}_2)_{kn} & & \text{(bulk, film, fiber)} & & \text{(improved strength)}
\end{array}$$

Figure 5.7 Process of silica gel formation.

temperature or the concentration of the sol, cross-linking of the nanoparticles leads to a highly viscous *lyogel*. The lyogel still contains solvent. In a subsequent *drying* step, the solvent can be completely removed. The porosity and the pore size of the now formed *xerogel* sensitively depend on the drying regime. The xerogel can be produced as monolithic bulk material, thin film, or thin fiber, depending on the processing regime. Thicker films ($>2\,\mu m$) are liable *to shrinkage crack formation*. Shrinkage cracks can be avoided by multiple depositions of thinner films or by polymeric additives. The mechanical properties of the xerogel can be improved by annealing at higher temperature. The whole process of gel formation can be summarized as shown in Figure 5.7.

The chemical and physical properties of the silica-based sol–gel materials can be modified by co-hydrolysis and co-condensation. This is done by mixing the SiO_2 precursors (a) with metal oxide or semimetal nanosols (e.g., using metallic components M = Al, Ti, Zr, and Sn or semimetals B and P) or (b) with alkoxysilanes with different organic residues R (e.g., R = alkyl groups, dyes, polymers, or biomolecules). Networks with mixed composition are formed (Figure 5.8) by covalent linkage in the mixed nanosols. The organically modified gels are also known under the name ORMOCERs, whereas xerogels with biological inclusions are called "biocers" in the following. The embedding of biological species in sol–gel matrices has received considerable attention in connection with the creation of novel biocomposite materials exhibiting the characteristic chemical and biochemical functionalities of enzyme, protein, and other biocomponents. Especially the immobilization of enzymes within modified silica matrices is of great scientific and practical interest for new biocatalysts and biosensors (Pierre, 2004; David, Yang, and Wang, 2011). By avoiding preparation

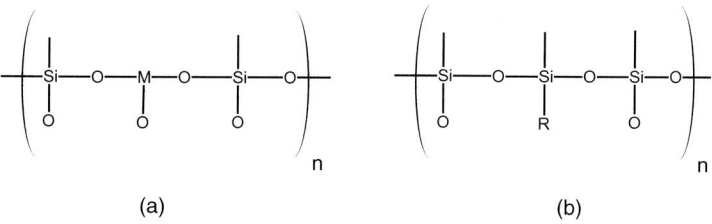

(a)   (b)

Figure 5.8 Co-hydrolysis and co-condensation by mixing the SiO_2 precursor with (a) metal oxides or (b) organic modified siloxanes.

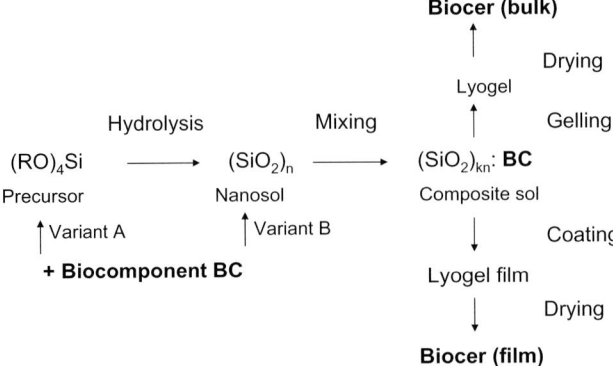

Figure 5.9 Variants of sol–gel entrapment of the biocomponent.

conditions that would lead to denaturation (extreme pH values and high levels of organic solvent in the nanosol), even cellular systems (such as bacteria, yeast cells, algae, etc.) can be embedded and still maintain their viability.

The flexibility of the sol–gel process offers a number of advantages and possibilities for an effective preparation of bioactive composites. Tetraethoxysilane (TEOS) is the commercially most important precursor for silica-based xerogel thin film production. Low materials cost, promising physical and chemical properties, and inexpensive processing techniques such as dip coating, spin coating, spray coating, or continuous casting make it suitable for various applications. One disadvantage appears, however, in connection with biohybrid silica materials: The formed alcohols R—OH during the hydrolysis of TEOS according to Eq. (5.3) can possibly denature the biomolecules or microorganisms entrapped in the gel unless precautions are taken to keep the R—OH concentration low, for example, by evaporation from the nanosol. The principal stages of the entrapment process are indicated in Figure 5.9.

To obtain biocomposites by the sol–gel method, the biocomponent (BC) can be added before or after hydrolysis of the precursor as shown in Figure 5.9. The selection of variant A or variant B depends mainly on the sensitivity of BC against the denaturing effect of R—OH. In variant B, the unfavorable effect of alcohol mentioned above can be avoided, for example, by removing the alcohol from the solvent with a bubbling inert gas stream before adding the BC to the nanosol.

Another interesting way is the use of the completely water-soluble precursor tetrakis(2-hydroxyethyl) orthosilicate (THEOS) (Figure 5.6) or poly(glyceryl silicate), which are obtained by substitution of the ethanol groups in TEOS with ethylene glycol or glycerin (Gill and Ballesteros, 1998). Their hydrolysis, other than that of water-insoluble TEOS, does not require organic cosolvents. In this way, the embedding of BC within the silica matrix can be realized under very mild and biocompatible conditions, especially since ethylene glycol or glycerin produced in hydrolysis are biocompatible water-soluble solvents. Thus, the biocomponents mix with the precursor and their intermediate products homogeneously up to the point of transition into the gel state. Hydrogen bonds are formed between the biopolymers (proteins and polysaccharides) already in the sol stage, which means that larger cross-linked aggregates of silica clusters and biopolymers are formed very early. Gel formation does not result in high shrinkage in this case, owing to the topological constraints, which means that shrinkage stresses remain low. This is a *one-stage process* since it is performed without applying a change of pH in between. The above-mentioned lyogel formation with TEOS is a *two-stage process* since the gel transition has to be triggered by pH change.

Drying of the gel with shrinkage strains of typically about 10% is the critical step in the sol–gel process. Careful process control avoids defolding of proteins and mechanical stress on microorganisms. Various modifications of the basic process have been developed in order to manufacture biocer materials with low shrinkage, high form stability, and various shapes (Figure 5.10). The addition of commercial alumina fibers (diameter 2–3 μm), for example, yields a *fiber reinforcement effect* (variant B in Figure 5.10). An advantage of biocers of the variant B is their excellent mechanical behavior: very high compactness, low shrinkage in drying, and high stability also with biocomponent contents below 5 wt%. Furthermore, the porosity of the material (about 50–60%) ensures a good accessibility of the embedded biocomponent. Residual water within the matrix enables the incorporated cells to live and show a high biocatalytic activity (see also Section 5.3.3). By a simple casting process and drying on air, various shapes with green-body character can be produced that are applicable as filler in bioreactor columns. In Figure 5.11, different forms of fiber-reinforced biocers are shown that have been used as packing material for the purification of industrial wastewater, for instance (see also Section 5.3.3). Because the quantity of biocomponent can be very small for

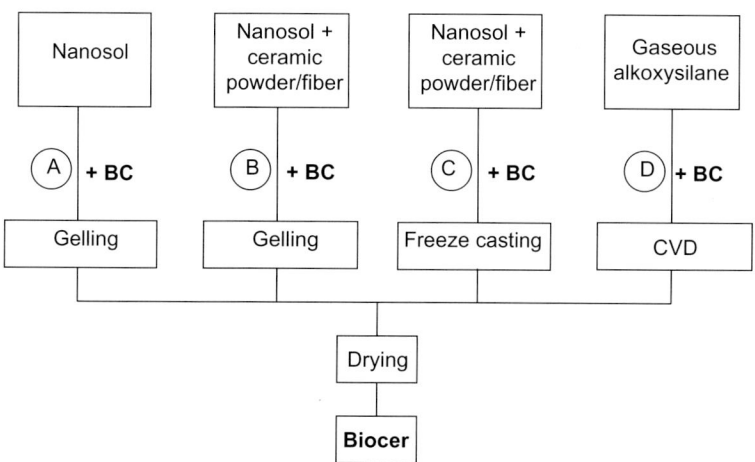

Figure 5.10 Biocer manufacturing: basic process (A) and variants (B, C, and D) with reduced shrinkage.

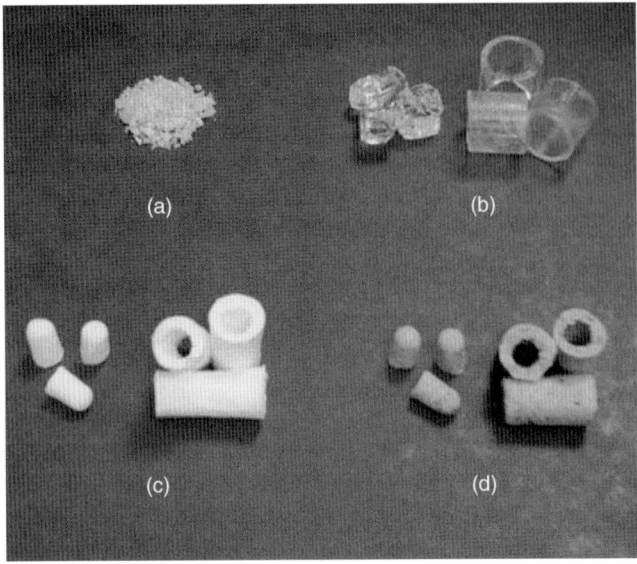

Figure 5.11 Packing materials for bioreactors. (a) Classified biocer powder: *Asp* in SiO_2. (b) Coated glass carriers. (c) Fiber-reinforced biocers: *Asp* in Al_2O_3/SiO_2. (d) Fiber-reinforced biocers: *Cga* in Al_2O_3/SiO_2 (*Asp: Aspergillus* spores; *Cga: Chlorella* green algae). (Reproduced with permission from Böttcher *et al.* (2004). the Royal Society of Chemistry.)

biocatalytic purposes, very inexpensive compositions can be produced by admixture of alumina or mullite powder.

Freeze gelation is another option of preparing complex shapes with almost zero shrinkage (variant C). This technique is also very suitable for biocer preparation. The process scheme is shown in Figure 5.12. A slurry made of an aqueous nanosol and

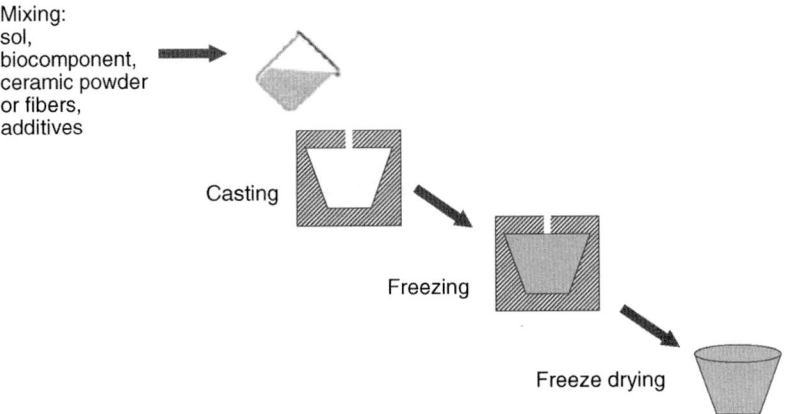

Mixing: sol, biocomponent, ceramic powder or fibers, additives

Casting

Freezing

Freeze drying

Figure 5.12 Preparation of freeze-gelated biocers.

biocomponents is poured into a mold. To improve the mechanical properties as well as the compatibility with the biocomponents, further components can be added to the slurry: ceramic powders or fibers for reinforcement, and nutrients and cryoprotectives for microorganisms. By freezing with liquid nitrogen (~77 K) the slurry solidifies, whereby the concentration of silica nanoparticles increases in the interstices between the ice particles, which favors their condensation. Subsequently, the material is heated for drying. In the final state, it has a continuous porosity of about 35%. The pore sizes and shapes are replicas of the ice crystals and thus depend on the velocity of the freezing front. This velocity cannot be preset since it slows down as the front moves. Three characteristic sizes and morphologies are observed (Figure 5.13): (i) fine, irregularly shaped pores at high cooling rates, (ii) unidirectional columnar pore channels at intermediate cooling rates with larger diameter at lower cooling rate, and (iii) coarse, irregularly shaped or dendritic large pores at low cooling rate. Well-developed columnar pore morphology is only found for intermediate front velocities . The pores are smallest near the surface where the front velocity is highest (b). The pores are larger and columnar below (b–d). The nanoporosity is seen to be not shaped after ice crystals (e and f). The nanostructure of the gel is made up of condensed globular particles (f). Judging from the aspect of the pore structure, freeze-gelated biocers are well suited for the exchange of external reagents and metabolic products of the entrapped microorganisms.

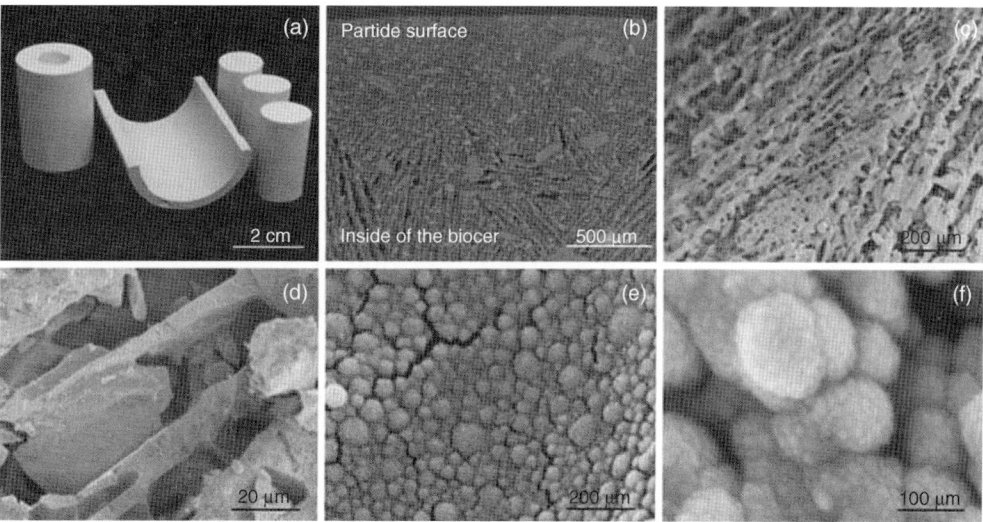

Figure 5.13 Photograph of freeze-gelated ceramics (a) and SEM images of the cross section of the wall of the hollow cylinder at increasing magnification (b–f). The material consists of 80 wt% mullite and 20 wt% SiO_2. (Reproduced with permission from Soltmann and Böttcher (2008). Copyright 2008, Springer + Business Media.)

Table 5.1 Change in the rate of phenol degradation by R. *ruber* embedded within a freeze-gelation matrix (Soltmann and Böttcher, 2008).

Operation in time (days)	Phenol degradation rate (mg/day/g biocer)
0	0.61
6	1.27
13	1.05
21	1.51
28	1.50

The large pore size gives enough space for cell division or germination of embedded spores. The possibility of extended proliferation of the embedded microorganisms in the freeze-gelated biocer is an important issue for the applicability of this technology. Freezing has a strong impact on the survival rate of the microorganisms. The formation of ice crystals inside the cells and the high osmotic pressure resulting from intercellular salt concentration due to freezing are problematic. For example, only about 1% of *Rhodococcus ruber* and 0.5% of *Pseudomonas fluorescens* bacteria survive the freeze gelation of TEOS (Soltmann and Böttcher, 2008). However, subsequent metabolic activity bears evidence of a remarkable vitality. For example, the increase of the glucose consumption by *P. fluorescens* bacteria embedded in a freeze-gelated biocer during the first 30 days can only be explained by a corresponding increase of the population. After 35 days, the surface of the biocer was cleaned, which reduced the population. Note also the long overall lifetime of the cells in the biocer. Similar dependences can be observed for the degradation efficiency of *R. ruber* bacteria, also embedded in a freeze-gelated biocer (Table 5.1).

Carturan *et al.* have developed a promising alternative path of biocer formation, the so-called Biosil method based on chemical vapor deposition (CVD) with a gaseous flux of alkoxysilanes in a moist atmosphere (variant D) (Campostrini *et al.*, 1996). The advantages of SiO_2 formation from gaseous precursors consist in the possibility of controlling the layer thickness by reaction time and alkoxide partial pressure, the immediate elimination of toxic by-products, and the deposition of a homogeneous SiO_2 layer in a single process step. It has been demonstrated that the Biosil process can be combined with other encapsulation techniques for the microencapsulation of enzymes and cells maintaining viability and function for virtually any biocomponent. For example, the cell embedding process has been successfully tested for the preparation of a bioartificial liver, devised particularly to preserve hepatocyte function and pancreatic islet viability and functionality.

5.2.2
Biological Properties of Silica-Based Biocers

At first view, silica-based biocers behave like polymer–ceramic composites. The overall mechanical properties correspond to the behavior of a pure silica xerogel.

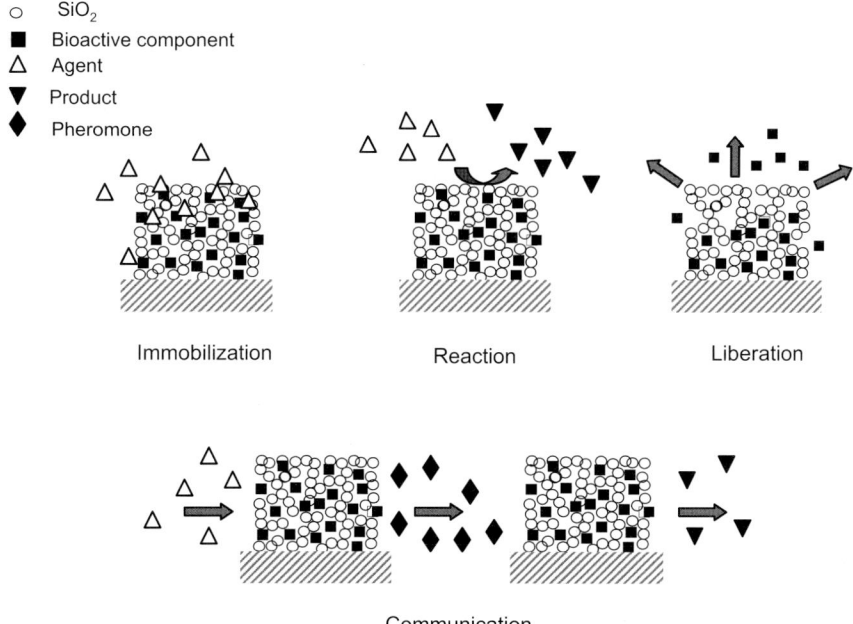

Figure 5.14 Possible functions to be fulfilled by biocers.

Young's modulus, strength, and hardness are similar to the corresponding data of the porous silica matrix. Only the fracture toughness is increased by the biomolecular inclusions. The essential difference of the materials' behavior is related to the biological and biochemical behavior. The biocer can be considered as a "living material." In this connection, we observe four new essential functions caused by the presence of living microorganisms (see also Figure 5.14):

i) Biologically controlled immobilization of inorganic or organic agents from aqueous solutions. A xerogel layer with entrapped biomolecules or microorganisms can serve as a substrate for *biogenic deposition processes*. This has been demonstrated by biomineralization and biotemplating directed by proteins or DNA, and also by biosorption controlled by biomolecules or microorganisms such as bacteria, yeast cells, or algae.

ii) Bioreactivity caused by immobilized enzymes or entrapped living microorganisms. Biocatalytic surface structures are notable examples for biologized silica-based xerogels. Entrapped enzymes have been shown to remain remarkably active and stable (Avnir *et al.*, 2006). Entrapped microorganisms facilitate the use of various natural metabolic processes for chemical reactions with inorganic or organic solutes in aqueous media. The entrapped cell can be considered as a highly specified microreactor. The biochemical reaction process of the microorganism is called *biotransformation*. Often the entrapped cells show a higher viability and stability in harsh environments.

iii) Liberation of biomolecules resulting from metabolic processes in entrapped cells. The open porosity of the silica gel allows the *exchange of metabolic products* with the environment.

iv) *Biological communication* by emission and reception of biomolecular signals. The entrapped microorganisms can communicate with others via emission and reception of specific signal molecules (for more details see Chapter 4).

Among all possible biological components for biocer preparation, bacterial spores are the most suitable ones. Biocers containing spores are long-term stable and less sensitive concerning preparation, drying conditions, and water content. Moreover, their reactivity is often comparable to that of other cell systems. The viability of the cells reduces with decreasing water content in the matrix (Figure 5.15). High water content, on the other hand, leads to low mechanical stability and promotes fouling. Hence, there is an optimum range of water content for a stable biocer. Additives such as polyols (glycerol, polyethylene glycol, and polyvinyl alcohol) or gelatin, which act as matrix softener, humidity preserver, and pore-forming agent (after leaching), suppress the cell lysis and prolong the viability of the embedded cells considerably. Coimmobilized nutrients improve the bioactivity. Under optimum storage conditions (low temperature and sufficient humidity), the viability and luminescent activity can be maintained for several months as has been demonstrated with *Escherichia coli* bacteria (Premkumar *et al.*, 2002).

Some microorganisms can undergo transformations from a stage of metabolic rest to a more active form, such as bacteria germinating from spores and fungus hyphae growing from spores or chlamydospores. For practical use of biocers, it would be useful to be able to switch on metabolic activity after storing the material for some time. It has been shown that embedded spores retain their metabolic functions in dried biocers. Figure 5.16 shows the budding of *Bacillus sphaericus*

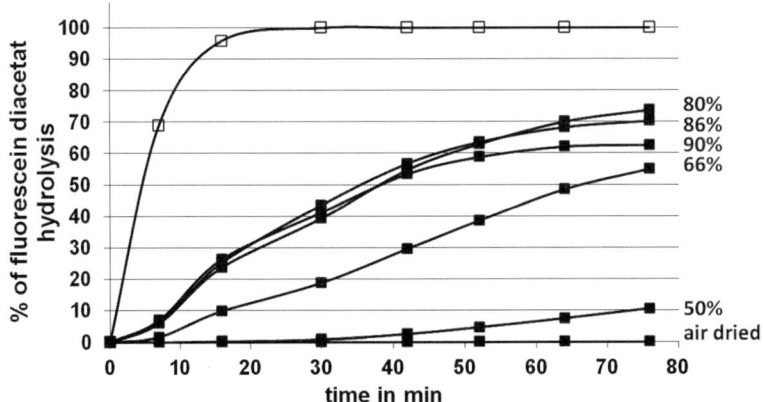

Figure 5.15 Impact of water content on the enzymatic cleavage of fluorescein diacetate by *B. sphaericus* cells. Upper curve: nonimmobilized cells; other curves:

biocers with different water content. (Reproduced with permission from Böttcher *et al.* (2004). Copyright 2004, the Royal Society of Chemistry.)

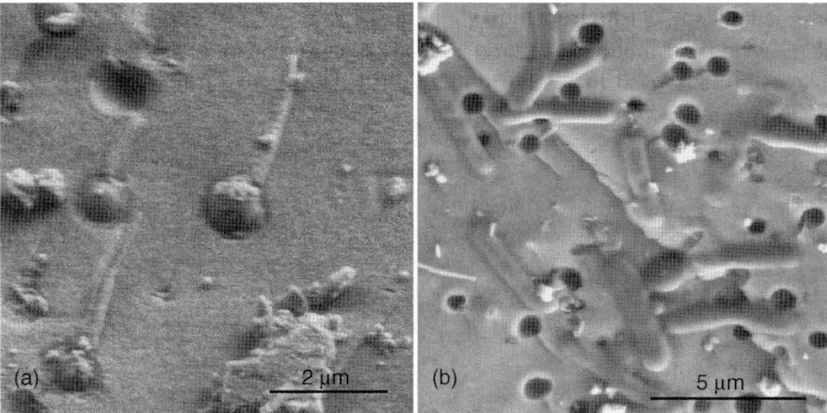

Figure 5.16 Germination of *B. sphaericus* spores embedded in a silica matrix. The spherical spores (a) are transformed into vegetative cells; (b) after 24 h exposure in nutrient broth (scanning electron micrographs). (Reproduced with permission from Böttcher *et al.* (2004). Copyright 2004, the Royal Society of Chemistry.)

spores after 24 h incubation in nutrient broth. The volume increase exceeding 100% during germination would not be an obstacle if the biocer could yield in some way. Usually the vegetative cells are found near the surface where the growing cells can find additional space by escaping through pore channels to the surface or by partial buckling or delamination of overlying thin silica films (Figure 5.16). Alternatively, also the large pore size in freeze-dried gels offers enough space for germination and proliferation in bulk materials.

5.3
Bioengineering

The incorporation of microorganisms or biomolecules into a ceramic matrix offers a wide range of applications. Two main routes can be distinguished (Figure 5.17). One group of materials exploits the sustained biological function of specific micro-organisms for technical use. Metabolic processes of the cell are the base for potential applications such as biocatalysis. In this connection, the entrapment of single catalytic biomolecules, the enzymes, in a porous silica matrix has some advantages in practical use, as we will see in the following. The high efficiency of metabolic process of entrapped microorganisms is used for the production of specific chemical compounds by biotransformation of appropriate nutrients. Microorganisms entrapped in porous silica find also first applications for bioremediation of polluted water and soil. Quite different options are provided by the cellular signaling and communication of entrapped microorganisms. There are already first applications of whole-cell biosensors and actuator systems where the entrapment of the living cells in a mechanically stable and bioinert matrix offers advantages. Also, the

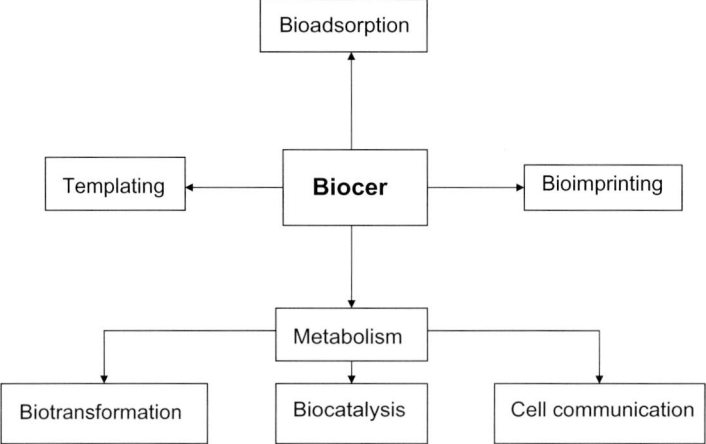

Figure 5.17 Possible routes for application of biocers.

immobilization of living cells opens a new field of tissue engineering for medical application. Bone regeneration, for example, requires scaffolds that enable the bone to be remodeled by resorbing cells, the osteoclasts, and the bone-forming cells, the osteoblasts. One option is the design of the scaffold starting from a porous silica–collagen composite (Heinemann *et al.*, 2009).

The second large group of biomaterials makes use of the embedded micro-organisms or their components as a passive structure with specific properties. Complementary to biotransformation, the bioadsorption of microorganisms or biomolecules immobilized on the surface or in the volume of porous silica materials can lead to low-cost and highly specific materials for remediation of polluted water. Nanostructured inorganic materials can be favorably produced by using immobilized microorganisms or biomolecules as templates for envisaged patterned inorganic structures on silica surfaces. The geometrical peculiarities of cells or biomolecules can also be used in a kind of bioimprinting for the preparation of porous silica matrices structured on the mesoscopic or nanoscopic scale.

In most of these approaches, the silica gel serves as a supporting matrix with sufficient strength and as a porous medium facilitating the exchange of chemical agents and biomolecules, the energy transfer, and the exchange of signal molecules with the environment. Examples for various successful developments are presented in the following.

5.3.1
Bioactive Sol–Gel Coatings and Composites

Biomaterials for the regeneration of bone make up a wide field of research and application. As explained in more detail in Section 6.2.8, bone is a highly complex living material. In the sense of mechanics, it is a hierarchically structured composite. Bone-forming cells, the osteoblasts, and bone resorbing cells, the osteoclasts, are embedded within a matrix consisting of the main components collagen and hydroxyapatite (HAP). These two cell types and various precursor cells communicate by exchanging signal molecules. This enables the bone structure to continuously adapt to loading, which means that the bone tissue is perpetually remodeled. The osteoclasts dissolve the HAP by local pH reduction and digest the collagen network. The osteoblasts express new collagen and initiate the nucleation of HAP at nucleation centers of the collagen fibrils. The extracellular matrix transmits the signals for the proliferation and differentiation of the various precursor cells of osteoclasts and osteoblasts. Knowledge of the processes going on inside living bone leads the way in the development of biomaterials for bone regeneration. There are two principal ways: (i) stable integration of the implant into the living tissue by *coating of load-bearing implants* with a bioactive layer and (ii) replacement of the implant by ingrowth of bone tissue. Materials based on *silicified collagen* have turned out promising for either option. From earlier developments of so-called bioglass, it was known that even plain silica is bioactive. Reactive OH groups on the silica surface reduce the HAP nucleation energy. The nucleation can be additionally promoted by adding certain dopants, such as P_2O_5, CaO, and Na_2O. The same applies to silica gel. Properties of several sol–gel layers for implant coating are shown in Figure 5.18. Besides silica gels (pure and with P_2O_5 addition), there are also two hybrid gels with gelatin entrapment. The behavior of a TiAl4V alloy as a well-established implant material is plotted for comparison. The good cell adhesion is the most interesting property of the various silica gel versions. Also, the low elastic modulus E can be considered as an advantage since it reduces the elastic misfit to the low-modulus bone tissue that avoids stress concentrations near the implant.

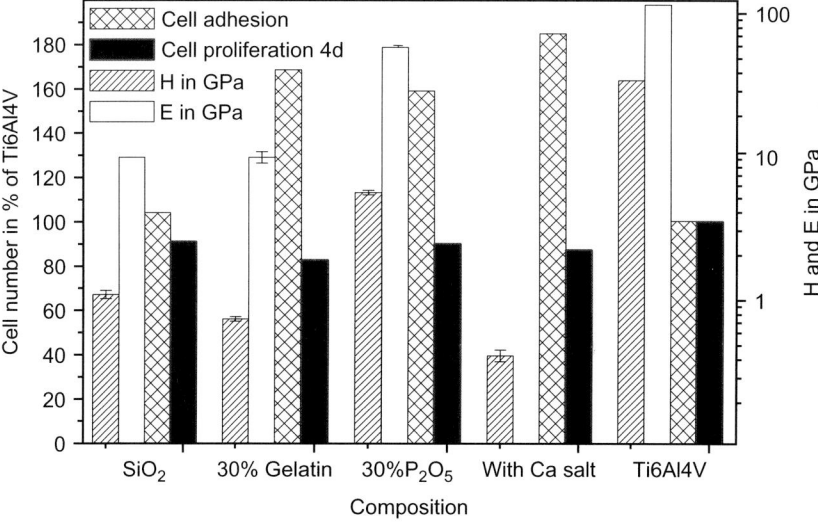

Figure 5.18 Properties of different sol–gel layers for implant coating. (Reproduced with permission from Böttcher (2000). Copyright 2000, Wiley-VCH Verlag GmbH.)

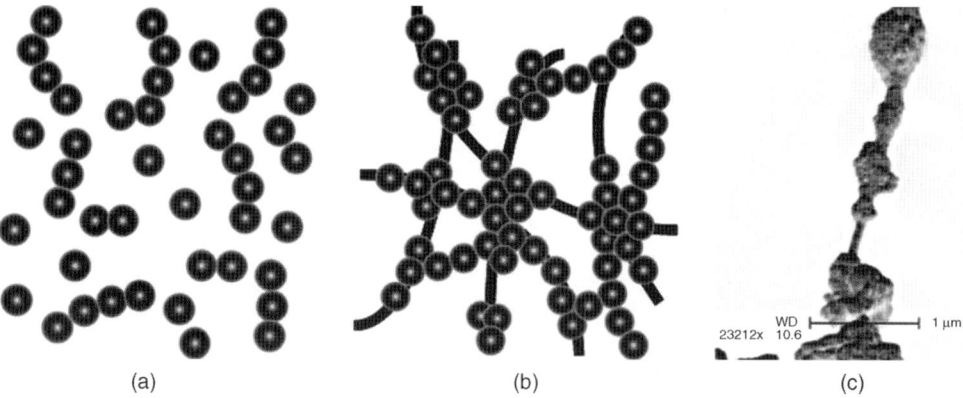

(a)   (b)   (c)

Figure 5.19 Polycondensation of silica particles leads to a three-dimensional inorganic gel network (a). The presence of an organic template, for example, fibrillar collagen, induces the formation of gel network consisting of silicified collagen (b). Single silicified collagen fibril (c, inverted SEM image). (Reproduced with permission from Heinemann *et al.* (2007). Copyright 2007, Wiley-VCH Verlag GmbH.)

For engineering of bone tissue, bulk-like collagen–silica composites are also considered as one promising option. Since a material to be implanted should have mechanical properties comparable to those of bone, the low strength of the porous gelatin-free silica gel would be an essential disadvantage for scaffold development. This problem can be overcome by additional entrapment of collagen in the silica gel. Collagen fibrils fulfill three functions in the biomaterial. First, electrostatic interaction between the negatively charged silica condensates and positively charged amino acids of the collagen favors the condensation of the silica clusters along the collagen fibrils (Figure 5.19). This leads to an accelerated gelation and a more open network structure. Second, the well-bonded collagen fibrils lead to higher strength and fracture toughness of the hybrid xerogel. Third, the collagen acts as a carrier of other biomolecules controlling the proliferation and differentiation of bone cells. Recently, Heinemann *et al.* (2009) have provided a nice example for the feasibility of a bone substitute made of silicified collagen. The chosen processing regime is shown in Figure 5.20. TEOS was chosen as a silica precursor.

Three kinds of materials were prepared for comparison: pure silica xerogel, silica–collagen composites, and silica–collagen composites with calcium phosphate cement (CPC) powder. The composition of the CPC was similar to the so-called biocement. It consists mainly of α-tricalcium phosphate and calcium hydrogen phosphate, two metastable phases that convert to hydroxyapatite in aqueous solution (for more details see Section 6.2.2). The entrapped collagen fibrils improved the mechanical behavior considerably. Stress–strain diagrams of the various materials under uniaxial compressive loading are shown in Figure 5.21. The large increase of compressive strength and fracture strain due to reinforcement by collagen is obvious. Note that the compressive strength of bone is typically in the range of about 30 MPa.

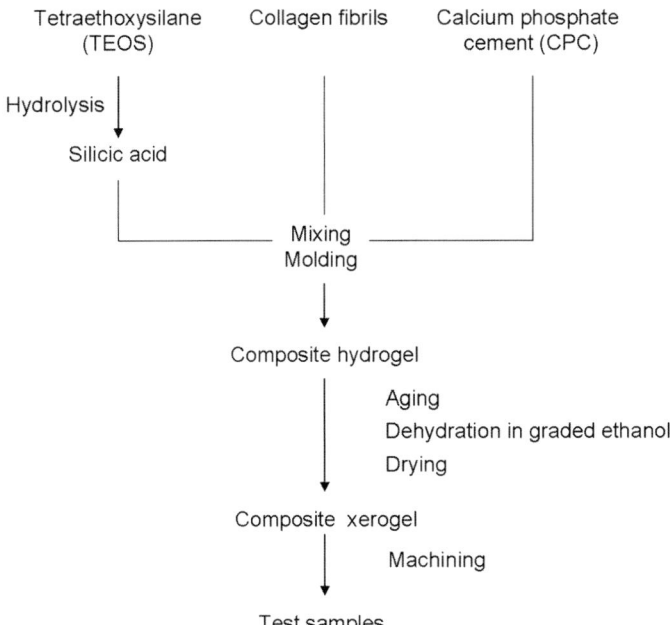

Tetraethoxysilane (TEOS) Collagen fibrils Calcium phosphate cement (CPC)

Hydrolysis

Silicic acid

Mixing
Molding

Composite hydrogel

Aging
Dehydration in graded ethanol
Drying

Composite xerogel

Machining

Test samples

Figure 5.20 Processing diagram for silica–collagen and silica–collagen–CPC composite xerogels. (Adapted from Heinemann *et al.* (2009).)

Also, the fracture faces show the reinforcing influence of the collagen fibrils (Figure 5.22). The almost smooth fracture face of the pure silica gel is roughened by the incorporation of collagen. The silicified collagen fibrils were pulled out of the bulk material. It means that near the tip of a propagating crack the fibrils bridge the crack, which leads to an unloading of the crack tip resulting in higher fracture

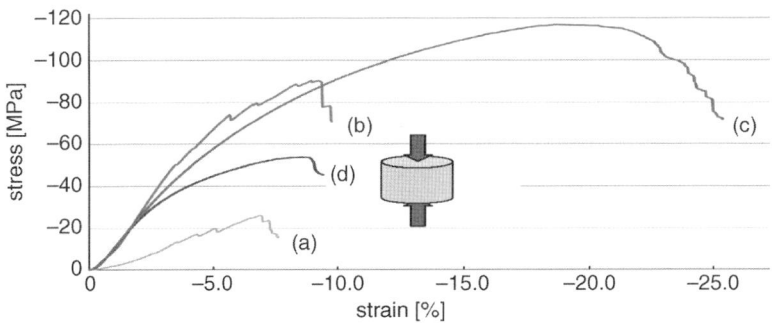

Figure 5.21 Stress–strain diagrams of silica–collagen and silica–collagen–CPC composite gels under uniaxial compressive loading: (a) 100% silica; (b) 85% silica, 15% collagen; (c) 70% silica, 30% collagen; (d) 52.5% silica, 22.5% collagen, 25% CPC. (Adapted with permission from Heinemann *et al.* (2009). Copyright 2009, Elsevier.)

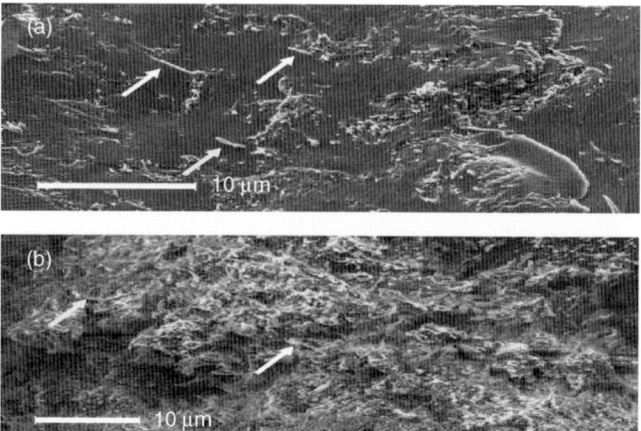

Figure 5.22 SEM images of fracture surfaces of the two-component xerogel 85/15 (a) and the three-component xerogel 52.5/22.5/25 with CPC (b). The white arrows point to pulled-out collagen fibrils. Scale bar: 10 μm. (Adapted with permission from Heinemann *et al.* (2009). Copyright 2009, Elsevier.)

strength and strain with increasing content of collagen. The metastable CPC phases have become partially transformed into a new phase during the gel formation. The effect of CPC on the strength can be detrimental. The observed roughening of the fracture surface is connected with an increase of fracture strength and strain. Larger untransformed agglomerates, however, can act as large strength-reducing flaws. Therefore, the addition of CPC always should be done in the presence of collagen fibrils to get a material with reliable mechanical behavior.

The bioactivity of the xerogel materials was tested by immersion in a simulated body fluid (SBF). The SBF contains calcium and phosphate ions to initiate the nucleation of calcium phosphate. The solution was replaced every day to ensure constant ion concentration and a pH of 7.4. These conditions enable the deposition of the bone mineral hydroxyapatite. Obviously, the reactive hydroxyl groups (Si—OH) at the surface of the silica xerogel yield the main contribution of nucleation centers. The EDX spectroscopy indicates that the Ca/P ratio on the surfaces of the pure silica gel and the silica–collagen composite xerogels is between 1.61 and 1.64, which is close to hydroxyapatite (1.67). The presence of CPC particles leads to further acceleration of HAP growth. After 7 days, a closed layer is formed where the Ca/P ratio of 1.50 can be explained by the presence of nontransformed CPC in the bulk of the xerogel. The pronounced bioactivity of the materials was proven by the behavior of bone cells seeded on the surfaces. It was shown that silica–collagen composites enable the adhesion, proliferation, and differentiation of human mesenchymal stem cells (hMSCs) into bone-forming osteoblast-like cells (Figure 5.23).

With the given material, it has been demonstrated that also the bone resorbing multinuclear osteoclasts can be grown by starting from human monocyte precursor

Figure 5.23 SEM micrographs of the silica–collagen hybrid xerogels 1 day after seeding hMSCs (a and b) and after 14 days (c and d) of cultivation of the osteogenically induced hMSCs. Scale bar: 200 μm (a and c) and 20 μm (b and d). (Reproduced with permission from Heinemann *et al.* (2007). Copyright 2000, Wiley-VCH Verlag GmbH.)

cells. The differentiation of the monocyte precursor cells into osteoclasts was induced by adding a mixture of glycerophosphate, dexamethasone, vitamin D3, and the cytokines M-CSF (macrophage colony stimulating factor) and RANKL (receptor activator of nuclear factor kappa B ligand) to the cell culture medium. Details of the role of cytokines in remodeling of bone tissue will be discussed in Section 6.2.8. In conclusion, the example shows that the bioactivity and the good mechanical behavior make silica–collagen composites manufactured via sol–gel processing a promising material for regenerative bone therapy.

5.3.2
Biocatalytic Sol–Gel Coatings

Enzymes entrapped in silica-based gels usually show better performance, such as higher activity and a longer lifetime, and in some cases higher thermal stability, compared to free enzymes. For instance, while free alkaline phosphatase is most active near pH 9.5, it still works well under very harsh conditions at pH as low as 0.9 when entrapped. This surprising behavior is caused by physical as well as chemical interactions of the xerogel matrix with the enzyme. The basic *physical effect* is mainly due to the confinement of the individual enzyme molecules to the pore spaces in the

Figure 5.24 Alkaline phosphatase shows an enhanced stability at extreme acidic environment when the molecule is entrapped in a sol–gel matrix. The two hydronium molecules (marked with arrows) in a cage containing 100 water molecules correspond to "pH 0." (Reprinted with permission from Frenkel-Mullerad and Avnir (2005). Copyright 2005, the American Chemical Society.)

xerogel. We can assume that, after shrinkage due to drying, the spaces have become so narrow that the shape of the enzyme molecule has become virtually fixed, which precludes denaturation by unfolding. Small conformation changes due to short-time temperature increase are reversed as soon as the temperature is lowered again. Also, the often observed stability of the enzymes at extreme pH may be a consequence of confinement. As explained by Frenkel-Mullerad and Avnir (2005), even large pH changes do not much affect the number of hydronium ions (H_3O^+) in the small volume available for water in the cage. The situation is schematically shown in Figure 5.24. It has been supposed that the enzyme is surrounded by one or two layers of water molecules in the pore of the gel. At pH 0, only two water molecules would be protonated to give hydronium H_3O^+, provided that estimates based on thermodynamics do still apply here (see Task 5.1). In a bulk solution, pH 0 would mean a strong bombardment of the protein with a large number of hydronium ions, which would lead to denaturation. Obviously, the presence of two hydronium ions would not essentially influence the protein structure with its multiple charge centers. Whereas a controlled physical confinement of enzymes is beneficial, extensive shrinkage of the xerogel during drying could cause undesirable conformation changes of the enzymes. Shrinkage stress can be reduced by incorporation of fibrous molecules such as hydroxyethyl carboxymethyl cellulose or collagen.

The *chemical influence* of the silica matrix can be controlled preferably via the choice of the precursor. As discussed above, the unfavorable influence of alcohol during the hydrolysis–condensation has to be excluded. This could be done in a two-step process where the alcohol is evaporated before the condensation step. A more suitable option is offered by the purely aqueous sol–gel process discussed in connection with THEOS as an example for a possible precursor. Also the covalent bonding of sugar molecules to the sol–gel matrix has proven successful. One problem is the formation of a tailored hydrophilic–hydrophobic silica interface in the xerogel since industrial syntheses are often performed in organic solvents. In these cases, the efficiency of the particular enzyme depends sensitively on the hydrophobic interaction. The interface between the enzyme and the xerogel can be optimized with mixtures of tetramethoxysilane ($Si(OMe)_4$, TMOS) and methyltrimethoxysilane ($MeSi(OMe)_3$, MTMS), for example. With a similar approach, Reetz, Zonta, and Simpelkamp (1995) were able to increase the activity of lipases entrapped in a hybrid silica matrix. Lipases are used for catalyzing hydrolysis of fat and oil into fatty acids and glycerol, and for esterification reactions in organic media. By hybrid precursors composed of mixtures of TMOS and silanes of the type $RSi(OMe)_3$ with $R = C_nH_{2n+1}$, a lipophilic environment can be created in the cages of the xerogel. A sharp increase of the activity of the lipases is observed for $n = 3$–5 up to about 100 times.

Besides their use in the white biotechnology, the stability and high activity of enzymes entrapped in a silica-based gel will find a wide field of application with biosensors. One example of such type of sensor is the glucose sensor based on the immobilization of glucose oxidase (GOD) in a sol–gel silica matrix, as one option, for catalysis of the oxidation reaction of glucose to gluconic acid and H_2O_2:

$$\beta\text{-D-glucose} + O_2 + H_2O \xrightarrow{\text{GOD}} \text{gluconic acid} + H_2O_2. \tag{5.6}$$

Different xerogel/GOD layer arrangements have been realized for electrochemical and optical sensors. The GOD activity depends strongly on the type of the sol–gel matrix. Incomplete hydrolysis of the sol and high alcohol concentration result in poor catalytic activity. The enzyme activity has been increased with so-called penetration agents, such as saccharides, coimmobilized within the sol–gel matrix, which increase the porosity by leaching. This increases the diffusive transport of the reactants within the layered structure. The glucose can be determined quantitatively by (i) measuring the change of the oxygen concentration, (ii) calorimetric measurement of the reaction heat, (iii) amperometric or voltametric measurement of H_2O_2, or (iv) optical sensing of the oxidation of a dye precursor by H_2O_2 catalyzed by peroxidase (Künzelmann and Böttcher, 1997). In a thin-film glucose sensor strip, the Pt working electrode was covered with a GOD/silica xerogel composite. Another design of such sensor was based on an enzyme cartridge with the GOD/silica composite deposited on zeolite beads.

5.3.3
Bioremediation

Biocers with embedded metal binding bacteria, yeasts, or algae that tolerate high concentrations of heavy metals are of special interest for the bioremediation of metal-contaminated wastewater. Immobilized microbial cells have already been used extensively for the pollutant biodegradation in water and soils. Immobilization offers two important advantages: protection of the embedded cells and usability of cells with slow division rates in bioreactors, decoupled from turnover time. There are two main approaches to bioremediation:

i) The active uptake of metals by the metabolic process. This process of *bioaccumulation* requires the viability of the immobilized microorganisms.

ii) The uptake of metals by complexation independent of the cell metabolism. For this *bioadsorption*, the viability of the microorganisms is not necessary. Dead cells or cell fragments such as bacterial cell surface protein layers (so-called S-layers) can be used. The advantage of biosorptive processes consists in high storage stability of the biocers, no nutrient addition, and multiple reusability by controlled desorption.

Promising examples of such approaches have been tested on the laboratory scale.

Sol–gel immobilized *yeast cells* accumulate hazardous heavy metal ions from aqueous solution, with affinity following the order $Hg > Zn > Pb > Cd > Co = Ni > Cu$ (Gadd and White, 1993; Al-Saraj *et al.*, 1999). By immobilization of *macro- and microalgae* in a silica matrix (content up to 30–50% (wt/wt SiO_2), porosity 40–60%), the high sorption efficiency of the biocomponent for heavy metals (Cr, Ni, Cu, and Pb) and the mechanical stability of the silica matrix have been combined to get a biocer with good performance (Soltmann *et al.*, 2010).

Microbial mats embedded together with nutrients in silica gel have been used for the removal of U(VI) from contaminated groundwater. In ecosystems, bioremediation is usually sustained by a heterogeneous consortium of microorganisms. Such consortium forms a multifunctional community that can realize its evolution more successfully than a single species. This principle has been proven also as a promising concept for the design of an optimized biocer for remediation of uranium-loaded water. A microbial mat grown under hostile conditions of a polluted medium has been taken as a pool for the selection of microorganisms to be immobilized in a silica-based biocer (Bender *et al.*, 2000). Uranium can exist in nature with different valences such as U(IV), U(V), and U(VI). U(VI) is highly soluble in water as UO_2^{2+} ions, whereas U(IV) has a very low solubility. U(V) usually occurs only in intermediate compounds of minor importance for remediation technology. Therefore, remediation of uranium-polluted water has to involve the sequestration of the UO_2^{2+} ions by reductive precipitation of U(IV) compounds. To produce such biocer, three microorganisms have been selected from a mat grown in uranium-polluted media: filamentous cyanobacteria *Oscillatoria*, dissimilatory metal-reducing bacteria *Rhodopseudomonas*, and dissimilatory sulfate-reducing bacteria. These microorganisms were immobilized in a silica gel forming silica mat particles (SMP). The interaction of the various constituents of the microbial consortia of the mats with silica particles leads to the high efficiency of remediation as summarized in Figure 5.25. The removal of U(VI) is initiated with sorption of UO_2^{2+}. At neutral pH, UO_2^{2+} ions are sequestered by Coulomb interaction at negatively charged surface sites of amorphous silica particles or at the cyanobacteria *Oscillatoria* immobilized in the silica gel. In the second step by metabolism of the metal-reducing *Rhodopseudomonas* and sulfur-reducing bacteria the adsorbed U(VI) is reduced to U(IV). This process is catalyzed by a broad spectrum of extracellular

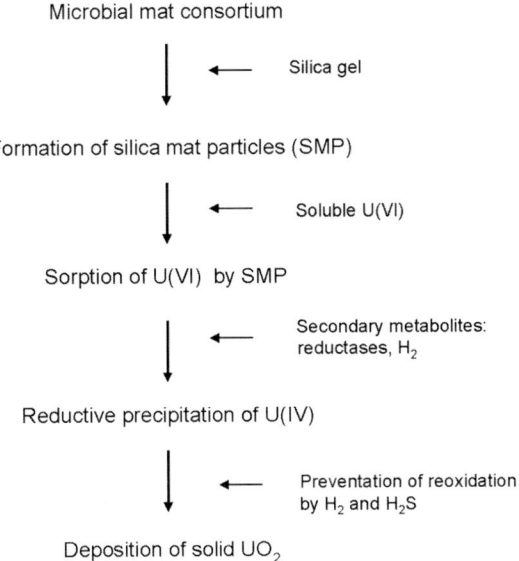

Microbial mat consortium

Formation of silica mat particles (SMP) ← Silica gel

Sorption of U(VI) by SMP ← Soluble U(VI)

Reductive precipitation of U(IV) ← Secondary metabolites: reductases, H₂

Deposition of solid UO₂ ← Preventation of reoxidation by H₂ and H₂S

Figure 5.25 Mechanisms causing removal of dissolved U by silica–microbial mat composites. (Adapted from Bender *et al.* (2000).)

reductase enzymes produced by *Rhodopseudomonas*. The low solubility of UO_2 leads to the precipitation of solid $UO_{2(s)}$ on cell walls or silica surfaces. The sulfate-reducing bacteria contribute to maintaining the reducing conditions of the active cells by the production of H_2 and H_2S. This also prevents the reoxidation of the UO_2 precipitates. Thus, the silica-based sol–gel process allows a successful mimicking of the synergism realized in microbial mats.

The efficiency of remediation technology can be increased by an appropriate tailoring of the xerogel structure as the following example shows (Raff *et al.*, 2003). Biocers based on cells of *B. sphaericus* JG-A12, isolated from a uranium mining waste pile, have been investigated with regard to their usefulness for the selective and reversible accumulation of heavy metals and radionuclides from drain waters of uranium wastes. The bioremediation process involves two steps: the biosorption,

$$SiO_2\text{-biocer} + U\text{-containing wastewater} \rightarrow SiO_2\text{-biocer/biosorbed U}, \quad (5.7)$$

as well as the following regeneration of the saturated biocer-based filters by washing with citric acid,

$$SiO_2\text{-biocer/biosorbed U} + \text{citric acid} \rightarrow SiO_2\text{-biocer} + U\text{-citrate}. \quad (5.8)$$

Differently manufactured biocers provide significant differences in the biosorption of uranium. Uranium is bound by both the biocomponent and the silica gel. High porosity due to freeze drying or addition of sorbitol during the preparation of the biocers leads to more rapid binding of the metals (Figure 5.26). Investigations

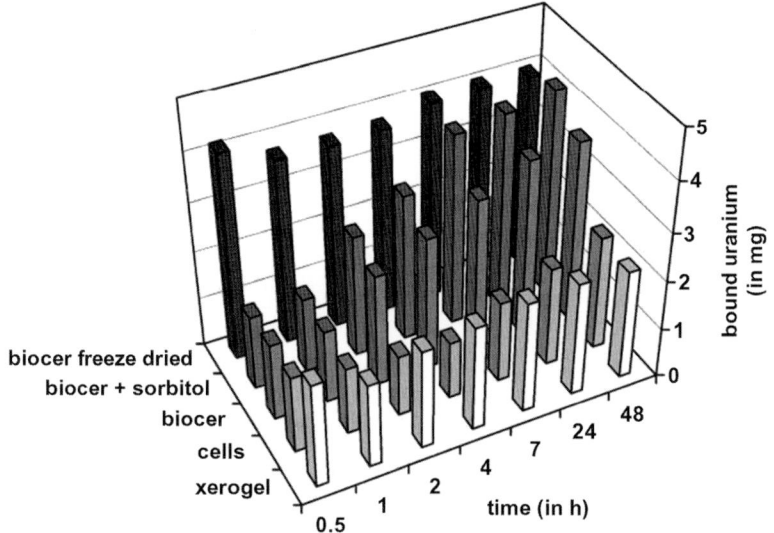

bound uranium (in mg)

biocer freeze dried
biocer + sorbitol
biocer
cells
xerogel

0.5 1 2 4 7 24 48

time (in h)

Figure 5.26 Uranium binding by different biocers based on *B. sphaericus* per 200 mg dry weight of biocers. (Reprinted with permission from Raff *et al.* (2003). Copyright 2003, the American Chemical Society.)

were carried out with 200 mg dry weight of sieved silica gel or biocer particles containing 36.4 mg of bacterial biomass corresponding to 2.6×10^{10} *B. sphaericus* cells. The metal binding capacity of the same amounts of free cells was measured as well. For a cost-efficient application as filter materials, a repeated regeneration of the uranium-saturated filters is necessary. Biosorbed uranium can be completely removed from the biocer with aqueous citric acid (Eq. (5.8)). Already a single desorption process step with 0.5 M citric acid at pH 4.5 for 24 h leads to an almost clean filter that can be used again for biosorption with nearly no loss of binding efficiency.

For biotransformation, the viability of the entrapped microorganisms is essential. Owing to the advantageous properties of biocers, such as low cost, biological and chemical inertness, mechanical stability, controllable porosity, and geometrical shape, there is a large potential for use of biotransformation in biocatalytic degradation of toxic substances from industrial wastewaters and soils. Decomposition processes have been realized with immobilized *Pseudomonas* strains for the biodegradation of the herbicide atrazine (Rietti-Shati, 1996) and polychlorinated biphenyls (Branyik *et al.*, 1998; Kuncova *et al.*, 2002), and with mixed bacteria and yeast strains for the continuous degradation of phenol in water (Branyik *et al.*, 2000). Phenol degradation was accomplished by biocers containing living *Rhodococcus rhodochrous* cells in a bubble column reactor (Fiedler *et al.*, 2004). The degradation was investigated in dependence on the residual water content of the silica matrix and pore-forming additives. The addition of sorbitol causes a higher activity of the embedded cells also in biocers with low water content (see Figure 5.27).

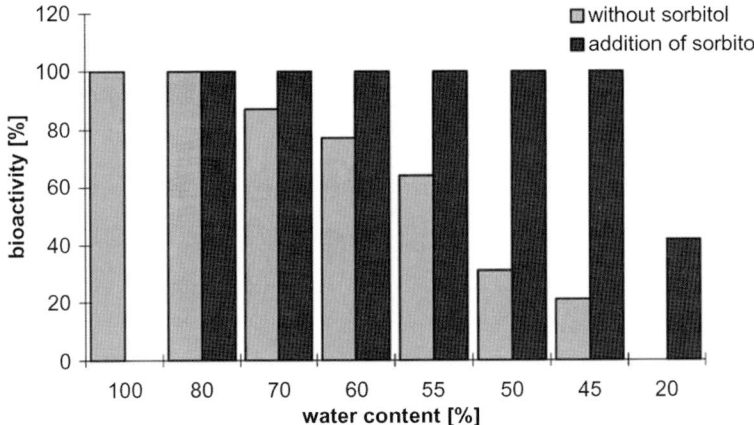

Figure 5.27 Phenol degradation by silica-embedded *Rhodococcus* sp. in dependence on water and sorbitol content. (Reproduced with permission from Böttcher *et al.* (2004). Copyright 2004, the Royal Society of Chemistry.)

Several glycols in salt-containing industrial wastewater have been found to be degraded very efficiently within silica-embedded *Aspergillus* spores (Fiedler *et al.*, 2004). To get a sufficient degradation rate, the spores have to be activated by additional nutrients such as glucose or casein and yeast extract. A comparison between the use of biocer-coated glass bodies (variant b) and fiber-reinforced packing material (variant c) as seen in Figure 5.11 reveals the high bioactivity of the reinforced biocer in a column bioreactor (Figure 5.28).

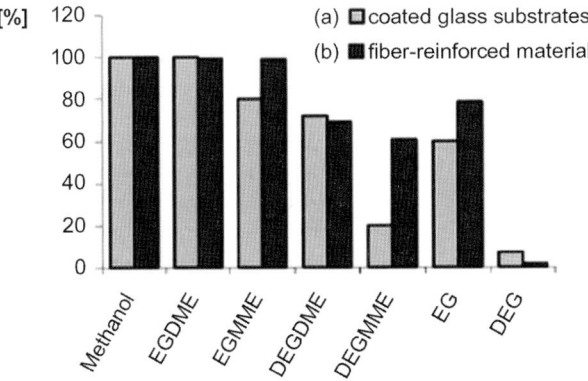

Figure 5.28 Degradation of different glycols after 21 days by silica-embedded *Aspergillus* spores (a) coated on glass and (b) fiber-reinforced packing materials. EGDME: ethylene glycol dimethyl ether; EGMME: ethylene glycol monomethyl ether; DEGDME: diethylene glycol dimethyl ether; DEGMME: diethylene glycol monomethyl ether; EG: ethylene glycol; DEG: diethylene glycol. (Reproduced with permission from Böttcher *et al.* (2004). Copyright 2004, the Royal Society of Chemistry.)

5.3.4
Cell-Based Bioreactors

The entrapment of living cells in silica-based xerogel enables metabolic processes to be used for biotechnical material synthesis (Carturan *et al.*, 1989; Pope, 1995; Nassif *et al.*, 2002; Avnir *et al.*, 2006; Soltmann and Böttcher, 2008). Immobilized cells have been exploited for the biotechnological production of amino and organic acids, antibiotics, enzymes, and other compounds. Increased metabolic activity and metabolite production, protection from environmental stresses and toxic pollution, and increased plasmid stability are possible advantages. Moreover, immobilized cells in the form of bulk materials or coated carriers can act in bioreactors as cell reservoir systems and thus can prolong the reaction time compared to free cell dispersions.

Therefore, some biocer compositions have already been successfully tested for the utilization in biotechnical material synthesis, beginning with the investigation of the fermentation process by which carbohydrates are converted into ethyl alcohol and CO_2, catalyzed by sol–gel entrapped yeast cells in silica (Carturan *et al.*, 1989; Gill and Ballesteros, 1998).

As the following example of astaxanthin shows, the entrapment of microalgae in a silica gel (variant A) offers new options for the design of the metabolite production process. Astaxanthin is a fat-soluble carotenoid pigment. It is mainly produced biologically by the microalgae *Haematococcus pluvialis* (Fiedler *et al.*, 2007). Astaxanthin is a secondary metabolite, which means it is not required in the living cell. Astaxanthin is used as a supplementary feed in fish farming and as a natural dyestuff in the food and cosmetic industries. Usually it is produced in large outdoor photobioreactors. In this process, all living algae are destroyed at the end of the astaxanthin production, and it is extracted from the biomass. Fresh cultivation of algal biomass is necessary. The aim of immobilization of the microalgae in silica gel is the realization of a continuous astaxanthin production. The astaxanthin production of the microalgae can be initiated by stopping the cell division under external stress, for example, by the combined application of Fe^{2+} compounds with NaCl or hydrogen peroxide as stress factors. This treatment strongly increases the astaxanthin formation within the microalgae embedded in the biocer (Figure 5.29). Combining the two stress factors, an optimum of astaxanthin production (indicated by formation of red colored cells) can be achieved. By additional stressing with 1 mM H_2O_2 and light irradiation, almost 100% of the cells take on strong red coloration. The dye can be extracted from the cells with various organic solvents without destroying the biocer. Owing to embedding of the cells in the porous xerogel combined with cell-protective additives, there is only a moderate decrease in viability after solvent extraction, which allows the immobilized microalgae to be recultivated. Thus, a continuous biotechnological process can be realized.

Successful examples of silica-based bioreactors prepared by the Biosil technique (variant D) have also been reported. Worth mentioning in this connection is the production of secondary metabolites such as umbelliferone and marmesin by immobilized plant cells (*Coronilla vaginalis*) (Campostrini *et al.*, 1996; Dal Monte,

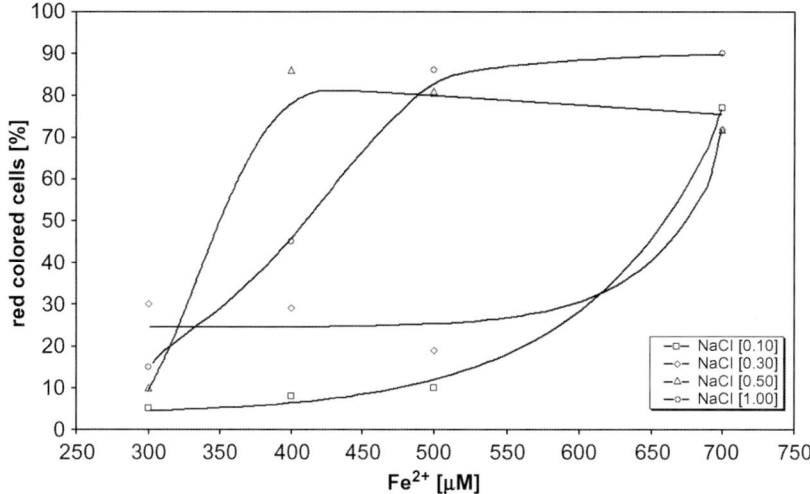

Figure 5.29 Fraction of astaxanthin producing (red colored) alga cells versus NaCl and Fe^{2+} concentrations. (Courtesy of D. Fiedler and U. Soltmann.)

2000), the alkaloid production by encapsulated *Catharanthus roseus* cells with metabolite productivity up to two orders of magnitude higher than that of free cells (Carturan *et al.*, 1998), and the production of the enzymatic complex invertase by immobilized plants cells (*Ajuga reptans*, selected from callus) with a 40-fold higher efficiency with respect to free cells as a result of suppression of the competitive process of proliferation (Pressi *et al.*, 2003). Advantages of this approach based on immobilization of the plant cells in porous silica matrices in comparison to production by free cells are continuous production of the secondary metabolites, control of the bioreactor, reduction of volumes and by-products, and the independence of metabolite production from season cycle and climatic trends or territorial habitat.

5.3.5
Silica-Based Controlled Release Structures

Silica xerogels are widely used as matrices for the manufacturing of drug release structures. The essential parameter controlling the release kinetics of infiltrated drugs is the open porosity of the silica matrix. The sol–gel process allows manifold modifications by alloying additives to the sol such as poly(D,L-lactide-*co*-glycolide) (PLGA), poly(dimethylsiloxane) (PDMS), or $CaO–P_2O_5$–poly(methyl methacrylate) (PMMA) in order to control the mesoporosity, increase the mechanical stability, or improve the biocompatibility, respectively. As a more challenging approach, the creation of implant materials consisting of embedded living cells that correct metabolism defects in the animal or human body is envisaged. Durable cell functionality, mechanical stability, well-controlled porosity, and high biocompatibility (no immunorejection) are the main preconditions for this purpose. Successful *in vitro* and *in vivo* results were obtained with silica gel-embedded pancreatic islets that catalyze the reduction of the blood sugar level of diabetic mice by insulin secretion (artificial pancreas) (Pope, Braun, and Peterson, 1997). With the Biosil technique (variant D), animal liver cells were fixed on a substrate and successfully tested for patients with hepatic defects (bioartificial extracorporeal liver) (Dal Monte, 2000).

The use of silica-based gels for manufacturing of drug release structures competes with the application of organic hydrogels and the layer-by-layer deposition of polyelectrolytes (see also Section 4.3). Recently, there is a growing research activity with the aim to combine the advantages of various materials into core–shell systems. For example, a cell embedded in hydrogel (e.g., alginate) with optimized biocompatibility can be coated with silica gel for higher mechanical stability.

5.3.6
Patterned Structures

The soft processing regime for silica xerogels is suitable for the creation of patterned structures by using biomolecules or cells as structure-forming units. Such kind of *bioimprinting* had first been studied for the preparation of mesoscopically patterned porous materials. Recently, it has been extended to nanostructured patterns. Inorganic materials such as SiO_2, TiO_2, and Al_2O_3 can be applied to transcribe the structure of the biomaterial. The transcription occurs via an organogel consisting of the organic component, the gel forming molecules, and the solvent. Electrostatic interaction of a cationic organic component with anionic gel forming constituents (e.g., silanolate groups $RSiO^-$) leads to a direct mapping of the organic template. After elimination of the organic component by heating, washing with organic solvents, or enzymatic decomposition, an inorganic mold of the organic pattern is obtained (van Bommel *et al.*, 2003). Among other organic components, microorganisms such as yeast cells (Chia *et al.*, 2000), *Bacillus subtilis* (Davis *et al.*, 1997), and the tobacco mosaic virus (Fowler *et al.*, 2001) have already been used as templates. The intention connected with these experiments is the manufacturing of mesoscopically porous ceramics. Thus, the biocer is only an intermediate stage in the processing route finally leading to an inorganic structure. Hereby, the role of the microorganisms is manifold. Their size and volume fraction determine the pore structure. In this way, silica tubes (inner diameter 20 nm, length about 1 μm) have been prepared by end-to-end self-aggregation of tobacco mosaic viruses (Fowler *et al.*, 2001) (see also Section 7.3). Macroscopic threads of *B. subtilis* have been used as templates for the manufacturing of silica ropes with 0.5 μm channels (Davis *et al.*, 1997). Monolayers of hexagonally packed yeast cells were embedded in silica films to create periodic two-dimensional pore arrays (pore size about 5 μm) (Figure 5.30). The high regularity of the pore pattern makes them candidates for structured substrates for gene screening libraries (Chia *et al.*, 2000).

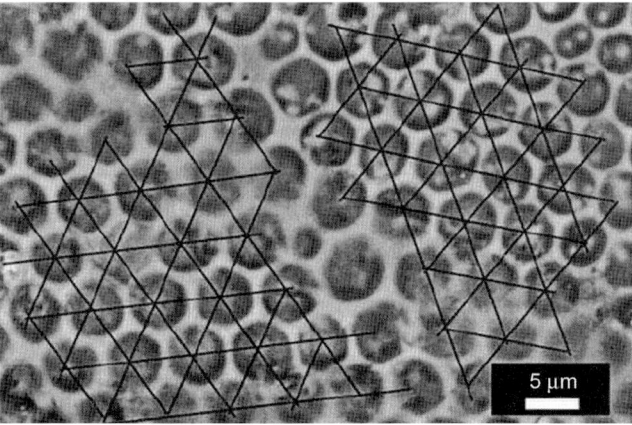

Figure 5.30 Silicate film containing a monolayer of hexagonal close-packed yeast cells. The ordered regions are indicated by lines. Defects caused by the presence of smaller cells are found in the center and upper left of the image. (Reprinted with permission from Chia *et al.* (2000). Copyright 2000, the American Chemical Society.)

5.4
Silicified Geological Biomaterials

Biocers with embedded microorganisms can also be found as geological material representing one of the oldest composite materials. Indications for the formation of natural composites composed by microorganisms and inorganic oxide (or carbonate)-based matrices can be observed in some microfossils found in Precambrian geological formations. The conditions in the Precambrian ocean – a biology dominated by bacteria and cyanobacteria existing in an environment enriched with silica, carbonates, and iron oxides – gave the natural setting for the formation of biocomposites.

There is ample evidence for natural biocomposite formation:

- Maritime sediment rocks with embedded layered cyanobacteria mats (stromatolites) at the west coast of Australia and South Africa.
- Proterozoic banded iron ores precipitated by cyanobacteria, with the latter being incorporated and still visible as fossils in Minnesota, USA.
- Cyanobacteria enclosed in silica stromatolite and cryptoendolithic communities found in hot spring sinters in Kenya (Renaut, Jones, and Tiercelin, 1998) and Iceland (Konhauser *et al.*, 2001). The microbial populations within the microstromatolite appear viable; that is, they still have their pigmentation, the trichomes are not collapsed, cell walls are unbroken, cytoplasm is still present, and they proved culturable.
- Permian chert with hematite and goethite decoration of microbial mats (Döhlen basin, Germany).

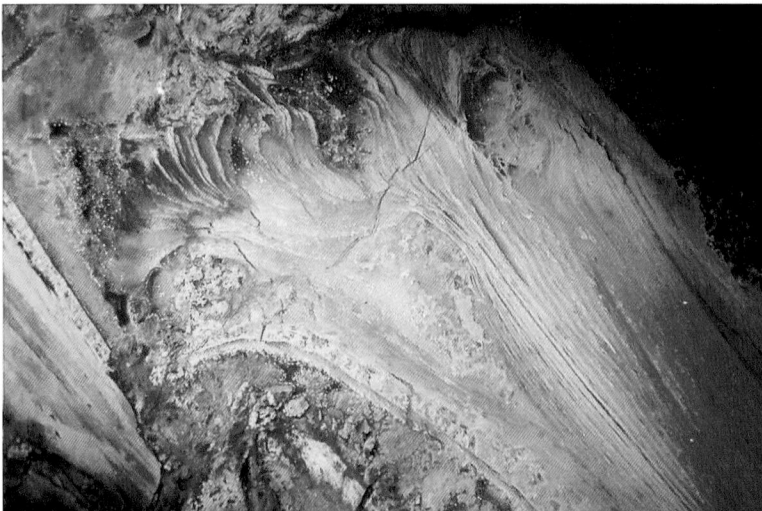

Figure 5.31 Iron oxides precipitated probably by cyanobacteria grown in very fine laminae before all turned into chert by silicification of the watery habitat. Width of the picture: 17.5 mm.

Large iron ore deposits known as banded iron formations were formed about 2×10^9 years ago as a result of photosynthetic oxygen production by cyanobacteria. This has been elaborated in Taylor, Taylor, and Krings (2009), where it is stated that "banded iron formations are laminated units containing iron oxides, sometimes in very fine laminae, separated by silica-rich layers." This can be vividly illustrated by much younger terrestrial formations as preserved in Lower Permian cherts (Figure 5.31).

The finely laminated iron oxide precipitates found in some of the red and yellow cherts from Döhlen basin look so similar to those from the Proterozoic banded iron formation, which are thought to be brought about by the physiological activity of cyanobacteria, that one can well assume that they, too, had originated in that way, although the involvement of cyanobacteria in the cherts considered here has never been proven directly. The present sample is a fragment of a chert layer of at least 23 cm thickness. Judging from the presence of foliage fragments and air-filled roots of the tree fern *Scolecopteris*, the colorful structures must have formed in a shallow swamp near the surface, since the cyanobacteria need water to live in and light for photosynthesis. This is compatible with the observation that the laminated precipitates are always disturbed as a result of some commotion in the swamp, possibly caused by strong winds, animals, and falling fern fronds. The disturbances occurred during different stages of silicification, as can be concluded from Figure 5.31. Obviously, the stack of layers had been torn into pieces while all was still fluid, and the individual limpid sheets separated themselves at the end of the piece seen here. Later, when the deformed stack fragment had obtained a jelly consistence, a crack ran across but no further.

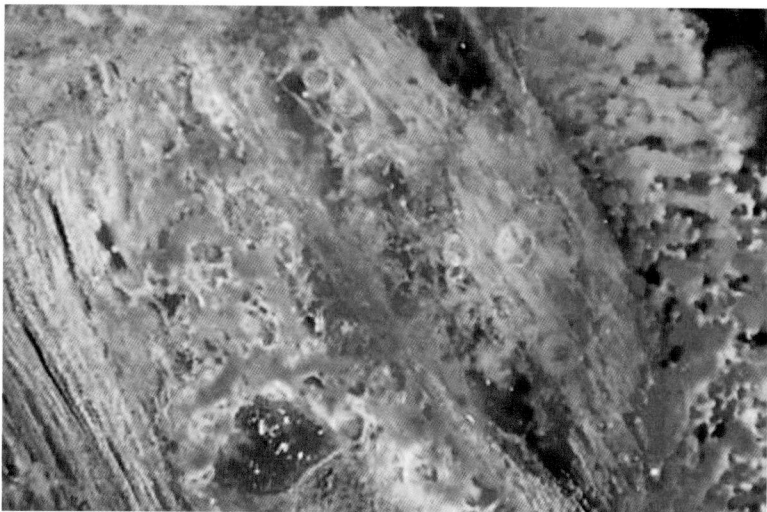

Figure 5.32 Chert with iron oxides precipitated on microbial formations grown among plant debris; fungus chlamydospores are faintly seen as transparent hollow spheres. Width of the picture: 5.5 mm.

This indicates that silicification combined with shrinkage was faster in the layer stack than in its vicinity. Microbes in water come more often as a multitude of floccules than as a stack of thin sheets. It is not known whether or not the sheets and floccules in Figure 5.32 (below left and above right) represent the same microbe species. Occasionally seen smooth transitions between the two forms seem to indicate that they do.

There are more features worth mentioning in Figure 5.32. A former cavity, probably a swamp gas bubble, is now filled with coarse quartz. A similar cavity with a level fill (not shown here) reveals the orientation of the sample during silicification so that it can be stated that the orientation of Figures 5.31 and 5.32 is the original one. Small bubble-like objects, 0.17–0.27 mm across, faintly seen randomly arranged along the inclined strip of decayed plant matter in Figure 5.32, are most probably "chlamydospores" of some fungus. They and the related hyphae are rare in the cherts of the Döhlen basin but abundant in the Rhynie chert. Besides the cyanobacteria or blue-green algae, their presence serves as additional evidence that this type of chert, like lots of others, was not formed by silicification of some sediment layer but by swamp water turning into silica gel that finally became chert. This has to be emphasized in order to contradict views adapted from outdated textbooks according to which cherts are generally formed by silicification of strata in the depth (Rösler *et al.*, 2007). It appears that by considering both the technology of artificial biocer preparation and fossil evidence from paleontology, either field of human aspiration may benefit from progress in the other one.

Task T5

5.1 Task 5.1: Estimate the pH in the small cavity of a silica gel in which an alkaline phosphatase molecule is entrapped. As shown in Figure 5.24, the cavity contains only 100 water molecules and two hydrogen ions (H^+) arranged in hydronium complexes (H_3O^+).

References

Al-Saraj, M., Abdel-Latif, M.S., El-Nahal, I., and Baraka, R. (1999) Bioaccumulation of some hazardous metals by sol–gel entrapped microorganisms. *Journal of Non-Crystalline Solids*, **248**, 137–140.

Avnir, D., Coradin, T., Lev, O., and Livage, J. (2006) Recent bio-applications of sol–gel materials. *Journal of Materials Chemistry*, **16**, 1013–1030.

Bender, J., Duff, M.C., Phillips, P., and Hilll, M. (2000) Bioremediation and bioreduction of dissolved U(VI) by microbial mat consortium supported on silica gel particles. *Environmental Science & Technology*, **34**, 3235–3241.

Böttcher, H. (2000) Bioactive sol–gel coatings. *Journal für Praktische Chemie*, **342** (5), 427–436.

Böttcher, H., Soltmann, U., Mertig, M., and Pompe, W. (2004) Biocers: ceramics with incorporated microorganisms for biocatalytic, biosorptive and functional materials development. *Journal of Materials Chemistry*, **14**, 2176–2188.

Branyik, T., Kuncova, G. et al. (2000) The use of silica gel prepared by sol–gel method and polyurethane foam as microbial carriers in the continuous degradation of phenol. *Applied Microbiology and Biotechnology*, **54** (2), 168–172.

Branyik, T., Kuncová, G., Páca, J., and Demnerová, K. (1998) Encapsulation of microbial cells into silica gel. *Journal of Sol-Gel Science and Technology*, **13**, 283–287.

Brunner, E., Richthammer, P. et al. (2009) Chitin-based organic networks: an integral part of cell wall biosilica in the diatom *Thalassiosira pseudonana*.

Angewandte Chemie – International Edition, **48** (51), 9724–9727.

Campostrini, R., Carturan, G., Caniato, R., Piovan, A., Fillipini, R., Innocenti, G., and Cappelletti, E.M. (1996) Immobilization of plant cells in hybrid sol–gel materials. *Journal of Sol-Gel Science and Technology*, **7** (1–2), 87–97.

Carturan, G., Campostrini, R., Diré, S., Scardi, V., and De Alteriis, E. (1989) Inorganic gels for immobilization of biocatalysts: inclusion of invertase-active whole cells of yeast (*Saccharomyces cerevisiae*) into thin layers of SiO_2 gel deposited on glass sheets. *Journal of Molecular Catalysis*, **57**, L13–L16.

Carturan, G., Dal Monte, R., Pressi, G., Secondin, S., and Verza, P. (1998) Production of valuable drugs from plant cells immobilized by hybrid sol–gel SiO_2. *Journal of Sol-Gel Science and Technology*, **13**, 273–276.

Chia, S., Urano, J., Tamanoi, F., Dunn, B., and Zink, J.I. (2000) Patterned hexagonal arrays of living cells in sol–gel silica films. *Journal of the American Chemical Society*, **122**, 6488–6489.

Dal Monte, R. (2000) The Biosil membrane: biomedical and industrial applications. *Journal of Sol-Gel Science and Technology*, **18**, 287–292.

David, A.E., Yang, A.J., and Wang, N. S. (2011) Enzyme stabilization and immobilization by sol–gel entrapment. *Methods in Molecular Biology (Clifton, NJ)*, **679**, 49–66.

Davis, S.A., Burkett, S.L., Mendelson, N.H., and Mann, S. (1997) Bacterial templating of ordered macrostructures in silica and silica–surfactant mesophases. *Nature*, **385**, 420–423.

Depagne, C., Roux, C. *et al.* (2011) How to design cell-based biosensors using the sol–gel process. *Analytical and Bioanalytical Chemistry*, **400** (4), 965–976.

Ehrlich, H. (2010) *Biological Materials of Marine Origin – Invertebrates*, Springer, Berlin.

Ehrlich, H., Deutzmann, R. *et al.* (2010) Mineralization of the metre-long biosilica structures of glass sponges is templated on hydroxylated collagen. *Nature Chemical Biology*, **2** (12), 1084–1088.

Fiedler, D., Hager, U., Franke, H., Soltmann, U., and Böttcher, H. (2007) Algae biocers: astaxanthin formation in sol–gel immobilised living microalgae. *Journal of Materials Chemistry*, **17**, 261–266.

Fiedler, D., Thron, A., Soltmann, U., and Böttcher, H. (2004) New packing materials for bioreactors based on coated and fiber-reinforced biocers. *Chemistry of Materials*, **16**, 3040–3044.

Fowler, C.E., Shenton, W., Stubbs, G., and Mann, S. (2001) Tobacco mosaic virus liquid crystals as templates for the interior design of silica mesophases and nanoparticles. *Advanced Materials*, **13**, 1266–1269.

Frenkel-Mullerad, H. and Avnir, D. (2005) Sol–gel materials as efficient enzyme protectors: preserving the activity of phosphatases under extreme pH conditions. *Journal of the American Chemical Society*, **127** (22), 8077–8081.

Gadd, G. M. and White, C. (1993) Microbial treatment of metal pollution: a working biotechnology? *Trends in Biotechnology*, **11** (8), 353–359.

Gautier, C., Livage, J. *et al.* (2006) Sol–gel encapsulation extends diatom viability and reveals their silica dissolution capability. *Chemical Communications (Cambridge, England)*, (44), 4611–4613.

Gill, I. and Ballesteros, A. (1998) Encapsulation of biologicals within silicate, siloxane, and hybrid sol–gel polymers: an efficient and generic approach. *Journal of the American Chemical Society*, **120**, 8587–8598.

Heinemann, S., Heinemann, C. *et al.* (2009) Bioactive silica–collagen composite xerogels modified by calcium phosphate phases with adjustable mechanical properties for bone replacement. *Acta Biomaterialia*, **5** (6), 1979–1990.

Heinemann, S., Heinemann, C., Ehrlich, H., Meyer, M., Baltzer, H., Worch, H., and Hanke, T. (2007) A novel bomimetic hybrid material made of silicified collagen: perspectives for bone replacement. *Advanced Engineering Materials*, **9** (12), 1061–1068.

Hildebrand, M., York, E., Kelza, J.I., Davis, A.K., Frigeri, L.G., Allison, D.P., and Doktycz, M.J. (2006) Nanoscale control of silica morphology and three-dimensional structure during diatom cell wall formation. *Journal of Materials Research*, **21**, 2689–2698.

Jeffryes, C., Campbell, J., Li, H., Jiao, J., and Rorrer, G. (2011) The potential of diatom nanobiotechnology for applications in solar cells, batteries, and electroluminescent devices. *Energy & Environmental Science*, **4**, 3930–3941.

Konhauser, K.O., Phoenix, V.R., Bottrell, S. H., Adams, D.G., and Headà, I.M. (2001) Microbial–silica interactions in Icelandic hot spring sinter: possible analogues for some Precambrian siliceous stromatolites. *Sedimentology*, **48**, 415–433.

Kuncova, G., Triska, J., Vrchotova, N., and Podrazky, O. (2002) The influence of immobilization of *Pseudomonas* sp. 2 on optical detection of polychlorinated biphenyls. *Materials Science and Engineering C*, **21**, 195–201.

Künzelmann, U. and Böttcher, H. (1997) Biosensor properties of glucose oxidase immobilized within SiO_2 gels. *Sensors and Actuators B*, **38–39**, 222–228.

Nassif, N., Bouvet, O. *et al.* (2002) Living bacteria in silica gels. *Nature Materials*, **1** (1), 42–44.

Pierre, A.C. (2004) The sol–gel encapsulation of enzymes. *Biocatalysis and Biotransformation*, **22** (3), 145–170.

Pope, E.J.A. (1995) Gel encapsulated microorganisms: *Saccharomyces cerevisiae*–silica gel biocomposites. *Journal of Sol-Gel Science and Technology*, **4**, 225–229.

Pope, E.J.A., Braun, K., and Peterson, C.M. (1997) Bioartificial organs I: silica gel encapsulated pancreatic islets for the treatment of diabetes mellitus. *Journal of Sol-Gel Science and Technology*, **8**, 635–639.

Premkumar, J.R., Sagi, E., Rozen, R., Belkin, S., Modestov, A.D., and Lev, O. (2002) Fluorescent bacteria encapsulated in sol–gel derived silicate films. *Chemistry of Materials*, **14**, 2676–2686.

Pressi, G., Dal Toso, R., Dal Monte, R., and Carturan, G. (2003) Production of enzymes by plant cells immobilized by sol–gel silica. *Journal of Sol-Gel Science and Technology*, **26**, 1189–1193.

Raff, J., Soltmann, U., Matys, S., Selenska-Pobell, S., Böttcher, H., and Pompe, W. (2003) Biosorption of uranium and copper by biocers. *Chemistry of Materials*, **15**, 240–244.

Reetz, M.T., Zonta, A., and Simpelkamp, J. (1995) Efficient heterogeneous biocatalysts by entrapment of lipases in hydrophobic sol–gel materials. *Angewandte Chemie – International Edition in English*, **34**, 301–303.

Renaut, R.W., Jones, B., and Tiercelin, J.-J. (1998) Rapid *in situ* silicification of microbes at Loburu hot spring, Lake Bogoria, Kenya Rift Valleys. *Sedimentology*, **45**, 1083–1103.

Rietti-Shati, M. (1996) Atrazine degradation by *Pseudomonas* strain ADP entrapped in sol–gel glass. *Journal of Sol-Gel Science and Technology*, **7**, 77–79.

Rösler, R., Zierold, T., Spindler, F., and Rudolph, F. (2007) Strandsteine. *Veröff Mus Naturkunde Chemnitz*, **30**, 5–24.

Ruiz-Hitzky, E., Ariga, K., and Lvov, Y. (eds) (2008) *Bio-Inorganic Hybrid Nanomaterials*, Wiley-VCH Verlag GmbH, Weinheim.

Scheffel, A., Poulsen, N., *et al.* (2011) Nanopatterned protein microrings from a diatom that direct silica morphogenesis. *Proc Natl Acad Sci USA* **108**, 3175-3180.

Schroeder, H.C., Boreiko, A. *et al.* (2006) Co-expression and functional interaction of silicatein with galectin: matrix-guided formation of siliceous spicules in the marine demosponge *Suberites domuncula*. *The Journal of Biological Chemistry*, **281** (17), 12001–12009.

Shimizu, K., Cha, J. *et al.* (1998) Silicatein α: cathepsin L-like protein in sponge biosilica. *Proceedings of the National Academy of Sciences of the United States of America*, **95** (11), 6234–6238.

Singh, S., Bhatta, U.M., Satyam, P.V., Dhawan, A., Sastry, M., and Prasad, B.L.V. (2008) Bacterial synthesis of silicon/silica nanocomposites. *Journal of Materials Chemistry*, **18**, 2601–2606.

Soltmann, U. and Böttcher, H. (2008) Utilization of sol–gel ceramics for immobilization of living microorganisms. *Journal of Sol-Gel Science and Technology*, **48**, 66–72.

Soltmann, U., Matys, S., Kieszig, G., Pompe, W., and Böttcher, H. (2010) Algae–silica hybrid materials for biosorption of heavy metals. *Journal of Water Resource and Protection*, **2**, 115–122.

Sumper, M. and Brunner, E. (2008) Silica biomineralization in diatoms: the model organism *Thalassiosira pseudonana*. *ChemBioChem: A European Journal of Chemical Biology*, **9** (8), 1187–1194.

Taylor, T.N., Taylor, E.L., and Krings, M. (2009) *Paleobotany*, Elsevier, Amsterdam.

van Bommel, K.J., Friggeri, A. *et al.* (2003) Organic templates for the generation of inorganic materials. *Angewandte Chemie – International Edition*, **42** (9), 980–999.

6
Biomineralization

6.1
Case Studies

Biologically Controlled Mineralization of Magnetic Nanoparticles in Magnetotactic Bacteria Magnetotactic bacteria had been discovered in marine marsh mud by R.P. Blackmore in 1975 and named *Magnetospirillum magnetotacticum*. The characteristic feature of magnetotactic bacteria is the existence of intracellular magnetic inclusions called magnetosomes (Figures 6.1 and 6.2). These are magnetic iron minerals (magnetite – Fe_3O_4 or greigite – Fe_3S_4) enclosed in membrane-bound vesicles. They are aligned in a chain and generate a resulting magnetic dipole moment. Until now, about 10 pure cultures of magnetotactic bacteria have been described. They give a beautiful example of the formation of nanosize functional materials in living organisms under *biological control*. It also demonstrates how, by way of evolution, the structure of these materials is optimized with respect to the functionality needed for particular conditions of life. Driven by flagella motors, the magnetotactic bacteria can move along the magnetic field lines. In marine areas they are found mainly at freshwater sites near the sediment–water interface, and in marine muds they are found under microaerobic to anaerobic conditions (Figure 6.3). The bacteria find the best growth conditions at the oxic–anoxic transition zone (OATZ). For this behavior, the term "magneto-aerotaxis" has been introduced. When the bacteria leave this zone for some reason, they can use the vertical component of the geomagnetic field as guidance toward the optimum depth. For this motion, the direction of the flagellar rotation is coupled to an aerotactic sensory system that acts as a switch when cells are at a suboptimal position.

Two different kinds of magneto-aerotactic bacterial strains have been found. There are strains called polar magneto-aerotactic which swim persistently in one direction along the magnetic field. There are other strains called axial magneto-aerotactic which swim in either direction along the magnetic field lines with frequent reversals of the swimming direction without turning around. In both cases, the existence of the magnetotactic system reduces the search for optimal oxygen concentration from three dimensions to one. Possibly, for the

Bio-Nanomaterials: Designing Materials Inspired by Nature, First Edition. Wolfgang Pompe, Gerhard Rödel, Hans-Jürgen Weiss, and Michael Mertig.
© 2013 Wiley-VCH Verlag GmbH & Co. KGaA. Published 2013 by Wiley-VCH Verlag GmbH & Co. KGaA.

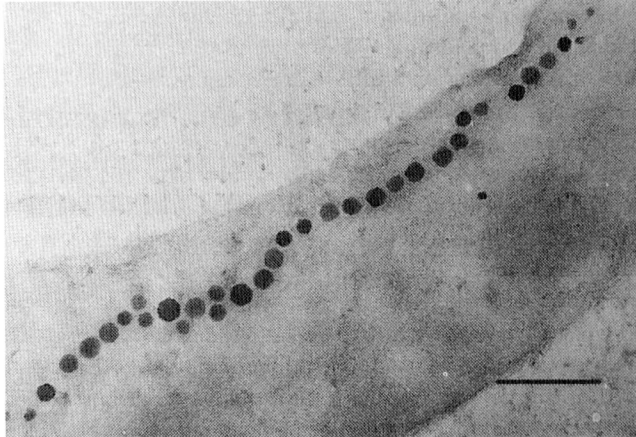

Figure 6.1 TEM of *Magnetospirillum gryphiswaldense*. Magnification of a cell with more than 30 cubo-octahedral crystals, maximal diameter 42–45 nm. Scale bar: 200 nm. (Reproduced with permission from Baeuerlein (2000), Copyright 2000, Wiley-VCH Verlag GmbH.)

sulfate-reducing bacteria which produce the ferrimagnetic greigite, there is an alternative magneto-chemotactic system responding to an optimum concentration of molecules as sulfide.

The OATZ also offers optimized conditions for the intracellular formation of magnetite nanocrystals. Free oxidized Fe(III) that is normally insoluble in water at biological pH can be transported to the cell as a Fe(III)–chelate complex. During crossing the membrane, it is reduced to Fe(II). Inside the cell, the Fe(II) ions are incorporated into vesicles aligned along the cell membrane. Within the vesicle

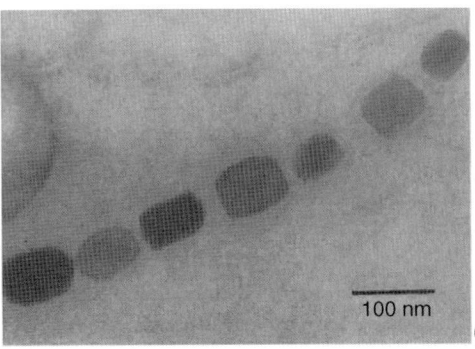
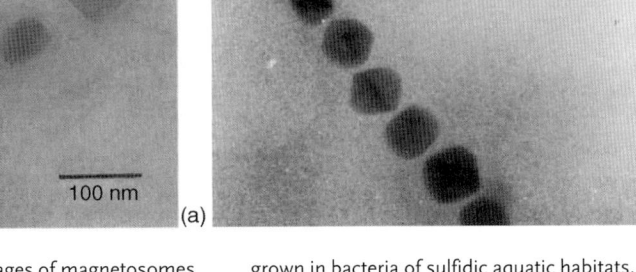

Figure 6.2 TEM images of magnetosomes. (a) Truncated hexahedral crystals of magnetite (Fe_3O_4) grown in marine vibroid strain MV-1. (b) Cubo-octahedral crystals of greigite (Fe_3S_4) grown in bacteria of sulfidic aquatic habitats. (Reproduced with permission from Bazylinski *et al.* (1994), Copyright 1994, John Wiley & Sons, Inc.)

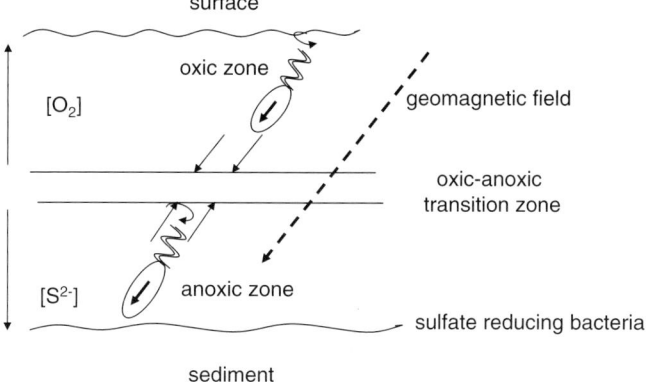

Figure 6.3 Schematics of the orientation of magnetotactic bacteria along the geomagnetic field on the Northern Hemisphere when moving toward the OATZ. There are optimized conditions for the intracellular formation of ferrimagnetic magnetite nanocrystals Fe_3O_4. Near the sediment, the enriched $[S^{2-}]$ concentration leads to the formation of ferrimagnetic greigite Fe_3S_4 by sulfate-reducing bacteria. (Adapted from Bazylinski and Frankel (2004).)

(magnetosome), amorphous hydrated Fe(III) oxide is precipitated first. In a following step, the amorphous phase transforms into the mixed valence magnetite Fe_3O_4. Magnetite is a ferrimagnetic material. The interaction of atomic magnetic moments leads to an arrangement of atomic moments in two sublattices of opposite orientation and different field strength. Therefore, they do not fully compensate and so an effective magnetic moment remains. In macroscopic crystals, there are randomly oriented magnetic domains with homogeneous magnetization. Below 5 nm size, thermodynamic fluctuations prevent the alignment of the atomic moments into a homogeneous domain. The fluctuations can be overcome by an external field known as *superparamagnetism*. Above a certain size of the crystal, the magnetic field energy of a single domain would be sufficiently high to cause a rearrangement of the atomic moments into multiple domains of random orientation. Hence, there is a critical size range of the nanocrystals with a magnetic single-domain structure. In magnetotactic bacteria, the growth of magnetite particles is restricted by the lipid cage of the magnetosome to a size range of 30–45 nm, where the single-domain structure is stable. A few morphological variants have been observed: cubo-octahedrons, elongated and flat hexagonal prisms, often also twinned single crystals (Figure 6.2). On average, there are 11–12 magnetosomes aligned as a chain along the cell wall. Their magnetization points in the same direction along the chain axis. Thus, an internal microcompass is formed within the magnetotactic bacteria, which enables them to use the geomagnetic field for orientation.

Freeze-Avoiding or Freeze-Tolerant Organisms   Droplets of pure water can be supercooled down to about $-40\,^{\circ}C$ without freezing. Only very small ice nuclei are formed and dissolved again in this metastable range below the freezing point. Near $-40\,^{\circ}C$, called supercooling point, nucleation temperature, or crystallization temperature, stable ice crystals are formed by homogeneous nucleation. The presence of an interface, such as dust grains or container walls, reduces the existing thermo-dynamic barrier for nucleation so that water practically freezes always at 0 °C. In every region of the globe where temperatures temporally sink below 0 °C, animals are confronted with the problem of freezing. Although some of them make use of thermally buffered microhabitats for hibernation, a larger part of them have developed one of two basic physiological adaptation mechanisms: *freeze avoidance* (the ability to prevent freezing) and *freeze tolerance* (the ability to survive body fluid freezing) (Duman, 2001). Already in the late 1960s, the biologist Arthur DeVries investigated glycoproteins that have been identified as antifreeze agents in the blood of Arctic fishes. Subsequently, a large group of *antifreeze proteins* (AFPs) were found in many different freeze-avoiding animals. In the same circle of studies, the so-called *protein ice nucleators* (PINs), which play a role in freeze-tolerant organisms, have also been found.

AFPs include a larger class of polypeptides that can be found in vertebrates, plants, fungi, and bacteria living in subzero environments. They bind at certain lattice planes of ice crystals and thus inhibit growth and recrystallization. The fairly high number of AFPs have different structures but same effect due to two reasons: a physicochemical and an evolutionary one. Ice crystals offer various faces that could be inhibited by polypeptides with different amino acid sequences. Inhibition can already be realized when the polypeptide binds at the lattice plane at a certain periodically distributed distances like a zipper. This means that not only the primary sequence of amino acids but also the tertiary structure of the folded protein governs the successful inhibition. Thus, five types of AFP differing in their primary sequences possess a similar tertiary structure inhibiting growth of ice crystals. Interestingly, also differences in the climate evolution in the Northern and Southern hemispheres seem to be the reason that two independent developments led to different results in the biological adaptation. Sea level glaciation dates back 1–2 million years in the Northern Hemisphere but 10–30 million years in Antarctica.

Quite another aspect of freeze tolerance is the ability of freeze-tolerant bacteria to serve at nuclei for ice formation and their impact on the ecosystem. They can be present in clouds and thus can be spread by rainfall to new habitants. Their nucleation effect is brought about by special bonding sites on their surface. Freeze-dried preparations of ice-nucleating bacteria are applied in the process of snow making at ski resorts. There is a serious debate concerning unwished effects on the ecosystem based on suspicions that this technology might prolong the duration of snow cover not only on the ski slopes but also in the vicinity.

6.2
Basic Principles

Characteristic sizes of inorganic constituents of biomaterials are often in the nanometer range. There are at least three reasons why evolution has brought forth such types of materials. First, *nanostructured inorganic materials* often show *exceptional functional properties* that cannot be realized with bulk materials. One example has been considered in Section 6.1 with the single-domain behavior of nanoscale magnetic particles used for magnetic navigation by bacteria. Second, *hierarchically structured nanocomposites* enable physical and chemical properties to be optimized simultaneously on the micro-, meso-, and macroscale. An instructive example is the remarkable mechanical properties of nanostructured calcium carbonate minerals in eggshells or nacreous layers and of calcium phosphate–collagen nanocomposites of hard tissue such as bone or tooth. It is really surprising that highly brittle materials can serve as components of macroscopic bodies with superior mechanical performance. Figures 6.4 and 6.5 show the hierarchical shell structure of *Strombus gigas*, the giant pink Queen Conch native to Caribbean habitats. It is composed of crossed crystalline lamellae of aragonite, a polymorph of calcium carbonate. Three orders of lamellae can be identified (Kuhn-Spearing *et al.*, 1996). The third-order lamellae with cross sections of about 100 nm × 250 nm and lengths of several micrometers serve as the basic building blocks. They are not homogeneous, but consist of twinned lattice domains of about 10–20 nm width. They are surrounded by a thin biopolymer film. Stacks of third-order lamellae are assembled into second-order lamellae with 5–30 μm in thickness and 5–60 μm in width. Stacks of second-order lamellae with the same orientation as the third-order stacks form a first-order lamella (5–60 μm thick, many micrometers wide). Staples of these first-order lamellae with differential orientations of the second-order lamellae are arranged in a three-layer structure, as shown in Figure 6.4. The layer thickness ranges between 0.5 and 2 mm. There are thin proteinaceous films between the lamellae of different orders and between the three layers. As a whole, the structure hierarchy of the mineral component comprises five length scales between the nanometer and the macroscale range. In this way, a hierarchy of strong components and weak interfaces of various orientations is obtained, which impedes the propagation of macrocracks (Figure 6.5). As discussed for bone fracture in Section 6.2.8.6, there are at least four

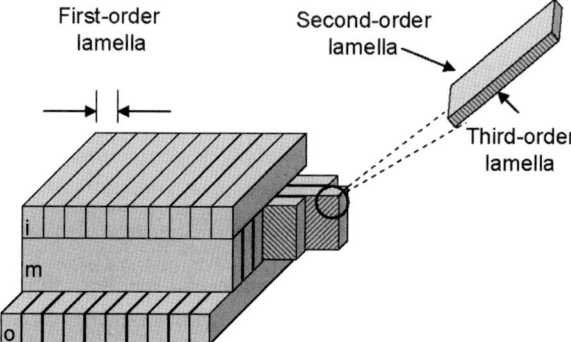

Figure 6.4 Cross-lamellar structure of the *S. gigas* conch shell (o: outer, m: middle, and i: inner layer). Each macroscopic layer is composed of first-, second-, and third-order lamellae. The third-order lamellae are twinned.

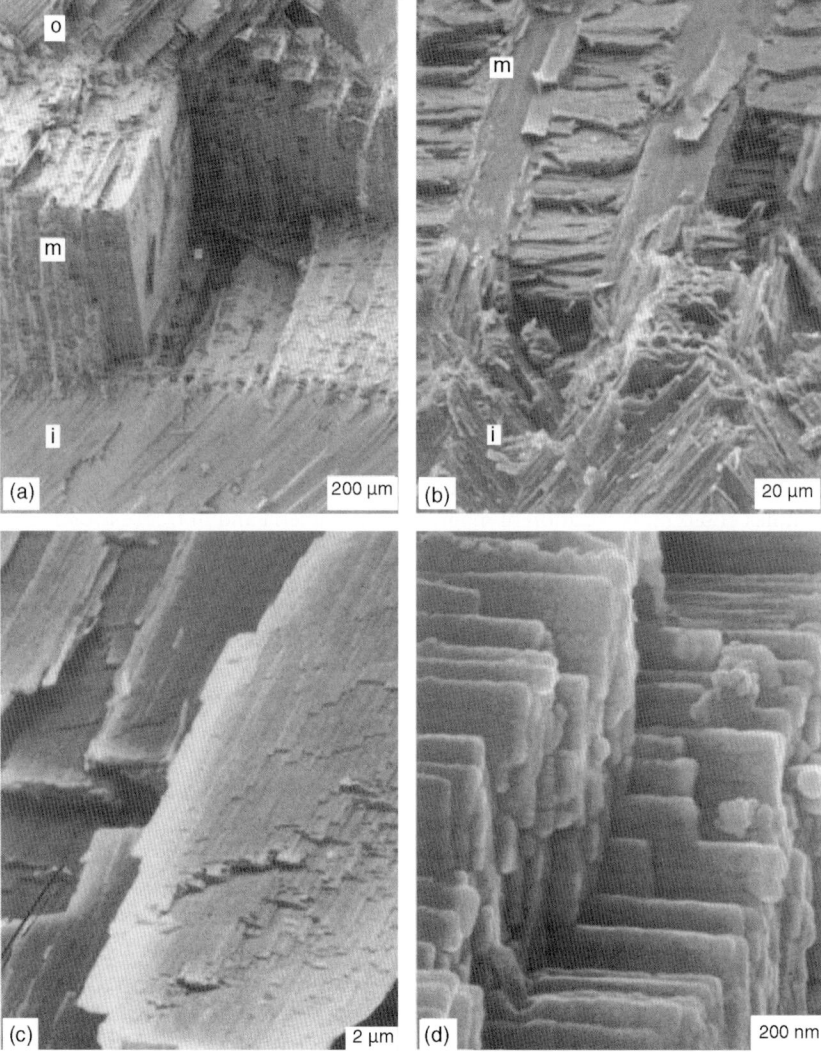

Figure 6.5 Fracture surface of a four-point bend specimen, showing extensive, delocalized, interfacial cracking. Note the increasing magnification. The crack propagated from the inner layer to the outer layer [bottom to top in part (a)], arresting temporarily at the inner–middle interface due to the change in microstructural orientation (orientation toughening). The first-order interfaces appearing smooth in part (a) appear "wood-like" at higher magnification (b). In part (c), the crack path within the second-order lamellae is directed along interfaces between third-order lamellae shown in part (d). (Reproduced with permission from Kuhn-Spearing *et al.* (1996), Copyright 1996, Springer Science+Business Media.)

microstructural mechanisms which also cause the nanostructured aragonite shell to be much tougher than the large-grained bulk aragonite (see also more data given in Kamat *et al.* (2000, 2004)).

Third, *nanostructures can change readily under external influences.* The large specific surface of nanoscale inorganic particles implies a strong thermodynamic driving force for dissolution–precipitation processes, transitions into other meta-stable states, or formation of amorphous phases with favorable consequences of high dissolution rates, high efficiency of mass transport in stable colloidal solution, and high growth rates of mineral phases. This has led to a variety of biomaterials with a particular ability for *remodeling and self-healing* at room temperature in a biological environment. On the other hand, there are numerous nanosized biominerals based on thermodynamically metastable phases with unexpectedly high stability. Apparently, biomineralization involves *two comple-mentary mechanisms: the deposition of the mineral phase and the growth inhibition on certain crystal faces by aggregation of some biomolecular ligands.* This can lead to structures differing from those grown in the absence of biomolecular ligands. The examples of biomineralization in magnetotactic bacteria and in freeze-avoiding or freeze-tolerant organisms discussed in the case studies have shown that the different size, shape, and composition of the biominerals are often connected with unusual functional properties.

The intriguing phenomena of biomineralization have inspired scientists from several professions to make thorough research. The explored basic principles gave a fundamental for establishing activities in various other scientific areas. First of all, the high impact on chemistry has to be mentioned. Bio-inspired materials chemis-try, an exciting new field of bioinorganic chemistry, has been shaped by Mann in the past 30 years (Mann *et al.*, 1993; Mann, 1996, 2000, 2001, 2009). This research has laid the base for innovative engineering applications of bionanotechnology, as there are new catalysts, magnetic and electric materials, photonic and plasmonic materi-als, and even improved structural materials for civil engineering. Another important contribution of bio-inspired materials science has been made to the engineering of materials for regenerative therapy of bone diseases. Furthermore, novel functional materials based on controlled biomineralization are becoming increasingly relevant for purification of polluted water and for bioremediation.

As to be discussed in the following, the selection process of natural evolution has brought forth biological materials and processing routes, some of being essentially superior to the conventional processing techniques of inorganic materials. This chapter is meant to give an overview of the basic principles of biomineralization. Beginning with the classical physicochemical concepts of mineralization in aqueous solutions, it will be shown how the presence of biomolecular structures causes remarkable changes in the reaction pathway and in the final outcome of mineral precipitation. By an appropriate choice of bimolecular scaffolds or ligands, the size, shape, and composition of the mineral can be tailored. Growth and remodeling of bone, which serves as an example of hierarchical architecture produced by bio-mineralization, will be discussed in detail.

6.2.1

Precipitation

Mineralization from the aqueous solution inside cells follows the well-known rules of thermodynamics, according to which any given amount of substance tends to get into a state of minimum free enthalpy provided that temperature and pressure are kept constant. Hence, the difference between the higher free enthalpy per amount of substance in the nonequilibrium state and the minimum free enthalpy per amount of substance, denoted here as Δg, is a measure for the tendency to get into the equilibrium state. Hence, Δg may be called the driving force of the process. This applies to processes like condensation of supersaturated vapor into liquids or solids, freezing of undercooled liquids, and crystallization of solvents in supersaturated solutions, as well as these processes running in the opposite way.

In the simplest case where the area of surfaces or interfaces does not vary in the process, as with vapor condensing onto a liquid surface in a pot, condensation of an amount of vapor into liquid of volume V changes the free enthalpy by $\Delta G = -\Delta g \cdot V$, where Δg is the free enthalpy difference per volume of liquid. (Note that here Δg is positive by definition.) If condensation goes along with increasing surface, as with the nucleation of tiny droplets or crystals in the absence of any preexisting surface, the full driving force Δg does not realize since the atoms or molecules on the surface are not as tightly bound as those in the bulk. Since they have been included into the term $-\Delta g \cdot V$ as if they were tightly bound, a correction is necessary, which is done by the additional term $\gamma \cdot A$ with opposite sign, where A is the surface area and γ is the free enthalpy related to the weaker bonding, per area of surface.

With this formalism, the difference between the free enthalpy of a precipitate and the free enthalpy of the same amount of substance in solution or as vapor is written as

$$\Delta G = -\Delta g \cdot V + \gamma \cdot A. \tag{6.1}$$

For simplicity, we consider spherical precipitates with radius r, in which case it is easily seen that ΔG is above zero for small precipitates with $r < 3\gamma/\Delta g$. This does not mean that precipitates of such size $\Delta G = 0$ necessarily dissolve. This becomes obvious by comparing two states of a given amount of substance: a precipitate with volume $V + dV$ and a precipitate with volume V and the incremental amount dV dissolved. Since the dissolved state has been chosen as the reference zero of the free enthalpy, it does not appear in the balance. Thus, the increment of ΔG related to an incremental change of volume can be written as

$$d\Delta G = -\Delta g \cdot dV + \gamma \cdot \frac{dA}{dV} \cdot dV. \tag{6.2}$$

Growing precipitate implies $d\Delta G < 0$ for $dV > 0$. It appears that $\gamma \cdot dA/dV$ acts as a retarding force that has to be overcome by the driving force Δg. For spherical precipitates again, (6.1) can be written in terms of radius and plotted as a curve (Figure 6.6) with a maximum, where $d\Delta G = 0$. With $dA/dV = 2/r$ for spheres, the

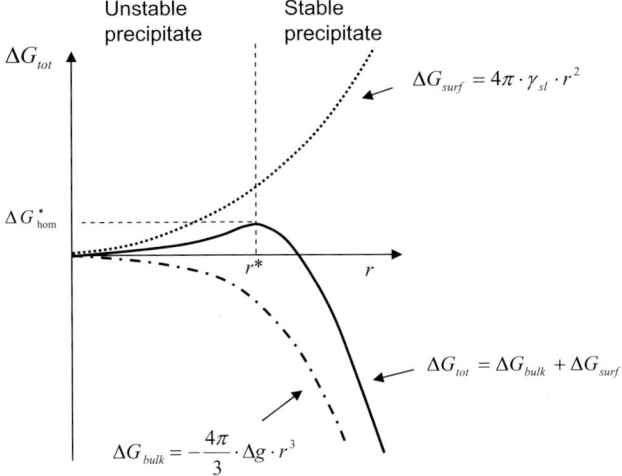

Figure 6.6 Difference in the free enthalpies of a substance in the solid and in the liquid solution, $\Delta G_{tot} = G_s - G_l$, versus radius of a spherical precipitate in a supersaturated solution.

position of the maximum, r^*, follows immediately from (6.2) as $r^* = 2\gamma/\Delta g$. Likewise, the growth condition $d\Delta G < 0$ becomes $r > 2\gamma/\Delta g$:

$$r^* = 2\gamma/\Delta g, \quad d\Delta G/dr < 0 \quad \text{for} \quad r > r^*. \tag{6.3}$$

It seems reasonable to assume that a similar relation is also valid for crystals, with some complication arising from the fact that different crystal faces have got differential γ values. From (6.2) follows that clusters with $r < r^*$ would never form from smaller ones by themselves. They have to be cobbled together by thermal fluctuations and are prone to decay. Among the clusters created by thermal fluctuations, only those with $r > r^*$ are able to grow, according to (6.3). Higher supersaturation implies higher driving force Δg, hence smaller critical cluster size r^*, so that thermal fluctuations produce clusters with $r > r^*$ at higher rate.

A relation between Δg and supersaturation is derived next.

6.2.1.1 Thermodynamics of Mineralization

Precipitation from a One-Component Solution Let us consider the formation of protein aggregates growing in a solution of monomers, as discussed in Chapter 7, in the particular case of the self-assembly of bacterial surface layer proteins. The free enthalpy of the solid phase depends on pressure and temperature, $G_s(p, T)$, whereas the free enthalpy of the solution $G_l(p, T, X_1)$ depends additionally on the concentration X_1 of the solute in the solvent. It is useful here to introduce the free enthalpy per molecule, also called its chemical potential: the chemical potential of the solid phase $\mu_s = \mu_s^*(p, T)$ depends on pressure and temperature, whereas the chemical potential of the solute μ_1 also depends on its concentration X_1. For calculating this dependence, we consider 1 mol of the solution, which means N_a particles altogether, $N_a \cdot X_1$ of the solute, and $N_a \cdot (1 - X_1)$ of the solvent. Avogadro's number $N_a = 6.022 \times 10^{23}$ is defined as the number atoms in 12 g of carbon. With this definition based on the atomic weight, N_a is also approximately the number of atoms in 1 g of hydrogen, or in 16 g of oxygen, and so on. Often the word *mole* is used as a synonym for Avogadro's number. A concentration of 1 mol/l is called one-molar solution, a unit denoted by the letter M. X_1 is called mole fraction of the solute. The total free enthalpy of the solution is given by

$$G_{solution} = N_a \cdot X_1 \cdot \mu_1(p, T, X_1) + N_a \cdot (1 - X_1) \cdot \mu_{solvent}(p, T, X_1). \tag{6.4}$$

Here, $\mu_1 \equiv \mu_{solute}$ and $\mu_{solvent}$ denote the chemical potentials of solute and solvent, respectively. In the simple case of an ideal solution, where the only effect of mixing is an increase of entropy by ΔS_{mix}, the total free enthalpy can be expressed by

$$G_{solution} = N_a \cdot X_1 \cdot \mu_1^*(p, T) + N_a \cdot (1 - X_1) \cdot \mu_{solvent}^*(p, T) - T\Delta S_{mix}. \tag{6.5}$$

ΔS_{mix} quantifies the additional randomness introduced by the presence of two types of molecules instead of one. It is defined as

$$\Delta S_{mix} = k_B \ln W, \tag{6.6}$$

where $k_B = 1.381 \times 10^{-23}$ J/K is the Boltzmann's constant and W is the number of distinguishable states that can be realized by permutation of the molecules among their fixed positions. In the trivial case of only one type of molecules, mixing would not have an effect, which means there would be only one state, $W = 1$, and zero entropy of mixing in agreement with (6.6). To calculate the number of states W in an ideal solution with $N_{solute} + N_{solvent} = N$ molecules, we have to keep in mind that the permutations of molecules of the same kind do not have any effect. Therefore, the $N!$ permutations of all molecules in the solution have to be divided by the permutations without effect in order to obtain the number of distinguishable macrostates:

$$W = \frac{N!}{N_{solute}! \cdot N_{solvent}!}. \tag{6.7}$$

(Porter and Easterling, 1992). For 1 mol of solution with the mole fractions X_1 of solute and $1 - X_1$ of solvent,

$$W = \frac{N_a!}{(N_a \cdot X_1)! \cdot (N_a \cdot (1 - X_1))!}. \tag{6.8}$$

With Stirling's approximation

$$\ln N! \cong N \ln N - N, \tag{6.9}$$

we get the contribution of mixing to the total free enthalpy from (6.8):

$$\Delta G_{mix} = -T\Delta S_{mix} = N_a \cdot k_B \cdot (X_1 \ln X_1 + (1 - X_1)\ln(1 - X_1)). \tag{6.10}$$

Comparison with Eq. (6.4) gives the chemical potential of the solute μ_1 in ideal solutions as

$$\mu_1 = \mu_1^*(p, T) + k_B \cdot \ln X_1. \tag{6.11}$$

Usually, solutions differ more or less strongly from ideal mixes, which means that mixing does not only affect the entropy but also gives rise to interactions between the components, resulting in either consumption or release of heat. Such interaction can be quantified in terms of enthalpy, known as the *quasi-chemical approach*:

$$\Delta H_{mix} = N_a \cdot \varepsilon \cdot X_1 \cdot (1 - X_1), \tag{6.12}$$

where ε is the difference between the bond energy of solvent–solute pairs of molecules and the average bond energy of solvent–solvent and solute–solute pairs:

$$\varepsilon = \varepsilon_{solvent,solute} - (\varepsilon_{solvent,solvent} + \varepsilon_{solute,solute})/2. \tag{6.13}$$

Note that the bond energies are negative quantities. Stronger bonds between solvent and solute than between the single species makes $\Delta H_{mix} < 0$, which stabilizes the solution. In the quasi-chemical approach, mixing of regular solutions is described by

$$\begin{aligned}\Delta G_{mix} &= \Delta H_{mix} - T\Delta S_{mix} \\ &= N_a \cdot \varepsilon \cdot X_1 \cdot (1 - X_1) + N_a \cdot k_B T \cdot (X_1 \ln X_1 + (1 - X_1)\ln(1 - X_1)).\end{aligned} \tag{6.14}$$

With the identity $X_1 \cdot (1 - X_1) = X_1^2 \cdot (1 - X_1) + X_1 \cdot (1 - X_1)^2$, the comparison with Eq. (6.4) shows that the chemical potential of the solute of a regular solution is given as

$$\mu_1 = \mu_1^*(p, T) + k_B T \cdot \ln X_1 + \varepsilon \cdot (1 - X_1)^2. \tag{6.15}$$

The last two terms in Eq. (6.15) can be combined into one by introducing the activity $\{X_1\}$ with

$$\ln\left(\frac{\{X_1\}}{X_1}\right) = \frac{\varepsilon}{k_B T} \cdot (1 - X_1)^2, \quad \mu_1(p, T, X_1) = \mu_1^*(p, T) + k_B T \cdot \ln\{X_1\}. \tag{6.16}$$

The change of activity with the molar fraction X_1 depends strongly on the bond energy ε, as shown in Figure 6.7. For a dilute solution with $X_1 \ll 1$, the activity can be approximated by Henry's law:

$$\frac{\{X_1\}}{X_1} = \exp\left(\frac{\varepsilon}{k_B T}\right) = \gamma_1, \tag{6.17}$$

where the activity coefficient γ_1 does not depend on the molar fraction X_1. As seen in Eq. (6.12), $\Delta H_{mix} > 0$ implies $\varepsilon > 0$, hence $\gamma_1 > 1$, which means the activity or the chemical potential can be interpreted as a measure of the tendency to leave the solvent. With increasing activity, the driving force for precipitation $\Delta\mu = \mu_1 - \mu_s$ increases.

With the known dependence of the chemical potential on the concentration, the condition for growth of a spherical precipitate can be formulated as $d\Delta G_{tot} < 0$, $dN_s > 0$ with

$$\begin{aligned}d\Delta G_{tot} &= (\mu_s - \mu_1) \cdot dN_s + \gamma_{sl} \cdot \frac{dA_{sl}}{dN_s} \cdot dN_s \\ &= (\mu_s - \mu_1) \cdot dN_s + \frac{2\gamma_{sl} \cdot v_s}{r} \cdot dN_s < 0,\end{aligned} \tag{6.18}$$

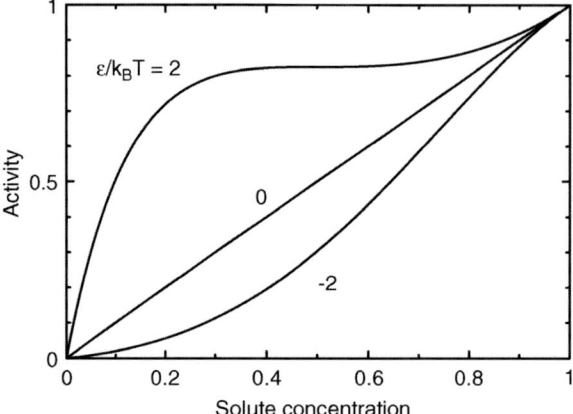

Figure 6.7 Activity $\{X_l\}$ versus solute concentration X_l for various values of the bond energy ε.

where $N_s = 4\pi r^3/(3v_s)$ is the number of molecules in the precipitate and v_s is the volume per molecule in the precipitate. In this formulation, the difference of the chemical potentials $\Delta\mu = \mu_l - \mu_s$ is the driving force for precipitation. In thermodynamic equilibrium, it is zero for large precipitates (with $2\gamma_{sl}/r \to 0$):

$$\Delta\mu = \mu_l(p, T, X_{eq}) - \mu_s^*(p, T) = 0. \tag{6.19}$$

With Eq. (6.16) we get

$$k_B T \ln X_{eq} + \varepsilon \cdot (1 - X_{eq})^2 = k_B T \ln\{X_{eq}\} = -\Delta\mu^*, \tag{6.20}$$

with $\Delta\mu^* = \mu_l^*(p, T) - \mu_s^*(p, T)$. In the case of low solubility, $X_{eq} \ll 1$, this reduces to

$$X_{eq} \cong \exp\left(-\frac{\Delta\mu^* + \varepsilon}{k_B T}\right). \tag{6.21}$$

For concentrations above X_{eq}, the driving force for precipitation, $\Delta\mu = \mu_l - \mu_s > 0$, is proportional to $\ln(X_1/X_{eq})$:

$$\Delta\mu = k_B T \ln\{X_1\} + \Delta\mu^* = k_B T \ln\left(\frac{\{X_1\}}{\{X_{eq}\}}\right) \cong k_B T \ln\left(\frac{X_1}{X_{eq}}\right). \tag{6.22}$$

The ratio $S_R = \{X_1\}/\{X_{eq}\} \cong X_1/X_{eq}$ is a measure for the relative supersaturation of the solution.

Chemically Controlled Dissolution–Precipitation Reaction   There are biomineralization processes where chemical reactions are involved in precipitation/dissolution. In the following, we will consider the formation of a solid phase $A_p B_q(s)$ by reaction of ions $A^{q+}(aq)$ and $B^{p-}(aq)$ in aqueous solution:

$$A_p B_q(s) \leftrightarrow p A^{q+}(aq) + q B^{p-}(aq). \tag{6.23}$$

In thermodynamic equilibrium, at constant pressure and temperature, the molar free enthalpy of the system keeps constant, while the molar fractions of the reaction components vary. This leads to the generalized form of Eq. (6.20):

$$k_B T \cdot \left(p \ln\{A_{eq}^{q+}\} + q \ln\{B_{eq}^{p-}\} \right) = \mu_s^* - (p\mu_A^* + q\mu_B^*) = -\Delta\mu^*, \qquad (6.24)$$

with the stoichiometric coefficients p and q. $\Delta\mu^* = p\mu_A^* + q\mu_B^* - \mu_s^*$ is called the standard Gibbs energy change for the reaction given in Eq. (6.24).

The equilibrium condition can be reformulated by introducing the solubility product K_{sp}:

$$\Delta\mu^* = -k_B T \ln K_{sp}, \qquad K_{sp} = \left\{A_{eq}^{q+}\right\}^p \cdot \left\{B_{eq}^{p-}\right\}^q. \qquad (6.25)$$

K_{sp} is a measure of the weighted activity product at equilibrium. Solubility products of typical biominerals cover a wide range, which means large differences of thermodynamic stability (Table 6.1).

The solubility product is only one of the parameters governing the formation of mineral phases in biomineralization. Usually it is meant to apply to large crystals.

According to (6.3), the driving force of precipitation includes a surface contribution and hence depends on the size of the precipitate. This implies that precipitates of unequal sizes cannot be in equilibrium with the solution and with each other. Under given conditions, there is only one equilibrium size. Smaller precipitates dissolve. They would be in equilibrium with higher concentrations, which is known

Table 6.1 Examples of solubility products for biominerals.

Mineral	Solubility product (log K_{sp})
Calcium carbonate	
Monohydrite ($CaCO_3 \cdot H_2O$)	−7.39
Vaterite ($CaCO_3$)	−7.60
Aragonite ($CaCO_3$)	−8.22
Calcite ($CaCO_3$)	−8.42
Calcium phosphate	
Brushite ($CaHPO_4 \cdot 2H_2O$)	−6.56
Octacalcium phosphate ($Ca_8H_2(PO_4)_6$)	−96.6
Hydroxyapatite ($Ca_{10}(PO_4)_6(OH)_2$)	−116.8
Fluorapatite ($Ca_{10}(PO_4)_6F_2$)	−118.0
Amorphous silica	−3.9
Iron oxides	
Ferrihydrate	−37.0
Goethite ($\alpha - FeOOH$)	−44.0
Gypsum ($CaSO_4 \cdot 2H_2O$)	−5.03

as the *Gibbs–Thomson effect*. Since precipitates are never of exactly equal size, they are never really in equilibrium. The larger precipitates get ever larger, while the smaller ones disappear. This phenomenon is known as *Ostwald ripening*. It is governed by the solubility product (6.25) extended by the surface term in (6.1):

$$\Delta\mu^* - \frac{2\gamma_{sl} \cdot v_s}{r} = -k_B T \ln K_{sp}. \tag{6.26}$$

In biomineralization, the surface effect is relevant for grain sizes below about 1 μm. There are various ways to control sizes of mineral precipitates, as pointed out in the following sections.

Controlling the solubility product can be a means for remodeling and healing of microdamaged structures. It is known that synthetic hydroxyapatite used in pastes or as implant coatings for bone repair defects shows an improved biodegradability for grain sizes below 100 nm range. Thus, nanosize biomimetic hydroxyapatite prepared near room temperature has some advantage over coarse-grain materials produced by plasma spraying or other high-temperature processing routes.

Analogous to the case of precipitation from a one-component solution, the driving force for precipitation $\Delta\mu$ can be expressed by the weighted activity product of a supersaturated solution:

$$AP = \{A^{q+}\}^p \cdot \{B^{p-}\}^q, \tag{6.27}$$

and the solubility product K_{sp}:

$$\Delta\mu = k_B T \ln AP + \Delta\mu^* = k_B T \ln\left(\frac{AP}{K_{sp}}\right) = k_B T \ln S_R, \tag{6.28}$$

where $S_R = AP/K_{sp}$ is defined as the relative supersaturation. For dilute solutions, the activities can be approximated by the corresponding expressions of the concentrations $[A^{q+}]$ and $[B^{p-}]$ similar to the last term in Eq. (6.22):

$$\Delta\mu \cong k_B T \ln\left(\frac{[A^{q+}]^p \cdot [B^{p-}]^q}{\left[A_{eq}^{q+}\right]^p \cdot \left[B_{eq}^{p-}\right]^q}\right). \tag{6.29}$$

6.2.1.2 Kinetics of Mineralization

Homogeneous Nucleation and Growth of Solid Particles in Liquids As already explained, precipitation proceeds in two stages: nucleation and growth of the precipitate (Figure 6.6). For a quantitative description of precipitation, we rewrite (6.1) with $\Delta g = \Delta\mu/v_s$ for a spherical precipitate (radius r) in the form

$$\Delta G = -\frac{4\pi}{3v_s} \cdot r^3 \cdot \Delta\mu + 4\pi \cdot r^2 \cdot \gamma_{sl}. \tag{6.30}$$

The radius r^* in (6.3) is rewritten in terms of v_s and $\Delta\mu$, and so is ΔG^*, the maximum of ΔG:

$$r^* = \frac{2\gamma_{sl} \cdot v_s}{\Delta\mu} \quad \text{and} \quad \Delta G^* = \frac{16\pi\gamma_{sl}^3 \cdot v_s^2}{3\Delta\mu^2}. \tag{6.31}$$

With Eqs. (6.22) and (6.29) for the thermodynamic driving force, the critical radius and the free enthalpy barrier can be written such that their dependence on the relative supersaturation S_R becomes apparent:

$$r^* = \frac{2\gamma_{sl} \cdot v_s}{k_B T \ln S_R} \quad \text{and} \quad \Delta G_{hom}^* = \frac{16\pi\gamma_{sl}^3 \cdot v_s^2}{3(k_B T \ln S_R)^2}. \tag{6.32}$$

Obviously, higher supersaturation makes lower critical particle size and lower activation barrier for homogeneous nucleation. As a condition for precipitation rate of practical relevance, the activation barrier for homogeneous nucleation ΔG_{hom}^* has to be so low that thermal fluctuations can create clusters exceeding the critical size r^* at a sufficiently high rate. With sufficiently high concentration of solute, n_0, the thermodynamic equilibrium distribution of subcritical clusters of size r can extend to the critical size r^* that enables the clusters to pass the activation barrier and become growing precipitates:

$$n^* = n_0 \cdot \exp\left(-\frac{\Delta G_{hom}^*}{k_B T}\right). \tag{6.33}$$

The stationary flow of clusters leaking from the pool of subcritical clusters over the G-threshold, J_{hom}, can be assumed to be proportional to n^*:

$$J_{hom} = f_0 \cdot n^*. \tag{6.34}$$

The factor f_0 has to be taken from the experiment. For practically relevant precipitation processes, it is about $f_0 = 10^{10} - 10^{11} \text{ s}^{-1}$. Relevant nucleation rates can be observed in homogeneous nucleation for $\Delta G_{hom}^*/k_B T < 60$. Growth of precipitates occurs beyond a supersaturation threshold that is proportional to γ_{sl}^3 (Figure 6.8).

Heterogeneous Nucleation   The activation barrier of homogeneous nucleation depends sensitively on the interface free enthalpy γ_{sl}. Therefore, the supersaturation needed for precipitation is much reduced by the presence of a template with lower interface energy $\gamma_{st} < \gamma_{sl}$. Let us consider a small isotropic nucleus at an isotropic template (Figure 6.9), which is in fact the well-known wetting problem.

The contact angle is determined by the balance of the interface stresses, which are manifestations of the interface free enthalpies:

$$\gamma_{tl} = \gamma_{st} + \gamma_{sl} \cdot \cos\theta. \tag{6.35}$$

Analogous to Eq. (6.30), the change of the free enthalpy is the sum of bulk and interface contributions:

$$\Delta G = -\Delta\mu \cdot N_s + \gamma_{sl} \cdot A_{sl} + \gamma_{st} \cdot A_{st} - \gamma_{tl} \cdot A_{st}. \tag{6.36}$$

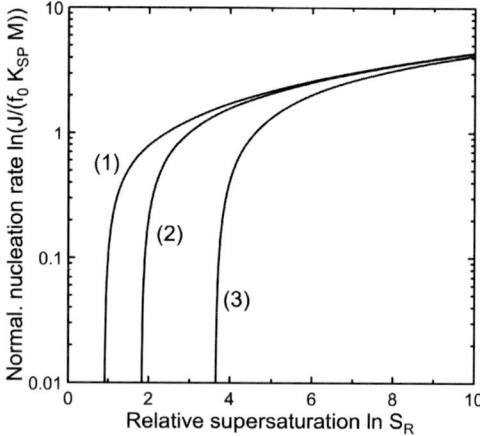

Figure 6.8 Estimate of the dependence of the stationary nucleation rate on the relative supersaturation S_R and the interface energy γ_{sl} of spherical precipitates. (1) $\gamma_{sl} = 0.025\,J/m^2$, (2) $\gamma_{sl} = 0.05\,J/m^2$, and (3) $\gamma_{sl} = 0.1\,J/m^2$, with $\nu_s = 1.186 \times 10^{-5}\,m^3$.

By an appropriate choice of the template, the activation barrier for nucleation can be much reduced. The geometry of the spherical cap is put into a function $S(\theta)$ that reduces (6.36) to a form similar to that of homogeneous nucleation:

$$\Delta G = \left(-\frac{4\pi}{3\nu_s} \cdot r^3 \cdot \Delta\mu + 4\pi \cdot r^2 \cdot \gamma_{sl} \right) \cdot S(\theta), \tag{6.37}$$

with $\quad S(\theta) = (2 + \cos\theta) \cdot (1 - \cos\theta)^2/4.$ $\tag{6.38}$

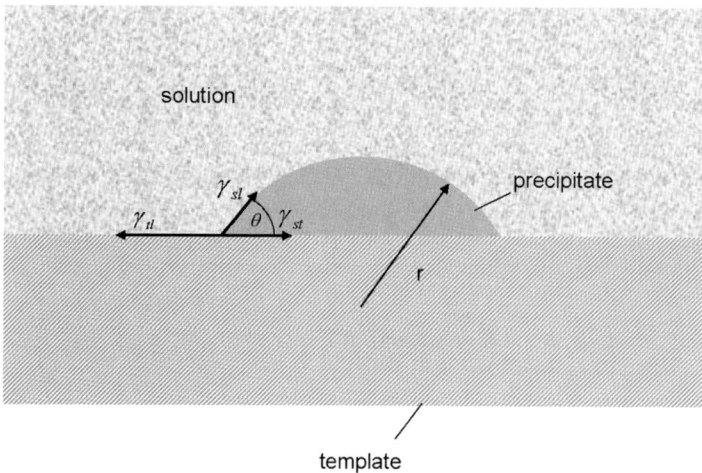

Figure 6.9 Schematic drawing of the equilibrium of the interfacial forces acting on a precipitate growing on a template by heterogeneous nucleation.

As ΔG_{het} has the form $\Delta G_{het} = \Delta G_{hom} \cdot S(\theta)$, it is immediately seen that the critical radius of a nucleating cap equals that of homogeneous nucleation, and the critical free enthalpy change is reduced by the factor $S(\theta)$:

$$r^* = \frac{2\gamma_{sl} \cdot v_s}{k_B T \ln S_R} \quad \text{and} \quad \Delta G^*_{het} = \frac{16\pi\gamma^3_{sl} \cdot v^2_s}{3(k_B T \ln S_R)^2} \cdot S(\theta). \tag{6.39}$$

The influence of the shape factor $S(\theta)$ can be very strong, especially for templates with small contact angle: $S(10^0) = 0.00017$ and $S(30^0) = 0.013$. Such reduction of the critical free enthalpy strongly favors heterogeneous nucleation, which therefore is a phenomenon highly relevant for biomineralization.

6.2.2
Phenomenology of Biomineralization

Natural solid biomaterials such as bones, teeth, and molluscs shells and also ancient biomineralized sediments such as stromatolites represent hybrid composites consisting of *inorganic and bioorganic components*. From the viewpoint of a materials engineer, it seems surprising that Nature chose only about 10 out of the 118 elements of the periodic table as the main ingredients for making about 60 different biomaterials: H, C, O, Mg, Si, P, S, Ca, Mn, and Fe. Typically, the grown minerals are nanocrystalline or amorphous phases. Among several polymorphic phases of a mineral, it is often not the stable one that is formed. The formation of *metastable phases* indicates that biomineralization does not simply follow the usual straightforward path toward the minimum of free enthalpy. Apparently, the minerals are guided along more complex reaction paths in the thermodynamic phase space by the presence of particular organic substances.

Calcium phosphates and calcium carbonates are the minerals that are most often found in organisms, also as metastable phases. The low-solubility product of phosphates and carbonates (Table 6.1) and the relatively high Ca level in extracellular fluids (typically 10^{-3} M) imply a high thermodynamic driving force for precipitation of these minerals. Calcium carbonate, for example, is known as calcite, aragonite, and vaterite. Such polymorphism is due to stepwise phase formation, as explained in the following section.

In biominerals, we usually observe the phosphate phases hydroxyapatite (HAP) and octacalcium phosphate (OCP). Amorphous calcium phosphate can be found in some organisms as a precursor phase. Hydroxyapatite is the most important biomineral in the group of calcium phosphates Together with matrix proteins (mainly collagen, amelogenin, and enamelin), it forms the complex structures of bone and teeth. The structure of biomineralized HAP, (Ca, Sr, Mg, Na, H_2O, [. . .])$_{10}$(PO$_4$), HPO$_4$, CO$_3$P$_2$O$_7$)$_6$(OH, F, Cl, H_2O, O, [. . .])$_2$, is usually characterized by a high density of crystal defects. The additional components and the lattice defects play an important role in the functionality of the biomineral in a given environment. For example, Na$^+$, NH$_4{}^+$, K$^+$, and F$^-$ affect the solubility of HAP. The solubility product of HAP with stoichiometric composition Ca$_{10}$(PO$_4$)$_6$(OH)$_2$ in water at 25 °C

Table 6.2 Solubility products of various calcium phosphate phases.

Ca/P ratio	Compound		$-\log (K_{sp})$ at 25 °C, pH = 7
0.5	Monocalcium phosphate monohydrate	$Ca(H_2PO_4)_2 \cdot H_2O$	1.14
1.0	Dicalcium phosphate dihydrate Brushite (DCPD)	$CaHPO_4 \cdot 2H_2O$	6.56
1.0	Dicalcium phosphate anhydrate Monetite (DCPA)	$CaHPO_4$	6.90
1.33	Octacalcium phosphate (OCP)	$Ca_8(HPO_4)_2(PO_4)_4 \cdot 5H_2O$	96.6
1.5	α-Tricalcium phosphate (TCP)	$\alpha\text{-}Ca_3(PO_4)_2$	25.5
	β-Tricalcium phosphate (TCP)	$\beta\text{-}Ca_3(PO_4)_2$	28.9
1.6	Hydroxyapatite (HAP)	$Ca_{10}(PO_4)_6(OH)_2$	116.8

is extremely low: $K_{sp} = [Ca^{2+}]^{10} \cdot [PO_4^{3-}]^6 \cdot [OH^-]^2 = 10^{-116.8}$. This high stability distinguishes HAP from various Ca phosphates in biomineralization and biomimetic formation of bone-like materials (Table 6.2).

Similar to the carbonates and phosphates, low-solubility products also lead to the formation of sulfate- and oxalate-based biominerals. As mentioned in Chapter 5, many unicellular organisms grow beautiful silica-based mineral structures from hydrated Si—O—Si complexes present in water. Their deposition as hydrated inorganic polymers is controlled by appropriate protein templates. Fe is the most important transition metal in connection with biomineralization. The various precipitation and dissolution reactions are governed by the redox behavior of the II/III oxidation states of Fe. Fe(II) is much more soluble in water than Fe(III). Therefore, Fe(III) is the dominating oxidation state in biominerals. Among them, ferrihydrite plays a crucial role. Ferrihydrite can be transformed into all other iron minerals such as goethite, lepidocrocite, and magnetite by dissolution–reprecipitation processes, including reductive dissolution–reprecipitation in the mixed valence state of magnetite. Similar to magnetite, the ferromagnetic greigite (Fe_3S_4) is formed in magnetotactic bacteria living in a sulfide-rich environment (see also Section 6.1). In the reducing early atmosphere, such bacteria were dominating life on Earth. For further details, see Mann (2001).

6.2.3
Basic Mechanisms in Biomineralization

Precipitation of the mineral phase in biomineralization may simply look like heterogeneous nucleation with the biomolecule serving as a substrate or template. If looked at more closely, it becomes obvious that the biomolecular structure affects the process in a much more sophisticated way than an anorganic solid does by way of its interface energy. The various weak interactions of biomolecules and their abilities of molecular recognition and self-assembly give rise to *site-specific interactions between the template and the growing mineral phase*. There are *additional geometric constraints* caused by the arrangements of the biomolecules such as cages, tubes, fibrils, sheets, frameworks, and so on, covering the size range from a few nanometers up to hundreds of micrometers. The mineralization process is also affected by the *coupling of the biological structures with metabolic processes of living cells*. This high complexity of biomineralization is also the reason for its fascinating structural richness.

When trying to analyze a complex problem, it is always useful to separate it, if possible, into partial problems that can be approached with simpler theoretical models. However, one has to keep in mind that understanding of the parts does not necessarily mean understanding of the whole.

In the following, we will try to classify the mineralization processes into categories, depending on the impact of biomolecular structures on the product phase. Processes that mainly follow the classical chemical synthesis route go here under *biologically mediated mineralization*. If nucleation, final structure, shape, and size of the mineral *depend mainly on biologically induced environmental conditions*, we speak of *biologically induced mineralization*. This is the case when a biopolymer affects crystal growth, for example, by site-specific adsorption.

Biologically induced mineralization usually concerns by-products of metabolic processes. If the mineral is unique in structure, shape, and size, the process is called *biologically controlled mineralization*. In this type of process, the mineral is the product of a *reproducible biological synthesis with specific crystallochemical properties*.

Biologically induced mineralization and biologically controlled mineralization differ with respect to their result: Size, shape, structure, and composition of the mineral particles vary with the former, but are highly uniform with the latter. Biologically controlled mineralization produces structures of high complexity such as the distribution of calcium phosphate nanocrystals in bone or the calcium carbonate crystal distribution in mollusc shells.

Any classification of biomineralization processes, of course, will be more or less arbitrary; nevertheless, the categories introduced above may be helpful in the search for biologically inspired synthesis routes of nanostructured inorganic materials.

6.2.4
Biologically Mediated Mineralization: the Competition between Inhibition and Growth

A conspicuous feature of mineral precipitates is their crystal habitus or habit that may vary between columnar and bladed for one substance, depending on various influences. Different crystal faces may have differential surface free enthalpies, and hence the habits are not equivalent with respect to free enthalpy. Under given conditions, there is one habitus with minimum free enthalpy and thus in stable equilibrium. Growing crystals, however, tend to deviate: the more the deviation from equilibrium, the faster they grow. So it has to be anticipated that any precipitate can be more or less out of equilibrium shape.

Mineral precipitates growing in the presence of organic solutes can be affected in another way: Organic molecules may bind loosely to the crystal surface and (temporarily) obstruct the diffusion paths of the mineral solute, thereby reducing its concentration at the crystal surface. Since the organics bind with unequal affinity with different crystal faces, they inhibit growth unevenly, which affects the habitus. So it appears that mineral precipitates obtained with biomineralization can find themselves in states far from equilibrium, also called metastable states.

6.2.4.1 Effect of Polypeptides on Precipitate Habitus

An experimental example of morphological control by biomolecule adsorption is presented by Tomczak *et al.* (2009) with the precipitation of ZnO crystals from equimolar aqueous solutions of $Zn(NO_3)_2$ and 1,3-hexamethylenetetramine (HMTA) (100 mM each) (Figure 6.10). Without any additional solvents in the aqueous solution, needle-like hexagonal crystals have been observed after incubating the samples at room temperature for 24 h, followed by a 72 h incubation at 65 °C. Mixing a specifically selected peptide with a concentration of 0.5 mg/ml to the solution, the same growth procedure yields hexagonal platelets, as shown in Figure 6.10.

High-resolution transmission electron microscopy revealed that the peptide does not disrupt the crystal structure. For a peptide concentration of 0.5 mg/ml, the ZnO nanostructures crystallized in a wurtzite lattice. With decreasing concentration of the peptide, the ZnO nanocrystals became elongated with hexagonal symmetry. The basic idea behind this experiment is sketched in Figure 6.11.

The peptide is adsorbed preferentially on the end faces of the hexagonal. By phage peptide display (see Section 2.3), a peptide with the amino acid sequence Gly—Leu—His—Val—Met—His—Lys—Val—Ala—Pro—Pro—Arg termed Z1 has been identified that specifically binds to ZnO nanoparticles (Figure 6.12). For technical reasons, a Gly—Gly—Gly—Cys tail was added to the C-terminus.

The density of zinc atoms is highest on the (002) lattice plane, but much lower on the (101) plane. The strong affinity of histidine and cysteine residues to zinc explains their preferred binding to the (002) surface.

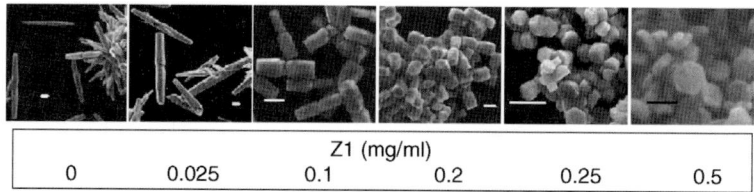

			Z1 (mg/ml)			
0	0.025	0.1		0.2	0.25	0.5

Figure 6.10 Effect of Z1 peptide on ZnO morphology. SEM micrographs of ZnO nanoparticles grown in the presence of Z1 peptide at the indicated concentrations. Scale bars 1 μm except for the image on the far right: 0.2 μm. (Reprinted with permission from Tomczak *et al.* (2009), Copyright 2009, Elsevier.)

Polypeptide

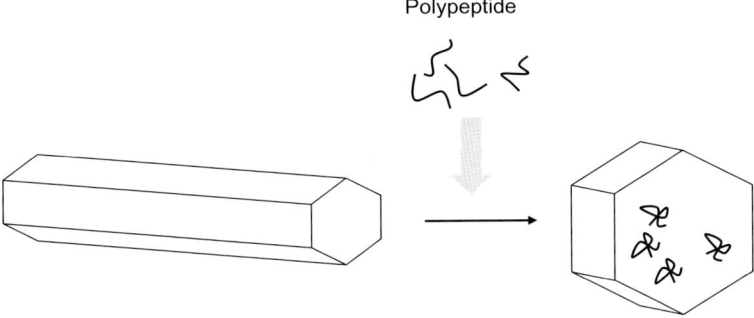

Figure 6.11 Effect of face-specific adsorption of polypeptide on habitus during crystal growth.

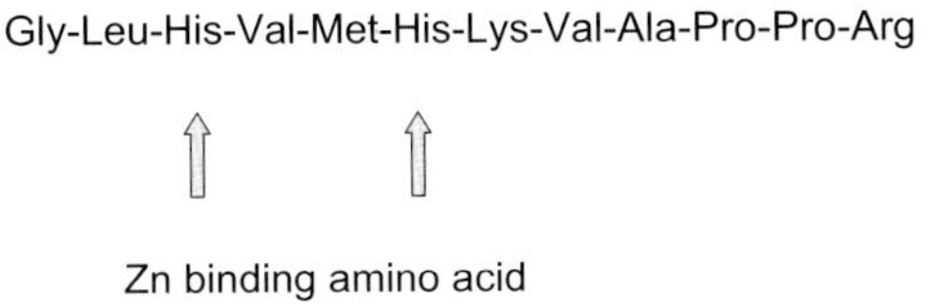

Gly-Leu-His-Val-Met-His-Lys-Val-Ala-Pro-Pro-Arg
Zn binding amino acid

Figure 6.12 Amino sequence of the Z1 peptide with two histidines that bind specifically to Zn atoms (Tomczak *et al.*, 2009).

Aspect ratio versus peptide concentration is shown in Figure 6.13 in the range of 0–0.5 mg/ml. The observation can be explained with the adsorption equilibrium according to the Langmuir adsorption isotherm. The fraction ϕ of occupied sites on the c plane is given as

$$\phi = \frac{K_L \cdot c_p}{1 + K_L \cdot c_p}, \tag{6.40}$$

Aspect Ratio (Length/Width)
10
9
8
7
6
5
4
3
2
1
0
0
0.1
0.2
0.3
0.4
0.5
Z1 (mg/ml)

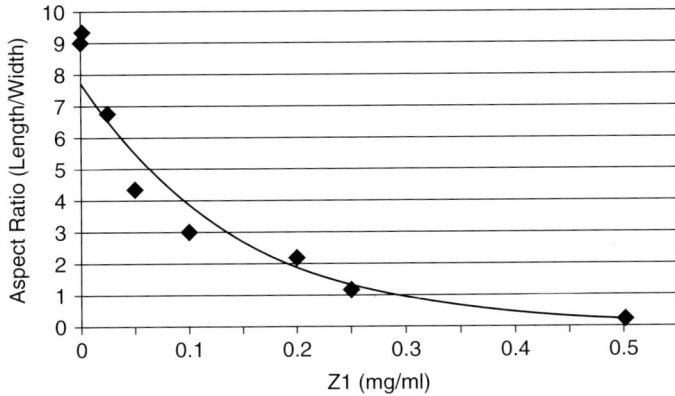

Figure 6.13 Effect of Z1 peptide on ZnO aspect ratio of ZnO hexagonal nanoparticles. (Reprinted with permission from Tomczak *et al.* (2009), Copyright 2009, Elsevier.)

where c_p is the peptide concentration in the solution and K_L is the Langmuir equilibrium constant. (See also Task 6.5.) The experimental data for the aspect ratio of the crystals yield a dependence

$$S(c_p)/S(0) = (1 - \phi) = (1 + K_L \cdot c_p)^{-1} \tag{6.41}$$

in good approximation (Muthukumar, 2009).

The importance of the face-specific peptide adsorption has been underlined by applying bovine serum albumin (BSA) in a control experiment. Adding 0.5 mg/ml BSA to the precursor solution does not affect the needle-like aspect because BSA is equally adsorbed on all faces.

6.2.4.2 The Formation of Metastable Polymorphs

As already explained, many of the known biominerals can exist in several different structural modifications. $CaCO_3$, for example, can be precipitated in the polymorphic crystallographic structures of calcite, aragonite, and vaterite with same stoichiometry. Polymorphic phases are also known for silica. Between the various calcium phosphate and iron oxide phases, there are only minor compositional differences. Since biomineralization proceeds in aqueous solution, amorphous phases and hydrated phases may be formed. Hence, there are a number of competing reaction channels for precipitation that can result in metastable phases as final reaction products. This is also known as *Ostwald step rule*. It says that it is usually not the most stable polymorph, but the least stable one that crystallizes first.

The selection of the dominating reaction can be governed by thermodynamic as well as kinetic principles. The thermodynamic potential changes driving the precipitation reactions differ by as little as 1–10 kJ/mol. Stable polymorphs of silica and calcium carbonate precipitation are seen in Figure 6.14. The differences of the enthalpy ΔH between the equilibrium phases and the other polymorphs are plotted for bulk samples. As the differences between the phases are small, any variation of the interfacial free enthalpy γ_{sl} can largely affect the precipitation path (Navrotsky, 2004).

Free enthalpy changes of two polymorphs plotted schematically versus particle radius in Figure 6.15 may serve as a simple example. It has been assumed here that the polymorph with the smaller $\Delta\mu$ gain (metastable bulk phase) has a smaller interfacial free enthalpy γ_{sl}. The graph shows that the smaller interfacial free enthalpy is connected with a smaller activation barrier in the nucleation kinetics.

Small interfacial free enthalpies are also typical for amorphous phases formed in aqueous solution, such as the incorporation of water molecules into the growing amorphous nucleus reduces the interaction of the surrounding solution with the solid precipitate. Nanosize amorphous particles have often been observed in early stages of crystallization. This has been assumed to be a kinetic effect, but the molecular dynamics simulations of the nucleation of calcium carbonate in aqueous solution (Raiteri and Gale, 2010) have shown that the favored precipitation of ultrafine amorphous calcium carbonate (ACC) below a critical diameter of about 3.8 nm is due to thermodynamics. This nonclassical nucleation pathway is brought

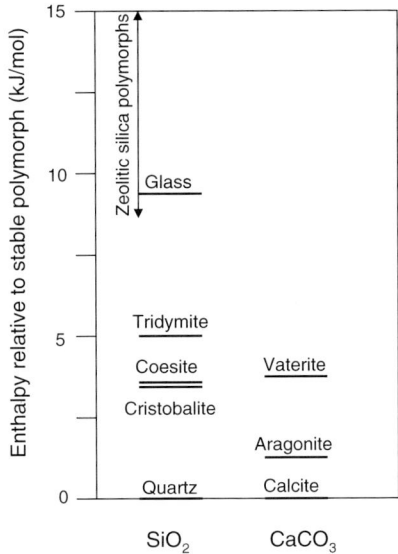

Figure 6.14 ΔH of polymorphs of silica and calcium carbonate in the bulk state. (Adapted from Navrotsky (2004), Copyright 2004, National Academy of Sciences, USA.)

about by the involvement of water in cluster growth. Calculations have shown that the addition of ion pairs of calcium cations and carbonate anions to an amorphous nucleus is always an exothermic process ($\Delta G \leq 0$). The ultrafine ACC nuclei are more stable than the crystalline calcite phase because of the lower enthalpy of the

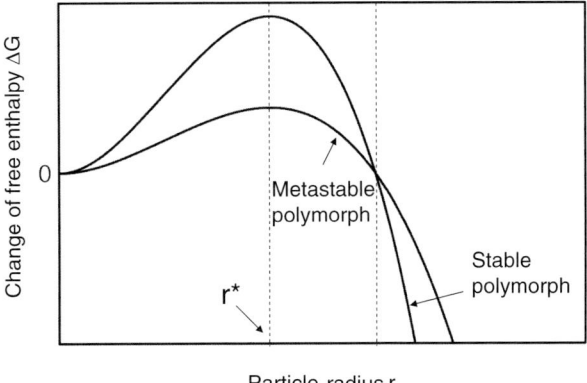

Figure 6.15 Free enthalpy of two polymorphs versus particle radius. The interfacial free enthalpy has been chosen such that for small particle size, the formation of the polymorph with the higher bulk free enthalpy is thermodynamically favored. (Adapted from Navrotsky (2004).)

ACC due to the incorporation of water. The stable ACC nuclei contain about one water molecule per calcium carbonate unit. With increasing cluster size, the thermodynamically favored fraction of water increases. Furthermore, the rough interface of the growing ACC nucleus offers a higher density of adsorption sites for the calcium carbonate ion pairs than the smooth surfaces of the crystalline phases. This leads to a faster growth rate of the ACC nuclei.

The simulation also shows that the incorporation of water into the ACC structure is connected with an entropy penalty due to additional constraints for the water molecules. This contribution to the free enthalpy implies that beyond 3.8 nm, the crystalline calcite phase is thermodynamically favored. In conclusion, based on thermodynamic arguments, a nonclassical nucleation pathway can be assumed (Cölfen and Mann, 2003). In an early stage, amorphous clusters are formed with a very small free enthalpy barrier, thus differing from the classical nucleation pathway of a crystalline calcium carbonate phase. The gain of free enthalpy Δg increases with increasing water content of the amorphous phase. From experimental observations, it can be concluded that the individual ACC clusters form larger aggregates in a later growth stage. With further growth, beyond the critical size of about 4 nm, the free enthalpy of the individual ACC particles is larger than that of the crystalline phases. Now a disorder–order transformation can occur that follows the rules of a classical nucleation process with a free enthalpy barrier. In biomineralization, this process can be furthered or delayed by proteins or other biomolecules with specific binding motifs.

Recently, it has been pointed out that liquid precursors induced by soluble biopolymers could be the first step for the biomineralization in aqueous solutions. Small anionic noncollagenous polymers, for example, can induce the formation of nanoscale liquid precursors in the hydroxyapatite mineralization in bone tissue (for more details, see Section 6.2.8). The schematic in Figure 6.15 also explains that the restricted growth of the particle in nanoscale cages can favor the amorphous phase or metastable polymorphs. This mechanism can be used for the production of highly reactive nanoparticles by biomineralization in preorganized small compartments. In conclusion, we see that variation of the interfacial free enthalpy can modify the crystallization pathways at the nanoscale.

6.2.5
Biologically Induced Mineralization: Role of the Epicellular Space and the Extracellular Polymeric Substances

Metabolic processes of cells can give rise to secondary reactions with the environment that lead to the *adventitious precipitation of inorganic compounds* on organic structures such as proteins, polysaccharides, or lipids. As indicated in Figure 6.16, reaction routes of biologically induced mineralization can involve the intracellular space, the epicellular space next to the cell wall, and the extracellular polymeric substances (EPS). Sometimes the chemical reaction of mineral formation can be activated by the presence of an enzyme. The particular chemical reaction parameters

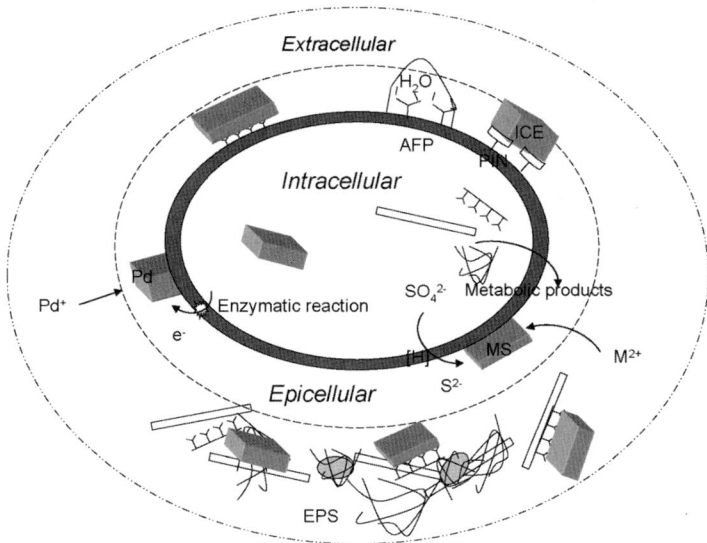

Figure 6.16 Representation of various mechanisms of biologically induced mineralization. EPS: extracellular polymeric substances, PIN: protein ice nucleator, AFP: antifreeze protein, M: metal, e^-, electron produced by an enzymatic reaction.

controlling the precipitation, such as supersaturation, pH, and ionic strength, are determined not only by the extraneous chemical conditions but also by the metabolic processes of the cell connected with the active transport of water, ions, and electrons across the cell wall.

The main domain for biologically induced mineralization is the epicellular space. Proteins or polysaccharides expressed from the cell are often immobilized at or in the cell wall. They can act as nucleators, catalysts, or also as inhibitors of mineralization. A very common case is the deposition of *calcium carbonate at the cell wall from supersaturated calcium bicarbonate solutions*. The calcium carbonate precipitation depends on three main process parameters: the activity product $\{Ca^{2+}\} \cdot \{CO_3^{2-}\}$, the pH, and the availability of nucleation sites.

With a sufficiently high concentration of Ca^{2+} ions, $CaCO_3$ is deposited according to

$$CaCO_3 \leftrightarrow Ca^{2+} + CO_3^{2-}. \tag{6.42}$$

The concentration of CO_3^{2-} is coupled to its equilibrium with dissolved CO_2 and HCO_3^-:

$$CO_2 + H_2O \leftrightarrow H_2CO_3. \tag{6.43}$$

$$H_2CO_3 \leftrightarrow H^+ + HCO_3^-. \tag{6.44}$$

$$HCO_3^- \leftrightarrow H^+ + CO_3^{2-}. \tag{6.45}$$

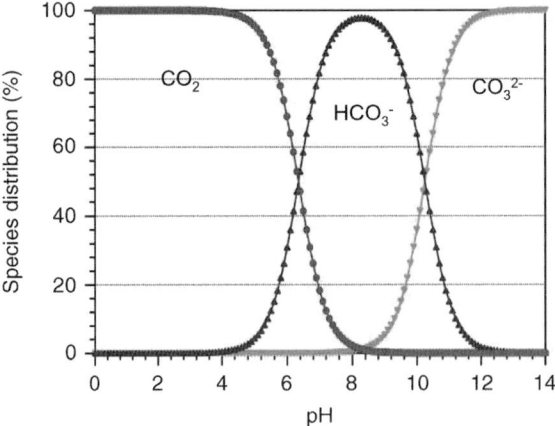

Figure 6.17 Fractions of CO_2, bicarbonate HCO_3^-, and carbonate CO_3^{2-} ions versus pH.

With the known equilibrium constants of the reactions, the concentrations of CO_2, HCO_3^-, and CO_3^{2-} can be calculated in dependence on the pH (Figure 6.17).

It appears that CO_3^{2-} necessary for $CaCO_3$ formation is available only above pH 8. This means $CaCO_3$ forms only if alkaline conditions are sustained by metabolic processes of the cell. Various metabolic processes, such as the dissimilatory sulfate reduction under anoxic conditions or the dissimilatory reduction of nitrate by heterotrophic bacteria, produce an alkaline environment in the epicellular space. Other metabolic processes produce CO_3^{2-} ions. Negatively charged groups on the cell wall can bind Ca^{2+} ions that bind CO_3^{2-} ions. Calcification goes often along with photosynthesis consuming carbon dioxide. The local decrease of the CO_2 concentration in aqueous solution with high concentrations of Ca^{2+} and HCO_3^- favors the precipitation of $CaCO_3$ (see also Section 6.2.9).

Another widespread biologically induced mineralization process is the microbial reduction of Fe(III), which plays a major role in the formation of iron-based biominerals. Oxides, oxyhydroxides, and oxyhydroxysulfates of iron are found in soil and sediments with a wide range of concentrations. The minerals exist in structures covering the range between amorphous ferrihydrite ($Fe_2O_3 \cdot nH_2O$) and crystalline phases as hematite (rhombohedral Fe_2O_3). There are various dissimilatory iron(III)-reducing bacteria (DIRB) (as *Geobacter sulfurreducens* and *Shewanella oneidensis*) that attack the Fe(III) mineral surface and increase the concentration of dissolved or sorbed Fe(II) by partial dissolution of the Fe(III) mineral. New biogenic phases as magnetite Fe_3O_4 are formed as a result of the metabolic activity of the bacteria. Electrons required for Fe(III) reduction are provided by a metabolic redox process. As this process of Fe(II) generation can be sustained by the bacteria quasi-continuously, the newly formed Fe(II) can also act as a mediator for reductive *in situ* transformation of many other reducible contaminants (Lloyd et al., 2008). Among them are critical pollutants such as

Cr(VI), U(VI), and Tc(VII). This mechanism plays a role, for instance, in bioremediation of wastewater.

Several *nonredox systems* involved in biomineralization have evolved along with the bacterial metabolism. They are based on the metabolic production of *precipitant ligands*, such as *sulfides, carbonates*, or *phosphates*, which can then react with metal ions in the epicellular or extracellular space. An example is the metabolic generation of hydrogen sulfide (H_2S) by the dissimilatory sulfate reduction. The sulfate-reducing bacteria reduce sulfate to oxidize organic matter and expel hydrogen sulfide. The hydrogen sulfide reacts with metal ions. *Desulfovibrio* bacteria, for example, preferentially produce pyrrhotite $Fe_{1-x}S$, which is hexagonal and para-magnetic for $0 \leq x \leq 0.11$, but monoclinic and ferromagnetic for $0.11 \leq x < 0.2$. Some pyrrhotite minerals are magnetically anisotropic: paramagnetic in one direc-tion and ferromagnetic perpendicular to it. Lipopolysaccharides (LPS), the major component of the bacterial cell wall of Gram-negative bacteria, is another relevant nonredox system involved in biomineralization. LPS protect the cell membrane against chemical attack. As described by Lloyd *et al.* (2008) for *Serratia* cells, the cells can form outer membrane vesicles (OMV) that contain LPS and an acid-type phosphatase.

Monophosphate groups of the lipid A backbone of the lipopolysaccharides serve as nucleation points for biomineralization of incoming metal ions such as Cd^{2+} or UO_2^{2+}. The process is catalyzed by an acid-type phosphatase (Figure 6.18). The biomineralization occurs in two steps. It is initiated by the formation of complexes between the metal ions and the monophosphate groups of the LPS. Simultaneously, the enzyme produces HPO_4^{2-}, which governs the growth of large metal phosphate crystals.

In Section 6.1, we have discussed the phenomena of freeze avoidance and freeze tolerance. These phenomena are related to *biologically induced formation of ice crystals*. Two groups of biomolecules have been identified as being involved in nucleation and inhibition of ice crystal growth. In the protein ice nucleator (PIN) and the AFP, also called ice structuring protein (ISP), specific molecular motifs regulate the freezing of water. In the following, we will consider examples of the two groups with respect to their function.

The protein ice nucleator (PIN) of overwintering freeze-tolerant queens of the hornet *Vespula maculata* is a hydrophilic 74 kDa protein containing about 20 mol% glutamate/glutamine. It is assumed that the hydrogen-bonding abilities of their hydrophilic side chains may be important for ice nucleation. In the freeze-tolerant larvae of the cranefly *Tipula trivittata*, a globular 800 kDa lipoprotein (LPIN) consisting of 45% protein, 4% carbohydrate, and 51% lipid has been identified as ice nucleator. The lipid is about 39% neutral lipids and 12% phospholipids. It has been suggested that one of the phospholipids, the phosphatidylinositol (PI), might arrange water in an ice-like pattern (Figure 6.19).

The hydroxyl groups of the inositol ring seem to be responsible for the nucleation process. The protein components of the LPINs assemble the PIs in chain-like structures of ice nuclei. It is assumed that the spherical LPINs arrange themselves in

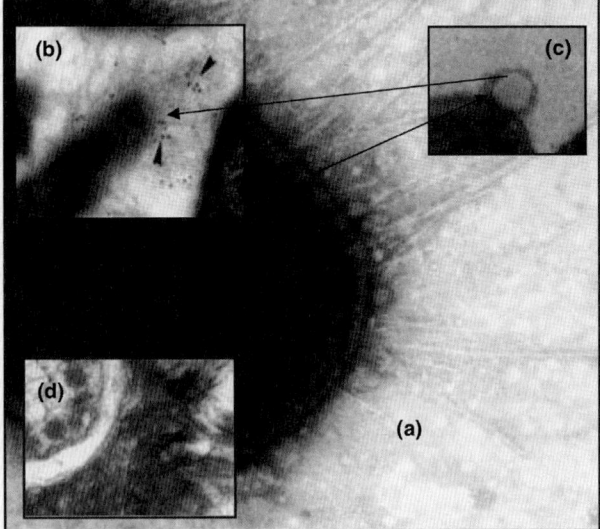

Figure 6.18 Deposition of metal phosphates by *Serratia* sp. (a) Surface features include flagella, pili, and outer membrane vesicles (OMV: shown enlarged in part (c)). Phosphatase (shown immunolabeled in part (b)) is tethered to material outside the cell. The OMVs also contain lipid A material derived from the outer membrane. They deliver the phosphatase to intercept incoming metal ions that initially coordinate with the lipid A phosphate groups followed by additional precipitation (shown as calcium phosphate deposition in part (d)). (Reproduced with permission from Lloyd *et al.* (2008), Copyright 2008, John Wiley & Sons, Inc., New York)

such a way that the inositol hydroxyls are organized as cooperative larger clusters (Figure 6.20). Thus, larger ice nuclei can be formed with the restricted amount of PI in the LPINs.

Structures of AFP have been first explored in Antarctic fishes. The best-documented structure is the AFP of the winter flounder. It consists of a single, long amphiphatic, alanine-rich protein with 37 amino acid residues. Four

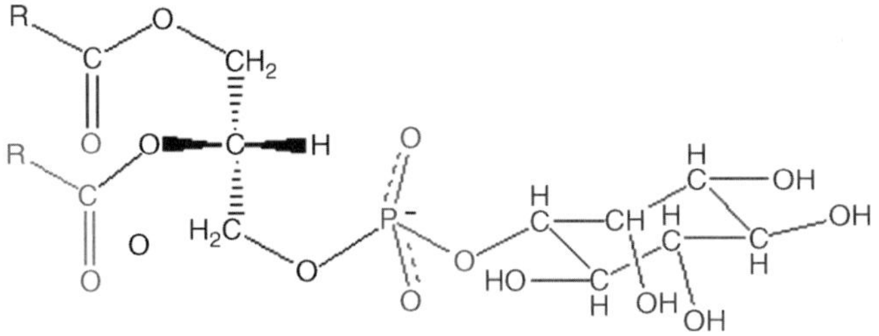

Figure 6.19 Structure of phosphatidylinositol. (Jag123, Wikimedia Commons (2007).)

Ice nuclei   Inositol hydroxyl regions

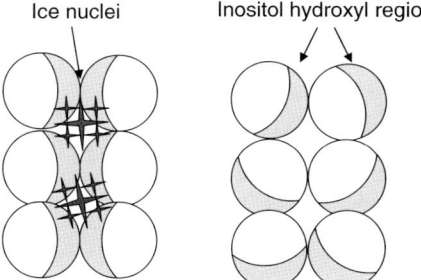

Figure 6.20 Schematic drawing of the supposed chain-forming assembly of *T. trivittata* LPINs. The dark areas represent the ice-nucleating inositol hydroxyls regions of an individual LPIN chain. The arrangement of the four LPINs in the left part allows the cooperative formation of a larger ice nucleus, whereas the arrangement in the right part would not suit for such structures. (Adapted from Duman, (2001).)

threonine residues, that is, Thr2, Thr13, Thr24, and Thr35, are decisive in the binding with ice crystals. By Monte Carlo simulation of the helix structure, it has been shown that in the energy-optimized structure, the hydroxyl groups of the four residues are aligned on almost the same line parallel to the helix axis with spacings of 1.61, 1.60, and 1.62 nm (Chou, 1992), which is close to the 1.66 nm periodicity along the $[01\bar{1}2]$ direction in ice (Figure 6.21).

The rather good match leads to the assumption that hydrogens of the threonine residues bind to oxygen atoms of the water molecules along the $[01\bar{1}2]$ direction in the ice crystal. A zipper-like connection of the AFP with the pyramidal planes of the ice crystal is formed. The presence of the small hydrophobic alanine residues in the regions between the threonine–ice contacts hinders the aggregation of additional water molecules. Later studies of AFPs of other organisms have shown similar inhibition mechanisms, among them being the hydrophobic interaction of the AFPs with the ice surface and the zipper-like hydrogen bonds provided by neutral polar residues such as serine, threonine, cysteine, asparagine, or glutamine (Duman, 2001).

There are various ways how the PINs and the AFPs contribute to freeze tolerance and freeze avoidance. Paradoxically, freeze tolerance of cells can be achieved by freezing in the extracellular space induced by PINs. The growing ice crystals extract water from the solution so that the concentration of solutes increases. The arising osmotic imbalance draws water from the cell so that the concentration of solutes in the cell fluid increases, which lowers the freezing point.

While ice nucleation is induced by PINs, the presence of AFPs can prevent the nanosize crystals from subsequent growth by recrystallization. Hence, they do not reach sizes that could harm the cell. (A tendency toward grain coarsening driven by free enthalpy is a general phenomenon of crystalline structures.) Therefore, it appears that the combined effect of AFP and PIN improves the freeze tolerance.

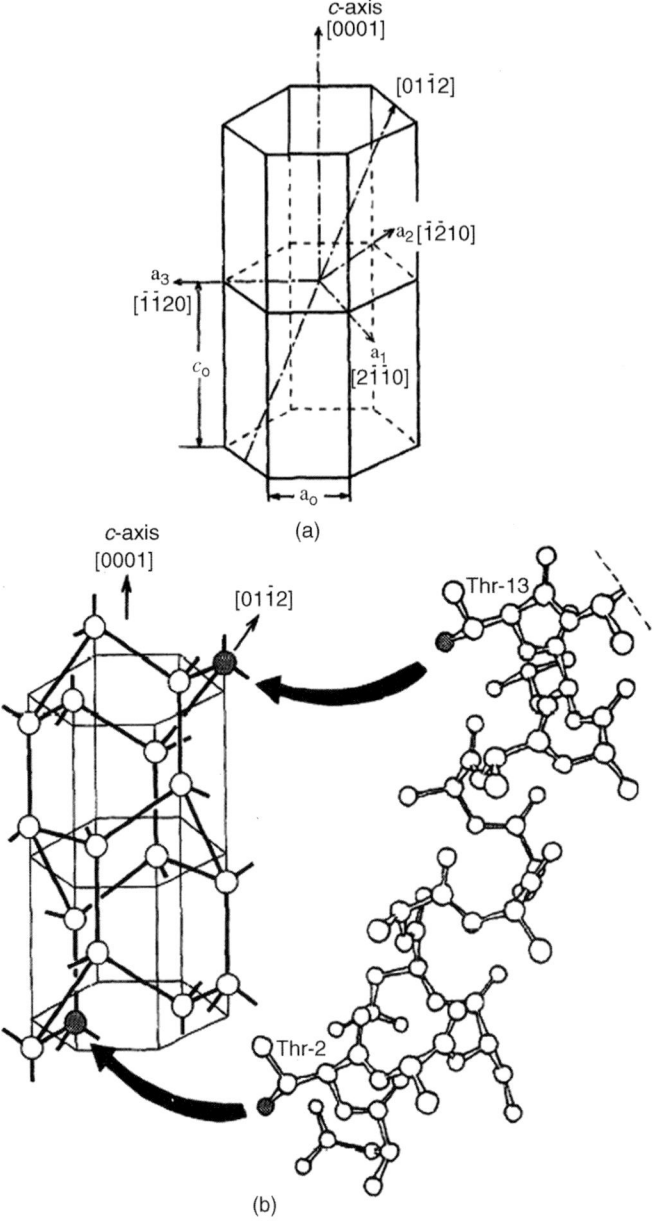

(a)

(b)

Figure 6.21 (a) The ice cell and its vector expression: $a_0 = 0.450$ nm and $c_0 = 0.732$ are the ice lattice parameters. Along the $[01\bar{1}2]$ direction, the periodicity unit is $d = 1.66$ nm. (b) Binding of the AFP helix to an ice nucleation structure along the $[01\bar{1}2]$ direction through hydrogen bonding. Only heavy atoms are shown. The shaded atoms are the hydroxyl oxygen atoms of the Thr residues and the oxygen atoms along the $[01\bar{1}2]$ direction in ice. They are associated with each other by an arrow indicating a hydrogen bond. (Reprinted with permission from Chou (1992), Copyright 1992, Elsevier.)

6.2.6
Biologically Controlled Mineralization: Molecular Preorganization, Recognition, and Vectorial Growth

The precipitation of nanocrystalline magnetite in magnetotactic bacteria described in Section 6.1 is an example of biologically controlled mineralization producing precipitates with nearly equal size and shape. It differs essentially from the random precipitation of iron minerals onto the cell wall of bacteria by dissimilatory reduction of Fe(III), as discussed in Section 6.2.5. Highly organized mineral structures of such type are always indicators of a possible biogenic history. As an interesting fact, there are also traces of iron oxide minerals in Martian meteorites that show morphologies very similar to those found in magnetotactic bacteria.

As pointed out by Stephen Mann, the study of magnetotactic bacteria and many other biomineralization processes has revealed several general principles behind their biological control, as compiled in Figure 6.22 (Mann *et al.*, 1993; Mann, 2000, 2001).

Let us consider first which tools for the control of chemical synthesis are available. Biologically controlled mineralization can proceed in separate vesicles or protein cages (intracellular), at the cell wall (epicellular), or among the so-called extracellular polymer substances (EPS) consisting of insoluble proteins, glycoproteins, or polysaccharides. In addition to these options, there is the particular intercellular mineralization between closely packed cells.

6.2.6.1 **Intracellular Mineralization**
Basic mechanisms that are typical for intracellular mineralization are shown schematically in Figure 6.23. The central entity of such process is a protein cage or vesicle. It forms the reaction space controlling the size, shape, and composition of the mineral. Very often such compartment is equipped with a

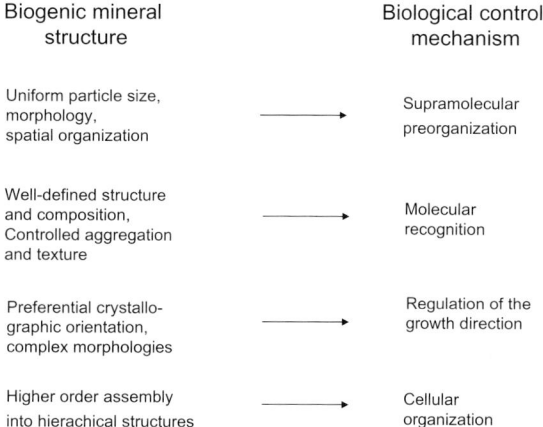

Figure 6.22   Basic mechanisms of biological control of mineralization.

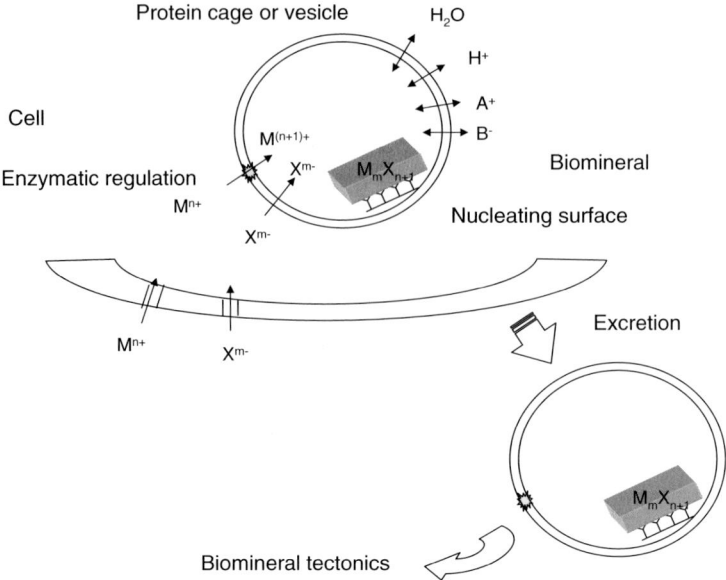

Figure 6.23 Schematic drawing of the various mechanisms of intracellular biologically controlled mineralization. (Adapted from Mann (2001).)

nucleating surface with a specific motif of molecular recognition structures. It controls the crystallographic structure and texture of the growing crystal. Moreover, we have to keep in mind that the shape of the compartment is not necessarily spherical. The shape of the compartment can be *mechanically controlled* in time by mechanisms such as deformation by cellular machines. This facilitates controlled vectorial growth leading to complex shapes. Furthermore, encapsulation of the growing crystal allows supersaturation of the reacting species M^{n+} and X^{m-} and the solubility of the product phase MX to be controlled. The substance transport through the cage membrane and the cell wall can be governed by diffusion-limited ion transport or chemical reaction. In the latter case, the transport can be facilitated via ion pumps also against concentration gradients. The supersaturation within the cage can be changed by direct and indirect mechanisms. The concentration of the reactants can be increased by pumping ions into the confined space or decreased by complexation with additionally present ligands (as citrate). Enzymatic reactions can shift the supersaturation significantly. Enzymatic redox reactions can change the valence of the reactants. For example, Fe(II) at ferroxidase centers (FOCs) turns into Fe(III) when iron ions move across the ferritin membrane (see below). The supersaturation can also be influenced indirectly by changes of the concentration of background ions A^+ and B^- because the ionic strength influences the activity product of the reactants. Similar effects can be caused by the efflux of water or by pumping of H^+ ions into the reaction space. Finally, the minerals can be arranged in a defined pattern inside the cell for a particular purpose, as for sensing elements in the magnetotactic bacteria.

Often the mineral is excreted, wrapped with a biomolecular envelope to enable its further deposition in the extracellular space. This can lead to large-scale hybrid organic–inorganic composites.

A very informative example of biologically controlled mineralization is the formation of ferrihydrate in ferritin and ferritin-like Dps proteins. As already mentioned in Section 6.1, iron can exist in two valence states, as Fe(II) and Fe (III). In the oxidized state, Fe(III) leads to the precipitation of insoluble Fe(III) oxides. The soluble Fe^{2+} ions are highly toxic in aerobic organisms by reducing molecular oxygen to the reactive superoxide anion radical (Eq. (6.46)) or by reacting with peroxides and generating the highly reactive hydroxyl radicals (Eq. (6.47)).

$$Fe^{2+} + O_2 \leftrightarrow \left\lfloor Fe^{2+} \cdots O_2 \leftrightarrow Fe^{3+} \cdots O_2^- \right\rfloor \leftrightarrow Fe^{3+} + O_2^{-*}. \tag{6.46}$$

$$\begin{aligned} Fe^{2+} + OOH^- &\leftrightarrow Fe^{2+} \cdots OOH^- \leftrightarrow OH^- + Fe^{3+} \cdots O^- \\ &\leftrightarrow OH^- + OH + Fe^{3+}. \end{aligned} \tag{6.47}$$

Therefore, the free Fe^{2+} ion concentrations are very low in the cytosol, about 10^{-8} M. Iron is stored in an insoluble but bioavailable form in iron-complexing proteins, or it can be linked as cofactor to enzymes. The best-known iron storage proteins are the ferritins. They form small spherical empty nanocompartments that allow iron oxide nanocrystals to be stored and iron ions to be released from the precipitate.

In eukaryotic cells, ferritin consisting of 24 monomers is the prevailing iron storage protein. The quarternary structure is a globular cage with cubic symmetry, formed by 12 dimers (Figure 6.24a). Its outer diameter is about 12 nm, and the width of the enclosure is about 8 nm. The overall symmetry of the quaternary structure is characterized by three- and fourfold symmetry axes. Along these axes, there are channels of about 0.3 nm width. Along the threefold channels, three aspartate and three glutamate residues represent hydrophilic surfaces, whereas the fourfold channels with three leucine residues are hydrophobic.

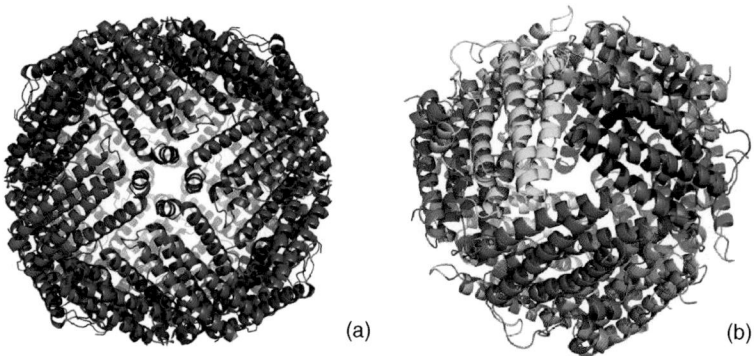

(a)   (b)

Figure 6.24 Comparison of the quarternary structures of ferritin (a) (Granier *et al.*, 2003) and ferritin-like Dps (b) (Ilari *et al.*, 2000).

In eubacteria, three ferritin-like protein subfamilies have been observed that can act as iron storage compartments. Two of them (the heme-binding bacterioferritin and the heme-free ferritin) have the same 24-mer architecture as the eukaryotic homologue.

Also, there are so-called ferritin-like Dps (DNA binding proteins from starved cells) consisting of 12 monomers. The Dps dodecamers are stress proteins that are overexpressed in starved cells in order to form tightly packed layered protein–DNA microcrystals. These sandwich-like structures are parts of a physiologically relevant survival strategy of the starved cells. In the Dps-like ferritin, the assembly of the monomers is highly similar to the 24-mer ferritin. Again, dimers form the quaternary structure, with tetrahedral symmetry as shown in Figure 6.24b. Its outer diameter is about 9 nm, and the width of the central cavity is about 4.5 nm. The channels along the four threefold symmetry axes are formed by similar structures as in the 24 mer ferritin. A hydrophilic pore channel extends along every one of these axes.

The basic mechanisms of immobilization of Fe(II) are very similar for the ferritin and the ferritin-like Dps. In these structures, the oligomeric protein shell is equipped with FOCs, where Fe(II) is oxidized into Fe(III) by O_2 in the 24 mer ferritins or by H_2O_2 in the 12 mer ferritins. In ferritin, the FOC mainly involves three glutamic acids at positions 27, 62, and 107, a glutamine residue at position 141, and 1 histidine residue at position 65, which bind the metal ion in a transition state (Figure 6.25) (Hempstead *et al.*, 1997).

In ferritin, two Fe(II) bound at this center are oxidized to Fe(III). The ferroxidase centers of the various ferritin-like Dps are structured in a similar way.

In most of the ferritin structures, the hydrophilic threefold channels are the favored pathways for diffusion of the Fe(II) toward the FOC. Probably there are also other transport channels that transverse the protein subunits and end above the ferroxidase centers. After oxidation at the ferroxidase center, the Fe(III) ions are ejected into the inner cavity. In ferritin, the iron-loading capacity can be up to 4500 iron atoms per protein. In the smaller Dps structures, the iron oxide core comprises around 500 iron atoms. There are three possible routes of precipitation under discussion.

Route I: Hydrated Fe(III) species are ejected from the FOCs into the aqueous inner compartment where a ferrihydrite crystal is grown by homogeneous nucleation in a supersaturated solution.

Route II: Ferrihydrite is deposited by heterogeneous nucleation on the inner surface of the protein shell. Three negatively charged neighboring glutamate residues located 0.7 nm apart on the cavity wall induce the deposition of the cationic Fe(III) ions by electrostatic interaction. Nearby there is another triple of glutamate residues along a groove toward the neighboring protein subunit. Thus, the cationic Fe(III) complexes are attracted by a surface site with six negatively charged surface groups. The electrostatic interaction lowers the activation energy of nucleation. By condensation reactions, oxo (Fe—O—Fe) and hydroxy (Fe—OH—Fe) bridges are formed, providing a nucleation layer for crystal growth

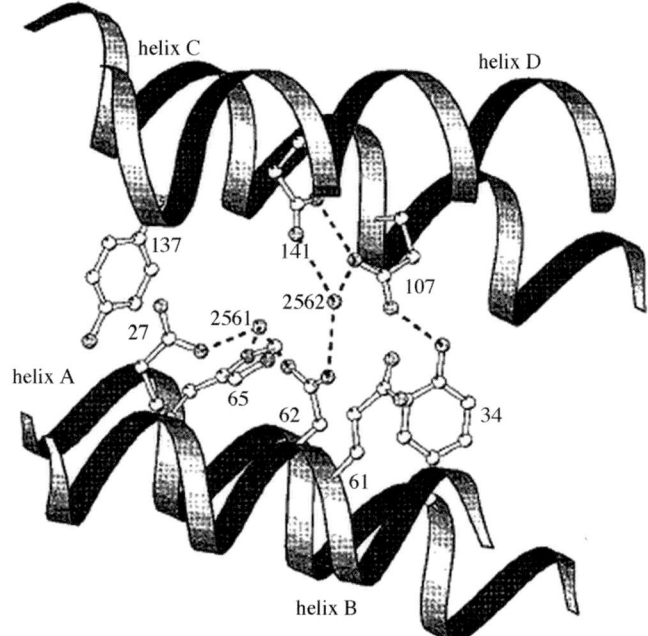

Figure 6.25 The ferroxidase center in the hydrophilic region of the H-chain subunit. The three glutamic amino acids at positions 27, 62, and 107, the glutamine at position 141, and the histidine at position 65 are binding the ferrous Fe(II). (Reproduced with permission from Hempstead *et al.* (1997), Copyright 1997, Elsevier.)

of ferrihydrite $Fe_2O_3 \cdot nH_2O$. Finally, one polycrystalline ferrihydrite grain can grow from multiple nucleation centers along the inner protein surface.

Route III: Fe(III) oxide species could have been formed at the ferroxidase centers and diffuse to the cavity wall where tiny clusters could be formed and then released to merge with the growing grain within the cage. In this case, the final product would be a coarser, polycrystalline material with higher reactivity. Such structures have been described in connection with biomineralization of iron-oxidizing bacteria (Banfield *et al.*, 2000).

The ferrihydrite precipitation in the ferritin cages (routes II and III) corresponds to a very common mode of mineral assembly. Often the nucleation is governed by electrostatic and stereochemical complementary fit at the inorganic–organic interface. The presentation of specific amino acid residues lowers the critical free enthalpy of the growing nucleus. In the above-mentioned ferrihydrite formation, the group of six glutamate residues at the inner surface of the protein shell is an example for the biomolecular control of the nucleation process. As an advantage over solid crystalline substrates, the soft peptide substrate avoids the elastic mismatch usually connected with heteroepitaxial growth of crystalline phases, which means another reduction of the free enthalpy barrier.

6.2.6.2 Epi- and Extracellular Mineralization

Cells can regulate the mineralization also by expression of biopolymers that are organized at the outer face of the cell wall or in the extracellular space. As a first helpful classification of the known mechanisms, we distinguish between the role of soluble and insoluble biopolymers in the control of mineralization. The main processes are compiled in Figure 6.26.

Insoluble biopolymers form the framework for mineralization in the extracellular space. A necessary precondition for the growth of a stable framework is a sufficiently high content of hydrophobic subunits in the biomolecules. Proteins and polysaccharides are the typical frameworks for macromolecules, as collagen in bone, dentine, and sponges, amelogenin in tooth enamel, chitin in mollusc shells and crab cuticle, frustulins in diatom shells, silicatein in sponges, and cellulose in plant cells. Acidic macromolecules that control site-specific nucleation events are often placed on the surface of the framework. The final shape and size of the mineral grains are governed by the nano-structure of the framework, often in interaction with additional water-soluble proteins.

A *soluble biopolymer* can interact with a growing nucleus of the solid mineral. Adsorption can be so intense that the mineral nucleus is completely wrapped so that growth is slowed down or fully suppressed. This can lead to the precipitation of very small particles with amorphous or paracrystalline structure with short- or medium-range order (similar to liquid crystals), but without the long-range order of the crystal lattice. On the other hand, the presence of biomolecules may further growth by acting as an enzyme accelerating the chemical reactions that yield the mineral.

There are indications of an interplay of insoluble and soluble biopolymers during biomineralization, as described in a *two-component model of the organic matrix*, which

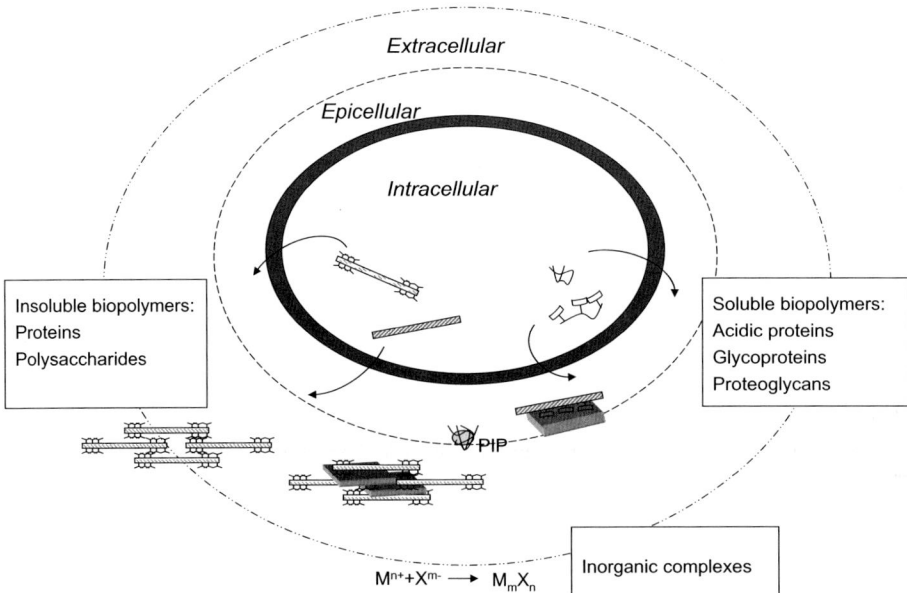

Figure 6.26 Schematic drawing of the various mechanisms of epi- and extracellular biologically controlled mineralization. (Adapted from Mann (2001).)

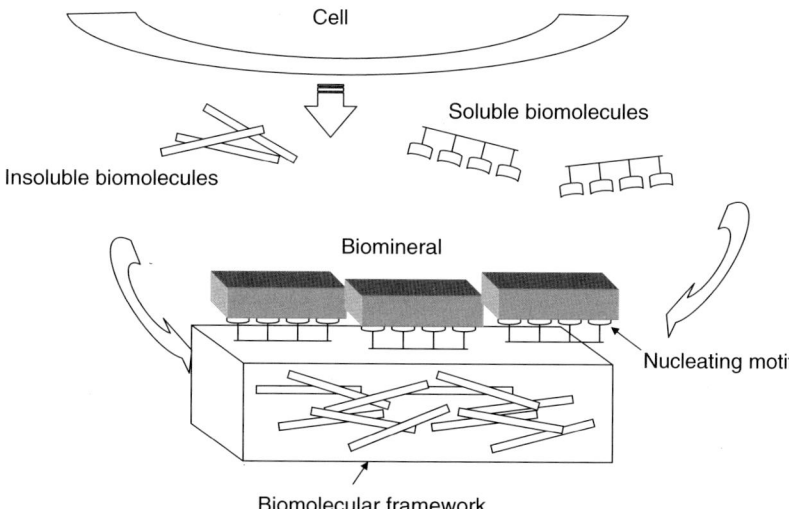

Figure 6.27 Schematic drawing of the interaction of soluble and insoluble biomolecules forming a nucleating surface of soluble biomolecules attached to a framework of insoluble biopolymers with some interfacial links.

comes in two variants. In one variant, the insoluble component forms the bio-mechanical framework stabilized mainly by hydrophobic interactions, whereas the soluble component acts as a *nucleating surface* attached to the framework with some interfacial links (Figure 6.27).

In a recently proposed other variant, the two components are supposed to perform more differentiated functions. There is increasing experimental evidence that the main function of the soluble biopolymers is the formation of a *nanosized precursor phase* shaped as nanodroplets (Figure 6.28).

As a possible pathway of biomineralization, the precursor already has the right chemical composition of the final biomineral by specific adsorption of the mineral-constituting ions. However, the ions are still assembled in a disordered amorphous structure. As we have seen in Section 6.2.4.2, an amorphous intermediate phase can lower the critical free enthalpy for nucleation. Moreover, the presence of the biopolymer can inhibit the growth of the crystalline nucleus. Sometimes these intermediate phases are called *polymer-induced liquid precursors* (PILP) of a mineral phase, although their structure can differ much from that of a liquid. Subsequent precipitation can occur via formation of loosely packed aggregates of nanodroplets. These colloidal aggregates can solidify at some bulky framework formed by insoluble biopolymers as, for example, the large collagen fibrils involved in bone growth (for details, see Chapter 7.) In case the framework presents a pattern of binding sites related to the final product phase, it can act as the guiding structure for the crystallization of amorphous nanodroplets.

The pathways of epi- and extracellular mineralization are displayed in Scheme 6.1.

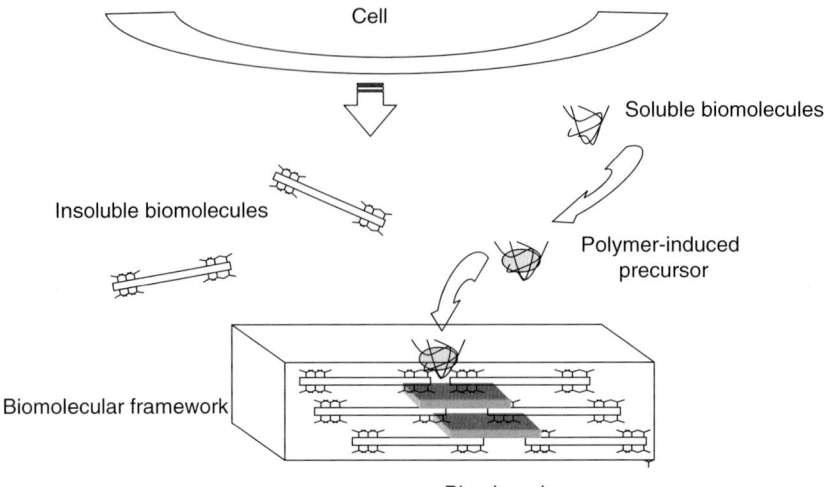

Figure 6.28 Two-component model of the organic matrix including a nanosized precursor phase. The soluble biopolymer induces the formation of a disordered precursor phase with the chemical composition of the final biomineral by specific adsorption of the mineral-constituting ions, whereas the insoluble component forming a framework controls the transition to the final ordered crystalline nanoparticles.

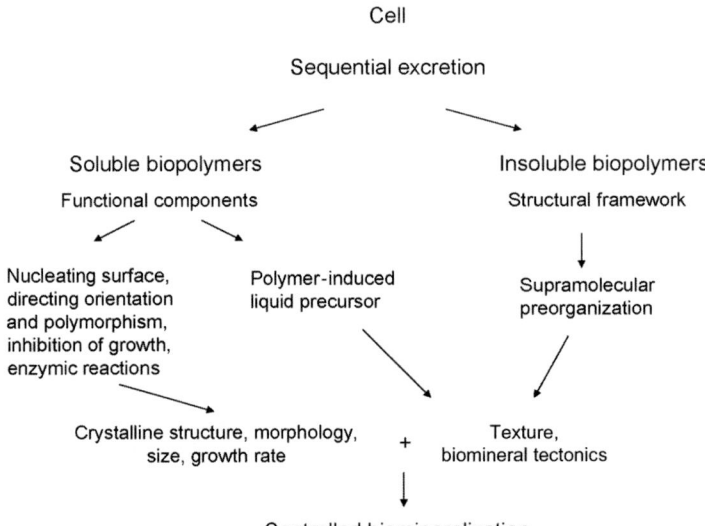

Scheme 6.1 Interplay of soluble and insoluble biopolymers in controlled epi- and extracellular mineralization.

Recently, Weiner and Addadi (2011) have pointed out that there are many examples of biomineralization in which a continuous transition from intracellular to epi- or extracellular mineralization can be observed. The unique process of these pathways is the formation of intracellular vesicles in which ions from the surrounding media (as seawater or body fluid) are concentrated temporarily. Examples are the seawater uptake by foraminifera, the calcite spicule formation by sea urchin larvae, the goethite formation in the teeth of limpets (Mollusca), and the guanine crystal formation in fish skin and spider cuticles (Weiner and Addadi, 2011). Aggregates of amorphous calcium phosphate enclosed in vesicles have been identified as precursors during bone formation in vertebrates (Mahamid *et al.*, 2011).

Inside the vesicles, the ions form a disordered phase that serves as precursor of the final mineral. The vesicle with the precursor phase is transported to the site of mineralization and destabilized there in order to release the mineral content. Three possible pathways are indicated in Figure 6.29.

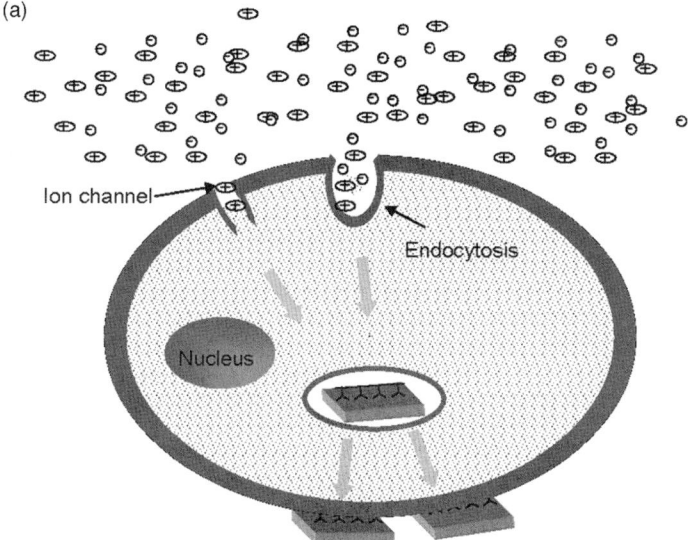

(a)

Ion channel

Endocytosis

Nucleus

Figure 6.29 Crystallization pathways involving (a) the formation of mature mineralized elements within a vesicle inside the cell, (b) a vesicle-confined space (syncytium), and (c) the formation of mature mineralized elements within a vesicle in the extracellular matrix. The ion-sequestering process occurs by endocytosis of seawater droplets and/or ion channels. (a) Transport within the cell to specialized vesicles in which the crystalline phase is formed. The mature mineralized product may remain within the cell or may be transported to the cell surface. (b and c) A first disordered mineral phase is formed in specialized vesicles. The disordered phase–bearing vesicles are transported into the syncytium or into the extracellular matrix. There the disordered phases are transformed into more ordered phases at the crystallization front. (Adapted from Weiner and Addadi (2011).)

(b)

(c)

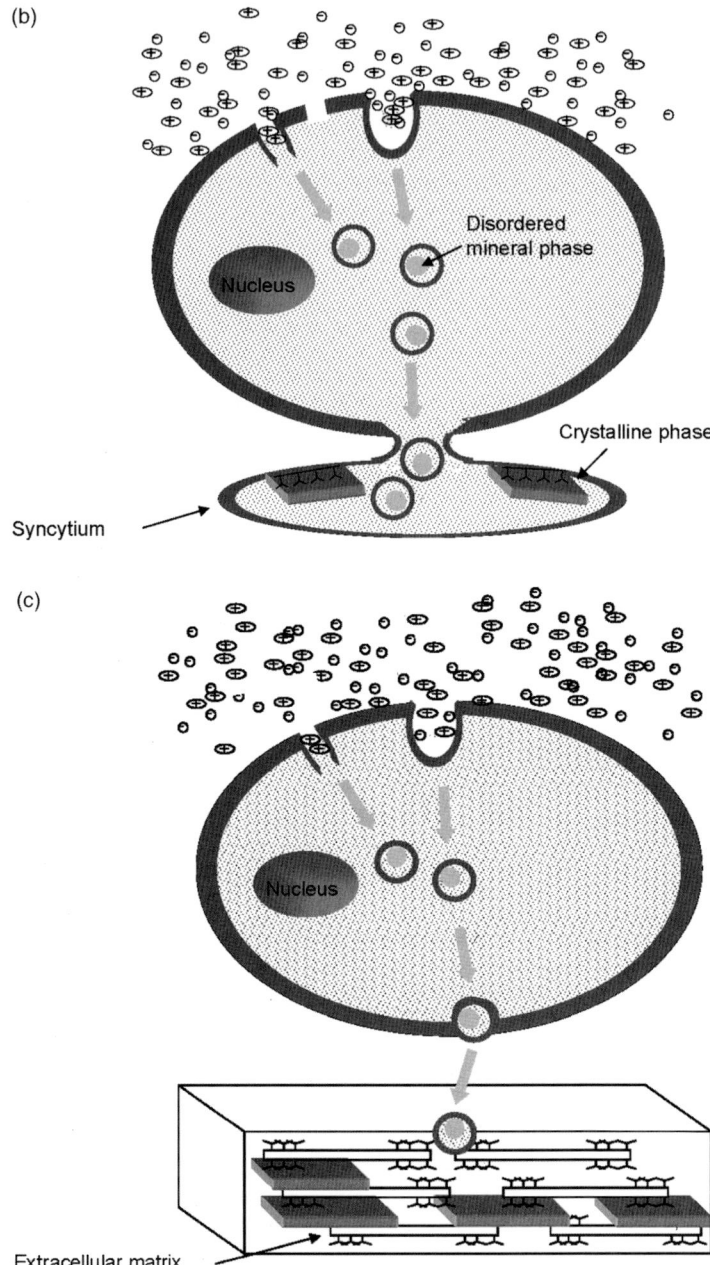

Figure 6.29 (*Continued*)

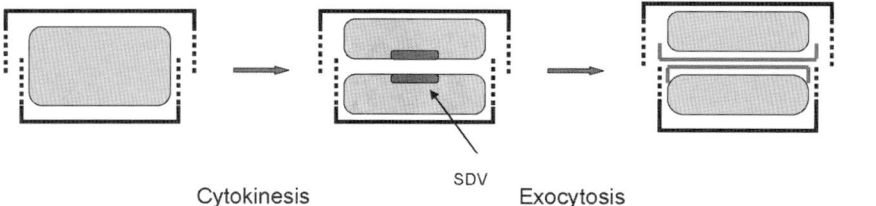

SDV

Cytokinesis   Exocytosis

Figure 6.30 Schematic illustration of the formation of new cell walls during cell division in the SDVs.

6.2.7
Mineralization of Diatom Shells: an Example of Unicellular Hierarchical Structures

As already mentioned in Section 5.1, the outer layer of the cell wall of diatoms is formed by two overlapping Petri dish-like mineralized parts, each consisting of a valve and several girdle bands. The parts tightly enclose the protoplast. They consist of amorphous silica organized in a highly regular structure. Tens of thousands of diatoms with different mineral structures are observed in Nature.

The formation of new mineral valves occurs during diatom cell division in the membrane-bound silica deposition vesicles (SDVs) (see also Figures 6.30 and 5.2). The complete mineral structure is exocytosed in a following process step to be a part of the outer cell wall. To fulfill these tasks, the SDV membrane, called silica lemma, provides some remarkable properties:

1) It withstands high osmotic pressure caused by the high silicic acid concentration during mineralization.
2) It contains transmembrane segments that enable a directed transport of silicic acid by so-called silicic acid transporters SITs. The SITs are carrier proteins with specific binding sites for silicic acid.
3) During mineralization, it changes its shape without loosing its integrity. It is shaped into long tubes or wide flat cylinders when finally the valves are deposited during cell division.

The grown mineral structures are characterized by distinct differences in the shape of the valves (macroscale), in their substructure (mesoscale), and in the size and distribution of pores (nanoscale in the range of 10–100 nm).

On the nanoscale, three classes of biomolecules govern the silica deposition: long-chain polyamines (LCPAs), posttranslational modified polypeptides (silaffins), and phosphopeptides (silacidins) (Sumper and Brunner, 2008). LCPAs have been found in all diatoms investigated until now. Representative examples are shown in Figure 6.31. Variations depending on species are observed in the chain length, the degree of methylation, the presence of secondary and tertiary amino functionalities, and the existence of quaternary ammonium functionalities.

Silaffins are composed of a polypeptide chain covalently bound to long-chain amphiphilic polyamines. The peptide chain is enriched in serine and lysine residues. All the serines are phosphorylated. The lysines are modified with polyamine chains. Therefore, silaffins are zwitterionic structures with many positive

Coscinodiscus granii

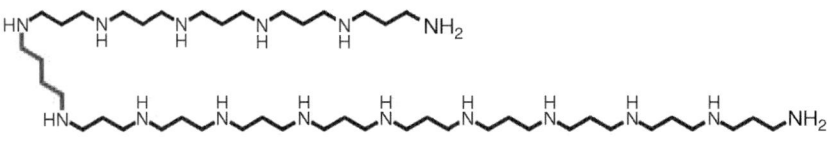

T. pseudonana

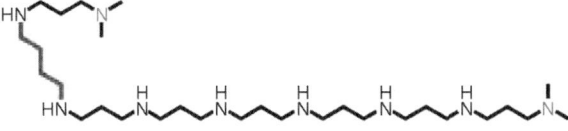

Coscinodiscus wailesii

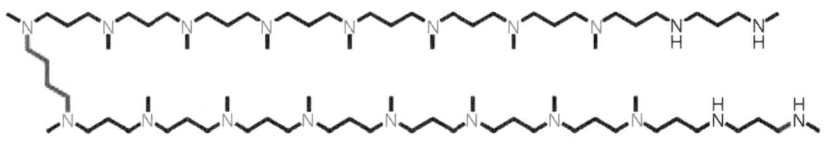

Cercartetus concinnus

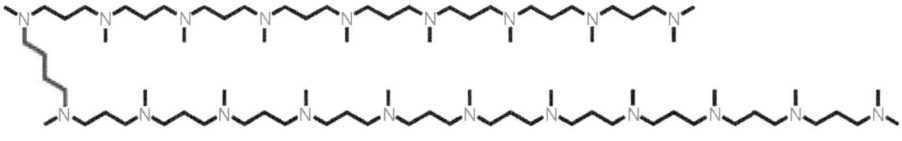

C. wailesii (minor component)

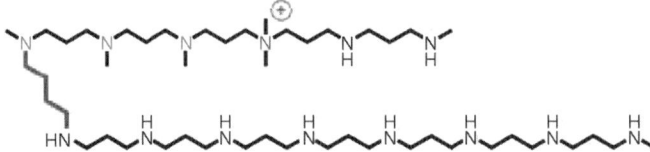

Figure 6.31 Representative structures of long-chain polyamines associated with biosilica from different diatom species. (Reproduced with permission from Sumper and Brunner (2008), Copyright 2008, Wiley-VCH Verlag GmbH.)

charges from the polyamines and negative charges introduced by phosphate groups. Figure 6.32 shows the primary structures of two silaffins from *Thalassiosira pseudonana*. Often the lysine residues are arranged in a definite K–(A/S/Q)–X–K tetrapeptide sequence (boxed in Figure 6.32). Furthermore, entirely polyanionic peptides have been identified. It is assumed that they may serve as cross-linking agents for the cationic polyamines. Silacidins found in *T. pseudonana* consist of

Sil 1L

LPGLTEMPTI SPTHEDYFFG

KSHKSHKSHK SKATKTLKVS KSGKSAKSSK SSG

*RRPL*FGVSQLSEGIAVGYAKSSGRSSQQAVGSWMPVAACILGALSFLN

Sil 3

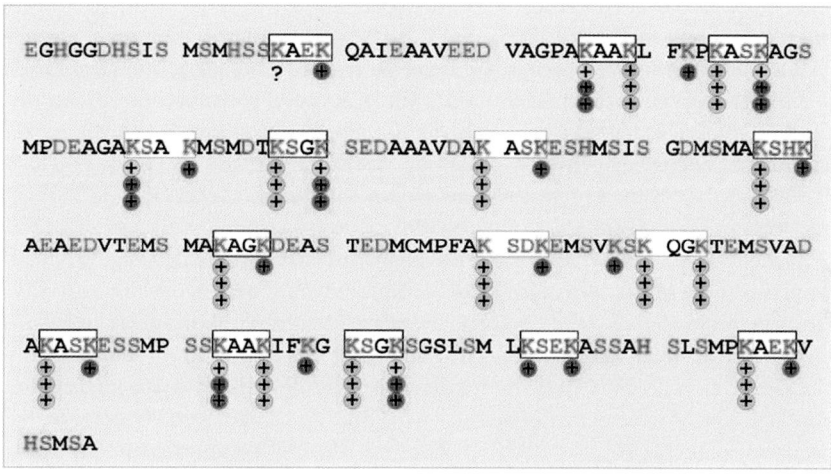

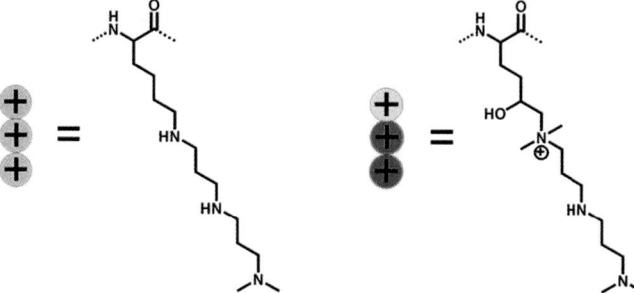

Figure 6.32 Primary structures of silaffins from *T. pseudonana*. Sil-1L: The amino acid sequence as derived from the corresponding gene contains a sequence RRPL that has previously been recognized as a signal for processing of silaffin precursor polypeptides. Therefore, the mature Sil-1L is likely to lack the C-terminal sequence shown in italics. Sil-3: The nature and distribution of complex lysine modification within the silaffin-3 polypeptide. The type of lysine modification is indicated by the symbols representing the chemical structures shown. Tentatively positive and negative charge carriers are shown as red and blue letters, respectively. (Reproduced with permission from Sumper and Brunner (2008), Copyright 2008, Wiley-VCH Verlag GmbH.)

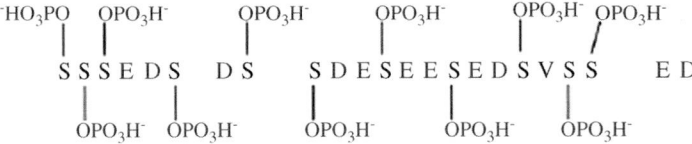

Figure 6.33 Primary structure and phosphorylation of silacidin A from *T. pseudonana*. (Reproduced with permission from Sumper and Brunner (2008), Copyright 2008, Wiley-VCH Verlag GmbH.)

serine phosphates and the acidic amino acids aspartic and glutamic acid. One example is shown in Figure 6.33.

All three kinds of biomolecules are involved in the formation of biosilica inside the SDVs. Interaction of orthosilicic acid with LCPAs and polycationic silaffins, also called catalytic silaffins, leads to the rapid growth of silica particles. In model experiments, it has been shown that densely packed aggregates of biosilica clusters are formed in mixtures of LCPAs and polycationic silaffins. With an increasing content of zwitterionic or polyanionic silaffins as well as silacidins, more open silica aggregates are produced. The polyanionic silaffins and silacidins form extended complexes with LCPAs or polycationic silaffins via Coulomb interaction.

At the positively charged regions of such complexes, monosilicic acid can polymerize into silica nanoparticles. However, the dominating content of negative charges (phosphate sites) inhibits the formation of densely packed aggregates of polysilicic acid. As the extent of silica polymerization is controlled in this way, the polyanionic sillafins are also called regulatory sillafins. Depending on the particular mixture of LCPAs and polycationic and polyanionic sillafins in the SDV, the structure of the amorphous silica varies in a large range.

The result of a model experiment studying the silica precipitation in polyamine–silaffin-3 mixtures is shown in Figure 6.34. There is an optimum polyamine–silaffin-3 ratio for the formation of extended silica aggregates. It correlates with the formation of an extended polyamine–silaffin-3 network in the SDVs.

The structure (density) of the intermediate silica aggregates governs the subsequent structure formation on the mesoscale, in particular the resulting pore size in the silica. Obviously, the mechanisms controlling the polymerization of silicic acid on the nanoscale cannot be responsible for observed structural organization on the mesoscale. As revealed in a study on valve formation in *T. pseudonana* by Hildebrand *et al.* (2008), the macromolecular components are involved in this process. By transmission electron microscopy of the valve, a radial distribution of branching silica ribs has been detected. Nanopores with an average size of about 20 nm exist between the ribs. Larger pores are located near the center. A thin, relatively smooth silica structure, the base layer, with nanopores is observed at the onset of mineralization. The rib structure is more pronounced at the distal face of the valve than at the proximal. With increasing growth time, the ribs built up to a ridge structure. The ribs at the distal surface consist of fused silica particles (diameter about 50 nm).

A possible model explaining this structure formation has been proposed by Mann (2000) and Hildebrand (2008), going back to the first ideas from Crawford and Schmid (1986) and Robinson and Sullivan (1987). They have assumed that proteins

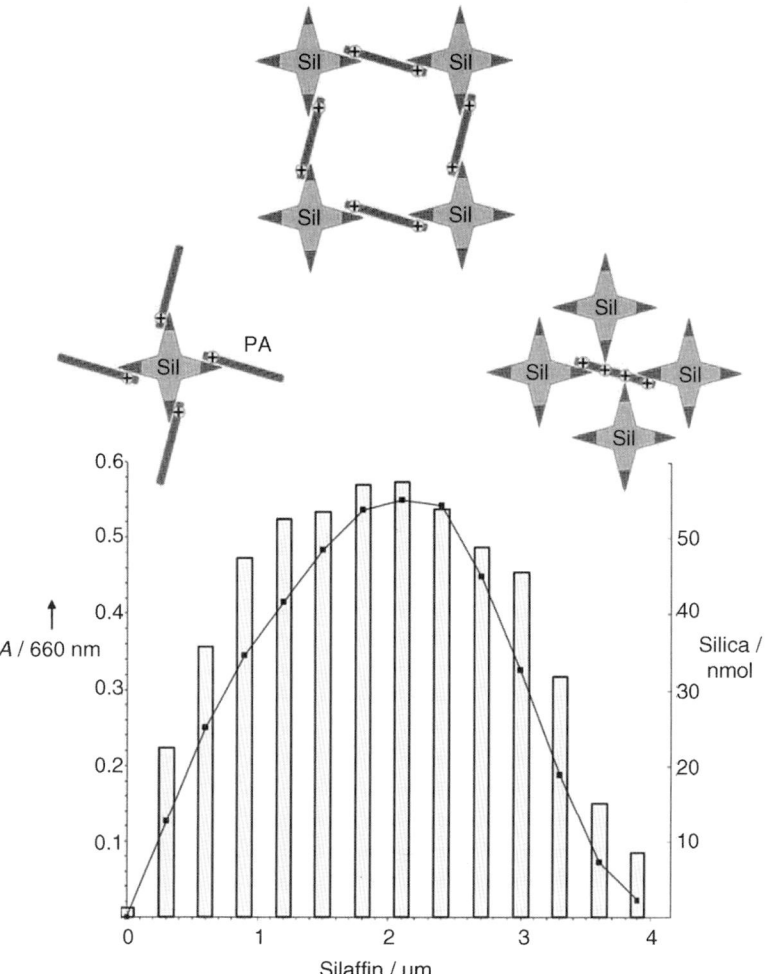

Figure 6.34 Aggregation behavior and silica-precipitating activity of polyamine/silaffin-3 mixtures. Increasing amounts of silaffin-3 were added to a constant concentration of polyamines (50 µM) buffered at pH 5.8 with sodium acetate. The resulting turbidity (A) was measured at 660 nm. The silica precipitation activities (gray bars, SiO_2 precipitated after 10 min) of different polyamine–silaffin-3 mixtures were determined by the addition of silicic acid. The positively charged polyamines (PA) are shown as black bars and the dark patches denote negatively charged clusters in silaffin-3 (Sil). (Reproduced with permission from Sumper and Brunner (2008), Copyright 2008, Wiley-VCH Verlag GmbH.)

from the cytoskeleton outside the SDV shape a pattern of silica-polymerizing proteins inside the SDV through transmembrane proteins (Figure 6.35). It has been proposed that internal proteins linked with the transmembrane proteins organize the distribution LCPAs and sillafins, which control the formation of the silica layer. The proposed model is supported by experimental evidence. Numerous microtubules have been observed next to the SDVs in the cytosol, precisely oriented parallel to the silica ridges in the SDV. The structure is controlled by the forces on the

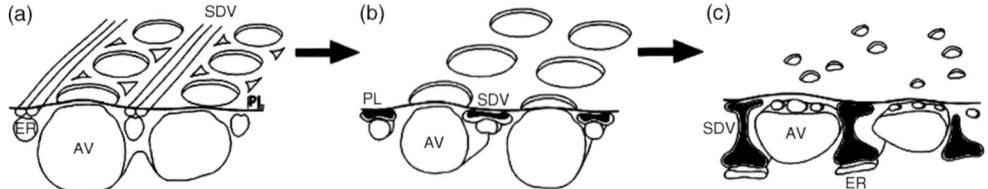

Figure 6.35 Stages in the formation of the siliceous diatom exoskeleton. (a) SDVs are preorganized with microtubules around the boundary space of large areolar vesicles (AR) attached to the plasmamembrane (PL). (b) The SDVs are mineralized with amorphous silica to give a patterned porous wall. (c) The mineralized wall is thickened by extension of each SDV in association with the endoplasmic reticulum (ER). In some diatoms, detachment and retraction of the areolor vesicles from the plasmamembrane result in infiltration with new SDVs and further mineralization of the pore space. (Adapted from Crawford and Schmid (1986) and Mann (2000), Copyright 2000, Wiley-VCH Verlag GmbH.)

SDV caused by the cytoskeleton. Microtubules are also attached to the leading edges of the SDV and pull it into final size. These microtubules often completely disappear after silification. The attachment of actin filaments has also been observed. It has been shown that they are involved in controlling the final diameter of the SDV.

Sumper and Brunner (2006) had proposed an alternative model to explain the formation of the mesoscale mineral structure. They assume that the structure is formed by a series of decomposition events in a solution composed of polysilicic acid, sillafin, and LCPA molecules. However, such model is not compatible with the size of the observed nanopores and their spatial distribution. Self-assembly of the mesoscopic structure by repeated decomposition of a preformed large sillafin microdroplet seems to be very unlikely, since tiny polysilicic acid clusters would already be stabilized in the solution, for instance, by the mobile LCPA molecules.

Recent studies have revealed that in addition to the nanostructure-controlling soluble biopolymers, such as silaffins in the diatoms already considered, insoluble biopolymers also seem to be involved in the formation of the mineralized cell wall. In *T. pseudonana* cells, for example, a chitin-based network has been identified. Its main function is the mechanical stabilization of the cell wall (Figure 6.36). It is involved in silica precipitation (Brunner *et al.*, 2009), as discussed in Section 5.1. Ogasawara *et al.* (2000) assumed that silicate ions and silica oligomers can interact with glycopyranose rings on the surface of the chitin nanofibril through polar and hydrogen bonds. This leads to a possible structure model of silica–chitin composites in diatoms and sponges as proposed by Ehrlich (2010).

6.2.8
Mineralization of Bone: an Example of Multicellular Biomineralization

There are biomineralized structures that extend over a range of hundreds of micrometers and beyond, such as the well-organized patterns of mussel shells or bone. This suggests that their formation is facilitated not only by reactions between some chemical constituents and biomolecules within individual cells but also by subtly controlled processes involving a multitude of interacting cells. Hence,

(a)

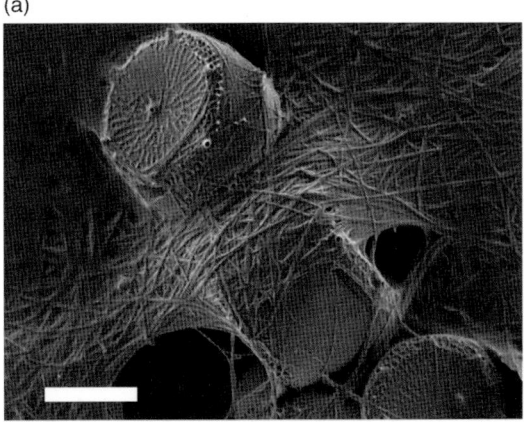

(b)

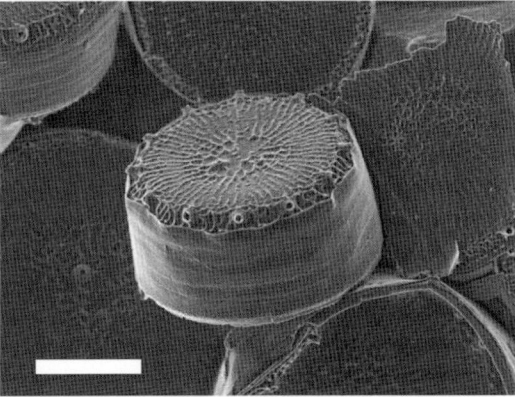

(c)

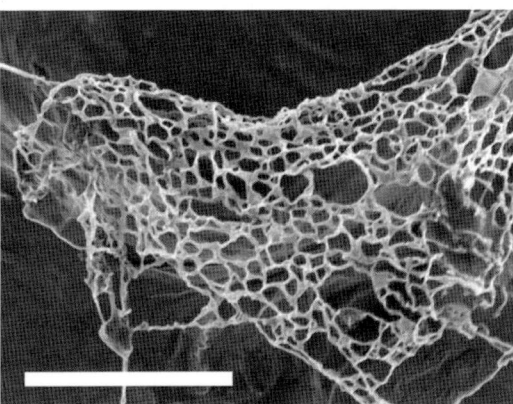

Figure 6.36 SEM images of *T. pseudonana* samples harvested with a filter (a) and a flow centrifuge (b). The majority of the siliceous cell walls withstand both the harvesting procedures and the subsequent treatment with sodium dodecylsulfate (SDS) and ethylenediamine tetraacetic acid (EDTA). (c) The image shows an organic scaffold extracted by NH4F treatment. Scale bar: 2 μm. (Reproduced with permission from Brunner *et al.* (2009), Copyright 2009, Wiley-VCH Verlag GmbH.)

the process as a whole has to be described on at least two structure levels: the nanoscopic level concerning the structure evolution of the nanomineral phases in a particular biopolymer matrix, and the mesoscopic level comprising larger objects, including cells and tissue, which is called, in the sense of Mann, biomineral tectonics of multicellular organisms. Undoubtedly, the processes on the mesoscopic and nanoscopic levels are closely interwoven. Nevertheless, one should try a practicable approach to the complex matter by treating the two levels separately first. The generation of the hierarchical structure of bone tissue is an example of multicellular control in biomineralization. Generally, we observe an interplay of intra- and extracellular processes as concisely indicated in the following:

i) Intracellular
  - Synthesis of soluble biopolymers.
  - Supramolecular preorganization leading to so-called procollagen molecules, which are already triple-helical structures. After excretion, by cutting off C- and N-terminal groups, the insoluble tropocollagen molecules are generated that form collagen fibrils (see Section 6.2.8.3).
ii) Excretion
iii) Extracellular
  - Supramolecular preorganization by large-scale cell-controlled resorption and alignment and nanoscale self-assembly of molecular bricks.
  - Site-specific mineralization by molecular recognition under cellular control.
  - Vectorial growth under local and global control.

The control of the different process steps and the temporal and spatial integration into the formation and remodeling of a hierarchically patterned mineral structure can occur only when the cells being involved in the processes have developed an appropriate way of communication. When we would like to integrate the natural biomineralization processes in the design of materials for bone regeneration and tissue engineering, then a bone substitute harboring one's own bone cells is expected to be very promising. Therefore, developments aiming at the creation of a suitable "living material" are under way.

6.2.8.1 The Mesoscopic Architecture of Bone

The biomineralization of bone is an outstanding example of synergistic cellular interaction mechanisms. Macroscopically, bone can be regarded as a hybrid composite composed of a matrix of insoluble collagen scaffolds, various soluble non-collagenous proteins, nanoscale bone minerals, and matrix forming and remodeling cells, the *osteoblasts* and *osteoclasts*, including various precursor cells. The macrostructure of mammalian bones shows a considerable variety, depending on the ontogenetic stadium and the particular location in the bone (Figure 6.37). For example, the volume ratio of organic and mineral components, the texture of the mineral phase, the porosity, and the number of cells embedded in the bone matrix may largely differ. Bone substance can be characterized by its density as cortical (compact) or cancellous (trabecular) (Figure 6.38). The porosity of cortical bone is 5–30%, with pore sizes in the micrometer range. Cortical bone amounts to

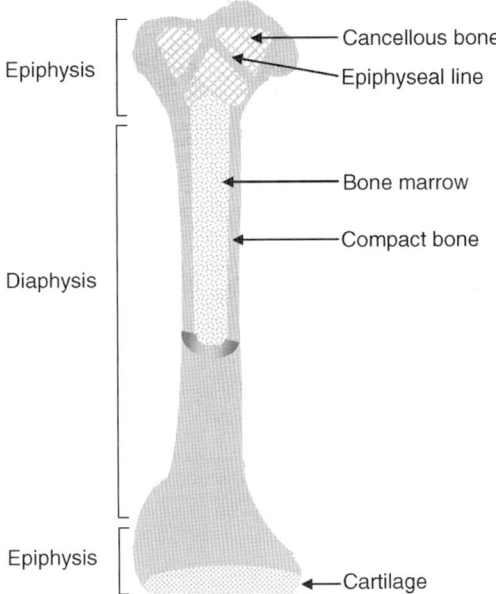

Figure 6.37 Schematic drawing of a long bone.

about 80 wt% of an adult skeleton. Cancellous bone with porosity of 30–90% and pore sizes up to a few millimeters, filling the interior of the individual bones, makes the remaining 20%.

Bone growth proceeds either by intramembranous ossification or by endochondral ossification. Intramembranous ossification is typical for flat bones. In intramembranous ossification, clusters of osteoprogenitor cells formed in a connective tissue differentiate into osteoblasts. They generate an extracellular matrix (ECM) of collagen type I and induce its mineralization. On the other hand, long bones are formed by endochondral ossification, where a cartilage scaffold is grown by chondrocytes first. After having deposited collagen and proteoglycans, they secrete alkaline phosphatase that controls the mineralization of the cartilage. After apoptosis of the hypertrophic chondrocytes, blood vessels can invade into the center of the long cartilage shaft. Hemopoietic cells and osteoprogenitor cells are transported with the blood into the cartilage. The hemopoietic and osteoprogenitor cells differentiate into osteoclasts and osteoblasts, respectively. Now the osteoblasts begin with the deposition of trabecular bone. The osteoclasts can break down the newly formed bone. The interplay of both processes facilitates a remodeling of the primary bone structure. Finally, a long bone is formed that consists of a middle part, the diaphysis, and the two structurally different ends, the epiphyses separated by cartilage zones, the epiphyseal plates, from the diaphysis (Figure 6.37). The outside of the bone is covered with a layer of connective tissue, the periosteum, and articular cartilage at both ends. The outer shell of long bones consist of compact bone.

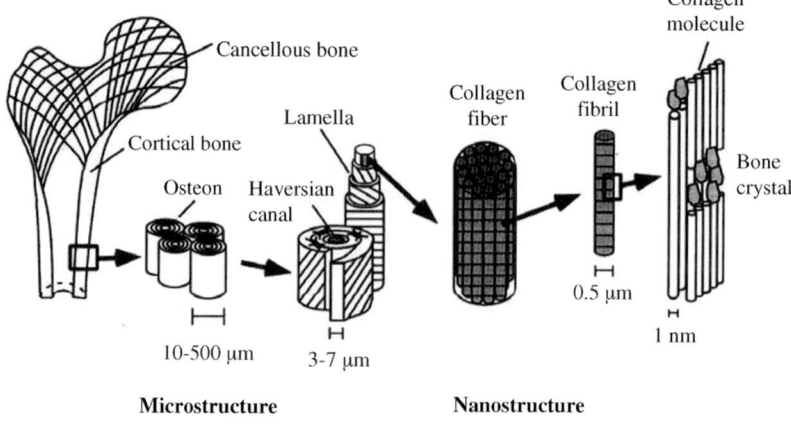

Microstructure

Nanostructure

Macrostructure   Sub microstructure

Sub nonostructure

Figure 6.38 Hierarchical organization of *cortical bone* from centimeter to nanometer scale. Tropocollagen filaments, a fibrillar collagen I with a diameter of about 1 nm and 300 nm in length, are the basic organic components of the ECM. By self-assembly, they are aligned in periodic structures with defined *hole zones*, which are nucleation sites for HAP nanoplatelets, about 45 nm in length, 20 nm in width, and 3 nm in thickness. The fibrils are arranged in alternating sheet-like lamellae with parallel fiber alignment. The *lamellae* form *osteons* in compact bone and *trabecules* in spongy bone. Blood capillaries (*Haversian canals*) are located in the center of the osteons. *Osteocytes*, immobilized osteoblasts, are located inside osteons: They are interconnected by the *canaliculi* pore system. (Reprinted with permission from Roh *et al.* (1998), Copyright 2012, Elsevier.)

Cancellous bone (spongy bone) fills the epiphyses. The cancellous bone and the interior part of the diaphysis are filled with bone marrow.

The compact bone has a well-organized microstructure. It is composed of osteons, which are layered cylindrical structures (diameter 10–500 μm) of mineralized collagen. The mineralized collagen fibrils are arranged in a plywood structure, with parallel orientation within every layer. The orientation rotates from layer to layer (Figure 6.38). An osteon contains *osteocytes* (mature osteoblasts) and a blood vessel in its center (*Haversian canal*). The osteocytes placed in liquid-filled *lacunae* are interconnected by the *canaliculi* pore system. The structure forming elements of the cancellous bone are the trabeculae that are layered structures of mineralized collagen with embedded osteocytes, also connected by *canaliculi*. The thickness of an individual trabecula is about 200 μm. The orientation of the mineralized collagen fibrils keeps near that of the axis of the trabecula. The large porous space of the cancellous bone is filled with marrow and other living cells.

There are two essentially different bone growth modes: *primary and secondary osteogenesis*. The primary bone growth is concentrated in the epiphyseal cartilage. The epiphyseal cartilage consists of loose fibril bundles of collagen. At the growth front, there is a high concentration of vesicles that deliver either nanocrystals or ions for the mineralization of an unorganized cancellous microstructure. The random deposition of defect HAP proceeds as *extrafibrillar mineralization* on the surface of collagen fibrils. The cancellous bone is characterized by low stiffness and mechanical strength. Subsequently, the primary bone structure is remodeled into a higher ordered structure (secondary bone formation). In this process, the bone-forming cells, the osteoblasts, secrete well-aligned collagen fibrils. The highly ordered structure of collagen fibrils harbors the preorganized sites for the textured precipitation of HAP crystals. Thick mineralized collagen fibrils with a diameter of about 80 nm are packed into concentric lamellae forming the osteons. The osteons are former osteoclastic resorption channels refilled with bone matrix. They include the osteocytes and the Haversian canal system. A Haversian canal (diameter about 50 μm) contains one or two blood vessels and nerves. It communicates with the osteocytes via the *canaliculi*.

6.2.8.2 Bone Remodeling and Bone Repair

Bone is an active tissue that is capable of adapting its structure to mechanical stimuli and of repairing structural damage by remodeling. In a growing skeleton, two distinct processes – bone modeling and remodeling – occur. *Bone modeling* is a process that alters the size or shape of the bone by selectively adding bone mass to a particular surface or by removing it from that surface, respectively. It involves either osteoblast activation with subsequent bone formation at the surface or osteoclast activation and bone resorption. However, both processes do not occur simultaneously at the same location. In a mature healthy skeleton, bone modeling is minimal, essentially restricted to exceptional cases such as response to certain diseases or prolonged weightlessness. Unlike modeling, *remodeling* is a sequential process that involves activation, resorption, and formation of new bone at the same locus by packet-wise replacement of old bone tissue with the new one. Remodeling or bone turnover serves for the renewal of bone tissue during the lifetime of the organism with only minor changes of shape as well as repair of microdamage. Bone remodeling is a multicellular process (Figure 6.39).

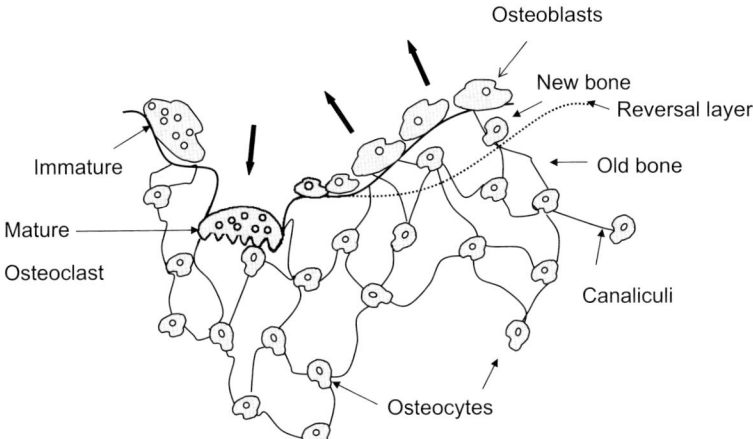

Figure 6.39 Cell activity during bone remodeling.

A basic multicellular unit (BMU) with the size of an osteon moves through the bone tissue. Osteoclasts, the bone resorbing cells, are arranged in the leading region of the BMU. They excavate a tunnel in the compact bone, roughly 250–300 μm across. The osteoclasts are followed by mononuclear cells that deposit a thin, mineral-deficient reversal layer of matrix that separates osteons from the surrounding interstitial lamellae. Osteoblasts, the bone-forming cells, adhere to the reversal layer behind these mononuclear cells. The osteoblasts deposit nonmineralized layers of a new osteoid centripetally. This process stops when only the space for the central Haversian canal is left. In trabecular bone, the remodeling occurs in the same sequence of cellular activity, with the difference that the osteoclasts do not form tunnel-like structures. Instead, scalloped packets of bones are excavated and refilled. In human cortical bone, the process of complete remodeling of a BMU takes approximately 120 days: about 20 days for excavation, 10 days for deposition of the reversal layer, and 90 days for the formation of the new osteon. The net amount of old bone resorbed and new bone deposited in the remodeling cycle, the so-called *bone balance*, is zero in mature healthy bone, but may be nonzero in case of disease, as with osteoporosis where it is negative. Bone remodeling and repair are the result of the synchronized action of bone-resorbing cells, the osteoclasts, and the bone-forming cells, the osteoblasts. A well-tuned communication is upheld between these two cell families during all the time between proliferation and differentiation of the precursor cells, immobilization in the bone matrix such as osteocytes, and programmed cell death (apoptosis). This communication enables the permanent remodeling of bone tissue, resulting in outstanding biomechanical properties such as self-healing of microdefects and shape adaptation to external loading. Before discussing this phenomenon in more detail, the cell lines from the stem cells via the committed cells to the osteoblastic and osteoclastic cells are briefly considered in the following (see also Figure 6.40).

The osteoblasts are derived from mesenchymal stem cells (MSCs) present in the bone marrow, spleen, and liver. The MSCs can differentiate into uncommitted osteoblast progenitors (OB_u). By specific signals such as growth factor (e.g., TGF-β) or other regulatory signals (e.g., the parathyroid hormone (PTH), vitamin D_3, interleukins, and corticosteroids), they can be changed into responsive osteoblasts (OB_p) and activated osteoblasts (OB_a). The active osteoblasts generate new osteons. At the end of this cycle, the osteoblasts are encapsulated in the new mineralized bone matrix as osteocytes, or they switch to the programmed cell death (apoptosis). In the compact bone, osteocytes serve as sensors for controlling the response of the bone tissue to mechanical loading.

The osteoclasts are derived from monocytes (hematopoietic progenitor cells). Stimulated by the macrophage colony-stimulating factor (MCSF), the osteoclast precursors (OC_p) develop from the uncommitted osteoclast progenitors (OC_u). The OC_p cells fuse together into multinuclear immature osteoclasts only in the presence of MCSF and the signal protein receptor actuator of nuclear factor kappa B (NF-κB) ligand (RANKL) present at the surface of osteoblasts. RANKL is liberated from the osteoblasts and interacts with the RANK receptor expressed on the surface of the osteoclast precursors. The interaction of the RANKL with the RANK receptor is

(a)

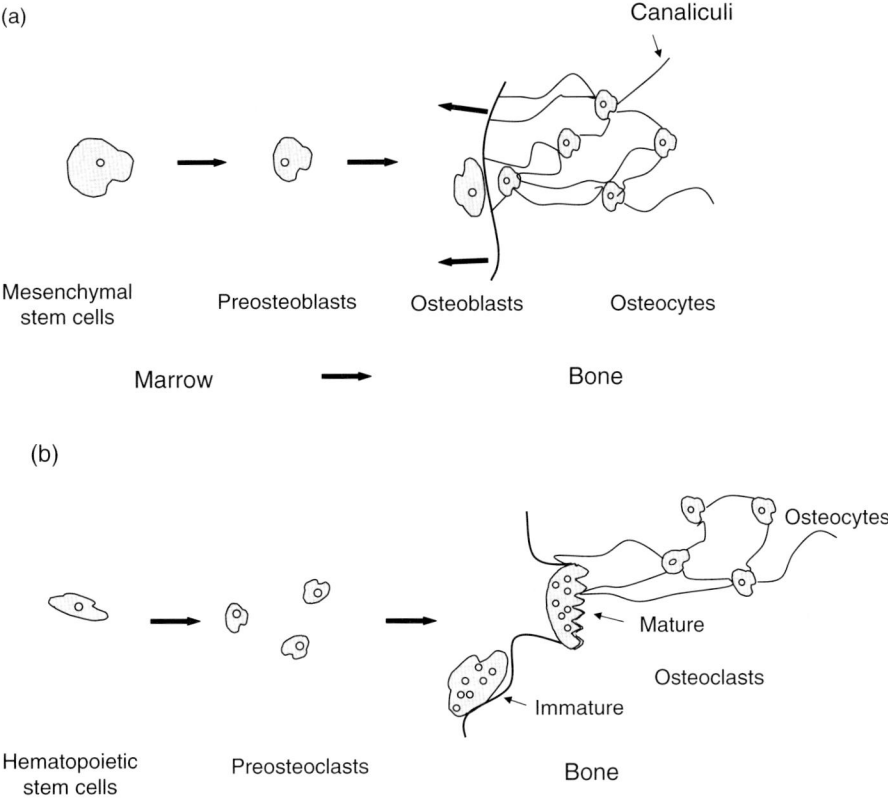

Canaliculi

Mesenchymal
stem cells

Preosteoblasts   Osteoblasts   Osteocytes

Marrow   →   Bone

(b)

Osteocytes

Mature

Osteoclasts

Immature

Hematopoietic
stem cells

Preosteoclasts   Bone

Figure 6.40 Lineage of osteoblasts (a) and osteoclasts (b).

additionally controlled by the soluble osteoprotegerin (OPG), which is produced by stromal/osteoblastic cells. RANKL together with a host of transcription factors is needed for the transformation into mature active osteoclasts. The active multi-nuclear osteoclasts perform the resorption process. Afterward, they switch into the apoptotic stage.

The main structure of the complex interaction between osteoblasts and osteoclasts is schematically shown in Figure 6.41.

The signaling pathway of the communicating osteoblasts and osteoclasts is based on the interaction of RANKL expressed by the cells of the osteoblastic cell line with its receptor RANK found on the surface of hematopoietic precursors of the osteoclastic cell line. As indicated in the scheme, the RANKL is formed on the surfaces of the responding osteoblasts and the active osteoblasts. It can be transferred into a soluble form by metalloproteases. Binding of RANKL at the RANK receptor causes the differentiation of the osteoclast precursors into active osteoclasts. The amount of RANKL available for the interaction with the RANK receptor is additionally controlled by the soluble ligand, osteoprotegerin (OPG). It is also produced by osteoblastic cells. Binding of free RANKL by OPG reduces the

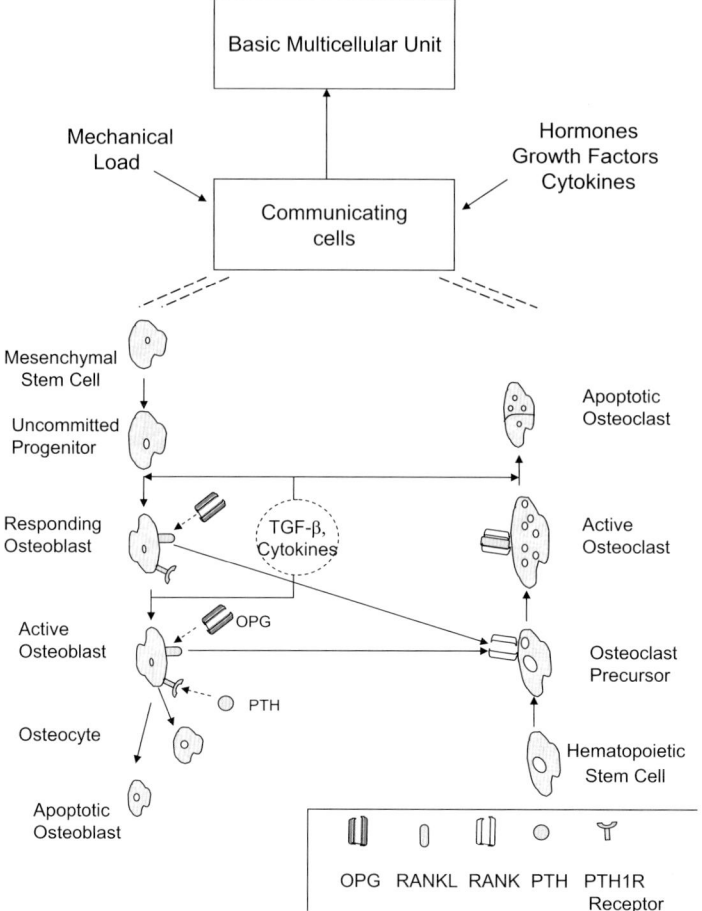

Figure 6.41 Communication of osteoblastic and osteoclastic cell lines in bone remodeling. Among the various existing signaling pathways, the RANK–RANKL–OPG pathway has been pictured in the scheme as one of the most relevant ones for cell differentiation. The PTH serves as an example of how cytokines and hormones control the balance of RANKL and OPG. The growth factor TGF-β is an example of functional factors released from the bone matrix during resorption activity of osteoclasts. As indicated, the process is additionally influenced by external biochemical signals and mechanical stimulation. (Adapted from Pivonka *et al.* (2008).)

possible activation of osteoclasts. The interplay of these proteins is noted as the RANK–RANKL–OPG pathway. The balance of RANKL and OPG is controlled by cytokines and hormones such as PTH, prostaglandins, interleukins, vitamin D_3, and corticosteroids. A PTH receptor expressed on osteoblasts only is shown in Figure 6.41. Binding of PTH leads to an increasing RANKL/OPG ratio that indirectly stimulates the osteoclasts. In addition to the described osteoblastic control of the osteoclasts, there is an essential feedback mechanism initiated by the active osteoclasts. Usually there are large amounts of growth factors, mainly from the TGF-β superfamily, stored in the ECM, which influence the differentiation of both osteoblasts and osteoclasts. When the active osteoclasts resorb the ECM, the growth factors are deliberated and interact with the surrounding cells. TGF-β, for example, promotes the formation of responding osteoblasts, but inhibits further differentiation into active osteoblasts. Thus, a big reservoir of responding osteoblasts is formed that can differentiate into active osteoblasts when the concentration of TGF-β decays. Furthermore, TGF-β interacts with active osteoclasts initiating their apoptosis. The described mechanisms demonstrate how the remodeling of bone is strictly controlled by biochemical signaling pathways, which makes it a very stable process in healthy bone tissue. Bone diseases as osteoporosis and also some aspects of bone metastasis formation are caused by alterations and imbalances of these pathways.

Bone remodeling is governed not only by biochemical control but also by mechanical loading. It is accelerated by both disuse and enhanced use. In case of disuse, bone resorption leads to a rapid loss of bone mass. In case of enhanced use, osteoclasts preferentially resorb bone tissue in the microdamaged zones. Afterward, new bone tissue is formed at these places. In case of overuse, the bone loss by microdamage and coupled bone resorption cannot be balanced by formation of new bone that finally leads to bone failure. Physiological loading gives rise to structural optimization of the load-bearing components of the ECM. In cancellous bone, the trabecular trajectories are roughly aligned along the direction of the maximum stress (*Wolff's law*), which means maximum strength with given amount of substance. In cortical bone, in regions under unidirectional tensile stress, the collagen fibers are favorably aligned along the loading direction, whereas in regions under uniaxial compressive stress the collagen fibers are oriented to a higher extent perpendicular to the loading direction, which is the direction of tensile strain. The cellular response to loading in connection with bone adaptation has been described by the following phenomenological rules (Robling *et al.*, 2006):

i) Dynamic but not static strain appears to be the primary stimulus of bone adaptation. The effect is essentially restricted to the frequency range of 0.5–90 Hz.
ii) During cyclic loading, the cells become desensitized and so the bone formation response decreases. After a period of rest, the cells become resensitized within seconds or hours, depending on loading stimulus.
iii) The response to an actual load may depend on previous loads.
iv) There is a threshold for local bone tissue response.

Among the cells of the osteoblastic lineage, there are cells specialized for sensing strain. This task is fulfilled by the osteocytes in the bulk of the compact bone and by the bone lining cells on the surface. Together their number in the bone tissue exceeds that of osteoblasts and osteoclasts by a factor of 20. The *canaliculi* pore system forms a fluid-filled network for the communication between the osteocytes and other bone cells. Probably the osteocytes respond indirectly to mechanical strain via extracellular fluid flow caused by deformation of the bone tissue. One possible pathway of mechanical signal transduction to the bone-forming cells seems to be the following. Fluid shear stress triggers the release of messenger molecules by the osteocytes, in particular prostaglandins and nitric oxide. These messengers inhibit RANKL expression and increase the OPG production. In this way the RANKL/OPG ratio in the communication pathway of osteoblasts and osteoclasts is changed (see above), which finally suppresses osteoclast differentiation. There are also other mechanisms involved in the mechanotransduction system of bone tissue. Bone lining cells respond to mechanical stimulation with dedifferentiation into active osteoblasts. Similarly, osteoblasts respond directly to mechanical stress in the absence of osteocytes. They secrete messengers (e.g., prostaglandins and nitric oxide) and express various growth factors under mechanical stimulation. It has been observed that mechanical loading reduces programmed cell death rates of osteocytes and osteoblasts. This extends the duration of bone matrix synthesis by the osteoblasts. Mechanotransduction in bone tissue, however, is still a matter of intense research. The available preliminary knowledge indicates that the stimulation of cell activity by applied external load may become part of future technologies for *ex vivo* engineering of bone tissue.

6.2.8.3 The Nanoscopic Structure of the Extracelluar Matrix of Bone

Bone as a hybrid inorganic–organic composite is distinguished by a well-organized microstructure of the ECM. The matrix is composed of about 33–43 vol% mineral phase HAP, 32–44 vol% organic substance, and 15–25 vol% water (Olszta *et al.*, 2007). The organic substance consists mainly of collagen fibrils that are composed of collagen type I monomers. The term "collagen" does not stand for a particular substance, but rather for a family of substances. Collagen is found in the extracellular matrices of most mammalian tissues with a particularly high content in connective tissues like skin (dermis), tendon, cartilage, and bone. About 30 known collagen types differ considerably in their composition and properties. However, there are some common features such as triple helices or triple-helix domains and the high hydroxyproline content. The amino acid sequence in a collagen molecule typically consists of repeating tripeptide units (glycine–X–Y), where the positions X and Y are often occupied by proline and hydroxyproline, respectively. Molecules of vertebrate collagen typically contain about 35% glycine and 21% proline and hydroxyproline. The content of any of the other amino acids is less than 5%. The collagen family includes molecules that form fibrils. Other types control fibril growth and are associated with fibrils. There are also types that form networks and filaments. In fibrillar collagen, three separated polypeptides, the so-called α-chains, form tropocollagen monomers that assemble into supramolecular micro- or macrofibrils. Three major types of fibrillar collagens are found in vertebrates. Collagen type I is part of the ECM of hard tissue such as bone and dentine, and it is also a major constituent of other tissues such as tendon, ligaments, and skin. Collagen type II is the scaffold-forming protein of cartilage, whereas collagen type III is the dominating structural protein in soft tissue.

Collagen type I acts as a tough matrix of the bone composite filled with stiff HAP nanoparticles, and it forms the supramolecular template for the biological control of the growth of the mineral phase. The proper combination of structural and functional properties is realized by a sophisticated hierarchic nano- and

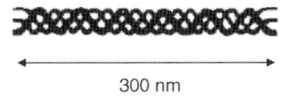

Tropocollagen

300 nm

Self-assembly

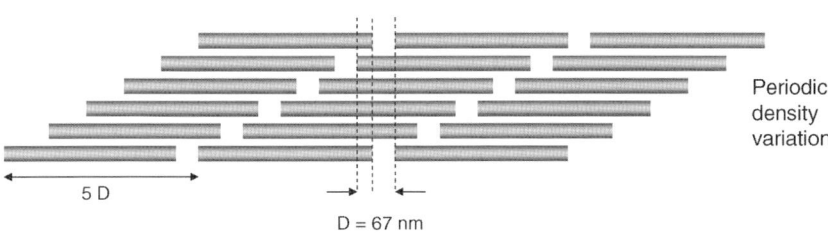

Periodic
density
variation

5 D

D = 67 nm

Figure 6.42 Self-assembly of collagen fibrils. After cleavage of short propeptide ends, tropocollagen molecules self-assemble into a fibrillar structure with periodically arranged hole zones. The staggering period $D = 67$ nm subdivides every filament into four long elements and one short overlap with 27 nm in length.

microstructure. The basic structural unit of collagen type I is a triple helix formed inside the osteoblasts. It consists of two identical $\alpha(1)$-chains and an $\alpha(2)$-chain arranged in a right-handed superhelix (Figure 6.42). The helix has a rise of 0.29 nm per amino acid residue and approximately 3.3 amino acids per turn. It is stabilized by hydrogen bonds and intrahelical water molecules. At the ends of the α-chains, there are propeptides that make the molecules soluble. After excretion, propeptides at the ends of the triple helix are cleaved off by enzymes. This leads to an insoluble tropocollagen molecule with about 300 nm in length and 1.5 nm in thickness. By self-assembly, the tropocollagen molecules are organized into long fibrils with a highly ordered nanostructure. Equally oriented molecules are arranged in a staggered pattern with a periodic shift of $D = 67$ nm. In this pattern, each filament is subdivided into five zones: four long zones of equal length (67 nm) and a short zone (27 nm). The shift by 67 nm maximizes the interfilament interactions. The periodicity leads to gap regions (or hole zones) between the aligned filaments of about 40 nm in length and 5 nm in width. Adjacent hole zones overlap and form grooves oriented in parallel rows perpendicular to the fibril axis (Figure 6.42). The tropocollagen assembles into fibrils with a diameter of about 100–200 nm, which finally can form thick fibers with a diameter up to 100 μm (Figure 6.43). *In vitro* self-assembly can be triggered by a change in pH and/or a rise of temperature. In these experiments, an interesting intermediate structure has been observed, the so-called microfibril (diameter 40 nm), composed of five tropocollagen molecules (Tuckermann, 2001).

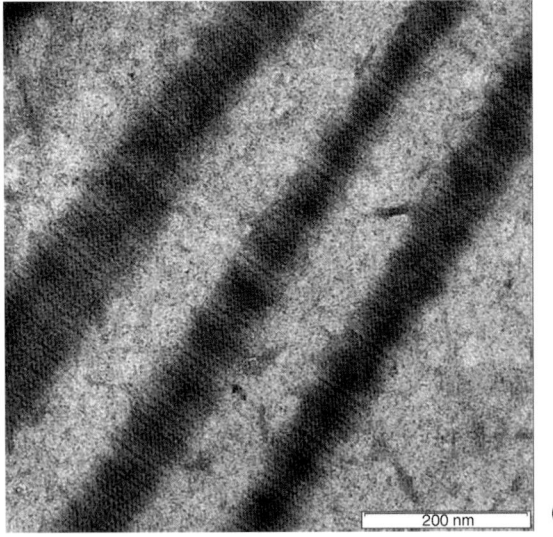

(a)

(b)

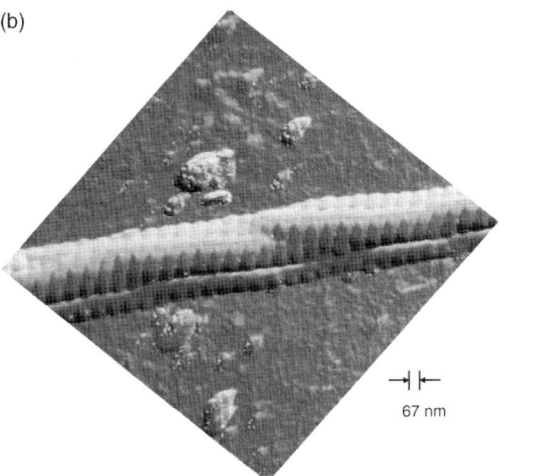

67 nm

Figure 6.43 (a) TEM image of stained collagen fibrils. Note the regular pattern within each of the fibrils. Scale bar: 200 nm. (Courtesy by A. Mensch.) (b) Atomic force microscopic image of a collagen type I fiber. (Courtesy by M. Tuckermann.)

In a mineralized collagen fibril, the hole zones are filled with HAP nanocrystals (Figure 6.44). These are plate-shaped particles, about 45 nm in length, 20 nm in width, and 3 nm in thickness. With its [001] axis, the crystals are aligned along the collagen fibril (Landis *et al.*, 1993). (See also cover image.) The platelets have some rotational disorder about the *c*-axis. The platelets are not coherently aligned with the *ab* plane. There is no biaxial symmetry in the diffraction pattern of transmission electron microscopy (Olszta *et al.*, 2007). This means the long accepted "deck of cards" model of bone's nanostructured architecture is not entirely supported by the

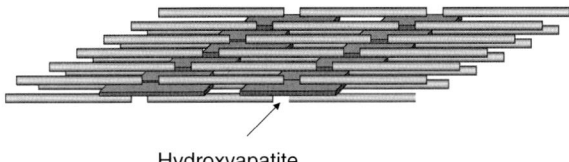

Hydroxyapatite

Figure 6.44 Mineralization process, based on *ex situ* observations using high-voltage TEM and tomographic reconstruction imaging (Landis *et al.*, 1993). Large and small HAP deposits occupy several hole zones. Preferential growth in *c*-axial direction follows the long axis of collagen. Two crystals formed in adjacent zones can fuse to create a large unit.

experimental observations. One missing link is yet a universally accepted 3D model of the supramolecular arrangement of the tropocollagen molecules in a collagen fibril (Ottani *et al.*, 2002).

It is surprising that the biomineralization of the collagen matrix, referred to as intrafibrillar crystallization, leads to such a close-packed interpenetrating structure of tropocollagen fibrils and nanocrystalline HAP platelets. Obviously, the shaping of the uniaxial crystal pattern needs structure-directing mechanisms. Besides the preorganized collagen microstructure, there is a larger fraction of water-soluble proteins and proteoglycans that are involved in these processes. The total content of noncollagenous proteins (NCPs) constitutes about 10–15% of the organic matrix. The main groups of anionic NCPs are compiled in Table 6.3.

The NCPs possess characteristic binding motifs composed of acidic amino acids (Figure 6.45). With a particular sequence of amino acids, the distance of the binding sites of the NCPs is adapted to a particular lattice plane of a growing crystalline phase.

The Gla proteins are small proteins characterized by the presence of amino acid motifs, including the γ-carboxyglutamic acid (Gla). With its twofold negatively charged residue, it offers a perfect binding site for the Ca^{2+} ions. An essential example of Gla proteins is osteocalcin. Osteocalcin is the most abundant NCP in bone (up to 20%). It is a highly specific osteoblastic marker produced during bone formation. The molecule contains 46–50 amino acids, depending on the species. Two antiparallel α-helices, the "Gla helix" (α1) and the "Asp—Glu helix" (α2), are framed by β-sheet structures. Because of three Gla residues of the α1-helix and the Asp—Glu residues of the α2-helix, osteocalcin has a strong binding efficiency to Ca^{2+} ions. By molecular modeling of the binding of osteocalcin to HAP, it has been shown that the three Gla residues together with the Asp residue can coordinate five Ca^{2+} ions on the prism face (100), with minor differences in the displacements needed for a coordination binding, and also on the secondary prism face (110) of HAP (Hoang *et al.*, 2003). This means that a preferred binding of osteocalcin at the prism faces could be one mechanism responsible for the favored elongation of the HAP platelets in the bone matrix. Besides its influence on the HAP mineralization, osteocalcin plays a decisive role in bone remodeling. The protein is secreted by the bone-forming cells, the osteoblasts. Its α-helical C-terminal sequence (Phe—Tyr—Gly—Pro—Val; α3-helix) seems to be responsible for a chemotactic

Table 6.3 Main types of water-soluble proteins and proteoglycans controlling the biomineralization of bone.

Macromolecules	Molecular mass	Structure-controlling units
Gla proteins		
Osteocalcin	6	γ-Carboxyglutamic acid (×3)
Matrix glaprotein	15	γ-Carboxyglutamic acid (×5)
Acidic glycoproteins		
Osteonectin	44 (Bovine)	Aspartate- and glutamate-rich
Sialoprotein II	200	Glutamate-rich
Osteopontin	33–55	Aspartate-rich/phosphorylated serine
Proteoglycans		
Bone proteoglycans	350	Chondroitin sulfate
Cartilage proteoglycans	1000	Chondroitin/keratan sulfate

Source: Adapted from Mann (2001).

activity of osteoclastic cells (monocytes, preosteoclasts, and osteoclasts) involved in the bone resorption process (see also Section 6.2.8.2).

The acidic glycoproteins are rich in aspartic acid (Asp) and glutamic acid (Glu) residues. Among the bone-relevant glycoproteins, the phosphoproteins are of

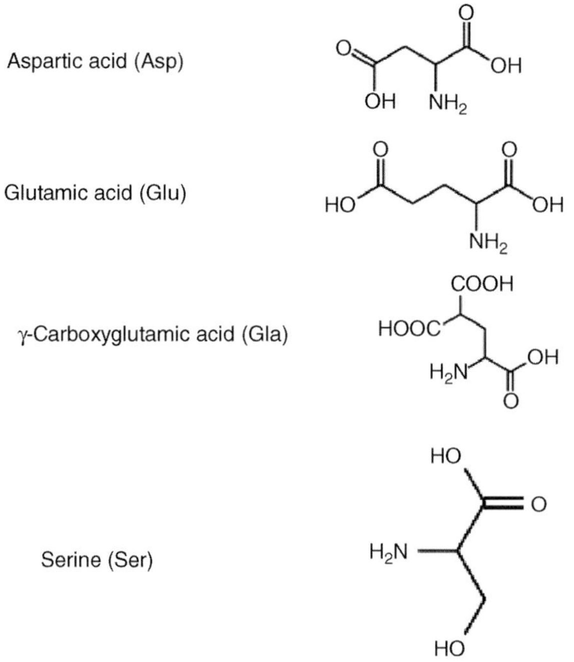

Figure 6.45 Amino acids mainly acting as components of structure motifs in biomineralization.

particular interest because of the presence of phosphorylated serine (PSer) in addition to aspartate and glutamate residues. One of the most abundant NCPs of the bone matrix is osteopontin. It has been found to be an effective inhibitor of HAP growth in the bone matrix. Osteopontin and phosphorylated serine bind preferably to the (100) and (010) faces of HAP, respectively. Osteopontin is also involved in the communication between osteoblasts and osteoclasts. It is expressed by the osteoblasts and osteocytes. Enrichment of osteopontin is observed at the interface of the bone matrix and also at interfaces between old and young bones, the so-called cement lines. There it activates the osteoclasts. The Gly—Arg—Gly—Asp—Ser motif has been identified as the one that controls the adhesion, spreading, and migration of osteoclasts. It initiates the process by which osteoclasts develop their ruffled borders to begin bone resorption.

Bone proteoglycans are mainly found in mineralized cartilage. In the bone matrix, they make only around 1 wt%. In the initial stage of bone growth, the cartilage forms the early shape of the final bone, whereas in mature bone, it forms compliant interfaces for a soft coupling between the rigid compact bones. The proteoglycans are organized along a central unit of repeating disaccharide units – the hyaluronic acid (Figures 6.46 and 6.47). A proteoglycan monomer is composed of a central core protein that serves as a carrier for about 80 glycosaminoglycan chains (GAGs). Two types of GAGs are known to be relevant for bone tissue development: the chondroitin sulfate (CS) and the keratan sulfate. These are polymers of disaccharide units with one acidic group on each saccharide ring (carboxylate and sulfate, respectively) (Figure 6.47). The negative charges are favored binding sites for Ca^{2+} ions.

Recently, growing evidence has been derived from experimental studies as well as molecular simulations that the face-specific interaction of proteins with inorganic crystals is mainly mediated by the *electrostatic interaction of charged subdomains* of the protein (Hunter *et al.*, 2010). This view differs from the one proposed by Addadi and Weiner (1985), according to which the *stereochemical interaction* governs the selection of the favored faces for protein adsorption. The stereochemical theory is based on three assumptions: (i) The crystal-binding region of the protein has a planar surface with charged groups projecting perpendicular to the plane. (ii) The spacing of the charged groups or the protein is commensurable with the spacing of cations in a

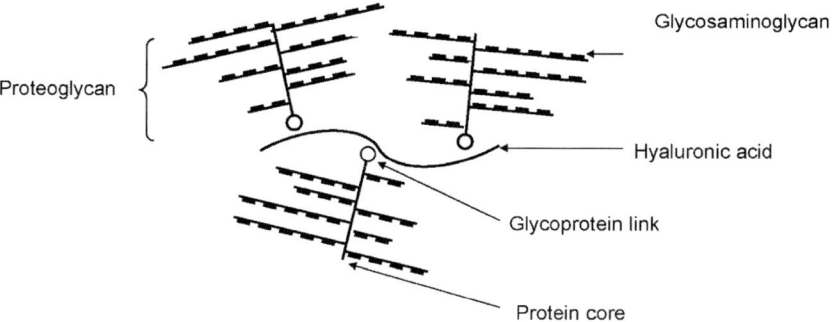

Figure 6.46 Proteoglycan aggregate.

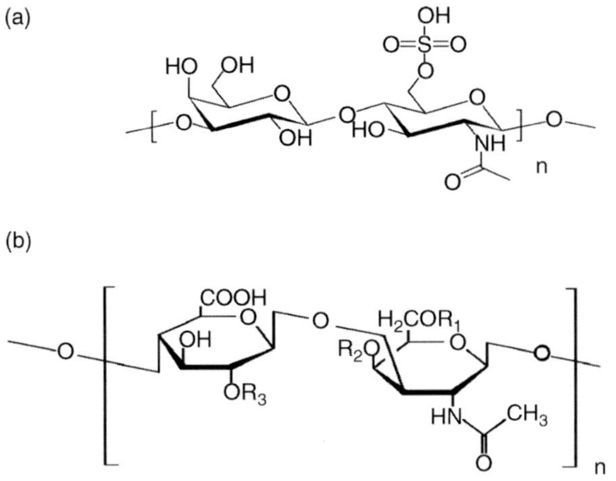

(a)

(b)

Figure 6.47 Glycosaminoglycans building together with proteins the proteoglycan aggregates. (a) Keratan sulfate. (b) Chondroitin-4-sulfate: $R_1 = H$, $R_2 = SO_3$, $R_3 = H$; chondroitin-6-sulfate: $R_1 = SO_3$, R_2, $R_3 = H$.

lattice plane of the crystal. (iii) The protein-binding face of the crystal has a particular orientation of anionic groups that optimizes the coordination of the cations shared with the protein.

The fact that proteins interacting with crystals usually behave like compliant polyelectrolytes is in disfavor of the stereochemical model in this case. This means that the flexibility of the protein allows the adsorption of charged subdomains of the protein with maximum conformational freedom to form electrostatic bonds between the anionic groups of the protein and the cations exposed to the crystal face. Reality seems to be between the two extremes: the high compliance of a polyelectrolyte and the rigidity and chemical specificity of a stereochemical pattern. The interpretation as a compliant polyelectrolyte opens up a new vista with regard to the role of GAGs in biomineralization. The high negative charge density and the compliance of the chains of GAGs explain their experimentally observed effect on mineral growth. Thus, we have to conclude that both concepts, the stereochemical model of protein–mineral interaction and the polyelectrolyte hypothesis, can serve as useful guides in the search for bio-inspired processing routes of well-defined nanostructured minerals.

6.2.8.4 The Polymer-Induced Liquid Precursor Process

For at least three decades, there was a debate going on over the question how platelet-like hydroxyapatite (HAP) nanocrystals can be formed inside the periodically arranged 40 nm long gap zones of the collagen fibrils. It was assumed that this process is governed by various acidic noncollagenous proteins. It was not clear whether the collagen matrix would play an essential role or only act as a passive scaffold. Furthermore, there was the problem of how the calcium and phosphate ions could reach the hidden gap zones. Simulation of such infiltration process by adding collagen fibrils into a solution of calcium and phosphate ions usually leads to the growth of larger HAP crystals on the surface of the fibrils. These observations finally gave rise to the idea that an amorphous mineral precursor phase grown on a soluble biopolymer could be the missing link in the above argumentation (Olszta et al., 2007; Nudelman et al., 2010). The basic concept of such process goes back to experiments on $CaCO_3$ precipitation in aqueous solution by Gower (1997). Addition of micromolar amounts of anionic polypeptides to a crystallizing solution of calcium and carbonate ions unexpectedly affected the crystallization pathway such that it differed completely from the usual pathway. A multistage process was identified: (i) With increasing supersaturation of calcium and carbonate ions, droplets of a highly hydrated precursor phase are found in the solution. The mineral ions are immobilized at the charged polypeptide. Any crystallization of a mineral is still inhibited. (ii) The droplets of the precursor accumulate and coalesce, forming films of a loosely packed amorphous $CaCO_3$ precipitate on substrates. (iii) Finally, the precursor phase crystallizes, under exclusion of water of hydration and most of the polymers, into the thermodynamically stable calcite phase. In this way, calcite films are obtained and with the help of moulds, tablets, nanofibers, and any patterns are also obtained. Gower pointed out that the anionic polypeptide acted as a *process-directing agent*. Therefore, he introduced the term polymer-induced liquid precursor (PILP) for this kind of precursor. The liquid character of the precursor is still a matter of debate. The chemical composition of the precursor droplets has been found to be nearly equal to that of calcite, except for the hydration. For simplicity, we prefer the term polymer-induced precursor (PIP) henceforth.

Gower proposed the hypothesis that such precursor phases could also play a role in biomineralization. This differs from processes where soluble anionic proteins govern crystal growth by selective interaction with crystallographic faces of a solid nucleus, as discussed in Section 6.2.4.1. In the latter case, the biopolymer can be regarded as a *structure-directing agent*. Several *in vitro* and *in vivo* studies of biomineralization of collagen led to the conclusion that acidic noncollagenous proteins act as *process-directing agents* for the nucleation of hydroxyapatite crystals in the collagen type I fibrils (Olszta, 2007). Figure 6.48 shows the main steps of the process supposed to be relevant for bone mineralization:

a) Acidic NCPs provide binding sites for calcium and phosphate ions. The NCPs also act as growth inhibitors for prenucleation clusters. By phase separation, nanodroplets of an amorphous precursor phase are formed. They form loosely packed negatively charged aggregates that are highly mobile.
b) The aggregates are attracted by positively charged regions in collagen preferentially located at the border of the gap and overlap zones.
c) Binding motifs for HAP formation at the collagen fibrils enable the transformation of the amorphous calcium phosphate into HAP crystals.

HAP nanoplatelets of length 25–100 nm were aligned preferentially with their c-axis [001] along the collagen fibril axis so that they were in mutual contact. Similar to the natural bone structure, there was some scatter of orientation about the c-axis. The observed structure of the mineralized collagen composite suggests the notions of a collagen fiber-reinforced mineral matrix or an *interpenetrating network structure*

(a)

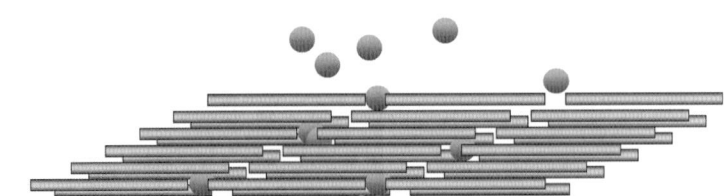

(b)

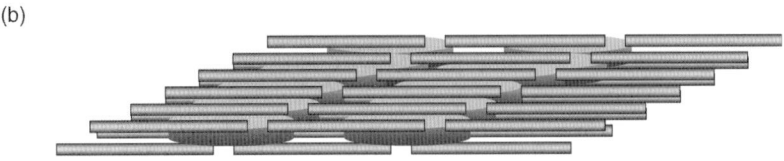

(c)

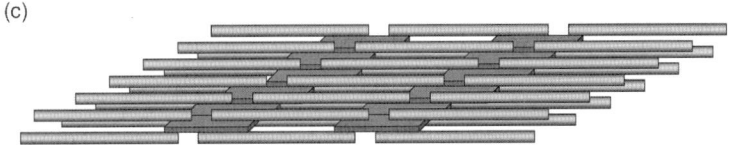

Figure 6.48 Proposed process of intrafibrillary mineralization of collagen via a polymer-induced precursor. (a) The negatively charged polymer sequesters ions and at some critical ion concentration, phase separation occurs within the solution, forming nanoscopic droplets of highly hydrated ACP. (b) Owing to their negative charge, droplet aggregates can be pulled into the positively charged hole zones and interstices of the collagen fibril. (c) The collagen fibril becomes fully imbibed with the amorphous mineral precursor. The amorphous precursor phase crystallizes, leaving the collagen fibril embedded with nanoscopic crystals of HAP. (Adapted from Olszta *et al.* (2007).)

of the mineral and collagen phases rather than that of a platelet-reinforced collagen matrix.

Results of a similar *in vitro* mineralization experiment by Nudelman *et al.* (2010) also supported the assumption of the existence of an amorphous precursor phase in biomineralization of collagen type I fibrils. The various early stages of bio-mineralization of collagen fibrils were studied with time-resolved cryogenic trans-mission electron microscopy. A collagen fibril was stained with uranyl acetate to reveal the ultrastructure of the collagen fibril. The mineralization was performed in a buffered solution containing $CaCl_2$, K_2HPO_4, and polyaspartic acid. After 24 h, ACP was observed outside the fibril at gap zone sites. First apatite crystals were detected after 48 h. They started to grow within a bed of ACP parallel to the fibril axis. After 72 h, elongated electron-dense HAP crystals were seen densely packed within the fibril. In other cases, the ACP bed became less densely covered. It has been shown that there is a correlation between the charge distribution in the ultrastructure of the collagen fibrils and the preferred infiltration path of the amorphous precursor. TEM investigation of nonmineralized fibrils stained with uranyl acetate indicated a positive net charge close to the C-terminal end of the collagen molecules located in a region of about 6 nm at the border between overlap and gap zones. High-resolution imaging of the mineralized fibrils proved that these areas of positive net charge of the collagen are also the locations to which the infiltrated polymer–ACP complexes bind and subsequently induce nucleation of HAP. Therefore, the collagen indeed plays an active role in the infiltration and alignment of the mineral precursors. In comparison to the NCPs, the *collagen appears to be the primary template* (Landis and Silver, 2009). Its charge distribution and geometric constraints control the oriented nucleation of the HAP crystals. Kniep and coworkers have studied the structure-guiding role of gelatin (denatured collagen) in fluorapatite mineralization. The existence of characteristic amino acid motifs directing nucleation and growth of the mineral phase has been revealed by these investigations (Kawska *et al.*, 2008; Kniep and Simon, 2008; Brickmann *et al.*, 2010; Kollmann *et al.*, 2010) (see also the discussion on fluorapa-tite–collagen composites in Section 6.3). It can be assumed that similar binding motifs control the hydroxyapatite crystallization in collagen-type I fibrils.

The polymer-induced amorphous precursor phase is related to the formation of a highly disordered solid phase in intracellular vesicles, discussed in Section 6.2.6.5. There seem to be two pathways of placing vesicles filled with highly disordered mineral phases in the extracellular space: (i) exporting vesicles filled inside the cell (Mahamid *et al.*, 2011; Weiner, 2011); (ii) exporting empty vesicles that become filled in the extracellular space (Mahamid *et al.*, 2011; Anderson *et al.*, 2005).

6.2.8.5 Scale-Dependent Mechanical Behavior of Bone

At first sight it appears surprising that so many different NCPs have been evolved. From *in vitro* experiments, it has been concluded that the *inhibition of extrafibrillar mineralization* may be the main function of the NCPs. Moreover, the NCPs play a major role in the scale-dependent mechanical behavior of the bone tissue. The specific binding of the NCPs with the HAP mineral as well as with the collagen fibrils essentially contributes to the *mechanical performance of mineralized bone* (Fratzl and Weinkamer, 2007). Bone combines strength and toughness in a remarkable way. It is difficult to combine high strength and high toughness in engineering materials: Those with high strength, such as certain steels and ceramics, are prone to brittle fracture and those with high fracture toughness, such as ductile steels and polymers, are less strong. In order to combine strength and toughness despite their partial incompatibility, the concept of composite materials has been promoted, which has turned out successful in a limited number of applications, such as fiber-reinforced polymers and sintered hard metals or cermets, for instance. Since the load transfer across the interface between the components is the central idea behind the concept, it is obvious that the properties of the interface are essential for the performance of the whole. The *control of the interface properties in the HAP–collagen composite* seems to be one function of the *NCPs*. Detailed investigations by Gupta *et al.* (2005, 2007) have revealed that a thin so-called glue layer between the mineralized collagen fibrils (Figure 6.49) is the decisive structure in this connection. It can be assumed that it consists of negatively charged proteins as osteopontin and fetuin, or proteoglycans, in a layer of only a few nanometers thickness.

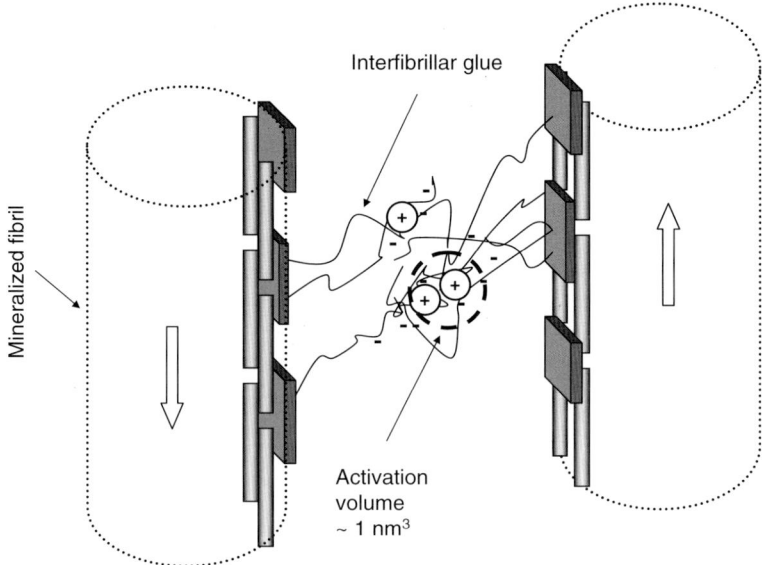

Figure 6.49 Putative structure of the glue layer formed by chains of negatively charged proteins or proteoglycans, interacting probably via multivalent cations, such as calcium. The arrows indicate the slip of the mineralized collagen fibrils. (Adapted from Fratzl and Weinkamer (2007).)

By micromechanical investigations of the deformation behavior of lamellar bone, the mechanism controlling the plastic behavior of the bone at the molecular level has been identified. The temperature and strain-rate dependence of the tensile behavior of microsamples led to the conclusion that the glue layer governs the stress transfer between the mineralized fibrils by extended plastic shear deformation. The elementary process causing the plastic slip in the glue layer is a stepwise opening and reconnecting of bonds. The individual rupture event is concentrated in an activation volume of about $1 \, nm^3$ and is connected with an activation energy of $1 \, eV$ (Gupta et al., 2007). Experiments by other authors have shown that the microscopic deformation of bone seems to be associated with calcium-dependent sacrificial bonds. It has been assumed that the positively charged Ca^{2+} ions create localized electrostatic bonds between the long chains of the negatively charged proteins or proteoglycans (see the schematic drawing in Figure 6.49). The interaction volume of overlapping fibrils of the glue layer indicated in the figure corresponds to the activation volume derived from the stress dependence of the strain rate. The bonds loosen under tensile load, and new ones form after a short slip. This assumption is supported by the observation of extensive self-healing behavior of lamellar bone under cycling loading in the inelastic regime.

Obviously, before applying the concepts of strength and fracture toughness taken from materials science to complex structures as bone, one has to make sure that such application is meaningful in an envisaged mechanical test. The mechanical behavior under higher load is the result of damage at various structure levels, as there are formation of microcracks, their accumulation into macrocracks, and the propagation

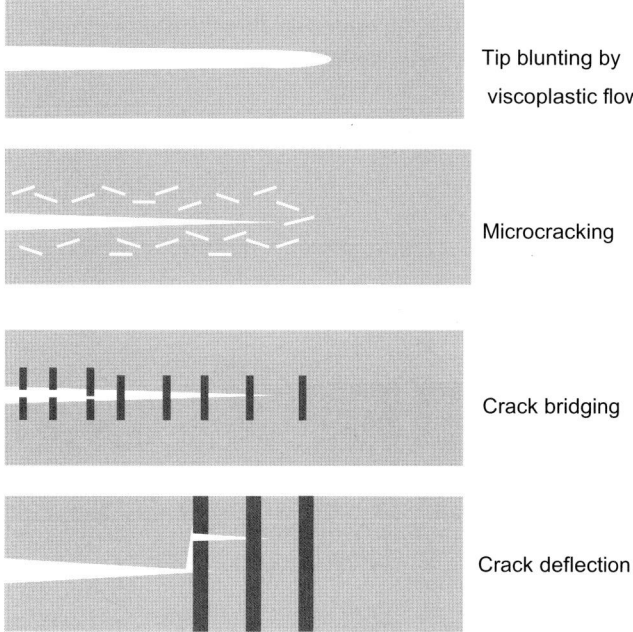

Tip blunting by
viscoplastic flow

Microcracking

Crack bridging

Crack deflection

Figure 6.50 The toughening mechanisms in bone. Different toughening mechanisms: blunting of the crack tip by viscoplastic flow, microcracking, crack bridging, and crack deflection. (Adapted from Peterlik *et al.* (2006).)

of the latter. A short overview of the main phenomena relevant on the macroscale is given in the following. The phenomenon of crack propagation in any material is adequately described by a formalism known as fracture mechanics: A given crack in a loaded material propagates if the stored strain energy released in propagation exceeds the mechanical work consumed in creating the new crack face. The increment of work δW consumed per crack face increment δA is called the crack extension energy or fracture toughness: $G_c = \delta W / \delta A$. G_c is a measure for the resistance of a material against macroscopic fracture. Contributions to G_c are provided by various energy-consuming processes that become activated in the field of high stresses near the crack tip (Peterlik *et al.*, 2006). Any process increasing G_c serves as a toughening mechanism. Various toughening mechanisms known from engineering materials have been found to be present in bone (Figure 6.50):

i) *Tip blunting by viscoplastic flow of the biopolymeric components of the bone.* The main contribution to the viscoplastic energy dissipation is probably due to slip of thin glue layers between the mineral-reinforced collagen fibrils, as already discussed.

ii) *Microcracking in the surrounding mineralized tissue.* A zone of microcracks around the main crack tip is known to make certain ceramics more fracture-resistant. In polymers such as polyethylene, crazes (small overstretched regions) instead of microcracks provide a toughening effect. The composite microstructure of bone has the potential for the formation of *microcracks and crazes*.

iii) *Localized bridging of the macrocrack.* Uncracked ligaments of biopolymer span the wake of the main crack, thus reducing the load on its tip, which is equivalent to increasing the fracture toughness.

iv) *Crack deflection and blunting of the crack tip.* The composite bone structure with a patterned sequence of weak and strong structural units leads to various deflections and blunting of the crack tip, influenced by the anisotropy of the textured bone microstructure.

The combined action of these mechanisms constitutes the high fracture toughness and fatigue strength of the bone. It also explains the strong anisotropy of the fracture toughness. In lamellar bone, for instance, the fracture toughness varies with the angle between crack propagation and collagen orientation from $375 \, \text{J/m}^2$ (crack propagation parallel to the lamellae) up to $9920 \, \text{J/m}^2$ (perpendicular to the lamellae) (Peterlik *et al.*, 2006; Fratzl and Weinkamer, 2007).

6.2.9
Ancient Evidence of Biomineralization

6.2.9.1 Stromatolites: the Oldest Fossils by Biogenic Mineralization

Biogenic extracellular limestone precipitates formed by photosynthetic microbial mats as dome-shaped laminated structures known as stromatolites (from Greek *stroma* = stratum) are among the oldest pieces of evidence of life on Earth, dating back to times about $3.5 \cdot 10^9$ years ago (Figure 6.51). In those times, excluding creatures feeding on microbes, microbial mats were quite abundant in shallow coastal waters. Recent stromatolite growth is restricted to a few ecological niches where photosynthetic

Figure 6.51 Stromatolites in the Hoyt Limestone (Cambrian) exposed at Lester Park, near Saratoga Springs, New York. (M.C. Ryel, Wikimedia Commons.)

microbial mats dominated by cyanobacteria can survive under conditions hostile to higher life forms, such as hypersalinity, extreme temperatures, and so on. It is assumed that the early biogenic stromatolites were also mainly formed by cyanobacteria. For their photoautotrophic way of life, all they need is light and CO_2 and a few of the omnipresent minerals from the seawater. The oxygen released by their photosynthetic activity was consumed by oxidation, which completely changed the composition of the early atmosphere, then consisting of methane, ammonia, and other gases, without oxygen. After the formation of the first stromatolites, it took about another 10^9 years before the cyanobacteria had turned the atmosphere from reducing to oxidizing, thus providing the environment for the evolution of higher forms of life.

The biogenic stromatolites are instructive examples of biologically induced mineralization. The calcite precipitation is induced by depletion of CO_2 in the microbial mat as a result of CO_2 consumption in photosynthesis (Mann, 2001). The photosynthesis can be concisely described by the reaction

$$CO_2 + H_2O \xrightarrow{hv} [CH_2O] + O_2 \qquad (6.48)$$

where $[CH_2O]$ stands for reduced carbon in organic molecules as sugar and carbohydrates. The depletion of CO_2 affects the dissolution equilibriums of CO_2 in water, as discussed in Section 6.2.5. As pointed out by Jansson and Northen (2010a), the efficiency of the calcification by cyanobacteria is unusually high because the CO_2-based photosynthesis combines an *intracellularcarbon-concentrating mechanism* (CCM) with the epicellular and extracellular deposition of $CaCO_3$). The main elements of the CCM of photosynthesis and its link with the calcification are schematically explained in Figure 6.52. The CCM is part of the so-called *Calvin cycle* that facilitates the assimilation of the inorganic carbon C_i to the organic world.

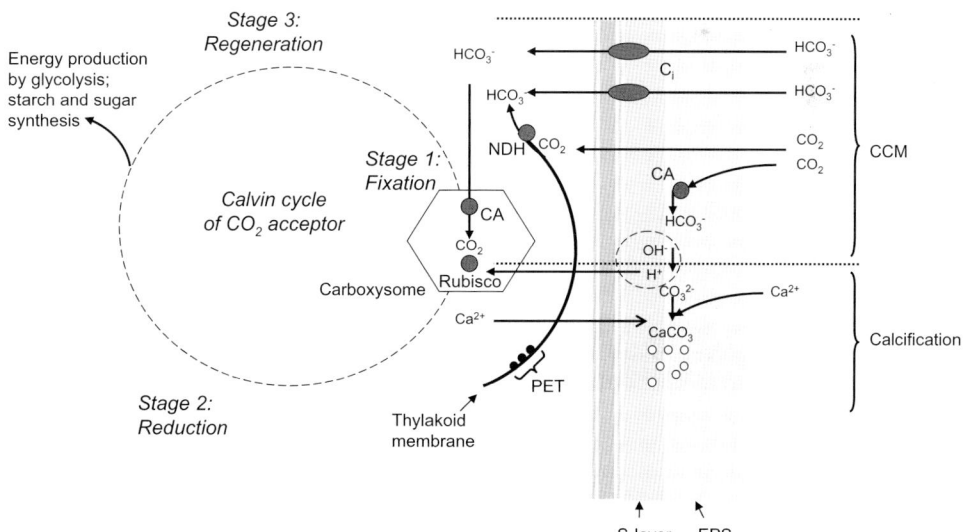

Figure 6.52 CCM and calcification mechanism in cyanobacteria. Inorganic carbon C_i is taken up by the cell.

In the particular case considered here, the inorganic carbon is provided by carbon dioxide dissolved in water as CO_2, H_2CO_3, HCO_3^-, and CO_3^{2-}.

The CCM leads to an essential increase of CO_2 concentration in the cyanobacteria (up to 1000-fold) over that of the extracellular medium. This occurs in the so-called carboxysome that relates to the first stage of the Calvin cycle of assimilation of CO_2 into organic carbon compounds: the fixation of CO_2 at the acceptor molecule, the ribulose 1,5-bisphosphate, which acts as a kind of carrier structure in the cycle. The Calvin cycle consists of three stages: (i) fixation of CO_2 at the ribulose acceptor; (ii) energy production by glycolysis, synthesis of starch, and sugar molecules; and (iii) regeneration of the ribulose acceptor (Nelson and Cox, 2008). The main uptake of the inorganic carbon occurs across the outer membrane and the plasma membrane via active transport of HCO_3^-, through HCO_3^-/Na^+ symports (simultaneous transport of two substrates across a membrane in the same direction) or through ATP-driven uniports (transport of only one substrate). Another way of inorganic carbon uptake is by diffusion of CO_2 that is transformed into HCO_3^- by NADPH hydrogenase (NDH) complexes on the thylakoids and plasma membranes. Inside the cyanobacteria, the HCO_3^- enters the carboxysome. In the carboxysome, the carbonic anhydrase (CA) catalyzes the conversion of HCO_3^- to CO_2. The highly enriched CO_2 is covalently attached to the ribulose acceptor. The attachment is catalyzed by the enzyme rubisco which can be activated by light irradiation. The active HCO_3^- transport is controlled by ATP produced by electron transport in the photosynthetic electron transport chain (PET) of the photosystem I (PSI). The CA activity in the carboxysome consumes H^+, which makes low concentration (high pH) in its vicinity. Hydrogen ions from outside move down the concentration gradient by diffusion so that the extracellular space, too, gets higher pH. It favors the formation of $CaCO_3$ nuclei via the reaction of Ca^{2+} ions with CO_3^{2-} at Ca^{2+} binding domains in the exopolysaccharide sheath (EPS) or on the proteinaceous surface layer (S-layer) that surrounds the cells. There are intracellular and extracellular Ca^{2+} sources.

6.3
Bioengineering

6.3.1
Bacteria-Derived Materials Development

6.3.1.1 Bio-Palladium: Biologically Controlled Growth of Metallic Nanoparticles

Mineral formations whose deposition was facilitated by microorganisms have attracted the interest of scientists ever since their nature had been recognized. Research in the subject has yielded a sizable amount of knowledge that serves as a basis for the efforts of microbiologists, chemists, and materials scientists to develop innovative functional materials. Promising results have been obtained in the latter three decades, such as the biologically controlled growth of functional nanoparticles mediated by bacteria, with the aim to facilitate low-cost large-scale production of nanocatalysts based on precious metals (PMs). It is well known that bacteria can use redox reactions involving metals (e.g., iron) as an energy source. There are also many bacterial strains that can reduce or oxidize trace elements existing in a certain biotope. This has been exploited for the development of bioremediation strategies for landscapes polluted with toxic metals or radionuclides (Lovley, 2003; Lloyd, Anderson, and Macaskie, 2005; Martinez *et al.*, 2007; Mkandawire and Dudel, 2009). Based on the same idea, a concept has been worked out to recover industrial wastewater loaded with PMs (Pd, Pt, Au, and Ag) by means of microbial redox systems (Korobushkina and Korobushkin, 1986; Lloyd *et al.*, 1998; Lloyd *et al.*, 2008; Deplanche *et al.*, 2011; Macaskie *et al.*, 2011). In particular, an efficient process for the formation of palladium nanoparticles has been worked out. As revealed by Lloyd *et al.* (1998), the biorecovery of PMs from aqueous solution, in the sense of our terminology, introduced in Section 6.2.6, is a biologically controlled mineralization process. As pointed out in the following, the biological control of the growth process produces nanoparticles whose structure differs from that of Pd nanoparticles produced by chemical synthesis, hence the term "bio-Pd nanoparticles."

The formation of transition metal clusters in aqueous solution has been widely studied in the past. The most interesting feature of this process is that such clusters form in solutions of transition metal salts often when a very mild reducing agent is added to the solution for reduction of the metal ions. It is known that cluster formation is an autocatalytic process, which is nucleation limited (Henglein, Ershov, and Malow, 1995; Watzky and Finke, 1997). The very first step of cluster growth is the decisive event. Therefore, it should not be astonishing that in biological systems, the deposition of transition metal nanoparticles can be observed often when in a cell such reducing conditions exist. We can assume that the resulting structures could show particularly interesting properties when the formation of the mild reducing conditions stays under biological control, which means that it is a part of some metabolic process. The formation of nanoparticles of Pd(0) at the cell wall of bacteria occurs by dissimilatory reduction. This is a redox process involving the reduction of metal ions coupled to the oxidation of an inorganic or organic substrate, for instance, molecular hydrogen or sodium formate (HCOONa). In a detailed study of Deplanche *et al.* (2010), the formation of bio-palladium nanoparticles in *Escherichia coli* by enzymatic oxidation of sodium formate was investigated. The electrons for the Pd(II) reduction are provided by the following reaction:

$$HCOONa + H_2O \rightarrow NaOH + CO_2 + 2H^+ + 2e^- \tag{6.49}$$

This oxidation reaction is catalyzed when hydrogenases are present in the cell. The reaction takes place at a bimetallic active site consisting of Fe atoms ([FeFe]-hydrogenases) or one Ni and one Fe atom ([NiFe]-hydrogenases), coordinated by biologically unusual CO and CN ligands. Hydrogenases are essential for sustaining anaerobic metabolism.

With the available electrons, the formation of palladium nanoparticles follows a reaction path that is different from the "classical" mechanism of homogeneous nucleation and growth, as described in Section 6.2.1. In the following, we consider an experimental situation as realized by Deplanche *et al.* (2010), starting with an aqueous solution of Na_2PdCl_4. In the picture of the "classical" model, the formation would occur as a two-step mechanism: (i) reduction of the bivalent Pd(II) to Pd(0), and (ii) subsequent aggregation of the zero-valent atoms to form metal clusters. However, as pointed out by Colombi Ciacchi (2002), there are experimental observations that cannot be explained by this nucleation model alone, for example, the formation of transition metal dimers that present higher oxidation states. This gave the motivation to study the very first events of the cluster formation by molecular simulation (Colombi Ciacchi *et al.*, 2001, 2003). Based on first-principles molecular dynamics (FPMD), the homogeneous nucleation of Pt clusters in a solution of K_2PtCl_4 by stepwise reduction of the salt complexes was investigated. The detailed analysis led to an essentially different picture of the complete nucleation process, within which the existing experimental observations can be interpreted. The results of these simulations and additional experimental observations lead to the following picture explaining the Pd cluster growth in aqueous solution, assuming that similar mechanisms govern the reduction of analogous Pt(II) and Pd(II) salts:

- The reduction of $PdCl_4^{-2}$ is faster when the solution has been aged to allow the water substitution of one or two chlorine ligands. Consistent with experimental observations for Pd salts (Wahl, 2003), it can be assumed that the electrons are favorably transferred to Pd(II) metal complexes only after hydrolysis to diaqua complexes $[PdCl_2(H_2O)_2]^0$:

$$[PdCl_4]^{2-} + H_2O \rightarrow [PdCl_3(H_2O)]^- + Cl^- \tag{6.50}$$

$$[PdCl_3(H_2O)]^- + H_2O \rightarrow [PdCl_2(H_2O)_2]^0 + Cl^- \tag{6.51}$$

The addition of one electron to $[Pd(II)Cl_2(H_2O)_2]^0$ leads to a stable $[Pd(I)Cl_2]^-$ complex. By adding one further electron, a stable $[Pd(0)Cl_2]^{-2}$ complex is formed.

$$[Pd(II)Cl_2(H_2O)_2]^0 + e^- \rightarrow [Pd(I)Cl_2]^- + 2H_2O \tag{6.52}$$

$$[Pd(I)Cl_2]^- + e^- \rightarrow [Pd(0)Cl_2]^{-2} \tag{6.53}$$

The three stable dichloro complexes form the pool for the subsequent nucleation events. In the earliest stage of the reduction process, the concentration of unreduced Pd(II) complexes is much higher than that of the reduced complexes. If the reduction rate is not too high (mild conditions), the formation of a bond between Pd(II) and Pd((I) complexes is a likely event. For the example of Pt cluster nucleation, the FPMD simulations have shown that this process occurs without an activation barrier.

- Addition of a further electron to the Pd(I)—Pd(II) dimer formed after a single reduction step leads to the loss of one chlorine ligand with strengthening of the Pd—Pd bond. A Pd(I)—Pd(I) dimer is formed. The Pd(I) atoms in both dimers are able to react with a second $[Pd(II)Cl_2(H_2O)_2]^0$ complex without waiting for a critical concentration of zero-valent Pd complexes. In Figure 6.53, snapshots of FPMD simulations of an analogous reaction leading to the formation of a Pt

(a) 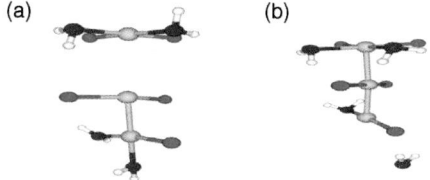 (b)

Figure 6.53 Snapshots of a FPMD simulation showing the formation of a Pt trimer after two reduction steps. (a) A Pt(II) complex approaches the Pt(I)–Pt(I) dimer. (b) The final trimer structure. Yellow: Pt, green: Cl, red: oxygen, white: hydrogen. (Courtesy of L. Colombi Chiacci.)

trimer after two reduction steps are shown. This may be considered as the first step of an autocatalytic cluster growth.

- FPMD simulations of the growth of larger Pt_nCl_m clusters (with $n > m$) have shown that unreduced $[Pt(II)Cl_2(H_2O)_2]$ complexes can be incorporated into small metal clusters at room temperature without noticeable energy barrier. The same should be valid for Pd cluster growth. By addition of Pd(II) complexes, the oxidation state and the nuclearity (number of metal atoms) of the cluster increase. This leads to enhancing of the electron affinity. Thus, reduction becomes more and more favorable during the growth process. At the same time, the reduction process is expected to cause desorption of chlorine ions from the cluster surface (Figure 6.54). New reactive sites for adsorption of Pd complexes are generated. The cluster growth proceeds as an autocatalytic process.

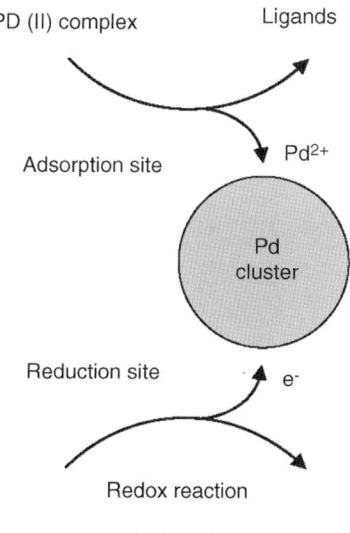

PD (II) complex   Ligands

Adsorption site   Pd²⁺

Pd cluster

Reduction site   e⁻

Redox reaction

Metabolism

Figure 6.54 Autocatalytic growth of bio-palladium clusters. (Adapted from Colombi Ciacchi (2002).)

In conclusion, the decisive contribution of the cell to the transition metal cluster growth is the creation of mild reducing conditions by the enzymatically controlled redox reaction. We expect that the reaction first leads to reduced Pd(I) complexes, which later aggregate into critical nuclei (not necessarily in the zero-valent state). The stepwise binding of further Pd(II) complexes and reduction enable stable autocatalytic cluster growth. The localization of enzymes in the cell wall influences the nucleation density and size distribution of the formed bio-Pd particles. It should be noted that the bacterial wall may also act as a template for the heterogeneous nucleation and growth of the particles, thus providing additional support for a preferential (accelerated) Pd particle development.

As shown by the experiments of Deplanche *et al.* (2010), in *E. coli*, three [NiFe] hydrogenases (Hyd-1, Hyd-2, and Hyd-3) govern the oxidation of sodium formate. Two of the hydrogenases (Hyd-1 and Hyd-2) are membrane bound with their catalytic subunits exposed to the periplasmic side of the cytoplasmic membrane. The third one (Hyd-3) is found at its cytoplasmic side (Figure 6.55).

A parent strain *E. coli* presenting all the three hydrogenases and four mutant strains have been used in the experiment. In three of these mutant strains, only one hydrogenase was present (be it Hyd-1, Hyd-2, or Hyd-3), whereas in the fourth strain all hydrogenases were lacking. The TEM micrographs clearly showed that all three hydrogenases were involved in the reduction process. Pd(0) particles grew on both sides of the cytoplasmic membrane of the parent cell. In cells of the three strains with only one type of hydrogenase, nearly equally sized Pd(0) particles (1–6 nm) were observed either in the periplasmic space or at the side of the cytoplasmic membrane facing inward. The Pd(0) nanoparticles were homogeneously distributed in the periplasmic space and/or along the cytoplasmic membrane. The narrow size distribution of the Pd(0) nanoparticles and their homogeneous distribution in a well-defined space are characteristic features of biological control by hydrogenase activity. The alternative case of biologically induced deposition was observed in the strain lacking all hydrogenases. Random nucleation of Pd(0) particles with large variation in particle size (up to 200 nm) was found.

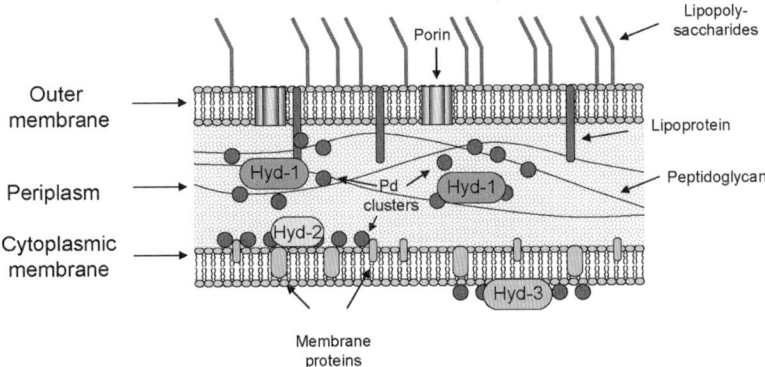

Figure 6.55 Arrangement of the NiFe hydrogenases in the periplasm and at the cytoplasmic membrane catalyzing the Pd cluster growth.

Extensive study of Pd(II) reduction by wild-type *Desulfovibrio fructosivorans* and three hydrogenase-negative mutants (e.g., mutant strains with deletions of NiFe hydrogenase, Fe hydrogenase, and Fe and NiFe hydrogenases) via H_2 has shown similar mineralization behavior. In the parent strain, Pd(0) clusters were grown at both cytoplasmic membrane and periplasmic loci. But in the double mutant and the mutant lacking the major periplasmic hydrogenase, the nucleation sites were confined to the cytoplasmic membrane. This is a strong evidence of the role of hydrogenases in Pd(0) deposition. Obtained preparation contained a subset of 4–10 nm Pd nanoparticles, associated with the membranes (Mikheenko *et al.*, 2008).

The very small size of the bio-Pd particles implies physical and chemical properties deviating from those of bulk Pd. With decreasing particle size, the influence of surface atoms on the overall material behavior increases, which affects, for example, the magnetic properties. Bulk palladium is paramagnetic, which means the magnetization is proportional to the applied magnetic field. With decreasing particle sizes below about 6 nm, an increasing contribution of ferromagnetism arises. The magnetization curve becomes a superposition of the saturated ferromagnetic and the paramagnetic components. Such behavior has already been observed for high-purity Pd nanoparticles produced by vacuum evaporation by Taniyama, Ohata, and Sato (1997). They explained this observation with a magnetic core–shell structure of ultrafine Pd particles consisting of a paramagnetic core and a ferromagnetic shell. Similar magnetization curves have been measured by Mikheenko *et al.* (2001), Macaskie *et al.* (2005), and Creamer *et al.* (2011) for bio-Pd (Figure 6.56). In this case, the bio-Pd nanoparticles were produced by means of the sulfate-reducing bacterium *Desulfovibrio desulfuricans*. After incubation of the cells in a solution of Na_2PdCl_4 at pH 2.3 for biosorption of Pd(II), H_2 was bubbled through the solution for ~30 min. Complete reduction of the Pd(II) due to the enzymatic oxidation of H_2 controlled by hydrogenase activity was observed. Two different samples of bio-Pd were prepared by using cells harvested at different stages of growth. Sample 1 was from an early stationary phase culture, while sample 2 was harvested in the midgrowth phase. In either case, a ferromagnetic contribution to the magnetization curve was observed. As sample 2 had a significantly larger nonlinear contribution, it can be assumed that cells of this growth stage produced a larger fraction of small Pd particles within the range of 1.2–6 nm. The structure of bio-Pd was revealed in that study by muon spin rotation spectroscopy (μSR) that allows the local magnetic environment in the sample to be probed. The measurements indicated an interaction of the muons with both the palladium electrons and the protons of the hydrogen that is trapped in the particles during the biologically controlled deposition. Hence, we can assume that the physical and chemical properties of bio-Pd nanoparticles depend on the stage of the cell culture. It is supported by an observation by Creamer *et al.* (2011) that the properties also depend on the composition of the growth medium.

Ultrafine PM clusters are of high interest for a variety of applications as catalysts.

As one important field of potential large-scale application of such catalysts, proton exchange membrane fuel cells (PEMFC), also known as polymer electrolyte fuel cells, have recently attracted attention. PEMFC could be one of the main power

(a)

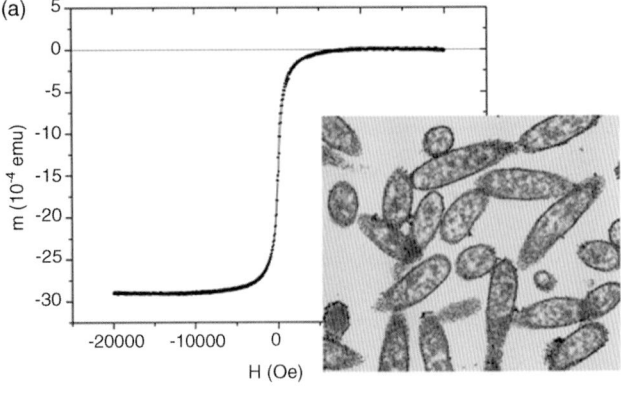

Pd(0) crystals in the bio-Pd powder with large ferromagnetic component

(b)

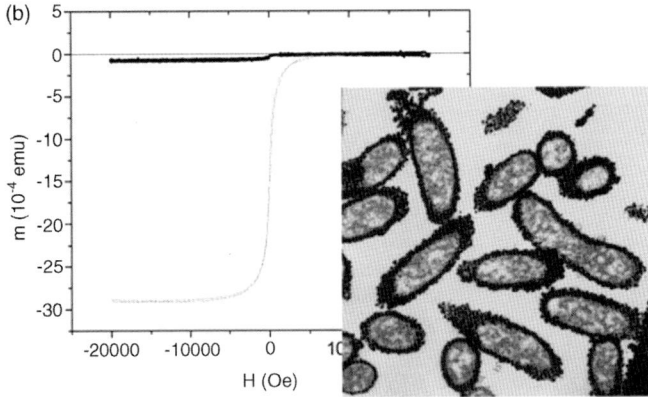

Pd(0) crystals in the bio-Pd powder with small ferromagnetic component

(c)

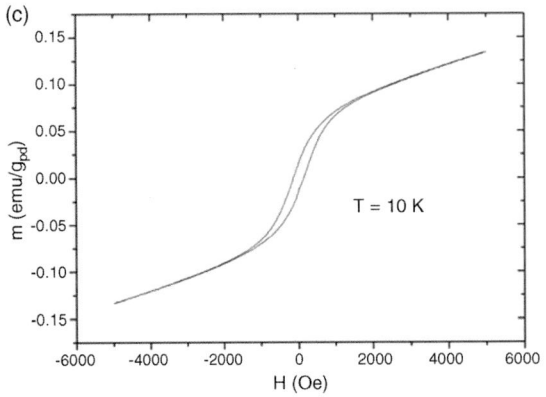

The presence of hysteresis in the magnetization curve of a bio-Pd sample is a direct proof of ferromagnetism in bio-Pd.

Figure 6.56 Illustration of bio-Pd made on *D. desulfuricans* in two ways to give (a) small deposits with a high ferromagnetic component or (b) larger deposits with a small ferromagnetic component as shown by the magnitude of magnetic moment versus applied magnetic field. In this way, it is possible to influence the catalytic efficacy of the bio-Pd. Part (c) shows proof of the ferromagnetic property of bio-Pd. (Reproduced with permission from Macaskie *et al.* (2005), The Biochemical Society.)

sources in future hydrogen-based technology for small power motor vehicles and generators and for small stationary applications. The basic elements of a PEMFC are two PM-based electrocatalysts (anode and cathode) and a proton-conducting membrane. H_2 is catalytically oxidized at the anode. The produced protons move across the proton-conducting membrane to the cathode. There they react catalytically with electrons in the presence of oxygen and water, which generates a voltage between anode and cathode. As mentioned before, microbial recovery of PMs by dissimilatory metal reduction is a promising technique for manufacturing the catalytic particles for such electrodes. The main advantages of bioreduction consist in the high production rate and low cost of the process of treating wastewater containing only very small concentrations of PMs of a few parts per million (ppm). Yong *et al.* (2010) have demonstrated how biorefining of PMs from wastes can be successfully used for manufacturing of inexpensive catalysts for fuel cells. Catalytically active bio-Pd was manufactured by *D. desulfuricans* and *E. coli* mutant IC007 (bio-Pd$_{D.\ desulfuricans}$ and bio-Pd$_{E.\ coli}$). Na_2PdCl_4 was used as Pd source. The dry biomass contained 25 mass% Pd. The electrodes were produced by coating carbon paper with heat-treated bio-Pd (at 700 °C under nitrogen atmosphere).

Applying the same process, electrodes were prepared by using PMs containing wastewater as metal source for the catalyst formation. Wastewater from industrial reprocessing with traces of PMs (2.8 mM Pt, 1.9 mM Pd, and 0.14 mM Rh) was treated with *E. coli* cells for metal deposition (bio-PM$_{E.\ coli}$). Furthermore, reference electrodes were prepared from commercial Pd powder produced by chemical synthesis (Pd$_{comm}$). The comparison of the maximum output power of fuel cells manufactured with electrodes made from the various powders gave the following values: bio-Pd$_{D.\ desulfuricans}$: = 141.4 mW, bio-Pd$_{E.\ coli}$: = 115.2 mW, bio-PM$_{E.\ coli}$: = 67.7 mW, and Pd$_{comm}$: = 99 mW. It has been proposed that the superior values for bio-Pd electrodes are the result of reduced aggregation of the bio-Pd nanoparticles because they are associated with carbon structures produced in the process of biomass incineration. The main advantage of the electrodes made from bio-PMs consists in the cost-efficiency of the powder production from cheap PM sources.

6.3.1.2 Biogenic Ion Exchange Materials

The formation of uranyl phosphate by precipitation in the epicellular space has been discussed in Section 6.2.5 as an example of biologically induced mineralization. The biogenic mineral has the remarkable property that it can be used as an ion exchanger for the removal of other radionuclides from nuclear reactor waste. In hydrogen uranyl phosphate (HUP), $H_3OUO_2PO_4 \cdot 3H_2O$, monovalent cations such as Na^+, NH_4^+, K^+, and Cs^+ and some divalent cations can be substituted for H_3O^+. Biogenic HUP can be generated by enzymatic cleavage of a phosphate donor molecule, for example, glycerol-2-phosphonate. Macaskie and coworkers have demonstrated that biogenic HUP produced with *Serratia* sp. cells facilitates cation exchange intercalation of Ni^{2+} into the interlamellar spaces of HUP (Bonthrone *et al.*, 1996; Amdursky *et al.*, 2010). HUP consists of UO_2PO_4 sheets separated by a two-level water layer. With the intention to develop an efficient remediation technology for water contaminated with nuclear fission and activated products, such as ^{137}Cs, ^{90}Sr, and ^{60}Co isotopes, bio-HUP has been applied for the removal of these isotopes by ion exchange. *Serratia* sp. biofilm, immobilized onto a polyurethane foam and mineralized with HUP, was applied in a flow reactor. The complete replacement of H$^+$ in the HUP matrix by the intercalating target metal species removed nuclides from the wastewater with high efficiency (Paterson-Beedle *et al.*, 2006). The demand for biogenic ion exchange materials could increase further as a result of the discovery that the biogenic zirconium phosphate can fulfill similar tasks (Mennan *et al.*, 2010).

6.3.2
Bio-Inspired Design of Mineralized Collagen and Bone-Like Materials

6.3.2.1 Biomimetic Growth of Apatite–Gelatin Nanocomposites

The apatite–gelatin system has turned out to be an ideal material for mimicking the mineralization of collagen type I. Particularly, the fundamental studies of Kniep and coworkers have shown that it is a suitable system for exploring the ability of collagen to direct structure formation processes in biomineralization of hard tissue (Kniep and Busch, 1996; Simon *et al.*, 2004b; Tlatlik *et al.*, 2006; Kawska *et al.*, 2008; Kniep and Simon, 2008; Paparcone, Kniep, and Brickmann, 2009; Kollmann *et al.*, 2010). At the same time, it is a promising material combination for the repair of bone and tooth dentine. The extracellular matrices of bone and dentine contain similar amounts of collagen ($\sim$25–40 and $\sim$20–30 wt%, respectively). The use of the soluble gelatin (denatured collagen type I) instead of collagen type I offers some advantages for mimicking the complex mineralization process. The increased mobility of the protein facilitates a better rearrangement of both the gelatin filaments and the growing mineral crystals. There are also advantages with respect to application: low cost, commercial availability, and well-defined physical and chemical properties. The minerals HAP ($Ca_{10}(PO_4)_6(OH)_2$) and fluorapatite ($Ca_{10}(PO_4)_6F_2$) have been chosen with respect to their possible use as repair materials for bone and teeth. Fluorapatite is favored for applications in dentistry as it is more resistant to acid attack than HAP. For this reason, toothpaste typically contains a source of fluoride anions.

Typically, apatite–gelatin nanocomposites can be produced by coprecipitation of apatite with gelatin in an aqueous solution or by a double-diffusion technique (Figure 6.57). The essential feature is the separate supply of the cations (calcium ions) and anions (phosphate and fluoride ions) to the reaction zone filled with gelatin. The morphogenesis of minerals precipitated in the gel depends on the concentration of Ca and phosphate ions; hence, they differ in the zones labeled P, C, and M. As shown in the time series of Figure 6.58, two types of structures develop within a few days. The two growth types start from a hexagonal prismatic seed and develop via dumbbell-like structures to spheres, but differ in the intermediate states: fan-like in the C bands (left columns in Figure 6.58), and cauliflower-like in the M and P bands (right columns). Fan-like growth is more distinctly realized with the crystal aggregate shown in Figure 6.59, where crystals are seen to have grown from the middle in diverging directions. A prismatic seed, not seen here, is supposed to be at the core.

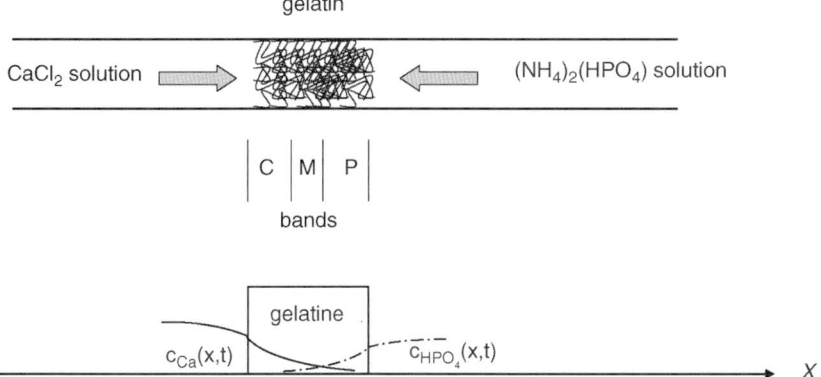

Figure 6.57 Principle of the double-diffusion technique, with arrows indicating the change of concentration profiles with time. The process results in HAP precipitates with different growth patterns: fan-like (C) and cauliflower-like (M and P).

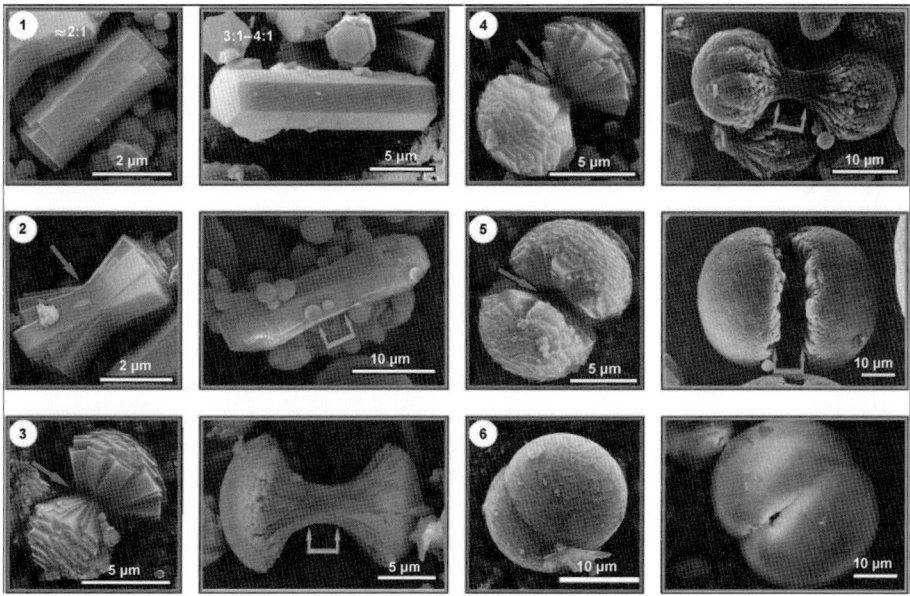

Figure 6.58 Scanning electron micrographs illustrating subsequent states (1–6) of morphogenesis of fluorapatite–gelatin nanocomposites for fan-like structures (left column in the series from state 1 to state 6 (C band)) and cauliflower-like structures (right column in the series from state 1 to state 6 (M and P bands)). (Reproduced with permission from Tlatlik et al. (2006), Copyright 2012, Wiley-VCH Verlag GmbH.)

Figure 6.59 TEM images of a fluorapatite–gelatin nanocomposite with fan-like aspect. Note the tiny white rods arranged in straight lines in the crystals seen on the enlarged section of the aggregate. They represent the (calcified) organic component (gelatin) of the composite. (Reproduced with permission from Tlatlik *et al.* (2006), Copyright 2012, Wiley-VCH Verlag GmbH.)

The minerals grown in the phosphate-rich M and P bands show a completely different structure. In Figure 6.60, the cross section of a fractured dumbbell is shown corresponding to subsequent states 3–5 (right column in Figure 6.58). In cauliflower-like pattern, new individual crystals grow on the basal face and near-end regions of the side faces of a central prismatic seed. The inset of Figure 6.60 shows a seed in an initial growth stage with first indications of crystals on the side faces and growing end faces. The cauliflower could be formed by repetition of such type of growth. The time sequence of the initial growth is indicated by a superposition of scanning electron microscopy (SEM) images of the initial growths stages in Figure 6.61. The growth starts with an elongated hexagonal prism ("young" seed). By outgrowth at both ends, a first dumbbell-like shape is formed ("mature" seed). The X-ray diffraction pattern of the mature seed indicates an overall 3D periodic structure of the inorganic component of the composite with the *c*-axis parallel to the long axis of the seed. By HRTEM, the crystallographic orientation of the inorganic phase has been determined in small

Figure 6.60 Cauliflower-like pattern formation by growth of fluorapatite–gelatin nanocomposites (SEM images). *Main picture*: Half of a dumbbell aggregate viewed along the central seed axis. *Inset*: Central seed exhibiting tendencies of developing of new crystals at both ends ("small" dumbbell). (Reproduced with permission from Tlatlik *et al.* (2006), Copyright 2012, Wiley-VCH Verlag GmbH.)

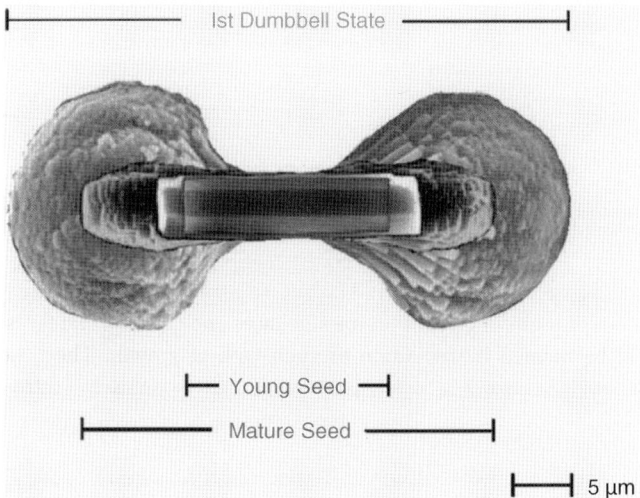

Figure 6.61 Superposition of SEM images of the initial growth stages of cauliflower-like fluorapatite–gelatin nanocomposite aggregates. (Reproduced with permission from Kniep and Simon (2008), Copyright 2012, Wiley-VCH Verlag GmbH.)

areas. It is consistent with the overall crystallographic orientation and independent of the embedded pattern of gelatin microfibrils.

The structures observed in mineralized gelatin gels resemble those known from fluorapatite precipitation from supersaturated aqueous solution without any fibrous organic material present (Prymak *et al.*, 2006). Figure 6.62 shows fluorapatite structures grown at 20 °C by immersing a polymer substrate into supersaturated aqueous solution. The pH was previously set and kept constant within 0.2–0.5 units during the experiment. Dumbbells were observed at pH 5–7 only. They look more fan-like at pH < 6, but rather bristly at pH > 6. At constant pH, a decrease of supersaturation led to fewer and smaller crystals. The presence of charged additives (cationic and anionic surfactants, poly(aspartic acid), and polylysine) inhibited the crystal growth. Below pH 5, compact spheres were found. At pH > 7, "coral-like" coatings were grown on the polymer substrate.

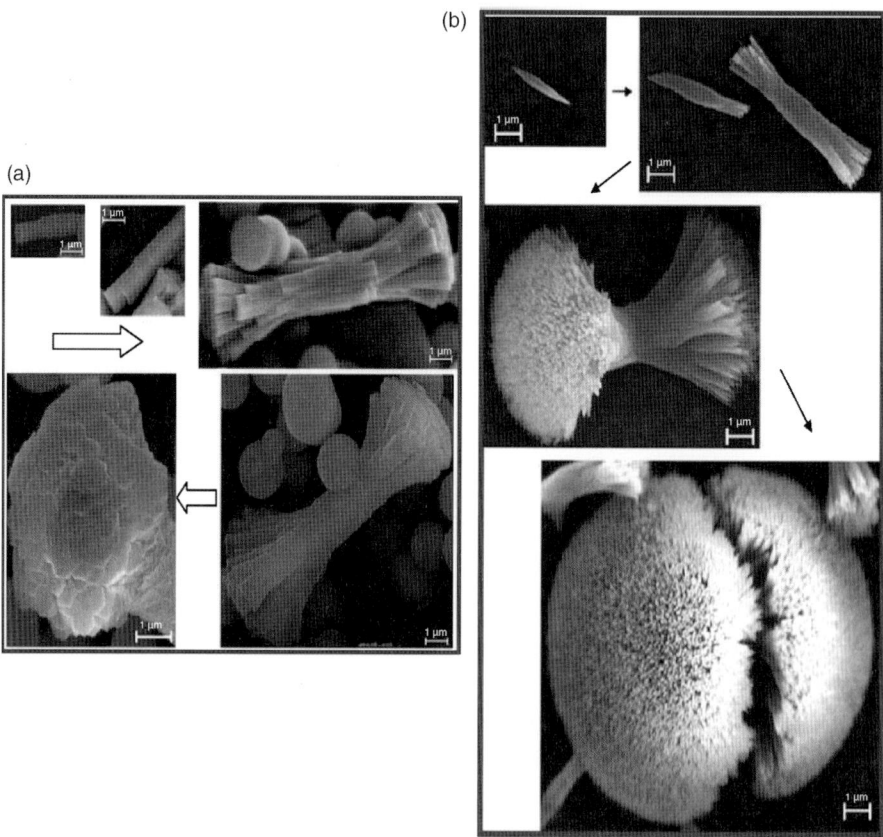

Figure 6.62 Dumbbell-like structures grown in a supersaturated solution from calcium acetate, KH$_2$PO$_4$, and KF on a polymer substrate (poly (ethylene-*co*-1,4-cyclohexane–dimethylene–terephthalate) (PETG)) after 24 h. (a) Grown at pH 5. (b) Grown at pH 6. Scale bars are always 1 μm. (Reprinted with permission from Prymak *et al.* (2006), Copyright 2012, American Chemical Society.)

The experimental findings were summarized in two geometrical models that could describe the sequential formation of the dumbbell structures (Prymak *et al.*, 2006). At pH 5–6, the crystal growth starts from a prism on which new layers develop. These layers finally form a closed sphere. At pH 6–7, the initial crystal has a "kayak-like" shape. New spikes are formed, which leads to branching into a dumbbell and finally to a closed sphere. It seems that the growth modes resulting in the different overall shapes are not yet fully understood. The formation of dumbbell-like crystal aggregates has been observed with a few more minerals: calcite dumbbells in the presence of Mg^{2+} and SO_4^{2-} (Tracy, Williams, and Jennings, 1998), magnesium calcite dumbbells in the presence of ammonium carbonate (Raz, Weiner, and Addadi, 2000), and calcite and $BaCO_3$ dumbbells in the presence of block copolymers (Cölfen and Qi, 2001; Yu and Cölfen, 2004).

The presence of the gelatin leads to some peculiarities of dumbbell formation caused by specific interaction of the fluorapatite nuclei with tropocollagen molecules, the basic constituent of gelatin. This conclusion has been derived from the detailed analysis of the microstructure in the young crystal seeds (Kniep and Simon, 2008). A texture faintly seen as dotted white lines on TEM micrographs of fan- and cauliflower-like crystal aggregates (Figures 6.59 and 6.63) gives the first impression of the gelatin

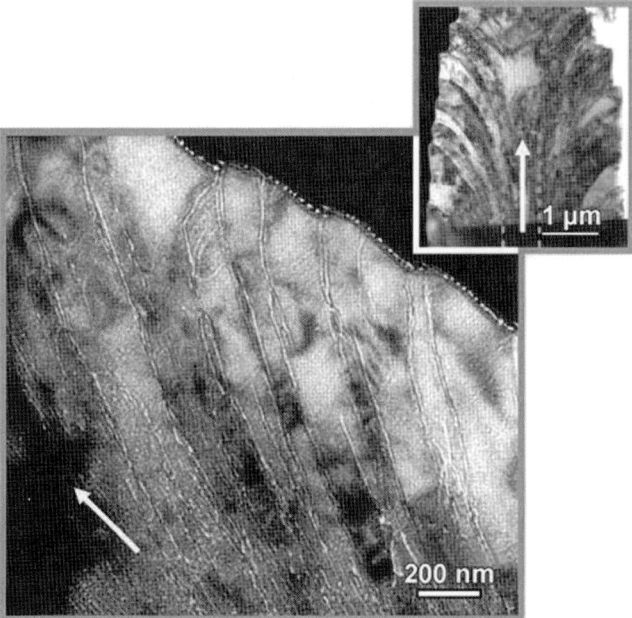

Figure 6.63 TEM images of fluorapatite–gelatin nanocomposites showing first tendencies of branching in the cauliflower-like growth series. (Bright arrow: orientation of the [001] axis of the seed crystal.) *Inset*: Overview of one-half of a prismatic seed with blossom-like development of the outgrowing secondary structure. *Main picture*: Dotted white lines are gelatin fibrils within the mineral structures. (Reproduced with permission from Tlatlik *et al.* (2006), Copyright 2012, Wiley-VCH Verlag GmbH.)

structure after mineralization. In the fan-like structures, gelatin microfibrils are observed that are aligned along the favored growth direction of the grown micro-crystals. Bent microfibrils are seen in the cauliflower-like structures.

More detailed investigations of the microstructure of the central prismatic seed of the cauliflower-like structure indicated that there is a second gelatin structure. HRTEM has revealed that there is a quasi-periodic array of cylindrical defects in the single-crystalline seed oriented along the [001] axis of the prismatic seed crystal (Simon *et al.*, 2004a; Simon, Schwarz, and Kniep, 2005). The diameter of the defects was found from a filtered image viewed along [001] of the apatite as about 3 nm. It is assumed that these defects are related to tropocollagen molecules (diameter about 1.5 nm). At mesoscopic magnification, the TEM images of the cross sections of the composite seeds reveal a superstructure caused by a patterned arrangement of the individual tropocollagen molecules. The fast Fourier transform (FFT) of a TEM image of a cross section tilted by 6°–8° with respect to [001] shows (001) reflections and diffuse streak-like reflections extending normal to the (001) reflections (Figure 6.64). The superstructure streaks show a periodicity of about 5 nm.

Crystals with an ordered array of enclosed biomolecules are known from matrix-mediated nucleation in biomineralization. They can result from inorganic nano-particles arranging themselves into crystals of sizes typically between tens of nanometers up to a few micrometers mediated by organic additives. The resulting superstructure with crystal habit can be formed via an intermediate structure called mesocrystal (Cölfen and Mann, 2003; Cölfen and Antonietti, 2005). The

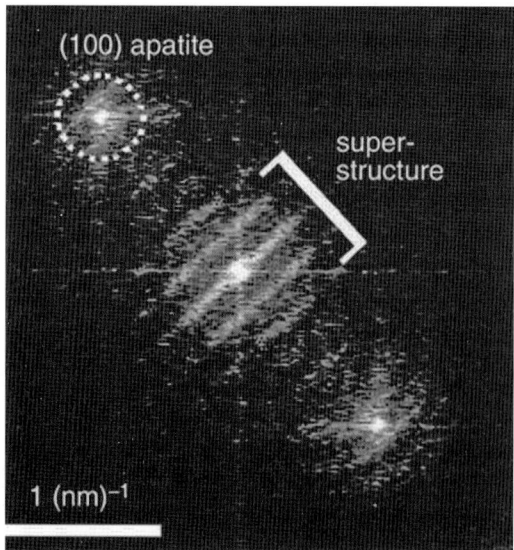

Figure 6.64 FFT along [001] of a bright-field TEM image of the seed lattice of cauliflower-like fluorapatite–gelatin composite. Diffuse streaks around the primary beam correspond to a periodicity of about 5 nm and multiples of this value. (Reproduced with permission from Simon *et al.* (2006), Copyright 2012, Wiley-VCH Verlag GmbH.)

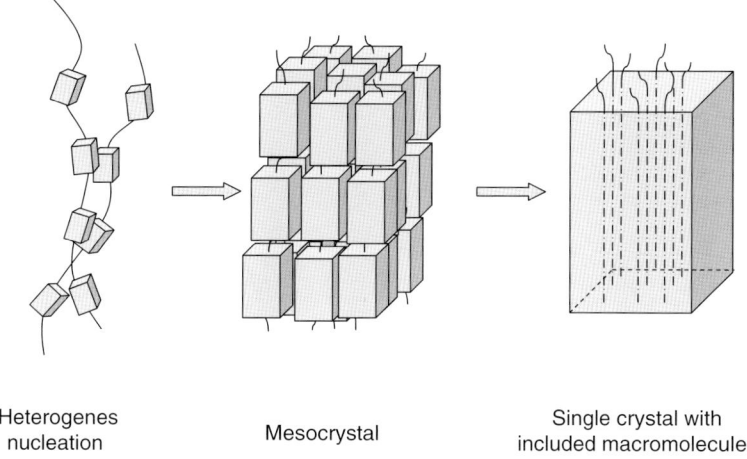

Heterogenes
nucleation

Mesocrystal

Single crystal with
included macromolecules

Figure 6.65 Single-crystal formation in a biomolecular matrix by a mesocrystal intermediate formed by nanoparticle self-assembly. (Adapted from Cölfen and Antonietti (2005).)

crystallographic interaction of the primary nanoparticles leads to a mutual alignment of inorganic building blocks. However, a small amount of lattice defects caused by the organic component is tolerated in the mesocrystal structure (Figure 6.65). Finally, the mesocrystalline structure can relax into a single crystal with enclosed macromolecules. Gels are favored matrices for the formation of mesocrystals because (i) high supersaturation can lead to a high density of small mineral particles and (ii) depending on the binding strength between the gel-forming molecules, their structure favors structural relaxation allowing efficient self-assembling of the inorganic particles.

For a model system with a well-defined structure, the tropocollagen helix, the influence of molecular motifs on the directed growth of the fluorapatite nanocrystals has been studied by atomistic simulation using empirical force fields for the interactions (Kawska *et al.*, 2008). The formation of Ca_3F stacks oriented along the tropocollagen helix has been revealed. These Ca_3F stacks are considered as nuclei for the formation of the apatite–gelatin nanocomposites (subunits of the seed). Governed by the lattice energy of the inorganic component, the tropocollagen filament is aligned parallel to the [001] axis of the seed crystal (Figure 6.66). Precondition for the formation of such structure in a gelatin matrix is a possible rearrangement of the random gelatin gel into a textured structure. The alignment of tropocollagen molecules in an apatite single crystal is connected with an interesting electric phenomenon. It is known that the triple helix of collagen is piezoelectric and pyroelectric (Lang, 1966; Lemanov, 2000). This behavior has been explained by the existence of internal electric dipoles of the various amino acids and the peptide bonds in a particular protein. For instance, glycine and proline as the main constituents of collagen exhibit strong dipole moments p (about $p_{glycine} = 15$ D, $p_{proline} = 13$ D, for comparison $p_{water} = 1.8$ D, with 1 D $= 3.336 \times 10^{-30}$ Asm). The existence of permanent dipoles in the collagen molecules gave the background for

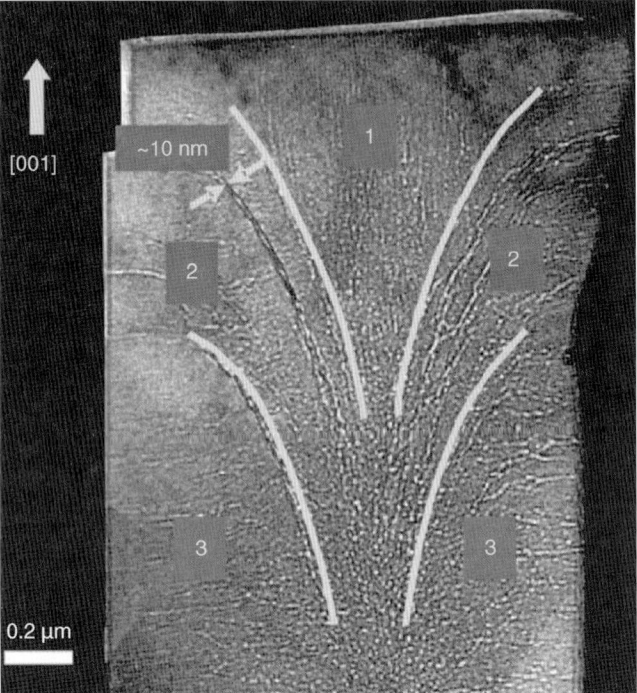

Figure 6.66 TEM image of the upper part of a FIB thin cut parallel [001] of a seed of cauliflower-like crystal. White lines indicate the borders between the distinct areas 1–3 representing different orientational characteristics of the microfibrils. (1) Central cone-like area containing microfibrils running along the *c*-axis direction. (2) Area with bent microfibrils opening like a flower and extending along the directions of the edges between basal and prismatic faces. (3) "Waist" area where microfibrils arrange themselves to finally adopt an orientation perpendicular to the prism faces. (Reproduced with permission from Brickmann *et al.* (2010), Copyright 2012, Wiley-VCH Verlag GmbH.)

the assumption that an ordered structure of collagen filaments in the young seed could form a resulting external electric field. The hypothesized electric field surrounding the seed crystal has been experimentally verified by electron holography (Simon *et al.*, 2006). The collagen filaments aligned in the primary seed make only a limited part of the gelatin matrix. The other part is organized as bent thicker microfibrils that are not oriented along the [001] axis of the seed, as shown in Figure 6.66. Kniep and coworkers proposed that their orientation is related to the electric field caused by the collagen filaments of the young seed (Paparcone, Kniep, and Brickmann, 2009). They call it an electric field-induced microfibril pattern. We have to remember that such mechanism is not observed with the fan-like structures. Apparently, the chemical interaction of the growing crystals controls the alignment of the microfibrils there. It cannot be excluded that such mechanism also contributes to the alignment of the microfibrils in the cauliflower-like structures.

With the possible explanation of patterning of the gelatin in cauliflower-like fluorapatite grown in the phosphate ion-rich region, the question arises why a similar mechanism cannot be observed with samples grown in the Ca^{2+}-rich region of the double-diffusion setup, where the fan-like structures are formed. At least two mechanisms seem to be involved: differential flexibility and effective dipole moment of the tropocollagen molecules. Microscopic observations led to the hypothesis that the triple helix of the tropocollagen is stiffened by contact with calcium ions, but softened and bent by contact with phosphate ions. This hypothesis has been supported by atomistic computer simulations of the interactions of calcium and phosphate ions with a "model molecule" built of three $(Gly—Pro—Hyp)_{12}$ poly-peptide strands (Tlatlik *et al.*, 2006). As explained in Section 6.2.8.3, $(Gly—Pro—Hyp)$ is the dominating motif of amino acids in the tropocollagen filament. Calcium and phosphate ions differ essentially with respect to their effect on the triple helix. As shown in Figure 6.67, the calcium ions form ionic bonds to the oxygen atoms of the carbonyl groups of the polypeptide backbones and to the side chains of proline and hydroxyproline, whereby the global shape of the triple helix is not affected. The Coulomb repulsion of the calcium ions bound at the backbone causes a straightening and additional stiffening of the tropocollagen molecule, thus increasing the persist-ence length. Furthermore, experiments indicate that the electric dipole field is disturbed by the Ca^{2+} ions. Electron holography did not produce evidence for a significant electric field inside seeds from the C zone (Simon, P., private communi-cation, Dresden, 2012). The influence of the phosphate ions is completely different.

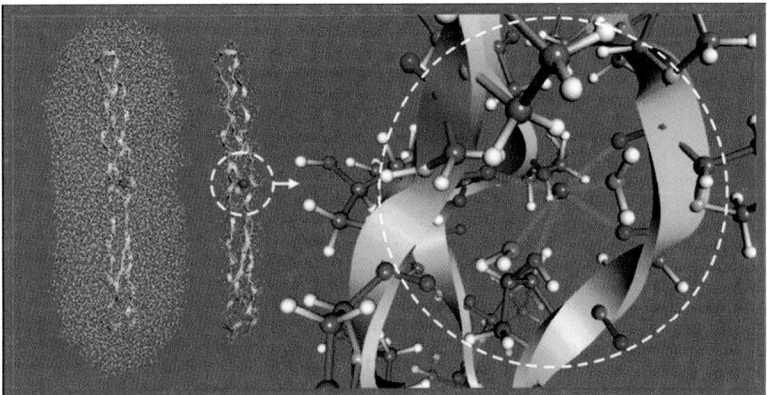

Figure 6.67 *Left*: Model of a triple-helical polypeptide including a solvent shell (about 5000 water molecules) of 1.2 nm minimum thickness together with one attached (included) calcium ion. Yellow ribbons represent the backbones of the three peptide strands of the protein filament. *Middle*: Same model without solvent molecules. *Right*: Enlargement of the binding site. Calcium ion forms ionic bonds (purple dashed lines) to carbonyl oxygen atoms of the protein ($d_{Ca-O} \approx 0.23$ nm). Further ionic interactions involve oxygen atoms (highlighted in green) of neighboring water molecules ($d_{Ca-O} \approx 0.27$ nm). H: white, O: red, C: gray, N: light blue, Ca: dark blue). (Reproduced with permission from Tlatlik *et al.* (2006), Copyright 2012, Wiley-VCH Verlag GmbH.)

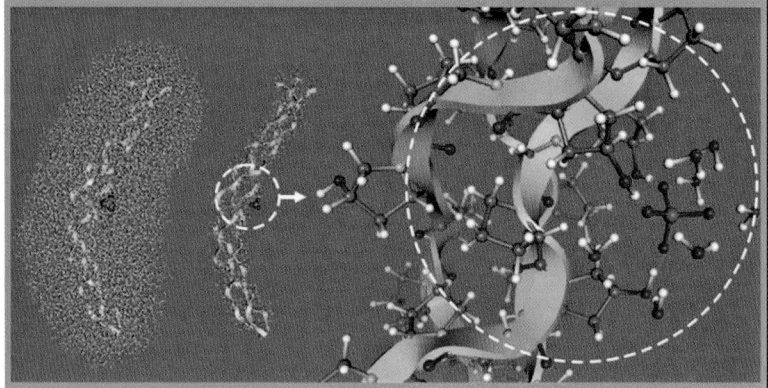

Figure 6.68 Same as Figure 6.67, but illustrating the association of a phosphate ion. For simplicity, only the hydrogen bonds between the phosphate ion and the protein filament ($d_{O-H} \approx 0.13$ nm) are depicted (dashed purple lines). Although three hydroxyproline–phosphate bonds are formed, only two strands of the triple helix are involved. This imbalance causes bending of the previously straight polypeptide. H: white, O: red, C: gray, N: light blue, P: purple. (Reproduced with permission from Tlatlik *et al.* (2006), Copyright 2012, Wiley-VCH Verlag GmbH.)

As shown in Figure 6.68, the phosphate ions are preferentially bound at the hydroxyproline side chains outside the triple helix. Two to three hydrogen bonds with the hydroxyproline residues and with amide groups are formed. Only two chains of the triple helix are linked with the phosphate ion. Most remarkable is the pronounced bending of the molecule. As the binding occurs at side chains, we can assume that it may cause softening of the triple helix.

The variable concentration of Ca^{2+} and phosphate ions in the double-diffusion process causes different *pretreatments of the gelatin molecules* in the three zones *before the mineral phase formation* starts, which finally could explain the observed mineralization of fan-like fluorapatite crystals in the Ca-rich zone and cauliflower-like mineral structures in the phosphate-rich zone. Differences in the stiffness and effective dipole moments of the gelatin molecules in the gel could explain the differential responses of the gelatin matrix to the growing fluorapatite.

6.3.2.2 Biomimetic Manufacturing of Mineralized Collagen Scaffolds

Mineralization of Collagen by Combined Fibril Assembly and Calcium Phosphate Formation The biomimetic synthesis of mineralized collagen fibrils is a basic step in the preparation of bone-like materials for regenerative therapies. As explained in Section 6.2.8.4, the incorporation of nanosize HAP crystals into the collagen fibrils in order to obtain intrafibrillar mineralization poses a problem. A solution is suggested by *in vitro* mineralization studies with chicken osteoblasts by Gerstenfeld *et al.* (1989), indicating that the *mineralization is a multistage process*. The collagen fibrils are grown first and noncollagenous proteins are produced during the following 10–20 days. In the final process stage, these proteins control the crystallization of the calcium phosphate (Boskey, 1989). The noncollagenous proteins contain large amounts of anionic amino acids (aspartate and glutamate). This suggested the use of polyglutamate and polyaspartate as a model system for the control of calcium phosphate mineralization (Hunter and Goldberg, 1994; Hunter et al., 1994; Gower, 1997; Bradt et al., 1999). More recently, it has led to the hypothesis that the noncollagenous proteins induce the formation of (polymer-induced) *mineral precursors* that are essential agents in the mineralization of collagen (Hunter et al., 2010). Typically, the penetration of the amorphous precursor into the collagen fibril structure and the subsequent crystallization take about 4 days. In order to accelerate this mineralization with the aim to achieve efficient processing, one should try to avoid the time-consuming transport of the precursor through the narrow intrafibrillar pores. This idea has been realized by simultaneous fibril assembly and mineralization within one process step (Bradt et al., 1999). In the presence of polyaspartate, tropocollagen molecules have been assembled into collagen fibrils simultaneously with the formation of calcium phosphate.

Information concerning the partial mechanisms of the process and their combined action can be derived from turbidity measurements on an aqueous solution of the reacting components. The turbidity versus time curves in Figures 6.69 and 6.70 visualize separately collagen fibrillogenesis, HAP formation, and their combined reaction, without polyaspartate influence. Tropocollagen, soluble at low pH, reconstitutes into fibrils by neutralization, which can be triggered by mixing with a buffer solution. Under optimized conditions, a three-dimensional network of collagen fibrils with gel-like properties is formed. The process has been monitored by

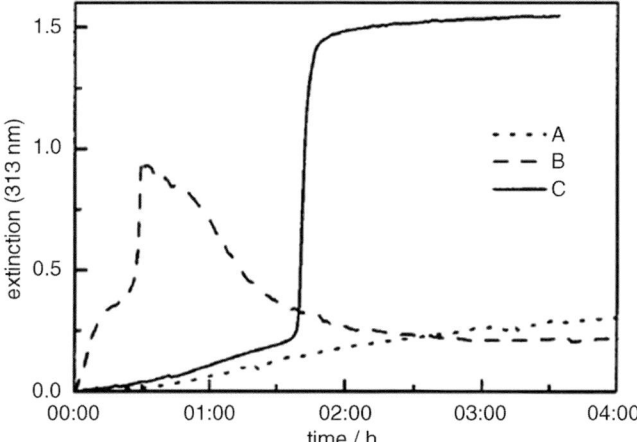

Figure 6.69 Turbidity during fibril assembly (a), calcium phosphate precipitation (b), and combined fibril formation and mineralization (c). The reactions were monitored through the extinction at 313 nm, with the value at the start of the reactions taken for reference zero. (Reprinted with permission from Bradt et al. (1999), Copyright 2012, American Chemical Society.)

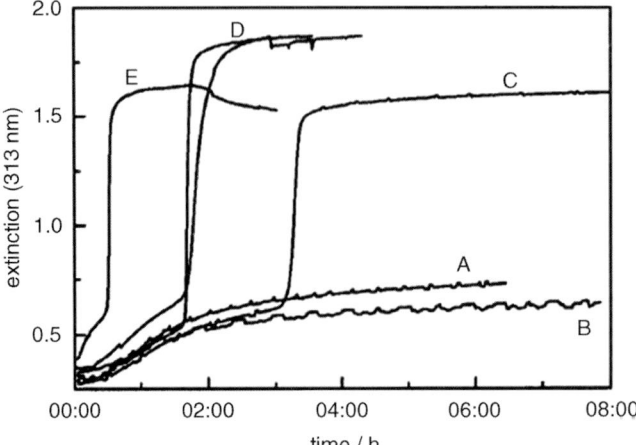

Figure 6.70 Effect of the calcium phosphate concentration on the kinetics of combined fibril assembly and mineralization. The reactions were monitored by measuring the extinction at 313 nm over time. Concentrations were (A) 7.5 mM Ca^{2+}, 4.5 mM phosphate; (B) 8 mM Ca^{2+}, 4.8 mM phosphate; (C) 8.5 mM Ca^{2+}, 5.1 mM phosphate; (D) 9 mM Ca^{2+}, 5.4 mM phosphate; and (E) 10 mM Ca^{2+}, 6 mM phosphate. Case D is represented by two curves in order to indicate the reproducibility of the experiment. (Reprinted with permission from Bradt *et al.* (1999), Copyright 2012, American Chemical Society.)

turbidity measurements [curve (A) in Figure 6.69]. HAP can be formed by mixing an aqueous solution of $CaCl_2$ (9.0 mM) containing 5.4 mM phosphate ions [curve (B) in Figure 6.69]. The turbidity increases in the first stage as a result of the precipitation of an amorphous phase and reaches a transient plateau value. After about 30 min, the transformation to the crystalline HAP phase causes a second steep increase of the turbidity, followed by a decrease due to sedimentation of the larger HAP clusters. By mixing an acidic $CaCl_2$ (9.0 mM) containing solution of tropocollagen (0.5 mg/ml) with the phosphate buffer (5.4 mM), the two reactions are initiated simultaneously in the mixture with pH 6.8 [curve (C)]. The first part of the turbidity versus time plot is similar to the fibril formation curve (A). After about 100 min, the turbidity rises rapidly to a plateau value caused by the combined turbidity due to collagen fibrillogenesis and mineralization. Infrared spectroscopy and SEM revealed an early precipitation of small calcium phosphate particles before the strong increase of turbidity. This suggests that the first smooth increase of turbidity is caused by fibril assembly and simultaneous formation of ACP. The following steep increase indicates the rapid transformation of the ACP into a crystalline phase. The comparison with curve (B) indicates that this transformation is delayed by collagen.

As a remarkable fact, the reaction sequence depends sensitively on the calcium and phosphate concentration (Figure 6.70). The sequence of reactions described above occurs only in a narrow window of concentrations. Lowering the concentrations by 10% leads to a complete suppression of mineralization [curve (B) in Figure 6.70], although the solution is highly supersaturated with respect to HAP. This indicates that the collagen stabilizes the amorphous precursor phase of the

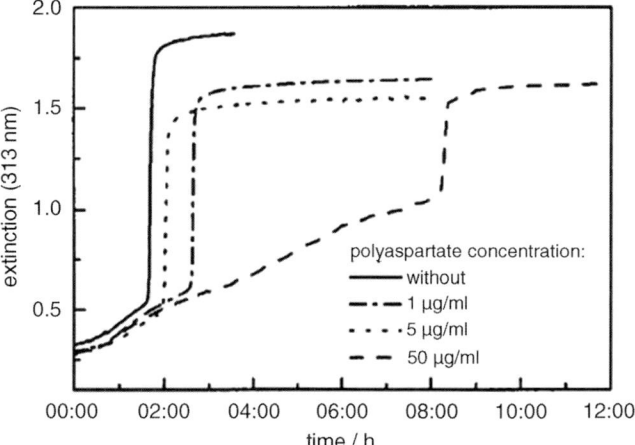

Figure 6.71 Effect of polyaspartate on the kinetics of combined fibril formation and mineralization. The calcium and phosphate concentrations are set at the optimum values (see Figure 6.70). (Reprinted with permission from Bradt *et al.* (1999), Copyright 2012, American Chemical Society.)

mineral. Otherwise, an increase of concentrations by 10% results in immediate precipitation of calcium phosphate after mixing [curve (E)]. Obviously, in this case, there is not enough time for the formation of collagen fibrils. The influence of polyaspartate on the whole process is twofold as the turbidity curves show (Figure 6.71). With increasing polyaspartate concentration, the amount of the amorphous phase grows, and the amorphous–crystalline transformation is delayed from about 100 min without polyaspartate to 8 h in the presence of 50 μg/ml polyaspartate. Obviously, this behavior corresponds to the model of a polymer-induced precursor phase, as discussed in Section 6.2.8.4. The nonmonotonous change of the transition point at low polyaspartate concentration can be explained by the additional influence of the polyamino acid on the fibril assembly. Fibril assembly can be accelerated as well as delayed depending on the applied concentration of the amino acid. Mainly the amorphous–crystalline transformation of the calcium phosphate is affected. Scanning electron micrographs show a significant influence of polyaspartate on the mineralized structures (Figure 6.72). Without polyaspartate, large aggregates of crystals are formed that seem to be only loosely bound to the collagen fibrils. With polyaspartate, separate crystals or only small clusters are formed on or in the fibrils. The improved attachment to the collagen fibril is probably caused by electrostatic interaction.

Transmission electron microscopy has revealed that the crystals are needles or very thin platelets between 15 and 80 nm in length and 3–4 nm in thickness (Figure 6.73). They are preferentially oriented parallel and are aligned along the collagen fibril that is bend by about 110° in this micrograph (Figure 6.73a). HRTEM (Figure 6.73b) shows a HAP needle with a diameter of 3.5 nm.

(a)

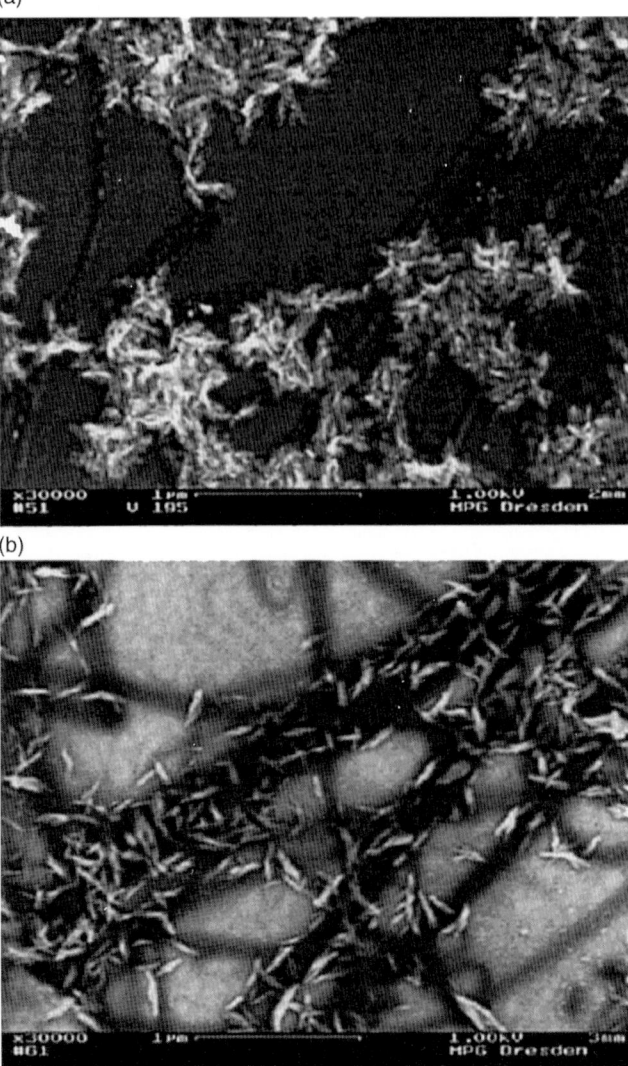

(b)

Figure 6.72 Scanning electron micrographs of mineralized collagen fibrils in the absence (a) and in the presence (b) of 50 µg/ml polyaspartate. (Reprinted with permission from Bradt *et al.* (1999), Copyright 2012, American Chemical Society.)

As we see, the process of synchronous fibrillogenesis and mineralization reproduces main features of the natural mineralized collagen. The advantage of such structure is its compositional and morphological similarity to the bone matrix. In particular, the nanodisperse HAP crystals are well biodegradable, which is an essential advantage for scaffold materials to be used in regenerative bone therapy. Larger scaffolds based on mineralized collagen are routinely produced now by

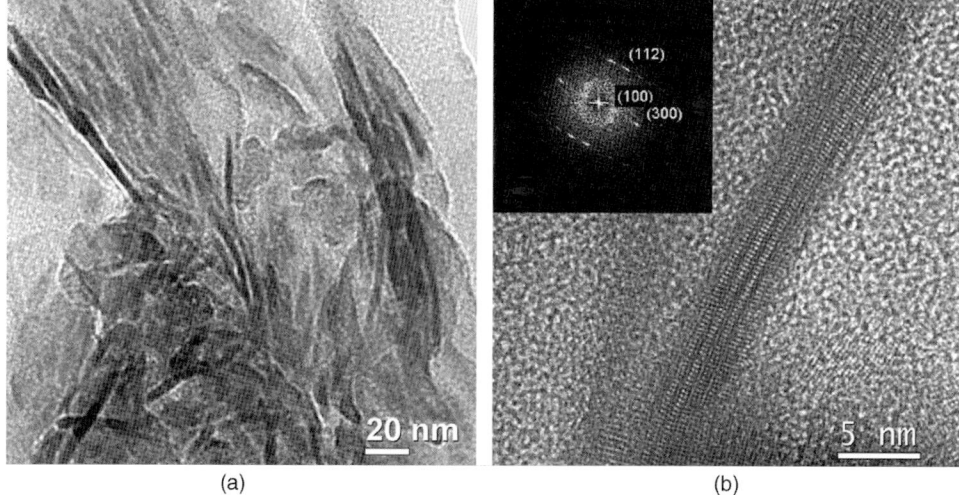

(a)   (b)

Figure 6.73 (a) Image showing individual needle- or platelet-like HAP crystals with typical length between 15 and 80 nm. (b) HRTEM of a HAP needle with a diameter of 3.5 nm. The (100) lattice planes of about 0.83 nm lattice distance are oriented along the needle axis. The inset displays the FFT of the crystal. Reflections of 0.83 nm (100), 0.275 nm (300), and 0.282 nm (112) of the hexagonal HAP crystal lattice are indicated. (Reprinted with permission from Gelinsky *et al.* (2008), Copyright 2012, Elsevier.)

means of various techniques. Freeze-drying of mineralized collagen suspensions has proven to be a flexible technique for preparing tailored scaffolds. By the freezing of suspensions cast into moulds, samples of any shape and size can be produced. By choosing an appropriate freezing regime, the pore size can be tuned. Freezing at temperatures between -20 and $-25\,°C$ generates a scaffold with interconnecting pores with diameters of about 200 μm (Figure 6.74). In order to obtain a scaffold that is stable under cell culture conditions, the mineralized collagen fibrils can be cross-linked with the carbodiimide derivative EDC (*N*-(3-dimethylaminopropyl)-*N'*-ethyl-carbodiimide) in 80% (v/v) ethanol. Such structure is best suited for bone cell cultivation (Gelinsky *et al.*, 2008).

The advantage of the freeze-drying process can also be used for the manufacturing of biphasic monolithic 3D scaffolds (Gelinsky, Eckert, and Despang, 2007). Such kind of materials is of interest for regenerative therapy of osteochondral defects. These are defects connected with deep lesions of the articular cartilage and damage to the underlying bone tissue. In such cases, special biphasic implant materials are needed that consist of a mineralized layer for the bony part of the defect and a nonmineralized layer for the chondral part. A promising solution of the problem has been worked out recently with the manufacturing of sandwiched scaffolds composed of mineralized collagen and hyaluronan–collagen (Figure 6.75). The mineralized collagen facilitates a stable connection with neighboring bone tissue, whereas a mixture of hyaluronan and nonmineralized collagen forms an appropriate matrix for chondrocytes, the cartilage cells. The duplex layer is obtained by starting from the

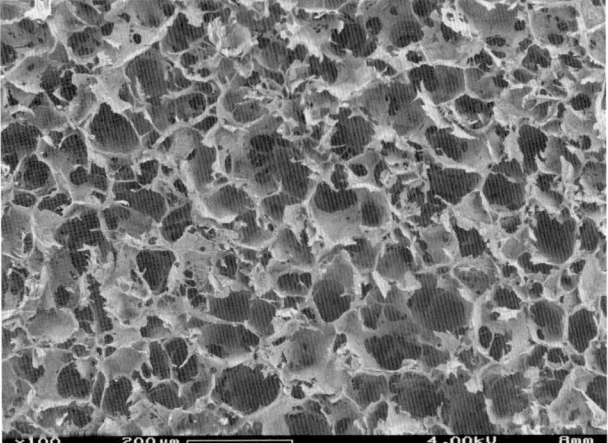

Figure 6.74 Scanning electron micrograph of a section through a three-dimensional scaffold made of biomimetically mineralized collagen. The interconnectivity of the pores is clearly visible. Scale bar: 200 μm. (Reprinted with permission from Gelinsky *et al.* (2008), Copyright 2012, Elsevier.)

liquid state, placing one liquid layer on top of the other, followed by freeze-drying and chemical cross-linking by means of covalent bonds (mainly amide bonds) between the collagen fibrils. The same process can be applied to the manufacturing of alginate-based osteochondral implants. The substitution of collagen by alginate allows the formation of calcium alginate–calcium phosphate composites with parallel capillary-like pores. The process will be explained in detail in the following section. Capillary systems are preferred in tissue engineering since they favor cell ingrowth and vascularization.

Mimicking the Anisotropic Structure of Bone Tissue When Heinrich Thiele started his studies on the formation of anisotropic inorganic and organic gels in 1947 at the

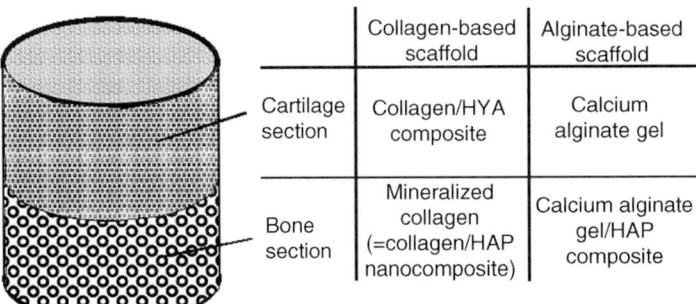

	Collagen-based scaffold	Alginate-based scaffold
Cartilage section	Collagen/HYA composite	Calcium alginate gel
Bone section	Mineralized collagen (=collagen/HAP nanocomposite)	Calcium alginate gel/HAP composite

Figure 6.75 Schematic representation of two types of biphasic, monolithic scaffolds for therapy of osteochondral defects. (Reprinted with permission from (Gelinsky, Eckert, and Despang, 2007), Copyright 2012, Carl Hanser Verlag, Munich.)

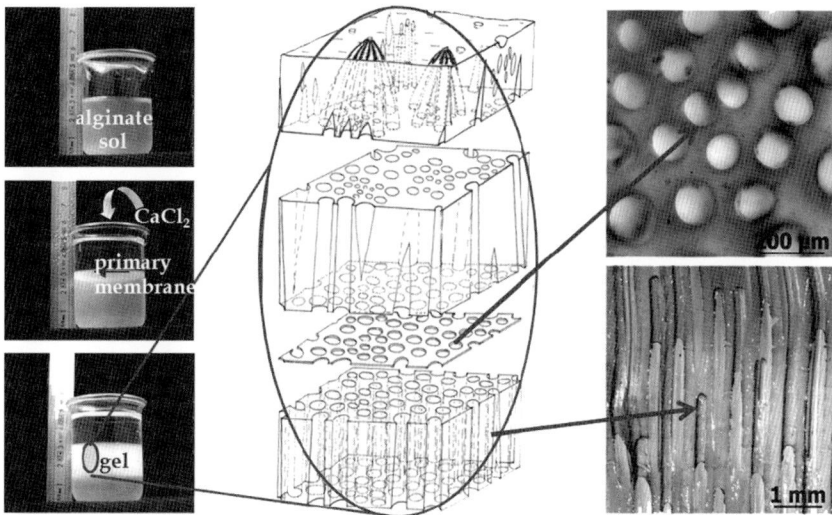

Figure 6.76 Sketch of the process of ionotropic gelation of alginate. The scheme in the middle was adapted from Wenger (1998). (Reprinted with permission from Despang, Dittrich, and Gelinsky (2011).)

chemical department of Kiel University, he soon realized that the new materials had a remarkable similarity with ordered collagen structures in decalcified bone. He noticed that *anisometric particles in colloidal solutions* align themselves during the sol–gel transition and generate an anisotropic gel if a diffusion front of electrolyte ions, such as Cu^{2+} or Ca^{2+}, is moving into the sol (Figure 6.76).

He explained this phenomenon by *ionotropy*, a concept used in materials chemistry for processes where ionic charge is locally accumulated at an organic surface. It is a potential mechanism for lowering the activation energy of inorganic nucleation on organic surfaces (Mann, 2001). Generally, a gel can be considered as a soft elastic material formed by cross-linking of the solute molecules in the sol. With increasing density of cross-links, the gel shrinks and its stiffness increases. As a common phenomenon, density fluctuations and phase transitions are observed beyond a critical density of cross-links (Elbert, 2011). The gel separates itself into two phases: a denser gel and a depleted solution. This kind of instability is called spinodal decomposition or syneresis in the case of phase separation in gel. This process is influenced by shrinkage stress (Onuki and Puri, 1999). The phase separation can be induced by exceeding thresholds of temperature, stress, or concentration of cross-linking ions, such as Ca^{2+} or Cu^{2+} in alginate solution.

Among the possibly organic colloids, the biopolymer alginate had been a favored system in the basic studies by Thiele. "Alginate" means alginic acid, an anionic polysaccharide found in the cell wall of brown algae. It is a copolymer, consisting of homopolymeric blocks of guluronic acid (G) and mannuronic acid (M) monosaccharide units covalently linked (Figure 6.77). The two monosaccharides possess the same carboxylic and hydroxyl groups, but differ in their configuration.

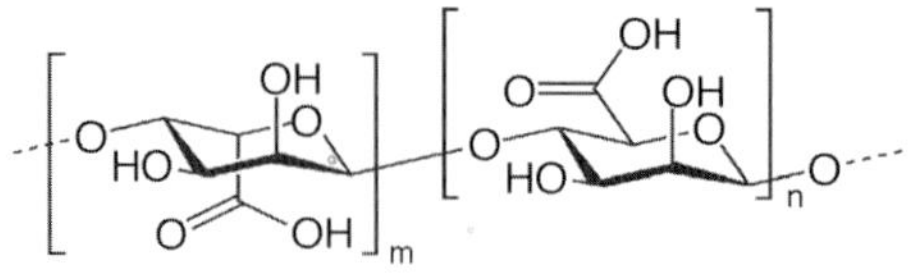

Guluronic acid   Mannuronic acid

Figure 6.77 Structural units of alginate copolymers consisting of homopolymeric blocks of guluronic acid (G) and mannuronic acid (M) monosaccharide units. Cu^{2+} or Ca^{2+} ions can create intermolecular complexes by coordinative bonds with carboxylic and hydroxyl groups between GG and MG acids.

Multivalent cations such as Cu^{2+} or Ca^{2+} can form intermolecular complexes by coordinative bonds with these functional groups. Such bonds can be formed between GG and MG sequences, whereas straight MM sequences offer no such binding sites for the cations. The presence of the cations results in the formation of the hydrogel. The particular composition of the alginate, which differs among the alga species, governs the viscoelastic properties of the hydrogel. G-rich alginates yield stiff hydrogels.

A droplet of alginate sol (alginate powder solved in a NaCl solution) in a $CaCl_2$ solution becomes immediately covered with a thin gel film. The gelation process induced by the diffusing ions can be emblematized by the nonstoichometric equation:

$$CaCl_2 + 2NaAlg \rightarrow CaAlg_2 + 2NaCl. \tag{6.54}$$

With Ca^{2+} diffusing into the droplet, the film becomes thicker until the droplet is completely transformed into a gel bead. Depending on the process conditions, tiny radial channels can form in the bead. In a planar configuration (Figure 6.76), the sol–gel transition initiated by a propagating diffusion front produces an anisotropic gel. In a first stage, a thin homogeneous gel membrane is formed by the reaction of the Ca^{2+} ions with the copolymer. The reaction front propagates as a planar diffusion front from the electrolyte overlayer into the alginate sol. With growing thickness of the gel membrane, its growth front can become unstable, resulting in water-filled parallel channels running almost through the whole length of the gel.

According to experiments, a well-developed capillary pattern forms under the following conditions:

i) The continuous membrane on the top has to reach a thickness of several micrometers.
ii) The concentration of cross-linking cations has to be beyond a critical value.
iii) The alginate concentration has to be below a critical value (about <5 wt%).

The diameter of the capillaries formed in the alginate gel is typically in the range of 30–460 μm. Depending on the intended application, this size can be tuned by variation of the molecular weight and concentration of the alginate and the kind and concentration of the electrolyte (Table 6.4).

Before discussing the next steps leading from the anisotropic alginate to a mimicked mineralized bone matrix, we will first consider the microscopic

Table 6.4 Alginate hydrogel scaffolds designed for different tissue engineering applications.

Target tissue	Pore size (μm)	Alginate concentration (wt%)	Molecular weight (kDa)	Electrolyte (M)
Bone	40–230	2	40–60	1 M CaCl$_2$
Embryonic stem cells	30	2	12–80	0.5 M CuSO$_4$
Neuronal tissue	27	2	100	1 M Cu(NO$_3$)$_2$
Vascularization	220–460	0.5–4	64–110	0.5–1.5 M CaCl$_2$

Source: Adapted from Despang, Dittrich, and Gelinsky (2011).

mechanisms responsible for the periodic capillary pattern of the gel. Two attempts to explain the experiments are mentioned here: diffusion–reaction-induced mechanism (Maneval *et al.*, 2011) and dissipative flow-induced patterning (Treml, Woelki, and Kohler, 2003). The basic of either model is the reaction of diffusing Ca^{2+} ions with alginate (Figure 6.78).

In the model of the *diffusion–reaction-induced mechanism*, it is proposed that the capillaries are formed due to anisotropic phase separation: The various steps of the phase separation are governed by the concentration profile of the diffusing Ca^{2+} ions. The structure development caused by indiffusion of Cu^{2+} as well as Ca^{2+} ions in an alginate sol has been studied by applying magnetic resonance (MR) microscopy. By MR microscopy, the spatially resolved measurement of distribution of water molecules and their diffusion coefficients is possible (Maneval *et al.*, 2011). From these measurements, a five-zone structure has been derived (Figure 6.79). With cation concentration increasing by diffusion, the alginate forms free-moving aggregates (zone 2) that turn into gel by cross-linking (zone 3, about 0.5 mm). Further cross-linking causes a separation into

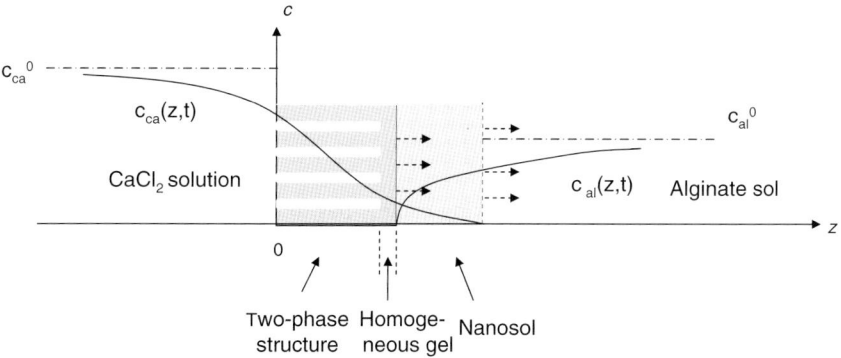

Figure 6.78 Diffusion–reaction zones in the gel front region during capillary formation. The calcium ion solution (extending far to the left with concentration c_{ca}^0) and the sodium alginate sol (extending far to the right with concentration c_{al}^0) are brought into contact at $z = 0$.

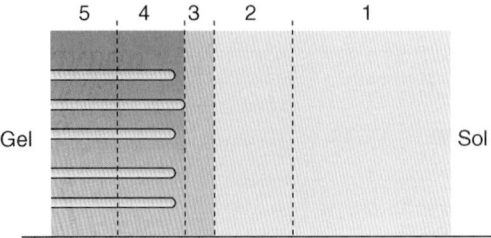

Figure 6.79 Structural model of capillary formation by the diffusion–reaction-induced mechanism as proposed by Maneval *et al.* (2011). The regions comprising the front are (1) sodium alginate sol, (2) dilute, pregel reaction zone where alginates react in small amounts with divalent cations, (3) homogeneous gel zone, (4) capillary formation zone, and (5) final heterogeneous gel formation zone.

denser gel and water-filled capillaries of well-defined shape, oriented in the direction of the diffusion front (zone 4). Subsequently, this structure undergoes an aging process (zone 5).

This is incompatible with the model of *dissipative flow-induced patterning of the gel* proposed by Kohler and coworkers, which is based on the assumption that the alginate chains, after getting attached with one end to the bulk gel, contract themselves toward the gel surface, thereby inducing a water flow via friction that gives rise to the formation of convective water rolls (Figure 6.80) (Treml and

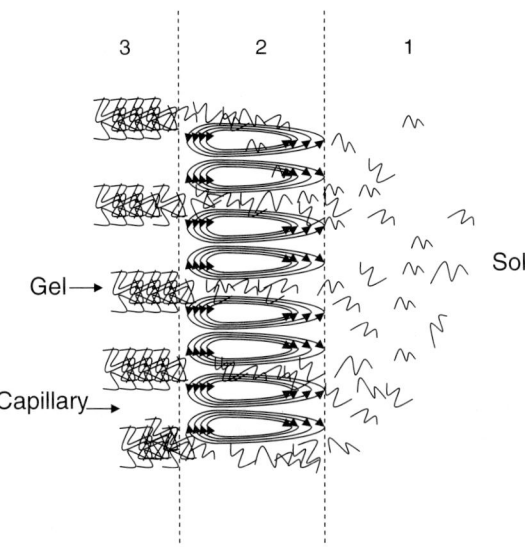

Figure 6.80 Structural model of capillary formation by dissipative flow-induced patterning of the gel proposed by Treml, Woelki, and Kohler (2003). The regions comprising the front are (1) sodium alginate solution, (2) contraction zone with roll cells, and (3) final heterogeneous gel formation zone.

Kohler, 2000; Treml, Woelki, and Kohler, 2003). As already mentioned, the MR measurements do not support the model involving liquid roll cells at the propagating gel front, but are compatible with the model known as *diffusion–reaction-induced mechanism*.

As an empirical fact, the ionotropic gelation of alginate sols is a very stable process that is not disturbed by the addition of other biopolymers as collagen type I, gelatin, or chitosan (Despang, Dittrich, and Gelinsky, 2011) (Figure 6.81). For bone tissue development, the incorporation of collagen type I or gelatin is of particular interest, but limited to small fractions only. In case of gelatin addition,

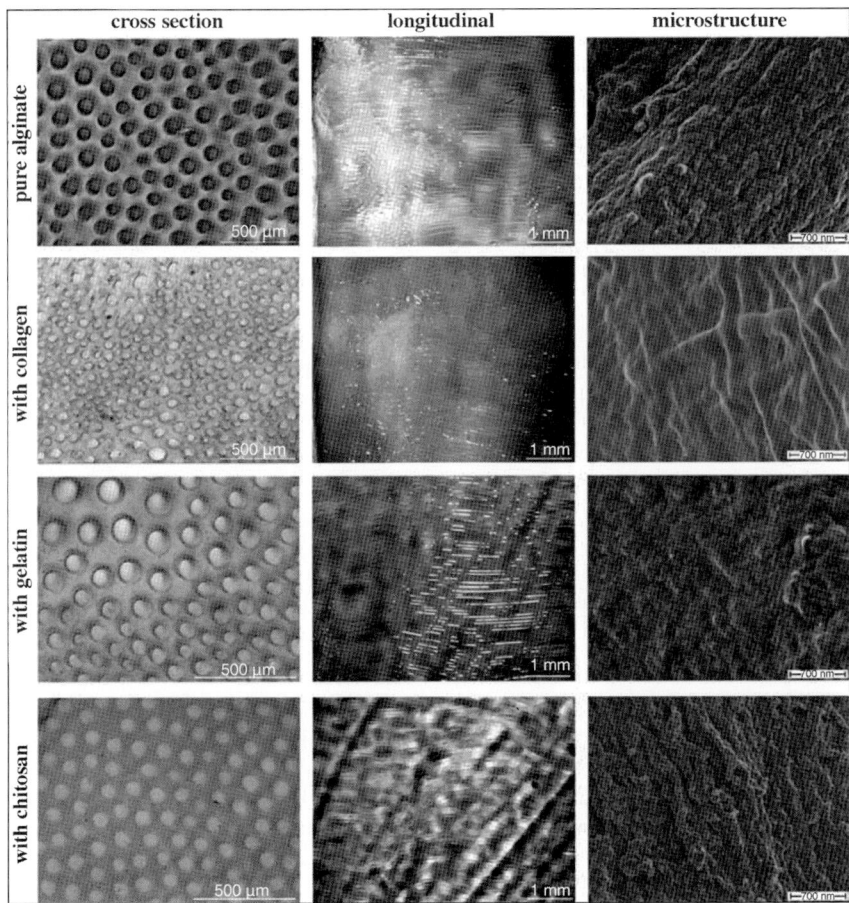

Figure 6.81 Polyelectrolytic hydrogels consisting of the negatively charged alginate and a second biopolymer. Note the significant decrease of diameter of the capillaries in case of addition of the long collagen type I molecules as well as of the positively charged chitosan. Scale bars: left column 500 μm, middle column 1 mm, and right column 200 nm. (Reprinted with permission from Despang, Dittrich, and Gelinsky (2011).)

isotropic gelation has to be avoided by keeping the process temperature above 30 °C. Very stable hydrogels have been prepared by adding chitosan. The positively charged polymer seems to form a stable network with the alginate molecules. The diameter of the capillaries is about half that in a gel of pure alginate. For hard tissue development, the good mineralization behavior of chitosan is an advantage.

There are various processing routes for mimicking bone tissue with anisotropic channel structure using gels consisting of pure alginate. HAP has been successfully incorporated into the ionotropic gel with the following processing routes:

- Immersion of the structured gel in simulated body fluid (SBF) containing calcium and phosphate ions needed for HAP growth (mineral content up to 11 wt%) Despang, Dittrich, and Gelinsky (2011),
- Precipitation of HAP by alternating diffusion of Ca^{2+} and phosphate ions into the prestructured gel (mineral content up to 50%) (Thiele and Awad, 1969).
- Simultaneous mineralization during the sol–gel transition by reaction of diffusing calcium and phosphate ions added to the alginate sol (mineral content 5–9 wt%) (Despang *et al.*, 2005).
- Addition of HAP powder into the sol before ionotropic gelation (mineral content more than 70 wt%) (Dittrich, Tomandl, and Mangler, 2002).

As shown by Dittrich, Tomandl, and Mangler (2002), the mechanical properties of the mineralized alginates can be improved by controlled heat treatment. By sintering at low temperatures, the mineralized gel can be transformed into a sintered ceramic without destroying the capillary pattern. Keeping in mind the intended use for hard tissue engineering, two conditions related to pore shrinkage and grain growth have to be met in heat treatment. The pore size for bone implants should be in the range of 100–300 μm. Temperatures beyond 1000 °C have to be avoided since HAP crystals exposed to such temperatures are no longer resorbable *in vivo* by osteoclasts (Despang, Dittrich, and Gelinsky, 2011). Heat treatment at 650 °C burns out the biopolymer without destroying the overall structure. In the so-called bisque, the pore diameter is in the range of 40–165 μm. In a mineralized alginate gel produced by addition of HAP powder into the sol before ionotropic gelation, the size of the HAP crystals is about 40 nm in the bisque. Its compressive strength of 4.5 MPa is comparable to that of cancellous bone (5–6 MPa). Sintering at 1200 °C leads to a further densification of the structure (pore diameter 30–115 μm) and larger crystal size (about 240 nm) (Despang, Dittrich, and Gelinsky, 2011). Scanning electron micrographs show the coarsening of the microstructure in the ceramic sintered at 1200 °C, which seems to be unfavorable for cell interaction in tissue engineering (Figure 6.82). *In vitro* studies with osteogenically induced human mesenchymal stem cells (hMSC) demonstrated the good biocompatibility of the bisque. Proliferation and differentiation of the hMSCs have been evaluated (Despang *et al.*, 2008). A typical marker for cell differentiation, the specific alkaline phosphatase (ALP) showed a maximum activity at day 14 after cell seeding. Within 3 weeks, the number of cells increased by a factor of 2.5. The cells adhered to the surface,

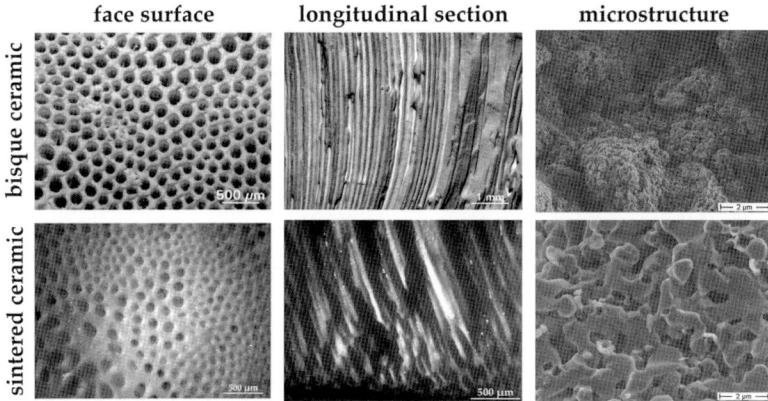

face surface longitudinal section microstructure

bisque ceramic

sintered ceramic

Figure 6.82 HAP bioceramic based on Ca–alginate–HAP–slurry with channel-like pores. *Top*: State after thermal treatment at 650 °C. *Bottom*: Sintered ceramic. The samples were prepared by addition of HAP powder into the sol before ionotropic gelation. Scale bars: left and middle column 500 μm and right column 2 μm. (Reprinted with permission from Despang, Dittrich, and Gelinsky (2011).)

propagated inside the pore channels, and grew to a confluent layer, as seen in Figure 6.83.

The presented example shows that the process allows the manufacturing of anisotropic HAP scaffolds with compressive strength similar to that of cancellous bone. The potential of ionotropic gels as precursors for many other artificial tissues becomes apparent in Figure 6.84, which gives an overview of structures achievable with this gelation technique. The overview shows typical structure elements covering a size range between nanometers and tens of micrometers.

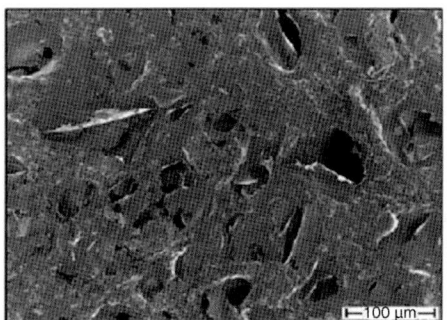 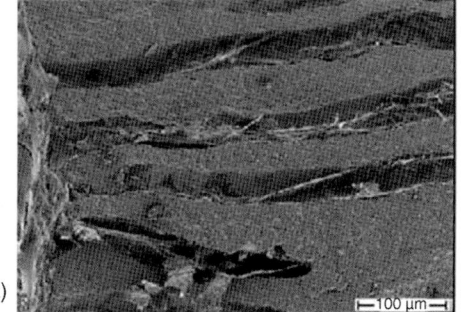

(a) (b)

Figure 6.83 Osteogenetically induced hMSCs after 14 days of *in vitro* cultivation on nanocrystalline HAP scaffolds in the state as bisque (Ca^{2+} gelled slurry, after heat treatment at 650 °C). (a) Surface. (b) Longitudinal section (scanning electron micrograph after supercritical drying). Scale bars are always 100 μm. (Reprinted with permission from Despang, Dittrich, and Gelinsky (2011).)

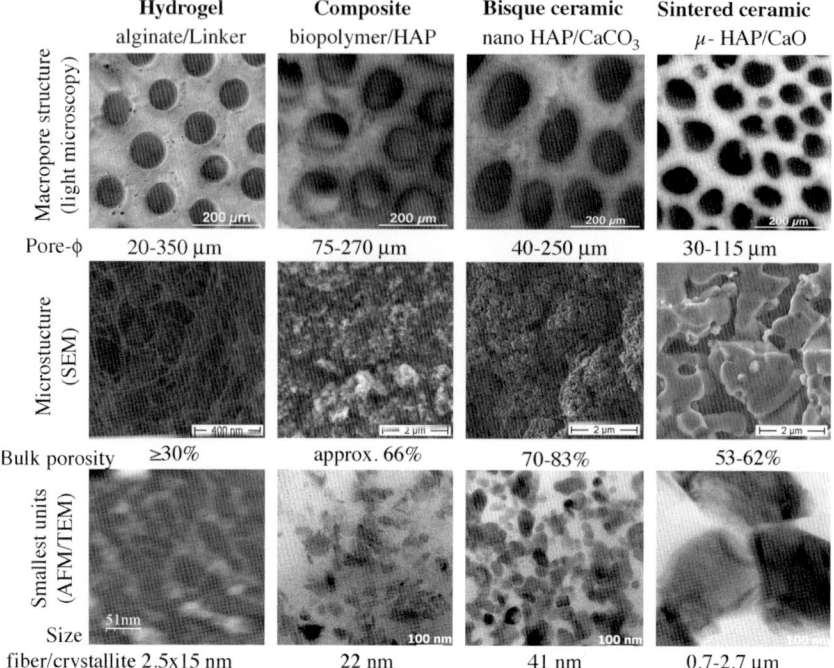

	Hydrogel alginate/Linker	**Composite** biopolymer/HAP	**Bisque ceramic** nano HAP/CaCO$_3$	**Sintered ceramic** μ- HAP/CaO
Macropore structure (light microscopy)				
Pore-ϕ	20-350 μm	75-270 μm	40-250 μm	30-115 μm
Microstucture (SEM)				
Bulk porosity	≥30%	approx. 66%	70-83%	53-62%
Smallest units (AFM/TEM)				
Size fiber/crystallite	2.5x15 nm	22 nm	41 nm	0.7-2.7 μm

Figure 6.84 Overview of anisotropic scaffolds consisting of different materials prepared via ionotropic gelation. (Reprinted with permission from Despang, Dittrich, and Gelinsky (2011).)

6.3.3
Biomimicking of Bone Tissue

The development of a "living" material suitable for the replacement of damaged bone can be regarded as an exotic novel branch of materials engineering. Such material should combine strength with the biological functions of different cell types present in living bone tissue. An ideal bone-like material should facilitate the morphogenesis of stem cells and bone precursor cells. It should offer the right conditions for homeostasis of the osteoblasts and osteocytes, and it should allow a complete remodeling into healthy natural tissue. The performance of a bone substitute is essentially characterized by three properties: (i) osteoconductivity, promoting the ingrowth of osteoprogenitor cells and blood vessels into the implant, (ii) osteoinductivity, stimulating the osteoprogenitor cells to form new bone tissue, (iii) osteogenicity, causing precursor cells to differentiate into active osteoblasts forming new bone. Depending on particular medical requirements, these properties have to be realized adequately. In cell-based therapy of pathogenic bone tissue, there are two alternative ways of integrating the artificial bone material into the natural bone tissue. As one variant, the morphogenesis of precursor cells or stem cells is induced in the bioengineered artificial bone matrix *ex vivo*, so that it acts as a carrier of cells to be transplanted together with the bone matrix. Alternatively, the transplanted artificial matrix material induces ingrowth and differentiation of cells from healthy tissue *in situ*. Depending on the strategy to be preferred, different material concepts have to be applied. In the *ex vivo* approach, the artificial matrix should contain biomolecular structures that act as signal molecules to direct the proliferation and differentiation of stem cells into precursor cells of the osteoblastic and osteoclastic cell lines. In the *in vivo* approach, the material to be implanted should favor the ingrowth of active osteoblasts and osteoclasts. Also, it should contain signal molecules that promote a fast remodeling into a natural ECM. Such concepts of scaffold design apply to tissue engineering in general and therefore are considered in Section 6.3.3.1 under more general aspects.

6.3.3.1 Natural versus Synthetic Biopolymers for Scaffold Design

Much effort has been devoted to the development of materials suitable as scaffolds for tissue engineering. Biologically derived as well as synthetic materials have been applied for this purpose. Natural materials, such as collagen from animal tissue or denatured bone, have the advantage of being structurally similar to the living tissue, which is relevant for the biological recognition as specific binding motifs of the interaction with ligands, the right composition for proteolytic degradation, and an optimum geometry for cell adhesion and migration. Problems are connected with necessary purification of the natural material and possible *immunogenicity* (critical response of the immune system). Synthetic materials offer a greater variety to choose from, better control of properties, and, as a major advantage, high purity. However, the incorporation of molecular recognition motifs needs additional efforts, and the number of suitable materials may be reduced by biodegradation and toxicity. Up to now, purified collagen type I is the first choice for scaffold design in regenerative bone therapy.

There are promising concepts for creating synthetic materials by combining various components in a systematic way (George *et al.*, 2011). First let us consider one informative example for such an approach, the use of leucine zipper polypeptides for hard tissue engineering (Gajjeraman *et al.*, 2008). The DNA-binding leucine zipper proteins contain a self-assembling leucine zipper domain. It consists of a repeating heptad motif *abcdefg*, where *a* and *d* are hydrophobic amino acids (at position *d*, leucine is common) and *e* and *g* are charged amino acids (mainly glutamic acid). The repeating domain has an α-helical structure. Driven by the hydrophobic interactions, coiled-coil dimers are formed. Depending on pH, temperature, and ionic strength, extended hydrogel network structures are formed by the interaction of the leucine zipper domains (see the scheme in Figure 6.94). The network can be functionalized by applying triblock copolymers composed of leucine zipper helix endblocks (LZ) and a water-soluble polyelectrolyte midblock. In the study by Gajjeraman *et al.* (2008), a polypeptide sequence of the C-terminal region from the dentin matrix protein 1 (DMP1) has been incorporated between the two LZs. It contains a HAP nucleating domain as well as an RGD motif for cell adhesion. In addition, the hydrogel network can be reinforced by chemical cross-links. For example, the leucine zipper construct can be stabilized by introducing several cysteine residues that form intermolecular disulfide bonds. Thus, the short

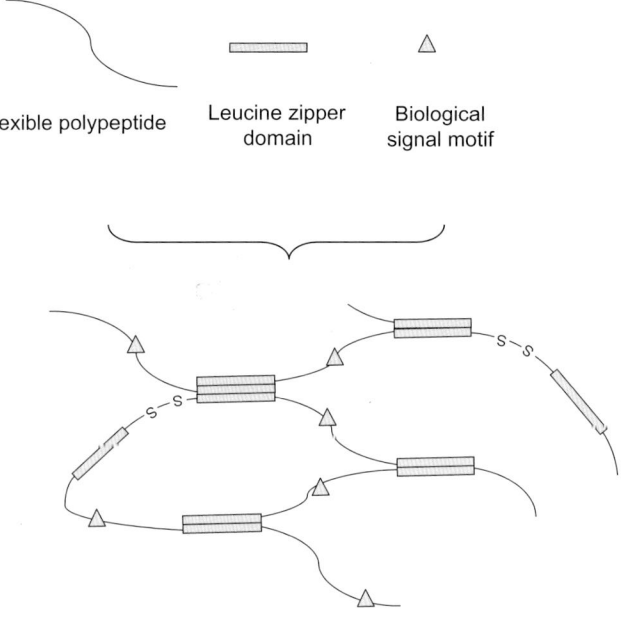

Flexible polypeptide   Leucine zipper   Biological
                           domain   signal motif

Figure 6.85 Schematic representation of the leucine zipper construct designed for bone regeneration. (Adapted from George et al. (2011).)

nanofibers are linked to high molecular weight polymers with additional cross-links in the network (Figure 6.85). *In vitro* nucleation experiments clearly demonstrated the functionality of the HAP nucleating domain. Figure 6.86 shows growth stages of calcium phosphate precipitates on a glass slide coated with the LZ-DMP1 copolymer. After 2 days, ACP precipitates had formed that transformed into well-crystallized needles within 8 days. During crystallization, the Ca/P ratio increased from 1.47 (amorphous phase) to 1.69 (well-crystallized needles). HRTEM and selected area electron diffraction showed that the crystals grow preferentially along the *c*-axis (Figure 6.86c). The lattice spacing of 0.345 nm corresponds to the (002) lattice plane of HAP. For comparison, mineralization experiments were performed on glass slides coated with LZ protein without DMP1 domain. After 8 days of growth, the deposited Ca phosphate appeared amorphous. The Ca/P ratio was 1.48. The comparison clearly shows the structure modulating function of the DMP1. The cell response is mediated by the RGD motif contained in the C-terminal DMP1 fragment. It is the same motif that has been identified in fibronectin, activating the cell surface integrins $\alpha_5\beta_1$ and $\alpha_v\beta_3$. The RGD (Arg—Gly—Asp) motifs of the LZ-DMP1 network cause an improved adhesion and migration of osteoblast-like cells that promote their incorporation and differentiation. Figure 6.87 shows the effect of a LZ-DMP1 coating on the response of preosteoblastic MC3T3-E1 cells plated on the glass. The cells spread extensively and expressed lamellipodial protrusions indicating the cell migration. The change of the cytoskeleton indicated signal transduction into the cell. After 48 h, the cells were stained for actin with a fluorescent dye

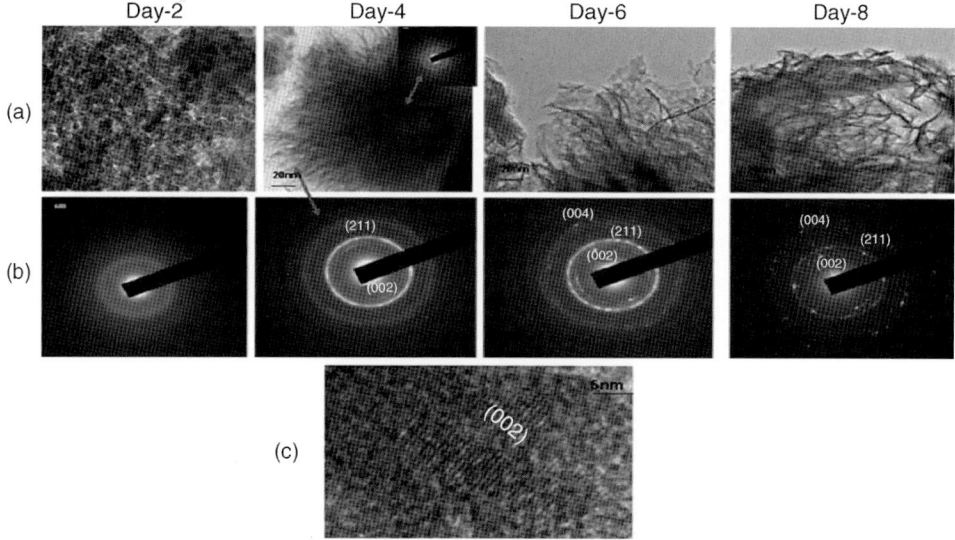

Figure 6.86 Characterization of the calcium phosphate deposits by TEM. (a) Time series of apatite crystallization. (b) Selected area electron diffraction of the obtained deposits. ACP precipitated within 2 days had transformed into well-crystallized HAP needles after 8 days. (c) HRTEM showing the lattice spacing of 0.35 nm corresponding to the (002) lattice plane of HAP. (Reproduced with permission from Gajjeraman *et al.* (2008), Copyright 2012, Wiley-VCH Verlag GmbH.)

(Oregon 514), and for focal adhesion sites with immunofluorescently labeled paxillin. Paxillin binds the cytoplasmic domains of the α_4, β_1, and β_3 integrins. The comparison of LZ-DMP1-coated and blank glass coverslips shows that the RGD domains promote the assembly of thick actin bundles inside the cell and the formation of focal adhesion sites mapped by the integrin assemblies.

6.3.3.2 Protein-Engineered Synthetic Polymers

When looking for a systematic way to mimic the complexity of the natural ECM of bone tissue, we can follow a concept worked out by Lutolf and Hubbell (2005) for engineering of artificial tissue. Tissue engineering means that the material chosen for the framework of the tissue to be grown has to be furnished with molecular structures that attract the cells from the surrounding living tissue to the scaffold as well as facilitate a transmission of signals from the scaffold to the surrounding tissue and vice versa. Two groups of so-called *effectors* are known to fulfill this task: (i) insoluble hydrated macromolecules carrying specific signal motifs (fibrillar proteins such as collagen, noncollagenous glycoproteins as fibronectin, and hydrophilic proteoglycans) (see Section 6.2.8.3) and (ii) soluble macromolecules (cytokines, among them chemokines and growth factors). Cytokines are molecular messengers that act between cells. They circulate in the tissue, but can act also over larger distances in the body. The chemotactic cytokines or chemokines induce directed chemically controlled migration of nearby responsive cells. Growth factors stimulate the cellular growth, proliferation,

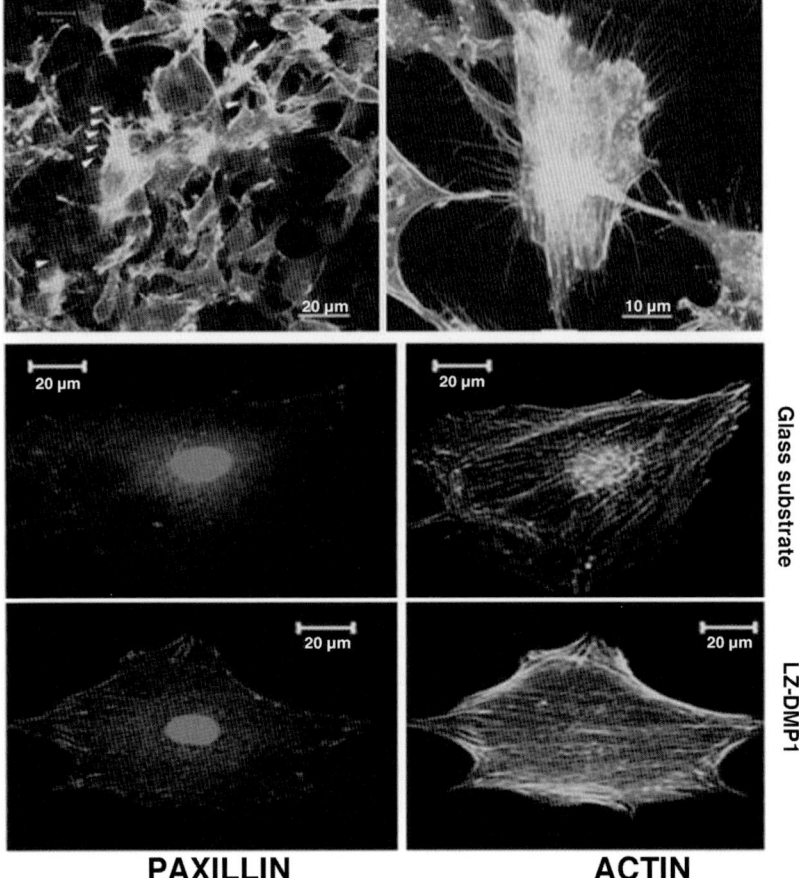

PAXILLIN   **ACTIN**

Figure 6.87 Fluorescent microscopy images of the effect of an LZ-DMP1 on the attachment of preosteoblastic cells at a glass substrate 48 h after cell plating. Paxillin binding to integrins was labeled with a fluorescent antipaxillin antibody. The actin cytoskeleton was labeled with a fluorescent dye (Oregon 514). In the first row, the expression of lammellipodial protrusions induced by the LZ-DMP1 coating of the glass substrate is shown. In the second and third rows, the organization of the cytoskeleton is compared for cells plated on a blank glass substrate and an LZ-DMP1-coated substrate, respectively. (Reproduced with permission from Gajjeraman *et al.* (2008), Copyright 2012, Wiley-VCH Verlag GmbH.)

and cellular differentiation. Among them, bone morphogenetic proteins (BMPs) and the transforming growth factor-ß (TGF-ß) are involved in the regulation of the osteogenic differentiation of hMSCs (see Section 6.2.8.2). Some of the cytokines inhibit cell growth or fulfill other tasks in the cell communication.

The example of leucine zipper polypeptides discussed in the previous section provides the outline of a general design concept for biofunctionalized scaffolds. In order to mimic the diverse properties of the natural ECM of a tissue, multiple domains with different functionalities should be included into the artificial scaffold. In Scheme 6.2,

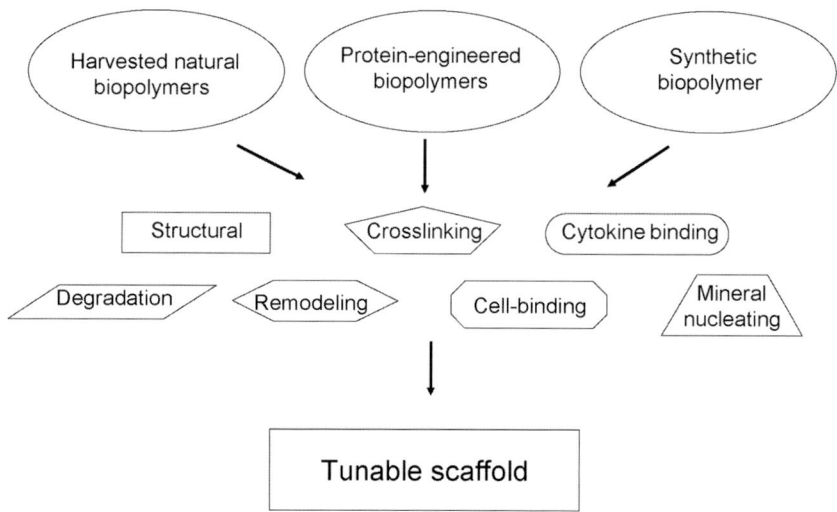

Scheme 6.2 Modular design of tunable scaffolds.(Adapted from Sengupta and Heilshorn (2010).)

relevant properties are summarized in terms of functional building blocks or domains: structural domains providing the shape of the tissue and its mechanical stability, cross-linking domains, cell binding domains, mineral-nucleating domains, domains binding cellular signal molecules, and domains regulating the scaffold remodeling and degradation.

When searching for means to meet this broad spectrum of demands, materials engineers can choose between two extremes to start with: Either they take a natural biomaterial such as collagen type I or a synthetic polymer such as poly(ethylene glycol) (PEG) as the basic material for the scaffold. Next, various functionalities have to be added by linking particular biomolecular units such as peptides, glycoproteins, proteoglycans, or polyaminoglycans with the main network structure. Starting from the genetically engineered DNA, proteins are generated that are composed of modular peptides, providing the desired functionalities. Combinations of individual modules allow the generation of tailored multifunctional biomaterials. Commonly, *E. coli* bacteria are used as expression system, because it is an inexpensive, robust, and fast growing cell system. Other suitable expression systems are yeast species like *Saccharomyces cerevisiae* and *Pichia pastoris*. These eukaryotic cells permit post-translational modifications of the recombinant proteins, for example, glycosylation, which can be essential for the functionality of the protein.

A protein-designed biomaterial should combine tunability of synthetic biomaterials with biocompatibility and minimum cytotoxicity of natural biomaterials. A comparison of advantages and challenges of the different classes of biomaterial scaffolds summarized in Table 6.5 shows that in this respect the protein-engineered biomaterials are most suitable. However, by taking into account aspects such as yield and cost-efficiency, it appears that combining the protein engineering of crucial functional modules with the manufacturing of synthetic or natural biomaterials is most promising.

Table 6.5 Comparison of biomaterial scaffolds.

	Synthetic biomaterials	Natural biomaterials	Protein-engineered biomaterials
Toxicity	Can be critical	Not critical	Not critical
Tunability	Good	Limited	Good
Bioactivity	Can be limited	High	High
Yield	High	Medium	Low
Cost-efficiency	High	High	Low

Source: Adapted from Sengupta and Heilshorn (2010).

The feasibility of manufacturing a mimetic ECM for *in situ* bone regeneration by making use of this concept has been demonstrated by Lutolf *et al.* (2003a, 2003b). PEG-based hydrogel is chosen here as the basic component to be conjugated with two functional peptides. The basic unit of the hydrogel was a vinylsulfone-functionalized PEG molecule. The incorporation of the intended biochemical functionalities was performed by conjugate addition reactions between conjugated unsaturations on the end-functionalized PEG macromers and thiol-bearing peptides (Figure 6.88). Two kinds of peptides have been prepared to mimic cell adhesion sites and a specific susceptibility to cell-triggered proteolysis. For inducing cell adhesion, pendant oligopeptide ligands containing a RGDSP motif have been chosen. It favors the adhesion of fibroblasts. As the second linker, a peptide has been selected that acts as substrate for matrix metalloproteinases (MMP). It is known that members of the

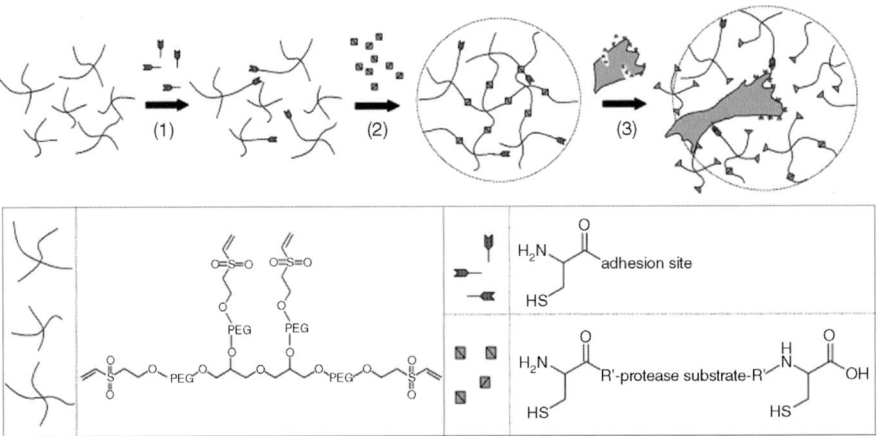

Figure 6.88 A Michael-type addition reaction between vinylsulfone-functionalized multiarm PEGs and monocysteine adhesion peptides (step 1, in high stoichiometric deficit) or biscysteine MMP substrate peptides (step 2, to come up to stoichiometric equilvalence) was used to form gels from aqueous solutions in the presence of cells. These elastic networks were designed to locally respond to local protease activity at the cell surface (step 3). (Reproduced with permission from Lutolf *et al.* (2003a), Copyright 2012, John Wiley & Sons, Inc.)

MMP family act as enzymes in denaturation of the bone matrix during remodeling. Therefore, the cell-triggered proteolytic degradation of the cross-links in the hydrogel network could be used for cell invasion and remodeling.

The success of that design concept was proven experimentally in bone regeneration using *in vitro* and *in vivo* assays. The invasion of human fibroblast cells (hF) was shown in case of an *in vitro* assay. Fibroblasts migrated out in three dimensions from a cell–fibrin cluster into an adhesive and MMP-sensitive synthetic matrix. The cells spread with an average velocity of about 7 μm/h. The invasion did not occur when the RGDSP motif in the network was missing or was substituted by a RDGSP sequence inactive in promoting cell adhesion. As the mesh size of the hydrogel was in the range of 30–50 nm, which is far below the cell size, migration was possible only when the network contained substrates for proteolytic digestion by MMPs. In presence of MMP inhibitors, hF invasion was significantly reduced, indicating that proteolysis was the main factor responsible for the outgrowth of cells. Furthermore, the hydrogel allowed a stable storage of human bone morphogenetic protein-2 (hBMP-2). This growth factor promotes the formation of new bone. It activates osteoprogenitor cells and induces their differentiation into active osteoblasts, the bone-forming cells (for more details about bone remodeling, see Section 6.2.8.2). The controlled release of hBMP-2 from the PEG network was triggered by exposing the network to MMP-2 enzymes. This means that under *in vivo* conditions, the hBMP-2 release occurs preferentially at bone defects connected with enhanced cell invasion by localized proteolytic activity. Therefore, the functionalized PEG-hydrogel makes the use of hBMP-2 in bone regeneration highly efficient.

6.3.3.3 Protein-Engineered Collagen Matrices

For bone tissue engineering, also the alternative approach based on the natural biopolymer collagen type I as matrix material has some advantages. As already mentioned, collagen type I is the appropriate template for the formation of nanoparticulate HAP that mimics the structure and shape of those in natural bone in a nearly perfect way. Moreover, the HAP nanocrystals offer binding sites for cytokines involved in the *in vivo* remodeling process or for genetically engineered peptides in case of *ex vivo* tissue engineering. As another advantage, the collagen fibrils as well as the HAP nanocrystals can be degraded by the osteoclasts (enzymatic digestion of the collagen and dissolution of HAP at low pH localized in the sealing zone of the adhering cell). Hence, with mineralized collagen type I as a starting material for scaffold manufacturing, an essential part of functionality of a bone-like extracellular matrix has already been realized at the very beginning. The following few examples demonstrate how scaffolds manufactured with mineralized collagen can be modified with functional components in order to mimic additional properties of the ECM of bone.

Addition of Binding Sites of Soluble Effectors   For *in situ* tissue engineering, it would be favorable if the scaffold to be implanted would contain components that could act as "catcher" structures immobilizing growth factors or other cytokines from the body fluids. Then, the scaffold could self-organize the functional domains needed for remodeling from the patient's own resources. In case of a scaffold made from mineralized collagen, a versatile structure with catcher properties can be generated by functionalization of the collagen–HAP scaffold with linear polysaccharides. Linear polysaccharide chains are an essential component in the ECM of bone tissue. In Section 6.2.8.3, GAGs have been introduced as the most important group of such polysaccharides. In connective tissue, there are variants of a few primary structural types: hyaluronic acid, keratan sulfate, and CS (Figure 6.47). As the chains are anionic, they can chelate Ca^{2+}. N-Acetylamido and hydroxyl groups can generate hydrogen bonds with phosphate ions, water, and OH^-. Therefore, the HAP nano-crystals of the mineralized collagen offer ideal binding sites for GAGs (Wise et al., 2007). One favored GAG to be used as a catcher structure is hyaluronic acid because it shows no immunogenic reactions. It promotes cell migration and cell adhesion. By introduction of sulfate groups, the biocompatibility with cells can be improved. Hintze et al. (2009) have studied the interaction of hyaluronic acid (Hya) and its sulfate modifications (high-sulfate sHya2.8 and low-sulfate sHya1.0) (Figure 6.89) with recombinant human bone morphogenetic protein-4 (rhBMP-4). For comparison, two other GAGs, CS and heparan sulfate (HS), have been included in the study. The interaction of the GAGs with rhBMP-4 is mainly electrostatic, caused by negatively charged disaccharides and positively charged residues of rhBMP-4. The binding efficiency of the various GAGs is evaluated with an enzyme-linked immunosorbent assay (ELISA) (Figure 6.90). The intensity of the colorimetric signal measured at 405 nm characterizes the binding of the BMP to GAGs immobilized in wells of a microtiter plate. The best binding is realized with sHya2.8, followed by sHya1.0. The binding is much weaker with unmodified Hya, CS, and HS. Well surfaces coated with 2% bovine serum albumin or untreated ones show no binding. The differences in binding affinity of the sulfated GAGs can be explained by the differences of the sulfate ion of the C-4 position in the N-acetylglucosamine unit. sHya2.8 is fully sulfated in the C-6 position and partially sulfated in the C-4 position. sHya1.0 and CS are primarily monosulfated. From the data it was concluded that sulfated hyaluronic acid is a promising "catcher" molecule for the immobilization of growth factors on/in scaffolds manufactured from mineralized collagen.

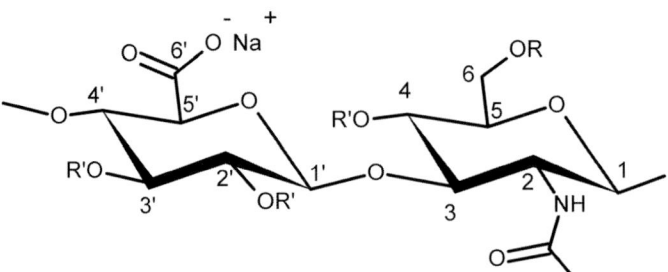

Figure 6.89 Chemical structures of differently sulfated hyaluronic acids (hyaluronic acid, Na salt (Hya): R, R′ = H; low-sulfate Hya (sHya1.0): R = SO₃Na, R′ = H; high-sulfated Hya (sHya2.8): R = SO₃Na, R′ = H or SO₃Na).

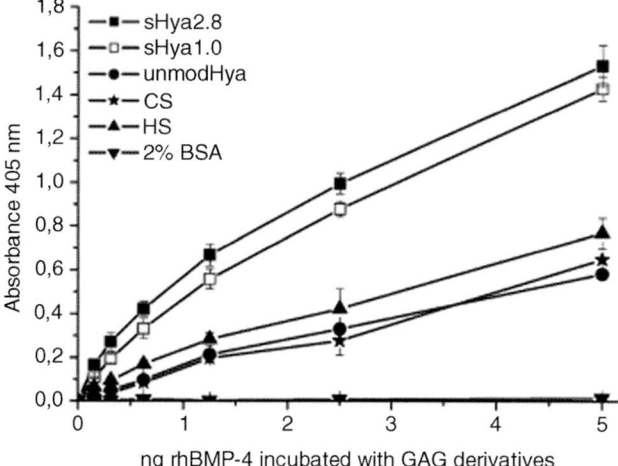

Figure 6.90 ELISA of up to 5 ng of rhBMP-4 with GAGs immobilized on wells of microtiter plates. Surfaces coated with 2% BSA or untreated wells served as negative or positive controls. The colorimetric reaction was followed by absorbance measurements at 405 nm. (CS = chondroitin sulfate, HS = heparan sulfate.) (Reprinted with permission from Hintze *et al.* (2009), Copyright 2012, American Chemical Society.)

Chemotactic Incorporation of Active Cells The use of synthetic bone tissue colonized with living cells *ex vivo* is limited to the treatment of small bone defects. The treatment of larger defects as those related to osteoporosis, tumor resection, or traumata encounters the problem of insufficient initial vascularization of the synthetic tissue. Cells placed in the inner regions of the new bone are not sufficiently supplied with oxygen and nutrients after implantation. As a possible solution, the initially cell-free ECM can be colonized with osteoblastic precursor cells *in vivo*, after implantation. The precursor cells should be attracted by effectors directing the cell migration into the inner region of the synthetic tissue. In case that bone marrow stromal cells (BMSCs) circulating in the bloodstream could be attracted, they would not only remodel the scaffold but also secrete factors (e.g., the vascular endothelial growth factor (VEGF)) that induce the ingrowth of blood vessels (angiogenesis). MSCs, a subset of BMSCs, can migrate directed by chemotactic interaction with various compounds such as fibronectin, platelet-derived growth factor (PDGF), and bone morphogenetic protein-2 (BMP-2).

An example of the induction of chemotactic interaction of BMSCs with mineralized collagen scaffolds has been worked out by Thieme *et al.* (2009). They used the well-characterized chemotactic interaction between the stromal cell-derived factor 1α (SDF-1α) and cells expressing its receptor CXCR4. This system has been applied to attract BMSCs from the surface into deeper structures of a porous scaffold manufactured from mineralized collagen (Section 6.3.2.2). In order to have a high density of CXCR4 receptors at the surface of the BMSCs, the CXCR4 was overexpressed in the cells by transient mRNA transfection. Chemotactic migration occurs when the cells detect a

(a)

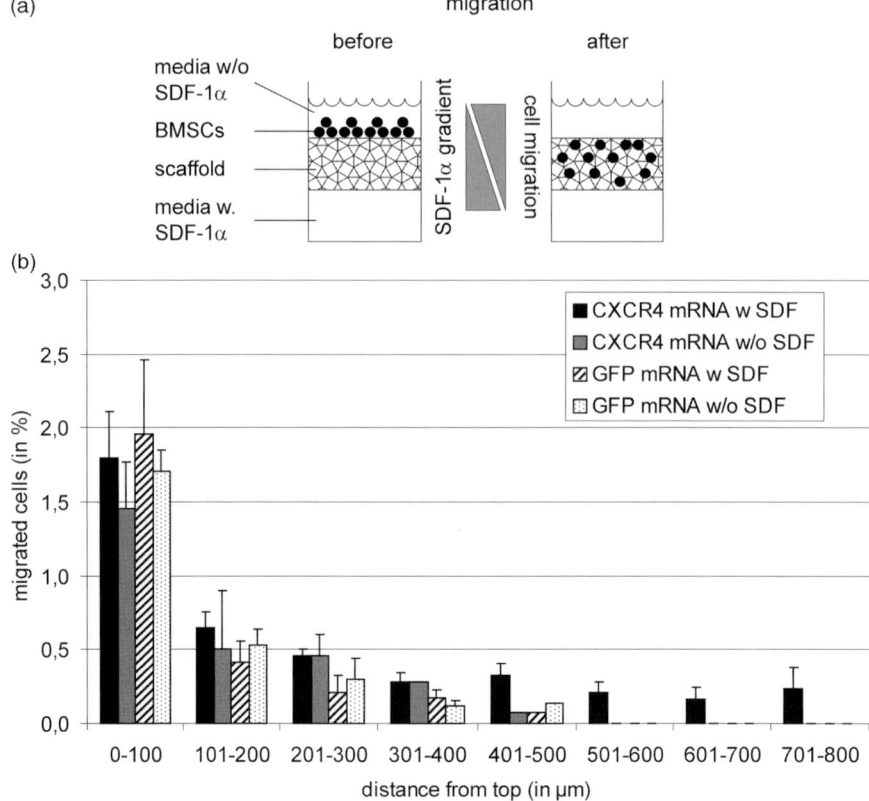

Figure 6.91 (a) Schematic diagram of the migration assay for 3D scaffolds. A fixed scaffold enables the application of stromal cell-derived factor-1α (SDF-1α) in the lower compartment of a plastic column. Different cell types are applied to the upper surface of the scaffold to examine their migration efficiency. (b) Depth distribution of the mRNA-transfected cells in the scaffold of mineralized collagen after 5 days. BMSCs had been transfected with CXCR4 (black and gray columns) or GFP (dashed and dotted columns). The bottom of the scaffold was either placed in media without SDF-1α (gray and dotted columns) or in media containing SDF-1α (black and dashed columns). (Reprinted with permission from Thieme *et al.* (2009), Copyright 2012, Mary Ann Liebert, Inc.)

gradient of the chemokine concentration guiding cells in the direction of maximum chemokine concentration. The experiment was carried out as shown schematically in Figure 6.91. BMSCs were put on top of the scaffold, whereas the bottom chamber contained SDF-1α. By diffusion, a gradient of chemokines was generated that induced cell migration in the opposite direction. After 5 days, the cell distribution was measured. For comparison, three other systems were also studied: CXCR4 transfected cells without SDF chemokine and cells transfected with green fluorescent protein (GFP) with and without SDF. In Figure 6.91b, the depth distribution of the cells reveals the chemoattraction of the CXCR4-transfected BMSCs toward the SDF-1α gradient down to a depth of 800 μm. It was shown that SDF-1α can also attract monocytes, which are the precursors of mature osteoclasts. This means that the SDF-1α–CXCR4 system is a promising chemoattractor couple for *in situ* bone tissue engineering.

In a similar way, chemotactic interaction can be induced by noncollagenous proteins as osteocalcin or osteopontin that we became acquainted with in connection with the control of HAP growth in Section 6.2.8.3. These two proteins not only possess specific amino acid motifs that cause a selective adsorption at specific lattice planes of the HAP nanocrystals but also present motifs interacting with specific bone cells. For example, osteocalcin, specifically binding on the (100) or (110) surface of HAP, has the peptide sequence (Phe—Tyr—Gly—Pro—Val) at its C-terminus, which induces a chemotactic interaction with monocytes as the precursor cells of osteoclasts (see Section 6.2.8.2). Osteopontin inhibiting HAP growth due to preferred binding of polyasparagine sequences at (100) surfaces of HAP has an additional binding motif (Gly—Arg—Gly—Asp—Ser) that interacts with osteoclasts via the $\alpha_V\beta_3$, $\alpha_8\beta_1$, or $\alpha_9\beta_1$ integrins. Therefore, the two proteins promote the remodeling of HAP–collagen composites through chemoattraction.

Immobilization of Osteogenic Factors   Porous mineralized collagen materials are favored matrices for the synthesis of bone tissue starting with hMSCs. For *in vitro* engineering, the scaffold has to be furnished with osteogenic effectors that induce differentiation of the hMSCs into active osteoblasts. For *in vivo* functionalization of the predesigned scaffold, one has to make use of the above-mentioned "catcher" properties. Following the general concept already explained, this can be realized by adsorption of soluble factors as cytokines or synthetic chemicals on or within the scaffold. Both components of the scaffold, the collagen filaments and the HAP nanocrystals, generate weak interactions (electrostatic and polar interactions, as well as hydrogen bonds) for immobilization of soluble factors. A further improvement of the binding efficiency can be achieved by adding linkers such as the GAGs already mentioned.

Evidence for the osteogenic differentiation of BMSCs is given in the work by Bernhardt *et al.* (2008) and Bernhardt *et al.* (2009), where the above approach is applied to 2D (tape) and 3D (cylinder) scaffolds. First, MSCs were isolated from bone marrow of healthy donors and expanded in a growth medium containing 10% fetal calf serum (FCS). Among other proteins, the FCS contains cell adhesion-promoting proteins, such as fibronectin and vitronectin. The mineralized collagen scaffolds are able to bind considerable amounts of serum proteins (Figure 6.92a). The adsorption characteristics show a continuous uptake of protein over the whole incubation time of 45 days. No repletion of the samples was observed. The absorbed protein layer promotes the conditions for cell adhesion and proliferation. The experiment demonstrates that the mineralized collagen acts as a highly effective "catcher" structure binding proteins from the surrounding media. Interestingly, there is also a continuous uptake of Ca^{2+} ions from the solution. As the HAP nanoparticles precipitated in the collagen–HAP scaffold have a calcium deficit, the scaffold acts as calcium sink. Furthermore, it has to be assumed that the simultaneously absorbed proteins bind the Ca^{2+} ions and act as nucleation sites for the growth of mineral phases.

The differentiation of stem cells *in vitro* requires specific growth media with chemical factors, growth factors, and hormones. BMSCs are multipotent, which

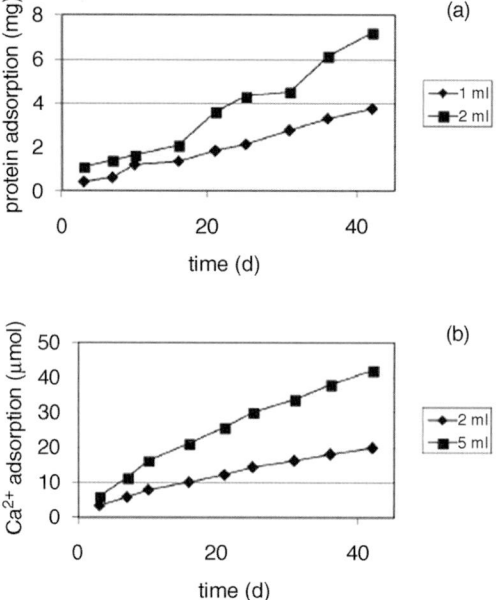

Figure 6.92 Adsorption characteristics of 3D collagen–HAP scaffolds incubated in the growth medium containing 10% fetal calf serum. (a) Protein adsorption. (b) Ca^{2+} adsorption. Scaffolds from mineralized collagen (10 mm diameter, 5 mm height) were incubated in 2 and 5 ml growth media. (Reproduced with permission from Bernhardt *et al.* (2009), Copyright 2012, John Wiley & Sons, Inc.)

means they are capable of self-replication and differentiation along various cell lineages. In the particular example, a combination of synthetic dexamethasone, β-glycerophosphate and L-ascorbic acid 2-phosphate [noted in the following as osteogenic supplement (OS)] was applied. The use of other growth media has been reported, for example, those containing the PTH (see also Section 6.2.8.2), the BMP-6, or the hepatocyte growth factor (HGF). As seen in the SEM micrographs of hBMSCs on tapes of mineralized collagen, the number of cells grows rapidly within a few days, and after 28 days the tapes were covered with a dense cell layer (Figure 6.93).

Microscopic evaluation of osteogenically induced hBMCS on a porous mineralized 3D scaffold has shown that the pore surfaces with time were covered with a thick cell layer. The differentiation of the cells toward the osteoblastic lineage can be followed by detection of characteristic osteogenic proteins, which are expressed in the different stages of differentiation. ALP is a relatively early marker of osteogenic differentiation being expressed during ECM maturation. Figure 6.94 shows the ALP expression at different time points of cell cultivation. The ALP concentration increases after stimulation of the osteogenic differentiation (OS+) in the 2D and 3D scaffolds. The differentiation has been confirmed by the detection of the upregulation of bone sialoprotein (BSP II) (Figure 6.95). BSP II is composed of various functional domains, including a hydrophobic collagen binding domain, a HAP-nucleating region of

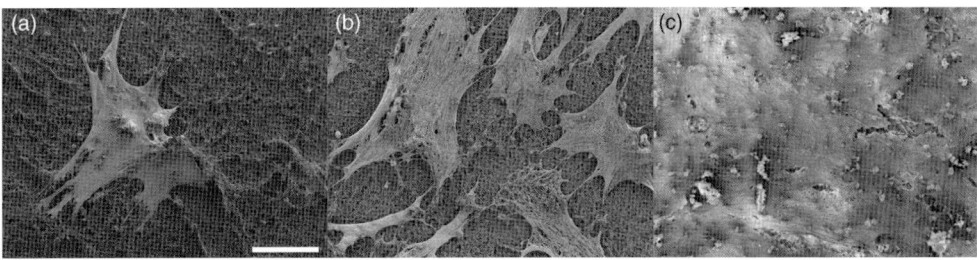

Figure 6.93 Scanning electron micrograph of hBMSC on tapes made of mineralized collagen cultivated for 1 day (a), 6 days (b), and 28 days (c) under osteogenic stimulation with dexamethasone, β-glycerophosphate, and ascorbate. Scale bar: 40 μm in all three images. (Reproduced with permission from Bernhardt et al. (2009), Copyright 2012, John Wiley & Sons, Inc.)

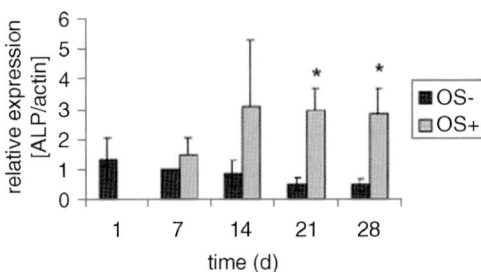

Figure 6.94 Specific ALP expression of induced (OS+) and noninduced (OS−) hBMSCs on 3D collagen–HAP scaffolds over 28 days. $n = 3$ (±standard deviation of the mean). Significant differences for OS+ at days 21 and 28, $^*p < 0.05$. (Reproduced with permission from Bernhardt et al. (2009), Copyright 2012, John Wiley & Sons, Inc.)

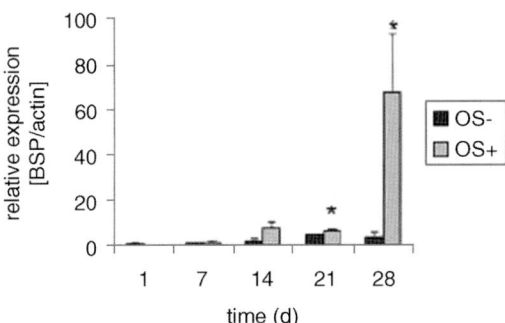

Figure 6.95 Specific BSP II expression of induced (OS+) and noninduced (OS−) hBMSCs on 3D collagen–HAP scaffolds over 28 days. $n = 3$ (±standard deviation of the mean). Significant differences for OS+ at days 21 and 28, $^*p < 0.05$. (Reproduced with permission from Bernhardt et al. (2009), Copyright 2012, John Wiley & Sons, Inc.)

glutamic acid residues, and an integrin-binding motif (Arg—Gly—Asp) near the C-terminus. The BSP II controls the HAP growth. It can act as a nucleation center of HAP crystals and is probably involved in the inhibition of crystal growth.

Remodeling by Tailored Cocultures   The original intention connected with tissue engineering of a bone implant is the combination of autologous (from the patient) MSCs with an artificial scaffold a few weeks before implantation with the aim to obtain a piece of bone tissue by *in vitro* cell cultivation. For this purpose, the differentiation of the MSCs has to be directed toward the osteoblastic lineage. Furthermore, active osteoclasts have to be generated by fusion of blood monocytes after implantation because the artificial scaffold can be transformed into a bone matrix only by the coordinated action of osteoclasts and osteoblasts. As a first step in this approach, an *in vitro* model has to be formed with a coculture of osteoblast and osteoclast precursor cells on a mineralized collagen scaffold. By adding appropriate effector molecules, the communication of the cells, the so-called cross talk, has to be tailored in such a way that natural bone is created by biodegradation of the artificial scaffold. The following example by Bernhardt *et al.* (2010) demonstrates essential process steps toward a realization:

i) *Selection of components*
    Tapes made from mineralized collagen have been applied as scaffolds. hMSCs from bone marrow and human peripheral blood monocytes have been chosen as precursor cells. The cell types were seeded on separate tapes.

ii) *Design of the coculture*
    The tapes were transferred into a well plate equipped with cell culture inserts of $0.4\,\mu m$ pore size. Tapes with seeded monocytes were placed in the lower cavities, and tapes with seeded hMSCs on top in the upper cavities. Thus, the different precursor cells were spatially separated. By this arrangement, a so-called indirect coculture was realized. The cells could communicate only by exchange of soluble cytokines. The main advantage of such a setup is the possibility to identify the relevant biochemical and structural changes of the different cell types. The growth medium of the cell culture was supplemented with soluble factors (dexamethasone, β-glycerophosphate, and vitamin D3) to induce osteogenic differentiation of hMSC into osteoblasts. Two cytokines were added for stimulation of the differentiation of the human monocytes into osteoclasts, the MCSF and the RANKL. As explained in Section 6.2.8.2, both factors are essential for the transformation of primary monocytes into active osteoclasts.

The promoting effect of the cell communication manifests itself in the higher proliferation rate of the osteogenically induced hMSCs (Figure 6.96a). A similar behavior was found with the specific ALP activity. The differentiation of the hMSCs into osteoblasts was accelerated by exchange of cytokines between the two cell types (Figure 6.96b). Scanning electron micrographs confirmed differences between the two growth regimes (Figure 6.97). The surface grown in coculture was rougher after a few days, but the difference vanished after a longer time. The tapes have become

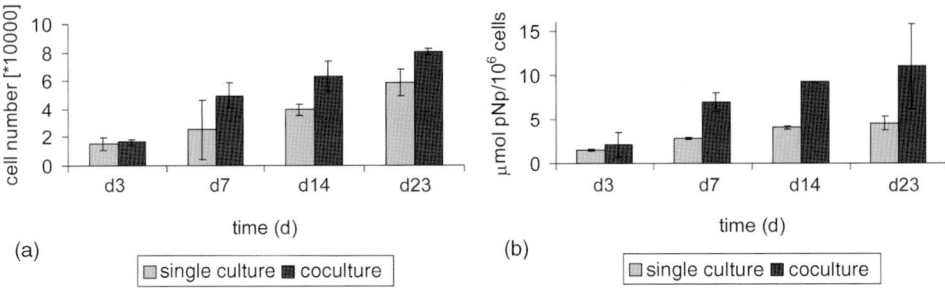

(a)

(b)

Figure 6.96 (a) Proliferation of osteogenically induced hMSC on mineralized collagen membrane in single culture and indirect coculture with human osteoclast-like cells. (b) Specific ALP activity of osteogenically induced hMSC on mineralized collagen membrane in single culture and indirect coculture with human osteoclast-like cells. (Reproduced with permission from Gelinsky et al. (2010), Copyright 2012, Springer Science+Business Media.)

covered by a nearly closed layer of elongated cells typical for active osteoblasts. Interestingly, the response of the osteoclastic cell type is different. Opening the communication path between the two cell types leads to a delayed differentiation into multinucleated osteoclasts. SEM revealed a lower density of multinucleated

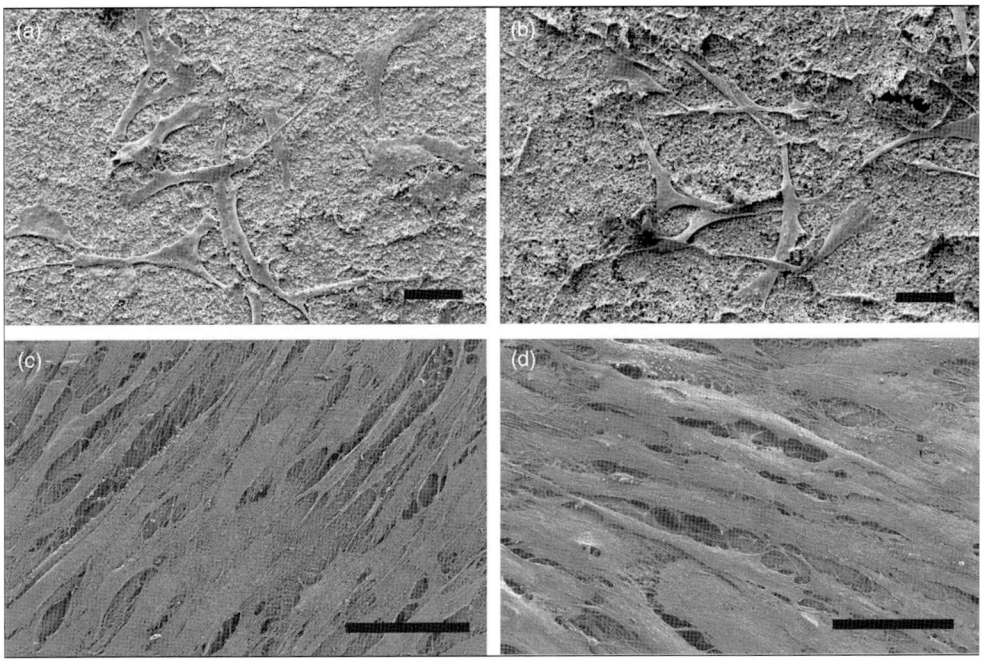

Figure 6.97 Scanning electron micrographs of osteogenically induced hMSCs on mineralized collagen tapes, cultivated separately (a and c) and in indirect contact with differentiating osteoclast-like cells (b and d) after 7 days of cultivation. Scale bars: 50 μm. (Reproduced with permission from Bernhardt et al. (2010), Copyright 2012, John Wiley & Sons, Inc.)

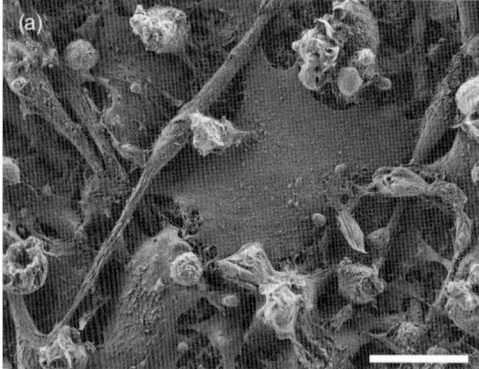
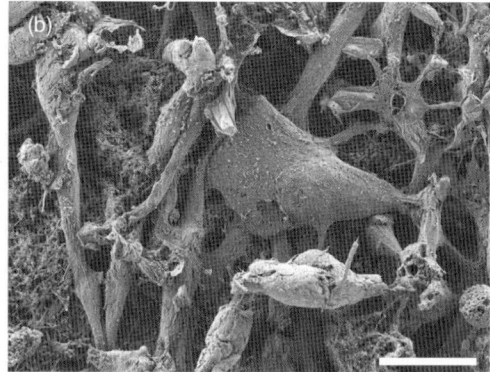

Figure 6.98 Scanning electron micrographs of osteoclast-like cells differentiated from monocytes on mineralized collagen tapes, cultivated separately (a) and in indirect contact with differentiating hMSCs (b) after 14 days of cultivation. Scale bars: 20 μm. (Reproduced with permission from Bernhardt *et al.* (2010), Copyright 2012, John Wiley & Sons, Inc.)

osteoclast-like cells in coculture (Figure 6.98). Experiments by Deyama *et al.* (2000) show that this is due to the production of extracellular phosphate, released by ALP during osteogenic differentiation.

The *in vitro* model worked out by Bernhardt *et al.* is a promising clue concerning the feasibility of bone tissue engineering by means of autologous cells. If this path could be followed until a practicable material for bone replacement is obtained, it would mean remarkable progress in regenerative bone therapy.

6.3.4
Microbial Carbonate Precipitation in Construction Materials

In civil engineering, construction materials as stone and concrete are usually exposed to the destructive influence of weathering. Accumulation of microdamage over a long time can impair the stability of large structures. Microdamage in living tissue typically does not accumulate, but gets repaired. It is well known, for instance, that microcracks in bone caused by overload trigger cellular activity closing the gaps. Such natural processes motivated materials engineers to search for innovative cost-efficient repair strategies in civil engineering. In particular, their interest was focused on microdamage of surface structures of buildings constructed with carbonate stones such as limestones, dolostones, and marbles, which deteriorated as a result of the progressive dissolution of the mineral matrix by weathering.

Traditional conservation treatments with water repellent agents or stone consolidants in order to regenerate the cohesion between grains in the damaged stone often lead to irreversible side effects accelerating the destruction of the stone due to the incompatibility of the agents with the microstructure. As an alternative approach, limewater (saturated solution of calcium hydroxide) can be applied with the aim to facilitate calcite formation. This had not been successful in an attempt to consolidate wall painting mortars. Insufficient penetration and the formation of soluble salts as reaction by-products limited the intended effect. Moreover, part of the small calcite crystallites grown by the treatment was not chemically bound to the internal structures and could not bridge the pores or microcracks in the damaged zones. At present, there is a promising new technique in the debate, based on microbially induced carbonate precipitation (MICP) (De Muynck, De Belle, and Verstraete, 2010). It is environment-friendly and well compatible with limestone. First examples of successful restoration of damaged ornamental stone have become known from several historic monuments in France, including Notre Dame de Paris.

As explained in the previous chapter, the key parameters of the carbonate precipitation are (i) the relative supersaturation $S_R = \{Ca^{2+}\} \cdot \{CO_3^{2-}\}/K_{sp,calcite}$ with the solubility product of calcite $K_{sp,calcite} = 4.8 \times 10^{-9}$, (ii) the pH, and (iii) the availability of nucleation sites. Microorganisms can produce high concentration of carbonate ions by bacterial metabolism (Section 6.2.5). They can generate an alkaline environment through physiological activities such as dissimilatory sulfate or nitrate reduction and by degradation of urea or uric acid. Furthermore, they offer nucleation sites as negatively charged groups at the cell wall, and they can create such sites also in the excreted EPS. The efficiency of the MICP process depends mainly on the so-called carbonatogenic yield of the biodeposition technique, which is the weight ratio of produced calcite and organic input. Highest yield, around 0.6, is obtained with *Bacillus cereus*. Various biodeposition techniques have been elaborated:

1) *The spraying of films of calcinogenic bacteria*
   In the first step, the surface to be protected is sprayed with a bacterial suspension culture. The deposited film has to be fed daily or every other day, typically restricted to five feedings. One alternative option is the use of dead cells or cellular fractions. This could give the advantage of avoiding uncontrolled bacterial growth.

2) *Exploiting microbial hydrolysis of urea*
   With the intention to speed up biomineralization, enzymatic hydrolysis of urea can be applied. It leads to an enhanced production of carbonate and an increase of the pH in the local environment of the microorganisms. The reaction equations of the intracellular hydrolysis shows that, in the total balance, one mole of urea yields two moles of ammonium ions and one mole of bicarbonate ions:

$$CO(NH_2)_2 + H_2O \rightarrow NH_2COOH + NH_3$$
$$NH_2COOH + H_2O \rightarrow NH_3 + H_2CO_3$$
$$2NH_3 + 2H_2O \leftrightarrow 2NH_4^+ + 2OH^-$$
$$2OH^- + H_2CO_3 \leftrightarrow CO_3^{2-} + 2H_2O$$
$$\text{(6.55)}$$
total reaction :
$$CO(NH_2)_2 + 2H_2O \rightarrow 2NH_4^+ + CO_3^{2-}.$$

In the presence of Ca^{2+} ions, local supersaturation can become high. Favored nucleation sites for the calcite deposition are the cell walls as they profusely offer negatively charged binding sites for the calcium ions. There are numerous bacteria strains that enzymatically degrade urea by expression of the enzyme urease. Among them, bacteria from the *Bacillus sphaericus* group show the highest proliferation rate.

3) *Cementitious materials*

For this technique, coatings consisting of a mixture of Portland mortar slurry, cell cultures of *B. sphaericus*, and an additional Ca source are prepared. As a Ca source, calcium acetate is favored over $CaCl_2$, avoiding chloride ions whose presence in the cement would be unfavorable. Variation of the Ca source leads to a change of the crystal morphology. Calcium chloride and calcium acetate favor the deposition of rhombohedral and spherulitic crystals, respectively. An interesting extension of this technique has been reported for the repair of larger cracks in concrete (De Muynck, De Belle, and Verstraete, 2010). In order to protect *B. sphaericus* from the alkaline cement matrix, the bacteria were embedded in a silica sol. A biocer precursor was prepared, including calcium chloride and urea as additional components.

4) *Self-healing concrete*

An innovative approach by Jonkers *et al.* (2010) aims at self-healing concrete to be obtained by embedding spore-forming microorganisms in the concrete matrix, where they are supposed to induce calcite precipitation in the wake of a propagating crack tip in the presence of humidity. To realize this idea, alkali-resistant spore-forming bacteria (e.g., *Bacillus pseudofirmus* DSM 8715 and *Bacillus cohnii* DSM 6307) were applied. Together with the spores, calcium lactate as a nutrient and calcium source was homogeneously distributed in the concrete matrix. Calcium lactate is converted by bacterial metabolism according to the reaction

$$CaC_6H_{10}O_6 + 6O_2 \rightarrow CaCO_3 + 5CO_2 + 5H_2O. \tag{6.56}$$

By reaction of the produced CO_2 with portlandite ($Ca(OH_2)$), an essential hydration product of cement, the amount of $CaCO_3$ is additionally increased:

$$5CO_2 + 5Ca(OH)_2 \rightarrow 5CaCO_3 + 5H_2O. \tag{6.57}$$

The expected effect is indicated in Figure 6.99. The bacterial spores enclosed in the cement matrix remain viable for a period up to 4 months. When a crack was formed, the increase of the humidity near to the crack wake causes the spores to germinate. The metabolic conversion of calcium lactate leads to precipitation of large calcium carbonate-based mineral particles bridging the gap. Decreasing viability with time after incorporation into the cement matrix limits the efficiency of the self-healing process. To overcome this drawback, further improvements have been worked out. Enclosing the bacteria and nutrients in clay capsules extended the viability considerably in lab tests. Now it is assumed that the encapsulated spores could germinate after lying dormant in the concrete structures for up to 50 years.

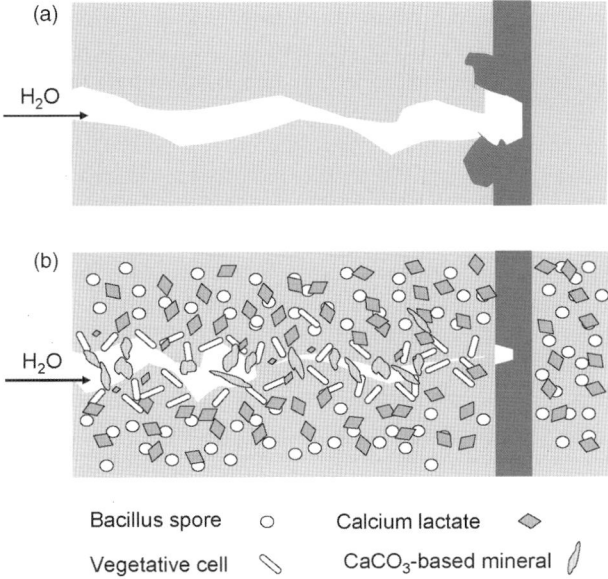

Bacillus spore ○   Calcium lactate ◇

Vegetative cell ⬭   CaCO₃-based mineral

Figure 6.99 Schematic drawing of conventional concrete (a) versus bacteria-based self-healing armored concrete (b). Crack ingress chemicals degrade the concrete matrix and accelerate corrosion of the reinforcement (a). The bacteria-based healing process activated by ingress water seals and prevents further cracking (b). (Adapted from De Muynck, De Belle, and Verstraete (2010).)

6.3.5
The Potential of Biomineralization for Carbon Capture and Storage (CCS)

The reduction of carbon dioxide emission is one of the major challenges of this century. One of the variety of scenarios proposed to meet this challenge is the search for strategies to develop efficient techniques for capturing the so-called point source CO_2 from power plants or other big industrial complexes. Injection of concentrated CO_2 into underground storages could be one solution of the problem. However, there are still uncertainties concerning longtime safety. Therefore, biological carbon sequestration suggests itself as a potential solution and has been included into the discussion (Jansson and Northen, 2010). In addition to reforestation, accelerated biomineralization is considered as a promising option. The basic idea is derived from the global calcification caused by microorganisms, in particular cyanobacteria in the oceans. As explained in Section 6.2.9, cyanobacteria can efficiently use solar energy to immobilize CO_2 as calcium carbonate. The CCM is an essential part of this biogenic calcification.

Halophilic cyanobacteria cultured in seawater or brine could provide a bio-based approach to CCS. Proposed calcium sources are agricultural drainage water, saline water produced from petroleum production, or calcium from gypsum or silicate minerals in connection with biologically accelerated weathering. There are still problems concerning the biological response of the cyanobacteria to the high CO_2 concentration from industrial sources. It can be assumed that with high CO_2 concentration, the cellular CCM can be neglected in comparison to diffusive transport. The conversion of CO_2 during the transport into the cell as well as its high concentration in the surrounding aqueous medium increases the H^+ concentration that finally impedes calcification. Therefore, research projects have been initiated to exploit genetic engineering of mutant cells that promote the uptake of HCO_3^- and inactivate the CO_2 uptake capacity. Alternatively, also porous materials could act as carriers for CA-like proteins that catalyze the conversion of the incoming CO_2 into HCO_3^- before it comes in contact with the cyanobacteria (Jansson and Northen, 2010b). It is obvious that such biomimetic solutions could be realized on a limited industrial scale only. Therefore, such research is focused on future niche applications of CCS such as for small-scale field power plants, municipal solid waste combustion, cement plants, and steel production.

Tasks T6

Task 6.1: The concentration of a solute in a real solution in equilibrium near the planar interface of a large precipitate is assumed to have the value $X_\infty = \exp\left(-(\Delta\mu^* + \Omega)/k_B T\right)$. Give the expression for the concentration in the solution near the interface of a small spherical precipitate with radius r. Derive from that formula an estimate of the concentration for higher temperatures. Calculate the increase of the concentration for a spherical calcium carbonate precipitate with a radius $r = 10\,\text{nm}$ compared to the value near the planar interface at a temperature of 500 K assuming for $\gamma_{sl} = 0.2\,\text{J/m}^2$ and for the molar volume $v_s = 10^{-5}\,\text{m}^3$.

Task 6.2: Derive the approximation for S_R given in Eq. (6.29) for dilute solutions. In this case, the activities very often can be expressed by the ion concentrations introducing the activity coefficients $\{A^{q+}\} = [A^{q+}] \cdot \gamma_A$, $\{B^{p-}\} = [B^{p-}] \cdot \gamma_B$. What has to be assumed for the activity coefficients in order that Eq. (6.29) is fulfilled?

Task 6.3: You have the task to prepare nanoparticles by homogeneous nucleation in an aqueous solution. The solubility product of the solute is $\log K_{sp} = -11$. How large should be the relative solubility S_R, which is needed to produce precipitates at room temperature with stationary nucleation rate of about $J_{hom} = 100\,\text{mM/s}$. For the aggregation frequency, we assume $f_0 = 10^{10}\,\text{s}^{-1}$. The molar volume of the precipitates should be $v_s = 10^{-5}\,\text{m}^3$ and their interface energy $\gamma_{sl} = 0.2\,\text{J/m}^2$.

Task 6.4: Equation (6.36) for the change of the free enthalpy ΔG_{het} during heterogeneous nucleation,

$$\Delta G_{het} = -\Delta\mu N_s + \gamma_{sl} A_{sl} + \gamma_{st} A_{st} - \gamma_{tl} A_{st}, \tag{6.36}$$

can be rewritten as

$$\Delta G_{het} = \left(-\frac{4\pi N_a}{3 v_s} r^3 \Delta\mu + 4\pi r^2 \gamma_{sl}\right) S(\theta), \tag{6.37}$$

with

$$S(\theta) = (2 + \cos\theta)(1 - \cos\theta)^2/4. \tag{6.38}$$

Equation (6.37) is similar to the corresponding one for homogeneous nucleation. Derive Eqs. (6.37) and (6.38) by using the following relationships for a spherical cap:

volume $\quad V_S = \pi r^3(2 + \cos\theta)(1 - \cos\theta)^2/3$

interfaces $\quad A_{sl} = 2\pi r^2(1 - \cos\theta), \quad A_{st} = \pi r^2 \sin^2\theta.$

The contact angle θ is given by the equilibrium condition for the interface energies γ_{sl}, γ_{st}, and γ_{tl}:

$$\gamma_{tl} = \gamma_{st} + \gamma_{sl}\cos\theta. \tag{6.35}$$

Task 6.5: As described in Section 6.2.4.1, the dependence of the aspect ratio of growing ZnO crystals on the peptide concentration c_p in the growth solution can be explained by the average coverage ϕ of the c surface with the peptide in equilibrium. The fraction ϕ of occupied sites on the c plane can be modeled by a Langmuir adsorption isotherm (Eq. (6.40)). Derive the Langmuir adsorption isotherm as the equilibrium solution of a linear reaction-controlled deposition of a sorbate monolayer.

References

Addadi, L. and Weiner, S. (1985) Interactions between acidic proteins and crystals: stereochemical requirements in biomineralization. *Proceedings of the National Academy of Sciences of the United States of America*, **82** (12), 4110–4114.

Amdursky, N., Molotskii, M., Gazit, E., and Rosenman, G. (2010) Elementary building blocks of self-assembled peptide nanotubes. *Journal of the American Chemical Society*, **132**, 15632–15636.

Anderson, H.C., Garimella, R. et al. (2005) The role of matrix vesicles in growth plate development and biomineralization. *Frontiers in Bioscience: A Journal and Virtual Library*, **10**, 822–837.

Baeuerlein, E. (ed.) (2000) *Biomineralization: From Biology to Biotechnology and Medical Application*, Wiley-VCH Verlag GmbH, Weinheim.

Banfield, J.F., Welch, S.A. et al. (2000) Aggregation-based crystal growth and microstructure development in natural iron oxyhydroxide biomineralization products. *Science*, **289** (5480), 751–754.

Bazylinski, D.A. and Frankel, R.B. (2004) Magnetosome formation in prokaryotes. *Nature Reviews Microbiology*, **2** (3), 217–230.

Bazylinski, D.A., Garratt-Reed, A.J. et al. (1994) Electron microscopic studies of magnetosomes in magnetotactic bacteria. *Microscopy Research and Technique*, **27** (5), 389–401.

Bernhardt, A., Lode, A. et al. (2008) Mineralised collagen–an artificial, extracellular bone matrix–improves osteogenic differentiation of bone marrow stromal cells. *Journal of Materials Science: Materials in Medicine*, **19** (1), 269–275.

Bernhardt, A., Lode, A., Mietrach, C., Hempel, U., Hanke, T., and Gelinsky, M. (2009) *In vitro* osteogenic potential of human bone marrow stromal cells cultivated in porous scaffolds from mineralized collagen. *Journal of Biomedical Materials Research Part A*, **90**, 852–862.

Bernhardt, A., Thieme, S. *et al.* (2010) Crosstalk of osteoblast and osteoclast precursors on mineralized collagen: towards an *in vitro* model for bone remodeling. *Journal of Biomedical Materials Research Part A*, **95** (3), 848–856.

Bonthrone, K.M., Basnakova, G. *et al.* (1996) Bioaccumulation of nickel by intercalation into polycrystalline hydrogen uranyl phosphate deposited via an enzymatic mechanism. *Nature Biotechnology*, **14** (5), 635–638.

Boskey, A.L. (1989) Non-collagenous matrix proteins and their role in mineralization. *Bone and Mineral*, **6** (2), 111–123.

Bradt, J.-H., Mertig, M., Teresiak, A., and Pompe, W. (1999) Biomimetic mineralization of collagen by combined fibril assembly and calcium phosphate formation. *Chemistry of Materials*, **11** (10), 2694–2701.

Brickmann, J., Paparcone, R. *et al.* (2010) Fluorapatite–gelatine nanocomposite superstructures: new insights into a biomimetic system of high complexity. *Chemphyschem*, **11** (9), 1851–1853.

Brunner, E., Richthammer, P. *et al.* (2009) Chitin-based organic networks: an integral part of cell wall biosilica in the diatom *Thalassiosira pseudonana*. *Angewandte Chemie: International Edition in English*, **48** (51), 9724–9727.

Chou, K.C. (1992) Energy-optimized structure of antifreeze protein and its binding mechanism. *Journal of Molecular Biology*, **223** (2), 509–517.

Cölfen, H. and Antonietti, M. (2005) Mesocrystals: inorganic superstructures made by highly parallel crystallization and controlled alignment. *Angewandte Chemie: International Edition in English*, **44** (35), 5576–5591.

Cölfen, H. and Mann, S. (2003) Higher-order organization by mesoscale self-assembly and transformation of hybrid nanostructures. *Angewandte Chemie: International Edition in English*, **42** (21), 2350–2365.

Cölfen, H. and Qi, L. (2001) A systematic examination of the morphogenesis of calcium carbonate in the presence of a double-hydrophilic block copolymer. *Chemistry*, **7** (1), 106–116.

Colombi Ciacchi, L. (2002) *Growth of platinum clusters in solution and on biopolymers: the microscopic mechanisms.* PhD thesis. Technische Universität, Dresden.

Colombi Ciacchi, L., Pompe, W. *et al.* (2001) Initial nucleation of platinum clusters after reduction of K(2)PtCl(4) in aqueous solution: a first principles study. *Journal of the American Chemical Society*, **123** (30), 7371–7380.

Colombi Ciacchi, L., Pompe, W., and De Vita, A. (2003) Growth of platinum clusters via addition of Pt(II) complexes: a first principles investigation. *The Journal of Physical Chemistry B*, **107**, 1755–1764.

Crawford, R.M. and Schmid, A.-M.M. (1986) *Biomineralization in Lower Plants and Animals*, Oxford University Press.

Creamer, N.J., Mikheenko, I.P. *et al.* (2011) Local magnetism in palladium bionanomaterials probed by muon spectroscopy. *Biotechnology Letters*, **33** (5), 969–976.

De Muynck, W., De Belie, N., and Verstraete, W. (2010) Microbial carbonate precipitation in construction materials: a review. *Ecological Engineering*, **36**, 118–136.

Deplanche, K., Caldelari, I. *et al.* (2010) Involvement of hydrogenases in the formation of highly catalytic Pd(0) nanoparticles by bioreduction of Pd(II) using *Escherichia coli* mutant strains. *Microbiology*, **156** (Part 9), 2630–2640.

Deplanche, K., Murray, A. *et al.* (2011) Biorecycling of precious metals and rare earth elements, in *Nanomaterials* (ed. M.M. Rahman), InTech.

Despang, F., Börner, A., Dittrich, R., Tomandl, G., Pompe, W., and Gelinsky, M. (2005) Alginate/calcium phosphate scaffolds with oriented, tube-like pores. *Materialwissenschaft und Werkstofftechnik*, **36**, 761–767.

Despang, F., Dittrich, R., Bernhardt, A., Hanke, Th., Tomandl, G., and Gelinsky, M. (2008) Channel-like pores in HAP-containing scaffolds for bone engineering: hydrogels versus ceramics. *Tissue Engineering Part A*, **14**, 819.

Despang, F., Dittrich, R., and Gelinsky, M. (2011) *Novel Biomaterials with Parallel Aligned Pore Channels by Ionotropic Gelation of Alginate: Mimicking the Anisotropic Structure of Bone Tissue*, InTech, Rijeka.

Deyama, Y., Takeyama, S. *et al.* (2000) Osteoblast maturation suppressed osteoclastogenesis in co-culture with bone marrow cells. *Biochemical and Biophysical Research Communications*, **274** (1), 249–254.

Dittrich, R., Tomandl, G., and Mangler, M. (2002) Preparation of Al_2O_3, TiO_2 and hydroxyapatite ceramics with pores similar to a honeycomb structure. *Advanced Engineering Materials*, **4**, 487–490.

Duman, J.G. (2001) Antifreeze and ice nucleator proteins in terrestrial arthropods. *Annual Review of Physiology*, **63**, 327–357.

Ehrlich, H. (2010) Chitin and collagen as universal and alternative templates in biomineralization. *International Geology Review*, **52** (7–8), 661–699.

Elbert, D.L. (2011) Liquid–liquid two-phase systems for the production of porous hydrogels and hydrogel microspheres for biomedical applications: a tutorial review. *Acta Biomaterialia*, **7** (1), 31–56.

Fratzl, P. and Weinkamer, R. (2007) Nature's hierarchical materials. *Progress in Materials Science*, **52**, 1263–1334.

Gajjeraman, S., He, G. *et al.* (2008) Biological assemblies provide novel templates for the synthesis of hierarchical structures and facilitate cell adhesion. *Advanced Functional Materials*, **18** (24), 3972–3980.

Gelinsky, M., Eckert, M., and Despang, F. (2007) Biphasic, but monolithic scaffolds for the therapy of osteochondral defects. *International Journal of Materials Research*, **98** (8), 749–755.

Gelinsky, M., Lode, A., Bernhardt, A., and Rösen-Wolff, A. (2010) *Stem Cell Engineering*, Springer, Berlin.

Gelinsky, M., Welzel, P.B., Simon, P., Bernhardt, A., and König, U. (2008) Porous three-dimensional scaffolds made of mineralised collagen: preparation and properties of a biomimetic nanocomposite material for tissue engineering of bone. *Chemical Engineering Journal*, **137**, 84–96.

George, A. and Huang, Ch-.Ch. (eds) (2011) Bioinspired strategies for hard tissue regeneration, in *Advances in Biomimetics*, InTech, Rijeka.

Gerstenfeld, L.C., Lian, J.B. *et al.* (1989) Use of cultured embryonic chicken osteoblasts as a model of cellular differentiation and bone mineralization. *Connective Tissue Research*, **21** (1–4), 215–223, discussion 224–225.

Gower, L.A. (1997) *The Influence of Polyaspartate Additive on the Growth and Morphology of Calcium Carbonate Crystals*. University of Massachusetts at Amherst, p. 119.

Granier, T., Langlois d'Estaintot, B. *et al.* (2003) Structural description of the active sites of mouse L-chain ferritin at 1.2 A resolution. *Journal of Biological Inorganic Chemistry*, **8** (1–2), 105–111.

Gupta, H.S., Fratzl, P. *et al.* (2007) Evidence for an elementary process in bone plasticity with an activation enthalpy of 1 eV. *Journal of the Royal Society Interface*, **4** (13), 277–282.

Gupta, H.S., Wagermaier, W. *et al.* (2005) Nanoscale deformation mechanisms in bone. *Nano Letters*, **5** (10), 2108–2111.

Hempstead, P.D., Yewdall, S.J. *et al.* (1997) Comparison of the three-dimensional structures of recombinant human H and horse L ferritins at high resolution. *Journal of Molecular Biology*, **268** (2), 424–448.

Henglein, A., Ershov, B.G., and Malow, M. (1995) Absorption spectrum and some chemical reactions of colloidal platinum in aqueous solution. *The Journal of Physical Chemistry*, **99**, 14129–14136.

Hildebrand, M. (2008) Diatoms, biomineralization processes, and genomics. *Chemical Reviews*, **108** (11), 4855–4874.

Hintze, V., Moeller, S. *et al.* (2009) Modifications of hyaluronan influence the interaction with human bone morphogenetic protein-4 (hBMP-4). *Biomacromolecules*, **10** (12), 3290–3297.

Hoang, Q.Q., Sicheri, F. *et al.* (2003) Bone recognition mechanism of porcine osteocalcin from crystal structure. *Nature*, **425** (6961), 977–980.

Hunter, G.K. and Goldberg, H.A. (1994) Modulation of crystal formation by bone phosphoproteins: role of glutamic acid-rich sequences in the nucleation of hydroxyapatite by bone sialoprotein. *The Biochemical Journal*, **302** (Part 1), 175–179.

Hunter, G.K., Kyle, C.L. *et al.* (1994) Modulation of crystal formation by bone phosphoproteins: structural specificity of the osteopontin-mediated inhibition of hydroxyapatite formation. *The Biochemical Journal*, **300** (Part 3), 723–728.

Hunter, G.K., O'Young, J. *et al.* (2010) The flexible polyelectrolyte hypothesis of protein–biomineral interaction. *Langmuir*, **26** (24), 18639–18646.

Ilari, A., Stefanini, S. *et al.* (2000) The dodecameric ferritin from *Listeria innocua* contains a novel intersubunit iron-binding site. *Nature Structural Biology*, **7** (1), 38–43.

Jansson, C. and Northen, T. (2010) Calcifying cyanobacteria: the potential of biomineralization for carbon capture and storage. *Current Opinion in Biotechnology*, **21** (3), 365–371.

Jonkers, H., Thijssen, A., Muyzer, G., Copuroglu, O., and Schlangen, E. (2010) Application of bacteria as self-healing agent for the development of sustainable concrete. *Ecological Engineering*, **36**, 230–235.

Kamat, S., Kessler, H., Ballarini, R., Nassirou, M., and Heuer, A.H. (2004) Fracture mechanisms of the *Strombus gigas* conch shell: II – micromechanics analyses of multiple cracking and large-scale crack bridging. *Acta Materialia*, **52**, 2395–2406.

Kamat, S., Su, X. *et al.* (2000) Structural basis for the fracture toughness of the shell of the conch *Strombus gigas*. *Nature*, **405** (6790), 1036–1040.

Kawska, A., Hochrein, O. *et al.* (2008) The nucleation mechanism of fluorapatite–collagen composites: ion association and motif control by collagen proteins. *Angewandte Chemie: International Edition in English*, **47** (27), 4982–4985.

Kniep, R. and Busch, S. (1996) Biomimetic growth and self-assembly of fluorapatite aggregates by diffusion into denatured collagen matrices. *Angewandte Chemie: International Edition in English*, **35**, 2624–2626.

Kniep, R. and Simon, P. (2008) "Hidden" hierarchy of microfibrils within 3D-periodic fluorapatite–gelatine nanocomposites: development of complexity and form in a biomimetic system. *Angewandte Chemie: International Edition in English*, **47** (8), 1405–1409.

Kollmann, T., Simon, P., Carrillo-Cabrera, W., Braunbarth, C., Poth, T., Rosseeva, E.V., and Kniep, R. (2010) Calcium phosphate–gelatin nanocomposites: bulk preparation (shape- and phase-control), characterization, and application as dentine repair material. *Chemistry of Materials*, **22**, 5137–5153.

Korobushkina, E.D. and Korobushkin, I.M. (1986) Interaction of gold with bacteria and formation of new gold. *Doklady Akademii Nauk SSSR*, **287**, 978–982.

Kuhn-Spearing, L.T., Kessler, H., Chateau, E., Ballarini, R., and Heuer, A.H. (1996) Fracture mechanisms of the *Strombus gigas* conch shell: implications for the design of brittle laminates. *Journal of Materials Science*, **31**, 6583–6594.

Landis, W.J. and Silver, F.H. (2009) Mineral deposition in the extracellular matrices of vertebrate tissues: identification of possible apatite nucleation sites on type I collagen. *Cells, Tissues, Organs*, **189** (1–4), 20–24.

Landis, W.J., Song, M.J. *et al.* (1993) Mineral and organic matrix interaction in normally calcifying tendon visualized in three dimensions by high-voltage electron microscopic tomography and graphic image reconstruction. *Journal of Structural Biology*, **110** (1), 39–54.

Lang, S.B. (1966) Pyroelectric effect in bone and tendon. *Nature*, **212**, 704–705.

Lemanov, V.V. (2000) Piezoelectric and pyroelectric properties of protein amino acids as basic materials of soft state physics. *Ferroelectrics*, **238** (1), 211–218.

Lloyd, J.R., Anderson, R.T., and Macaskie, L.E. (2005) Bioremediation of metals and radionuclides, in *Bioremediation* (eds R. Atlas and J. Philp), ASM Press, Washington DC, pp. 293–317.

Lloyd, J.R., Pearce, C.I. *et al.* (2008) Biomineralization: linking the fossil record to the production of high value functional materials. *Geobiology*, **6** (3), 285–297.

Lloyd, J.R., Yong, P. *et al.* (1998) Enzymatic recovery of elemental palladium by using sulfate-reducing bacteria. *Applied and Environmental Microbiology*, **64** (11), 4607–4609.

Lovley, D.R. (2003) Cleaning up with genomics: applying molecular biology to bioremediation. *Nature Reviews Microbiology*, **1** (1), 35–44.

Lutolf, M., George, R., Zisch, A., Tirelli, N., and Hubbell, J. (2003a) Cell-responsive synthetic hydrogels. *Advanced Materials*, **15** (11), 888–892.

Lutolf, M.P. and Hubbell, J.A. (2005) Synthetic biomaterials as instructive extracellular microenvironments for morphogenesis in tissue engineering. *Nature Biotechnology*, **23** (1), 47–55.

Lutolf, M.P., Weber, F.E. *et al.* (2003b) Repair of bone defects using synthetic mimetics of collagenous extracellular matrices. *Nature Biotechnology*, **21** (5), 513–518.

Macaskie, L.E., Baxter-Plant, V.S. *et al.* (2005) Applications of bacterial hydrogenases in waste decontamination, manufacture of novel bionanocatalysts and in sustainable energy. *Biochemical Society Transactions*, **33** (Part 1), 76–79.

Macaskie, L.E., Mikheenko, I.P. *et al.* (2011) Today's wastes, tomorrow's materials for environmental protection, in *Comprehensive Biotechnology, Volume 6: Environmental Biotechnology and Safety* (eds B. Moo-Young*et al.*), Pergamon, New York, pp. 719–725.

Mahamid, J., Addadi, L. *et al.* (2011) Crystallization pathways in bone. *Cells, Tissues, Organs*, **194** (2–4), 92–97.

Maneval, J.E., Bernin, D. *et al.* (2011) Magnetic resonance analysis of capillary formation reaction front dynamics in alginate gels. *Magnetic Resonance in Chemistry*, **49** (10), 627–640.

Mann, S. (ed.) (1996) *Biomimetic Materials Chemistry*, Wiley-VCH Verlag GmbH, New York.

Mann, S. (2000) The chemistry of form. *Angewandte Chemie: International Edition in English*, **39** (19), 3392–3406.

Mann, S. (2001) *Biomineralization: Principles and Concepts in Bioinorganic Materials Chemistry*, Oxford University Press, Oxford.

Mann, S. (2009) Self-assembly and transformation of hybrid nano-objects and nanostructures under equilibrium and non-equilibrium conditions. *Nature Materials*, **8** (10), 781–792.

Mann, S., Archibald, D.D. *et al.* (1993) Crystallization at inorganic–organic interfaces: biominerals and biomimetic synthesis. *Science*, **261** (5126), 1286–1292.

Martinez, R.J., Beazley, M.J. *et al.* (2007) Aerobic uranium (VI) bioprecipitation by metal-resistant bacteria isolated from radionuclide- and metal-contaminated subsurface soils. *Environmental Microbiology*, **9** (12), 3122–3133.

Mennan, C., Paterson-Beedle, M. *et al.* (2010) Accumulation of zirconium phosphate by a *Serratia* sp.: a benign system for the removal of radionuclides from aqueous flows. *Biotechnology Letters*, **32** (10), 1419–1427.

Mikheenko, I.P., Mikheenko, P.M. *et al.* (2001) Magnetic testing of Pd loaded bacteria, in *Biohydrometallurgy: Fundamentals, Technology and Sustainable Development. Part B: Biosorption and Bioremediation* (eds V.S.T. Ciminelli and O. GarciaJr.), Elsevier, New York, pp. 681–688.

Mikheenko, I.P., Rousset, M. *et al.* (2008) Bioaccumulation of palladium by *Desulfovibrio fructosivorans* wild-type and hydrogenase-deficient strains. *Applied and Environmental Microbiology*, **74** (19), 6144–6146.

Mkandawire, M. and Dudel, E.G. (2009) Uranium in the water of abandoned uranium mines: ecotoxicology and bioremidiation implications, in *Uranium: Compounds, Isotopes and*

Applications (ed. H.G. Wolfe), Nova Science Publishers, New York.

Muthukumar, M. (2009) Theory of competitive adsorption–nucleation in polypeptide-mediated biomineralization. *Journal of Chemical Physics*, **130** (16), 161101.

Navrotsky, A. (2004) Energetic clues to pathways to biomineralization: precursors, clusters, and nanoparticles. *Proceedings of the National Academy of Sciences of the United States of America*, **101** (33), 12096–12101.

Nelson, D.L. and Cox, M.M. (2008) *Lehninger: Principles of Biochemistry*, W.H. Freeman and Company, New York.

Nudelman, F., Pieterse, K. *et al.* (2010) The role of collagen in bone apatite formation in the presence of hydroxyapatite nucleation inhibitors. *Nature Materials*, **9** (12), 1004–1009.

Ogasawara, W., Shenton, W., Davis, S.A., and Mann, S. (2000) Template mineralization of ordered macroporous chitin–silica composites using a cuttlebone-derived organic matrix. *Chemistry of Materials*, **12**, 2835–2837.

Olszta, M.J., Cheng, X. *et al.* (2007) Bone structure and formation: a new perspective. *Materials Science and Engineering R*, **58**, 77–116.

Onuki, A. and Puri, S. (1999) Spinodal decomposition in gels. *Physical Review E*, **59** (2), R1331–R1334.

Ottani, V., Martini, D. *et al.* (2002) Hierarchical structures in fibrillar collagens. *Micron*, **33** (7–8), 587–596.

Paparcone, R., Kniep, R., and Brickmann, J. (2009) Hierarchical pattern of microfibrils in a 3D fluorapatite–gelatine nanocomposite: simulation of a bio-related structure building process. *Physical Chemistry Chemical Physics*, **11**, 2186–2194.

Paterson-Beedle, M., Macaskie, L.E. *et al.* (2006) Utilisation of a hydrogen uranyl phosphate-based ion exchanger supported on a biofilm for the removal of cobalt, strontium and caesium from aqueous solutions. *Hydrometallurgy*, **83**, 141–145.

Peterlik, H., Roschger, P. *et al.* (2006) From brittle to ductile fracture of bone. *Nature Materials*, **5** (1), 52–55.

Pivonka, P., Zimak, J. *et al.* (2008) Model structure and control of bone remodeling: a theoretical study. *Bone*, **43** (2), 249–263.

Porter, D.A. and Easterling, K.E. (1992) *Phase Transformations in Metals and Alloys*, Chapman & Hall, London.

Prymak, O., Sokolova, V., Peitsch, T., and Epple, M. (2006) The crystallization of fluoroapatite dumbbells from supersaturated aqueous solution. *Crystal Growth & Design*, **6** (2), 498–506.

Raiteri, P. and Gale, J.D. (2010) Water is the key to nonclassical nucleation of amorphous calcium carbonate. *Journal of the American Chemical Society*, **132** (49), 17623–17634.

Raz, S., Weiner, S., and Addadi, L. (2000) Formation of high-magnesium calcites via an amorphous precursor phase: possible biological implications. *Advanced Materials*, **12**, 38–42.

Robinson, D.H. and Sullivan, C.W. (1987) How do diatoms make silicon biominerals. *TIBS*, **12**, 151–154.

Robling, A.G., Castillo, A.B. *et al.* (2006) Biomechanical and molecular regulation of bone remodeling. *Annual Review of Biomedical Engineering*, **8**, 455–498.

Roh, J.Y., Kuhn-Spearing, L. *et al.* (1998) Mechanical properties and the hierarchical structure of bone. *Medical Engineering and Physics*, **20**, 92–102.

Sengupta, D. and Heilshorn, S.C. (2010) Protein-engineered biomaterials: highly tunable tissue engineering scaffolds. *Tissue Engineering: Part B*, **16** (3), 285–293.

Simon, P., Carrillo-Cabrera, W., Formanek, P., Göbel, C., Geiger, D., Ramlau, R., Tlatlik, H., Budera, J., and Kniep, R. (2004a) On the real-structure of biomimetically grown hexagonal prismatic seeds of fluorapatite–gelatine-composites: TEM investigations along [001]. *Journal of Materials Chemistry*, **14**, 2218–2224.

Simon, P., Lichte, H. *et al.* (2004b) Electron holography of non-stained bacterial surface layer proteins. *Biochimica et Biophysica Acta*, **1663** (1–2), 178–187.

Simon, P., Schwarz, U., and Kniep, R. (2005) Hierarchical architecture and real structure in biomimetic nano-composite of fluorapatite with gelatine: a model system for steps in dentino- and osteogenesis? *Journal of Materials Chemistry*, **15**, 4992–4996.

Simon, P., Zahn, D. *et al.* (2006) Intrinsic electric dipole fields and the induction of hierarchical form developments in fluorapatite–gelatine nanocomposites: a general principle for morphogenesis of biominerals? *Angewandte Chemie: International Edition in English*, **45** (12), 1911–1915.

Sumper, M. and Brunner, E. (2006) Learning from diatoms: nature's tools for the production of nanostructured silica. *Advanced Functional Materials*, **16**, 17–26.

Sumper, M. and Brunner, E. (2008) Silica biomineralization in diatoms: the model organism *Thalassiosira pseudonana*. *Chembiochem: A European Journal of Chemical Biology*, **9** (8), 1187–1194.

Taniyama, T., Ohata, E., and Sato, T. (1997) Ferromagnetism of Pd fine particles. *Physica B*, **237–238**, 286–288.

Thiele, H. and Awad, A. (1969) Nucleation and oriented crystallization apatite in ionotropic gels. *Journal of Biomedical Materials Research*, **3** (3), 431–441.

Thieme, S., Ryser, M. *et al.* (2009) Stromal cell-derived factor-1alpha-directed chemoattraction of transiently CXCR4-overexpressing bone marrow stromal cells into functionalized three-dimensional biomimetic scaffolds. *Tissue Engineering: Part C, Methods*, **15** (4), 687–696.

Thomas-Keprta, K.L., Clemett, S.J. *et al.* (2002) Magnetofossils from ancient Mars: a robust biosignature in the Martian meteorite ALH84001. *Applied and Environmental Microbiology*, **68** (8), 3663–3672.

Tlatlik, H., Simon, P. *et al.* (2006) Biomimetic fluorapatite–gelatine nanocomposites: pre-structuring of gelatine matrices by ion impregnation and its effect on form development. *Angewandte Chemie: International Edition in English*, **45** (12), 1905–1910.

Tomczak, M.M., Gupta, M.K. *et al.* (2009) Morphological control and assembly of zinc oxide using a biotemplate. *Acta Biomaterialia*, **5** (3), 876–882.

Tracy, S.L., Williams, D.A., and Jennings, H.M. (1998) The growth of calcite spherulites from solution. I: experimental design techniques. II: kinetics of formation. *Journal of Crystal Growth*, **193**, 374–388.

Treml, H. and Kohler, H.-H. (2000) Coupling of diffusion and reaction in the process of capillary formation in alginate gel. *Chemical Physics*, **252**, 199–208.

Treml, H., Woelki, S., and Kohler, H.-H. (2003) Theory of capillary formation in alginate gels. *Chemical Physics*, **293**, 341–353.

Tuckermann, M. (2001) *Struktur-Eigenschafts-Beziehungen in Kollagen und Leder*. VDI Fortschrittsberichte, vol. 17, p. 204.

Wahl, R. (2003) *Reguläre bakterielle Zellhüllenproteine als biomolekulares Templat*. Thesis. Technische Universität, Dresden.

Watzky, M.A. and Finke, R.G. (1997) Transition metal nanocluster formation kinetic and mechanistic studies. A new mechanism when hydrogen is the reductant: slow, continuous nucleation and fast autocatalytic surface growth. *Journal of the American Chemical Society*, **119**, 10382–10400.

Weiner, S. and Addadi, L. (2011) Crystallization pathways in bimineralization. *Annual Review of Materials Science*, **41**, 21–40.

Wenger, Th. (1998) Herstellung gerichtet-strukturierter Keramiken. Thesis. University of Regensburg

Wise, E., Maltsev, S., Davies, E., Duer, M., Jaeger, Ch., Loveridge, N., Murray, R., and Reid, D. (2007) The organic–mineral interface in bone is predominantly polysaccharide. *Chemistry of Materials*, **19**, 5055–5057.

Yong, P., Mikheenko, I.P. *et al.* (2010) Biorefining of precious metals from wastes: an answer to manufacturing of cheap nanocatalysts for fuel cells and power generation via an integrated biorefinery? *Biotechnology Letters*, **32** (12), 1821–1828.

Yu, S.H. and Cölfen, H. (2004) Bio-inspired crystal morphogenesis by hydrophilic polymers. *Journal of Materials Chemistry*, **14**, 2124–2147.

7
Self-Assembly

7.1
Case Study

The cell envelope of bacteria can be regarded as a multi-purpose structure. It has to keep the cell in shape, protect it against damage from mechanical and chemical impact, maintain a fairly constant chemical composition, and facilitate the selective import of nutrients and export of waste products. Any damage to the cell envelope could be fatal. Therefore, it is necessary that the cell envelope has a *high potential for structure regeneration*. Obviously, natural selection has led to unique structures which can meet these demands. They are found even with prokaryotic microorganisms, which represent life forms dating back to 3.5 billion years ago. The cell envelope of prokaryotic organisms (archaebacteria and eubacteria) is composed of outer cell wall and inner cytoplasmic membrane. The cell wall of almost all archaebacteria is a crystalline proteinaceous layer termed S-layer. Also, in hundreds of walled eubacteria, such S-layers form the outer cell wall.

The Vienna group of Sleytr has explored in a pioneering work the beauty and richness of many of these S-layer structures and opened a wide spectrum of possible applications for bioengineering (Dieluweit *et al.*, 1998; Sleytr and Beveridge, 1999a; Schuster *et al.*, 2001; Moll *et al.*, 2002; Sleytr *et al.*, 2003a, 2003b; Horejs *et al.*, 2011). One exciting message for materials scientists is the finding that there is one basic property, the ability of self-assembly of protein monomers into two-dimensional lattice structures. In Figure 7.1, two examples of such cell walls are shown. As it can be seen, a periodic lattice layer covers the whole cell surface. The proteins can be arranged in periodic lattices of various symmetries. Figure 7.2 summarizes the possible symmetries which have been found. There are lattices with oblique ($p1$, $p2$), square ($p4$), or hexagonal ($p3$, $p6$) symmetries. The size of the unit cell of the various lattices varies between 3 and 35 nm for the various strains. About 30–70% of the unit cell consists of pores of identical size and shape. Often in one unit cell, there are two or more pores of distinct geometry with diameters in the range of 2–8 nm.

In Figures 7.3–7.5, characteristic examples are shown of S-layers with $p1$, $p4$, and $p6$ symmetry, respectively. The unit cells are composed of two, four, and six

Bio-Nanomaterials: Designing Materials Inspired by Nature, First Edition. Wolfgang Pompe, Gerhard Rödel, Hans-Jürgen Weiss, and Michael Mertig.
© 2013 Wiley-VCH Verlag GmbH & Co. KGaA. Published 2013 by Wiley-VCH Verlag GmbH & Co. KGaA.

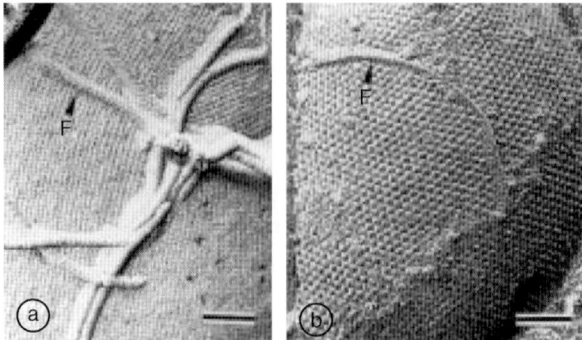

Figure 7.1 Electron micrographs of freeze-etched preparations of bacterial cells from (a) *Aneurinibacillus thermoaerophilus* DSM 10155 showing a square (*p4*) S-layer lattice and (b) archaeon *Methanomicrobium mobile* DSM 1539 covered with a hexagonal (*p6*) array. F: flagella. The bar corresponds to 100 nm. (Reproduced with permission from Sleytr *et al.* (1999b). Copyright 1999, Wiley-VCH Verlag GmbH.)

identical subunits, respectively. The high-resolution electron micrographs of the unit cells show the different nanopores very nicely. In Figure 7.5, with two alternative methods, it is demonstrated how the lattice constant of the hexagonal lattice can be determined by evaluating either the filtered reverse fast Fourier

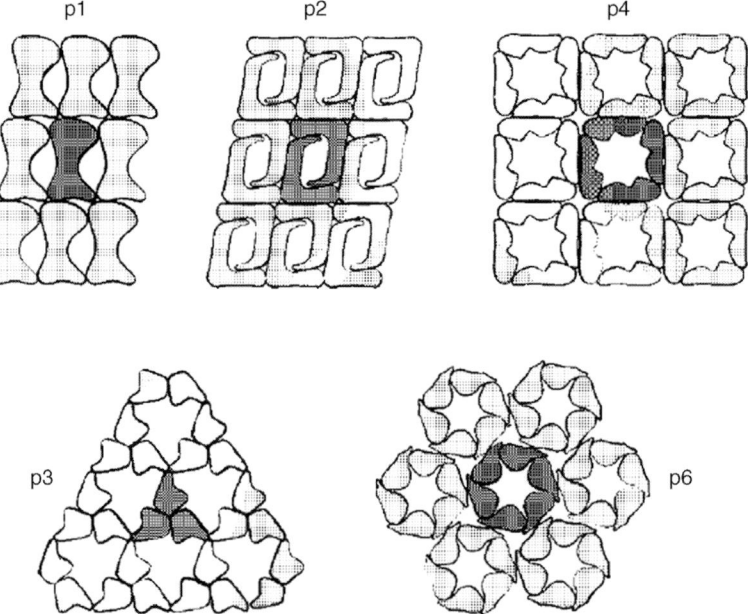

Figure 7.2 Schematic drawings of the S-layer lattice types. S-layer unit cells can be arranged in lattices with oblique (*p1*, *p2*), square (*p4*) or hexagonal (*p3*, *p6*) symmetry. (Reproduced with permission from Sleytr *et al.* (1999b). Copyright 1999, Wiley-VCH Verlag GmbH.)

Figure 7.3 Three-dimensional model of the protein mass distribution of the S-layer of *Bacillus stearothermophilus* NRS 2004/3a/V2 (outer face) (a). The S-layer is about 8 nm thick and exhibits a center-to-center spacing of the morphological units of 13.5 nm. The protein meshwork shows one square-shaped, two elongated, and four small pores per morphological unit (*p4*). The model was obtained after recording several tilt series of a negatively stained preparation by transmission electron microscopy, performing Fourier-domain computer image reconstructions over each individual tilted view and combining all processed views to a three-dimensional data volume. Computer image reconstruction of the scanning force microscopic images of the topography of the inner face of the S-layer lattice from *B. sphaericus* CCM 2177 (b) and *Bacillus coagulans* E38 ± 66/V1 (c). The images were taken under water. The surface corrugation corresponding to a gray scale from black to white is 1.8 nm. The center-to-center spacing of the morphological unit of the square lattice (*p4*) (b) is 14.5 nm. The S-layer lattice in (c) shows an oblique unit cell ($a = 9.4$ nm, $b = 7.5$ nm, and base angle $= 80.8°$) (*p1*). The bars correspond to 10 nm. (Reproduced with permission from Sleytr *et al.* (1999b). Copyright 1999, Wiley-VCH Verlag GmbH.)

transform (FFT) of the power spectrum of the original TEM micrograph or the nonfiltered autocorrelation analysis. The lattice constant of 18.1 ± 0.7 nm has been obtained for the hexagonal lattice of *Thermoanaerobacterium thermosulfurigenes* EM1.

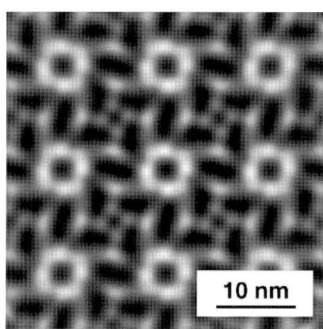

10 nm

Figure 7.4 Structure of the S-layer of *Sporosarcina ureae* ATCC 13881 with *p4* symmetry. Projection map of the negatively stained native S-layer after correlation averaging with a resolution of 1.7 nm using a Fourier-filtered TEM image as a reference. Protein appears white, while the heavily stained pores appear dark. (Reproduced with permission from Mertig *et al.* (1999). Copyright 1999, Springer Science+Business Media.)

(a)

(b)   (c)

Filtered Fast Fourier Transform (FFT)

(d)   (e)

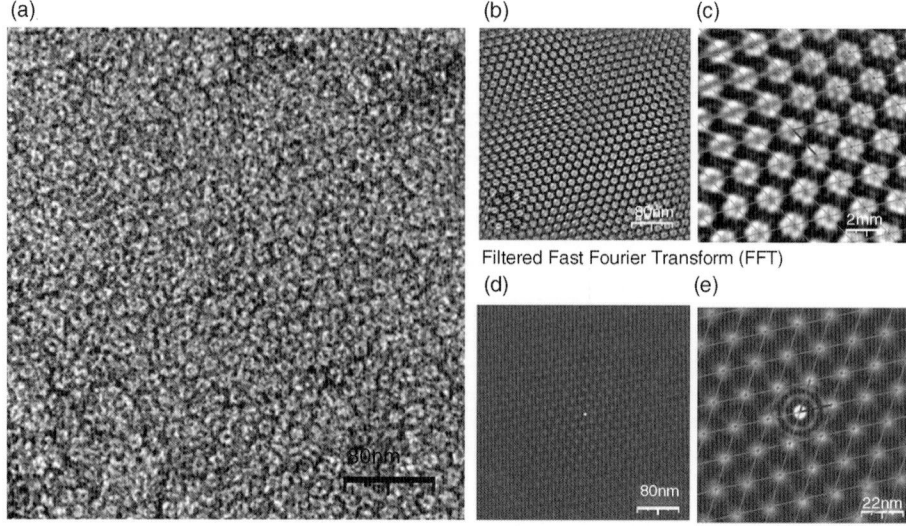

Nonfiltered autocorrelation analysis

Figure 7.5 Transmission electron micrograph of the S-layer of *T. thermosulfurigenes* EM1. The sample was negatively stained with uranyl acetate. (a) Original image of the assembled protein layer after staining with uranyl acetate. The power spectrum (insert) reveals a *p*6 symmetry. (b) By filtered reverse FFT of the power spectrum, a lattice constant of 18.1 ± 0.7 nm has been obtained. (c) Enlarged section from (b) for determination of the lattice constants. (d) Result of a nonfiltered autocorrelation analysis of (a). (e) Enlarged section of (d) for the determination of the lattice constant. (Reproduced with permission from Wahl (2003), and Blüher (2008).)

The S-layer is composed of a single protein or glycoprotein. Their molecular weight is in the range of 40–170 kDa for various species. The proteins are typically weakly acidic (pI ~ 4–6). One finds large amounts of glutamic acid, aspartic acid (~15 mol%), and hydrophobic amino acids (~40–60 mol%) and a high lysine content (~10 mol%). Hydrophobic and hydrophilic amino acids do not form extended clusters. These structure features constitute that one can expect attractive as well as repulsive interactions between the building blocks of the membrane with a well-defined equilibrium separation and reversible association–dissociation of the building blocks in assembly. Such reversible aggregation kinetics is most important, so the assembly can approach to the lowest-energy structure leading to high regularity of the final structure over a large scale. The *in vivo* observation of the cell growth kinetics has shown that the subunits of the lattice are secreted, folded, and self-assembled onto the underlying cell membrane. During cell generation (about 20 min), the cell wall is formed. This means about 500 subunits per second must be synthesized.

On the inner surface of the S-layer, domains have been identified that control the connection with the underlying cell wall layers (Figure 7.6). Three kinds of binding

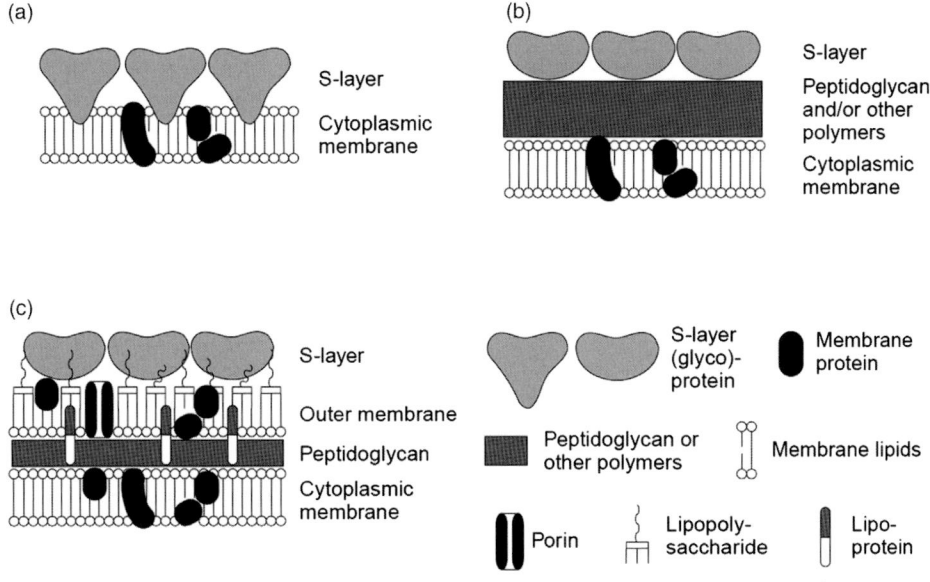

Figure 7.6 Schematic drawing of the supramolecular architecture of the three major classes of prokaryotic envelopes containing crystalline bacterial cell surface layers (S-layers). (Reproduced with permission from Sleytr *et al.* (1999b). Copyright 1999, Wiley-VCH Verlag GmbH.)

have been observed: (a) A cell envelope structure of Gram-negative archaea with S-layers as the only cell wall component external to the plasma (cytoplasmic) membrane. (b) The cell envelope of Gram-positive archaea and eubacteria. In eubacteria, the rigid wall component is primarily composed of peptidoglycan. In archaea, other wall polymers (e.g., pseudomurein or methanochondroitin) are found. (c) Cell envelope of Gram-negative eubacteria, composed of a thin peptidoglycan layer and an outer membrane. If present, the S-layer is closely associated with the lipopolysaccharide of the outer membrane (Sleytr and Beveridge, 1999a). The terms "Gram-positive" and "Gram-negative" characterize compositionally specific color changes of the cell wall due to staining. Gram-positive bacteria are those that are stained dark blue or crystal violet by Gram staining. This is in contrast to Gram-negative bacteria that cannot retain the violet stain. With a counterstain (safranin or fuchsine), they appear red or pink. The high content of peptidoglycan in Gram-positive bacteria makes the stable crystal violet stain.

With regard to possible bio-inspired materials development, there is a remarkable feature of archaebacteria. They can be found in very different habitats all around the earth and probably contribute up to 20% of its biomass. Most interesting is the larger part of extremophiles among them. There are thermophiles that can live in hot springs such as geysers, black smokers, and oil wells, growing at temperatures 45–100 °C. In black smoker from the Gulf of California, *Methanopyrus kandleri* was discovered at a depth of 2000 m at temperatures of 84–110 °C. In a hydrogen–carbon dioxide-rich environment, it acts as methane former. Halophiles have been observed in salt lakes. Alkaliphiles and acidophiles live under extreme chemical conditions. For example, it has been reported that *Picrophilus torridus* grows at pH 0, which is equivalent to 1.2 molar sulfuric acid. Such properties initiated some speculations that archaebacteria could be a source of extraterrestrial life. As their habitats are similar to conditions existing on Mars, they could be transferred between planets via meteorites.

7.2
Basic Principles

7.2.1
Basic Phenomena of Self-Assembly and Self-Organization

Objects of any size can agglomerate if there are attracting forces between them. The term "self-assembly" does not refer to this most general phenomenon, but is used in a narrower sense: It is usually restricted to the assemblage of equal elements into a regular structure. Crystallization is the most commonly known example of such phenomenon, but for reasons of convention rather than logic it is usually not covered by "self-assembly." The use of this term is still more restricted: It is mainly used in connection with macromolecular or macroscopic elements. The formation of bubble floats with crystal-like arrangement of millimeter-sized bubbles is suitable for the demonstration of self-assembly on a macroscopic scale. Here, the attraction is mediated by surface tension. On the molecular level, of course, the forces known from chemistry are at work. There do not seem to be strict rules for the use of the related terms. The self-assembly of viruses into a regular lattice is often called crystallization. "Self-assembly" is often used synonymous with "self-organization," although the latter term covers a wider range of phenomena, including spontaneous pattern formation, an example of which are the Liesegang fringes. In the same way, as the term "crystallization" allows for the incorporation of various faults into the crystal lattice, the meaning of "self-assembly" has been extended to allow for the admixture of other substances, including bonding agents.

Self-assembly processes observed in biomolecular systems offer promising bio-inspired approaches for bottom-up assembly of nanostructures. The phenomena are closely related to present topics in supramolecular chemistry. Therefore, we will apply here the terminology proposed by Lehn (1995) in his pioneering work on supramolecular chemistry:

> "*Supramolecular self-assembly* concerns the spontaneous association of either few or many components resulting in the generation of either discrete oligomolecular supermolecules or of extended polymolecular assemblies such as molecular layers, films, membranes etc. The formation of super-molecules results from the recognition-directed spontaneous association of

well-defined and limited number of molecular components under inter-molecular control of the non-covalent interactions that hold them together."

The assembly of complex structures with the help of structured substrates is denoted here by the term *template-directed assembly*. The substrate itself may have been produced by self-assembly, as in the example presented in Section 7.3.2. Often the binding of a biomolecule with a second one is associated with a conformation change. As a result of changed conformation, the binding of a ligand to one site influences the binding properties of another site on the same protein. This can make binding either easier (positive cooperativity) or more difficult (negative coopera-tivity). In case of positive cooperativity, an increase of growth rate and stability of the assembly can be observed with increasing size.

7.2.2
Self-Assembly of Protein Filaments: the Cytoskeleton

The cytoskeleton is an example of self-assembly of functional components of a living cell. It is a three-dimensional network of protein filaments stabilizing the cell shape and the structure of the cytoplasm. It is composed of three types of filaments: the actin filaments, the microtubules, and the intermediate filaments (Table 7.1). Every type of filament is assembled from individual small protein subunits that associate by non-covalent bonds with long uniform multistranded filaments. The actin monomer is a globular protein that contains a nucleotide – ATP or ADP – in a deep cleft. The nucleotide regulates the polymerization and depolymerization. The subunits of the microtubules are $\alpha\beta$-tubulin dimers. They contain in a cleft GTP or GDP. The formation of the various filaments of the cytoskeleton can be understood as a polymerization of these protein subunits. The size of the individual protein subunits ranges between a few nanometers and a few tens of nanometers. The final size distribution of the filaments is the result of the association and dissociation kinetics of a finite number of protein subunits described by the total concentration of the protein subunits n_t.

A remarkable feature of the various filaments is the large range of length, they span from the molecular level to the cell level. The filaments are dynamic structures. Their length and arrangement in the cell can change very quickly. Cell motion and mitosis are connected with dramatic changes of the whole cytoskeleton. For instance, actin filaments are involved in the formation of protrusions at the leading

Table 7.1 Components of the cytoskeleton (Howard, 2001).

Filament type	Subunit	Subunit size	Filament length
Actin filament	Actin monomer	5.5 nm	35 nm–100 μm
Microtubule	$\alpha\beta$-Tubulin dimer	8 nm	1–100 μm
Intermediate filament	Wrapped protofibrils	~10 nm	In the micrometer range

edge of a motile cell (see Chapter 3). Microtubules vary their length greatly while separating the chromosomes during the later phases of mitosis. The typical growth rates and shrinkage rates of microtubules are about 1 to 10 μm/min, respectively. They can switch from growth to shrinkage within a few minutes. The quick response of the cytoskeleton implies that large numbers of protein subunits have to polymerize or depolymerize within minutes. This requires highly efficient mechanisms of self-assembly and dissociation. Whereas the self-assembly kinetics of the intermediate filaments is not yet well understood, the processes driving the self-assembly of actin filaments and microtubules have been revealed. It has been found that the filament length is not set by a static equilibrium with the monomer concentration (passive polymerization), but by a dynamic equilibrium sustained by a permanent supply of energy released in hydrolysis of nucleotides (active polymerization). The advantage of dynamic equilibrium is its flexibility. It enables filament length extension or reduction as required by physiological processes to be efficiently controlled via energy supply. Before discussing the active polymerization in more detail, we will consider the passive polymerization as a reference mechanism suitable as a model process describing the self-assembly under equilibrium conditions. For this purpose, we follow a presentation given by Howard (2001).

Passive Polymerization of Protein Filaments   Protein filaments are usually multi-stranded aggregates (e.g., microtubules) or helically arranged chains (e.g., actin filaments) of individual protein subunits. The advantage of a multi-stranded aggregate in self-assembly of long filaments can be explained when we compare the equilibrium of association and dissociation processes of a single-stranded and a double-stranded protein filament, as schematically shown in Figure 7.7.

The number of available binding sites makes the difference between the two growth modes. In the single-stranded model, the so-called Einstein polymer (Hill, 1987), only one binding site is occupied by association of one protein unit with the chain, whereas in the double-stranded chain two binding sites are involved.

(a)

(b)

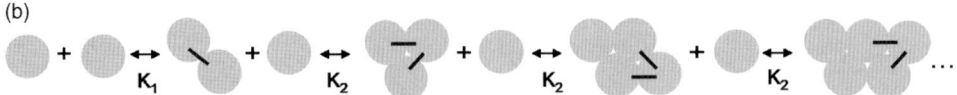

Figure 7.7 Aggregation of (a) single-stranded and (b) double-stranded protein chains. K_i ($i = 1, 2$) are the dissociation constants for the single- and double-bonded subunits, respectively. For large total concentration of the protein subunits n_t, the equilibrium concentration of free monomers n_1 approaches the value of the dissociation constants K_1 and K_2 in case of single- and double-stranded filaments, respectively. (Adapted from Howard (2001).)

Therefore, the dissociation constant K_2 of proteins at the end of a double-stranded chain is much smaller than that at the end of a single-stranded chain ($K_2 \ll K_1$). Thus, in equilibrium, the average number of protein units N_{av} assembling into one polymer filament is much larger than that assembling into a single-stranded chain.

Growth Kinetics of Single- and Double-Stranded Polymer Chains We consider the growth of *double-stranded aggregates* grown by association–dissociation reactions of monomers. The kinetic equations are given as

$$\frac{dn_1}{dt} = -2\sigma_1 D n_1 n_1 - \sum_{i=2}^{\infty} \sigma_i D n_1 n_i + 2\lambda_2 n_2 + \sum_{i=3}^{\infty} \lambda_i n_i,$$

$$\frac{dn_i}{dt} = \sigma_{i-1} D n_1 n_{i-1} - \sigma_i D n_1 n_i - \lambda_i n_i + \lambda_{i+1} n_{i+1}, \quad i > 1. \tag{7.1}$$

Here, $n_i(t)$ stands for the time-dependent concentration of filaments of size i. D is the diffusion coefficient of the monomers, σ_1 and $\sigma_i \equiv \sigma$ if ($i \geq 2$) are the collision cross sections for aggregation of a monomer with another monomer and with an i-mer ($i \geq 2$), respectively, and λ_i is the dissociation rate of an i-mer. We assume that the dissociation rates λ_i are size independent with exception of λ_2:

$$\frac{dn_1}{dt} = -2\sigma_1 D n_1 n_1 - \sum_{i=2}^{\infty} \sigma D n_1 n_i + 2\lambda_2 n_2 + \sum_{i=3}^{\infty} \lambda n_i,$$

$$\frac{dn_2}{dt} = \sigma_1 D n_1 n_1 - (\sigma D n_1 + \lambda_2) n_2 + \lambda n_3,$$

$$\frac{dn_i}{dt} = \sigma D n_1 n_{i-1} - (\sigma D n_1 + \lambda) n_i + \lambda n_{i+1}, \quad i > 2, \tag{7.2}$$

$$\frac{dn_f}{dt} = \sigma D n_1 n_{f-1} - \lambda n_f, \quad i = f.$$

The term f is the maximum size of filaments corresponding to the total number of subunits involved in the assembly process. With a given initial distribution of the filament concentrations $n_i(t = 0)$, Eq. (7.2) allows us to calculate the time dependence of the assembly caused by association and dissociation of monomers. In Figure 7.8, an example is shown of a solution under the simplifying assumptions $\sigma_1 = \sigma$ and $\lambda_2 = \lambda$. This assumption leads from the growth model of a double-stranded polymer chain to the description of the growth model of a single-stranded chain. Then, Eq. (7.2) can be transformed in a normalized shape:

$$\frac{dn_1}{d\tau} = -2K_1^{-1} \cdot n_1 n_1 - K_1^{-1} - \sum_{i=2}^{\infty} n_1 n_i + 2n_2 + \sum_{i=3}^{\infty} n_i,$$

$$\frac{dn_i}{d\tau} = K_1^{-1} \cdot n_1 n_{i-1} - (K_1^{-1} \cdot n_1 + 1) n_i + n_{i+1}, \quad i \geq 2, \tag{7.3}$$

$$\frac{dn_f}{d\tau} = K_1^{-1} \cdot n_1 n_{f-1} - n_f, \quad i = f,$$

with $\tau = t \cdot \lambda$ and $K_1^{-1} = \sigma D/\lambda$. In the example, an initial distribution of monomers was assumed with a supersaturation $n_1(t = 0)/K_1 = 100$. The simulation shows that the size distribution of filaments relaxes to a stationary distribution with

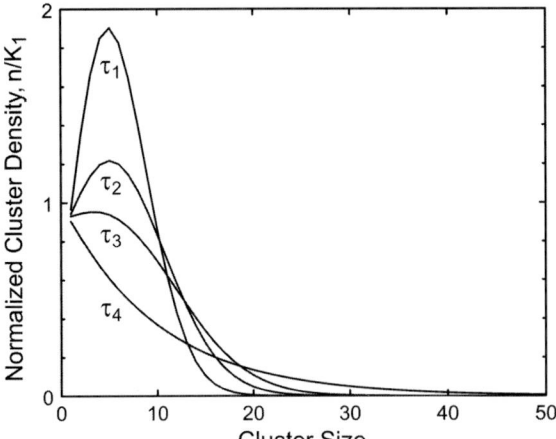

Figure 7.8 Growth kinetics of single-stranded polymer chains. Time dependence of the cluster size distribution normalized to the dissociation constant K_1. (Time steps t_i normalized to the inverse value of the dissociation rate λ^{-1} referred to in $\tau = t \times \lambda$ with $\tau_1 = 1.0 \times 10^3$, $\tau_2 = 2.0 \times 10^3$, $\tau_3 = 3.2 \times 10^3$, $\tau_4 = 1.0 \times 10^5$.) For $t = 0$, a monodisperse distribution of monomers with the supersaturation n_1 $(t = 0) = 100 K_1$ has been assumed.

exponential decay. Assuming the dissociation constant $K_1 = 1\,\mu\text{M}$, the collision cross section $\sigma = 1\,\text{nm}$, and the diffusion coefficient $D = 500\,\mu\text{m}^2/\text{s}$, the equilibrium is reached in about 5 min.

For the *general case of growing double-stranded filaments*, described by Eq. (7.2), the stationary solutions with $dn_k/dt = 0$ for all k from 1 to f give the equilibrium concentrations of the filaments:

$$-2\sigma_1 Dn_1^2 - \sigma Dn_1 \sum_{i=2}^{\infty} n_i + 2\lambda_2 n_2 + \lambda \sum_{i=3}^{\infty} n_i = 0,$$

$$\sigma_1 Dn_1^2 + \lambda n_3 - (\sigma Dn_1 + \lambda_2)n_2 = 0, \tag{7.4}$$

$$\sigma Dn_1 n_{i-1} + \lambda n_{i+1} - (\sigma Dn_1 + \lambda)n_i = 0, \quad i > 2,$$

$$\sigma Dn_1 n_{f-1} - \lambda n_f = 0, \quad i = f.$$

With the dissociation constants $K_1 = \lambda/\sigma_1 D$ and $K_2 = \lambda/\sigma D$, we define normalized equilibrium concentrations $\tilde{n}_1 = n_1/K_1$, and $\tilde{n}_i = n_i/K_2$, with $i \geq 2$. An iterative solution of Eq. (7.4) can be given by

$$\tilde{n}_i = \frac{K_1}{K_2}\tilde{n}_1 \cdot \tilde{n}_{i-1} = \frac{K_2}{K_1} \cdot \exp\left(-\frac{i}{n_0}\right), \quad 1 \leq i \leq f, \tag{7.5}$$

with

$$n_0^{-1} = \ln\left(\frac{K_2}{n_1}\right). \tag{7.6}$$

The polymer chains show an exponential decay of the size distribution in the stationary equilibrium.

For the *growth of single-stranded filaments*, we get with $\sigma_1 = \sigma$, $\lambda_2 = \lambda$, and $K_1 = K_2$:

$$\tilde{n}_i = \tilde{n}_1^i = \exp\left(-\frac{i}{n_0}\right), \quad \text{with} \quad n_0^{-1} = \ln\left(\frac{K_1}{n_1}\right), \quad 1 \leq i \leq f. \tag{7.7}$$

Note that the decay constant n_0^{-1} of single-stranded filaments is larger than the constant of double-stranded filaments for same n_1 due to the larger dissociation constant K_1. It means that double-stranded polymers form longer chains in equilibrium. As shown by Howard (2001), for large $f \gg 1$, the average length of the polymer (average number N_{av1}) of subunits per chain follows for single-stranded filaments:

$$N_{av1} = \sum_2^\infty i p_i = \sum_2^\infty i \frac{\tilde{n}_i}{\sum_2^\infty \tilde{n}_k} = 1 + \frac{1}{1 - \tilde{n}_1}. \tag{7.8}$$

The size distributions of the filament concentrations given in Eqs. (7.5) and (7.7) depend on the equilibrium concentration of the free monomers n_1. This value can be determined when the total concentration of subunits is known.

For *single-stranded filaments*, the normalized total concentration of subunits $\tilde{n}_t$ is

$$\tilde{n}_t = \sum_1^\infty i \tilde{n}_i = \frac{\tilde{n}_1}{(1 - \tilde{n}_1)^2} \tag{7.9}$$

or

$$\tilde{n}_1 = 1 + \frac{1}{2\tilde{n}_t} - \sqrt{\frac{1}{\tilde{n}_t} + \frac{1}{4\tilde{n}_t^2}}. \tag{7.10}$$

Always $\tilde{n}_1 < 1$ is fulfilled. For, $\tilde{n}_t \gg 1$, we get

$$\tilde{n}_1 \cong 1 - \frac{1}{\sqrt{\tilde{n}_t}}, \quad n_0 \cong \sqrt{\tilde{n}_t}, \quad N_{av1} \cong \sqrt{\tilde{n}_t}. \tag{7.11}$$

This means, for large values of the total concentration of protein subunits n_t, the concentration of free monomers n_1 is nearly equal to the dissociation constant K_1:

$$n_1 \cong K_1, \quad \text{for} \quad n_t \gg K_1. \tag{7.12}$$

In Figure 7.9, examples of normalized equilibrium concentrations are shown for different values of the total concentration of protein subunits. Note that for small initial supersaturation, the equilibrium concentration of the free subunits n_1 is significantly smaller than the dissociation constant K_1.

For *double-stranded filaments*, we get the following stationary solution for $\tilde{n}_t$ (Howard, 2001):

$$\tilde{n}_t = \tilde{n}_1 + \frac{K_2}{K_1}\tilde{n}_1^2 + \frac{K_2}{K_1}\frac{\tilde{n}_1^2 \cdot (2 - \tilde{n}_1)}{(1 - \tilde{n}_1)^2}. \tag{7.13}$$

$$N_{av2} = 2 + \frac{1}{1 - \tilde{n}_1}. \tag{7.14}$$

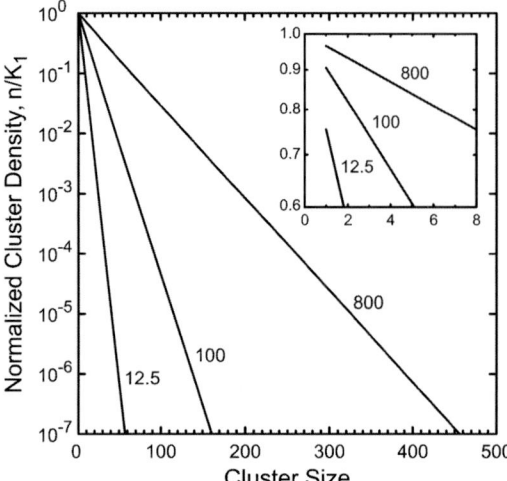

Figure 7.9 Equilibrium cluster size distributions with dependence on the initial monomer concentration normalized to the critical cluster concentration in an infinite system for an assembly of single-stranded filaments. The distributions are the asymptotic values of simulation runs for large time. Three initial monomer concentrations with different relative supersaturation $S_R = n_1 (t=0)/K_1$ have been assumed with $S_R = 800$, 100, and 12.5, respectively.

For $\tilde{n}_t \gg 1$, it yields

$$\tilde{n}_1 \cong 1, \quad N_{av2} \cong \sqrt{\frac{K_1}{K_2}} \cdot \sqrt{\tilde{n}_t}, \tag{7.15}$$

and

$$\tilde{n}_i = \frac{K_2}{K_1}, \quad \text{with} \quad i \geq 2 \quad \text{provided} \quad i \ll n_0. \tag{7.16}$$

It means that the average number of subunits per double-stranded filament increases by a factor K_1/K_2 with respect to the value of a single-stranded filament:

$$N_{av2} \cong \frac{K_1}{K_2} \cdot \sqrt{\frac{n_t}{K_1}} \cong \frac{K_1}{K_2} \cdot N_{av1}. \tag{7.17}$$

Following Howard (2001), an estimate for the dissociation constants of actin filaments yields $K_1 = 100\,\text{mM}$ and $K_2 = 1\,\mu\text{M}$. Hence, for a total concentration n_t equal to 10 times the dissociation constant K_2, the mean size is $N_{av2} \cong 10^3$, which corresponds to an actin filament length of about 2.75 μm.

Active Polymerization of Protein Filaments The hydrolysis of the nucleotide triphosphate complexes (NTPs) bound to the subunits enables the active

Growing phase   Shrinking phase

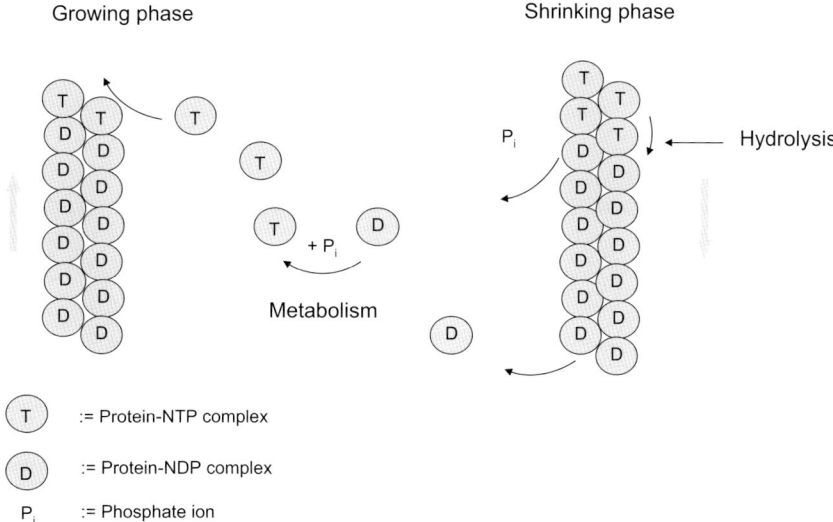

Metabolism

Hydrolysis

T := Protein-NTP complex

D := Protein-NDP complex

P$_i$:= Phosphate ion

Figure 7.10 The polymerization cycle for actin and tubulin. (Adapted from Howard (2001).)

polymerization of the polymer chains. The active polymerization of actin filaments and microtubules is characterized by the following four mechanisms, as schematically shown in Figure 7.10:

1) Active polymerization is based on the much higher association rate of actin–ATP complexes and tubulin–GTP complexes with the end of a corresponding filament in comparison to that of the nucleotide diphosphate (NDP) complexes.

2) When the NTP complexes are bound to the filaments, hydrolysis leads to the formation of NDPs with the dissociation of one phosphate ion P_i. The key for understanding the high assembly rate of actin filaments and microtubules is the large increase of the hydrolysis rate of the NTPs when the complexes are bound to the actin filament and microtubule. If incorporated into the filaments, the hydrolysis rate of ATP increases to $\sim 0{,}05\,\mathrm{s}^{-1}$ and that of GTP to $>0.2\,\mathrm{s}^{-1}$. It means that actin filaments older than a few minutes and microtubules older than several seconds consist of NDP complexes only.

3) After transformation in NDP complexes, the depolymerization of the filaments is favored as the dissociation constants $K = k_{\mathrm{off}}/k_{\mathrm{on}}$ for NDP complexes are much larger than those for NTP complexes. The experimental data show that the critical concentration at one end of the filament differs from that at the other end. It is connected with a structural polarity of actin filaments and microtubules. The end with the faster growth kinetics (lower critical concentration) is called the plus end. When the concentration of the NTP–monomer complexes is between the critical concentrations of the two ends, then the plus end grows and the minus end shrinks. There is one value of the monomer concentration in this range

where the growth of one end balances the shrinkage of the other end. This effect is called treadmilling.

4) The polymerization cycle of the filament is completed by converting the dissociated NDP–monomer complexes back to NTP–monomer complexes via metabolic processes. The exchange of NTP for NDP can occur only after the subunit has dissociated. In the case of tubulin, the rate of this process is about 1/s. In the case of actin, the exchange is accelerated by profilin, a so-called exchange factor. The activity of the metabolic process controlling the exchange of NTP complexes for NDP complexes governs the growth or shrinkage of the filament.

7.2.3
Self-Assembly of β-Sheets: the Amyloid Fibrils

About a quarter of all proteins generated in cells do not correctly fold in the process of their synthesis. Usually the misfolded proteins are destroyed in the cells. When this process is disturbed, then the misfolded protein is excreted and converts into an insoluble extracellular amyloid fibril through a sequence of transition states (Figure 7.11). In some cases, these fibrils can cause serious diseases.

The width and length of amyloid fibers are 7–10 nm and a few micrometers, respectively. Their shape is given by a bundle of highly ordered filaments consisting of stacked ß-sheets. The secondary structure of these β-sheets is oriented perpendicular to the filament axis in high regularity. Many of the amyloid-forming proteins have a higher fraction of aromatic residues in the core region of the β-sheet. Their interaction stabilizes the stacked structure. The other parts of the protein fold outside the core region correctly. The main part of the domains outside the core region is cleaved off by proteases later. The amyloid fibrils can aggregate into larger plaques that are deposited on a particular tissue. For instance, a large trans-membrane protein called β-amyloid precursor protein (APP) found in most human

native stage   unfolded intermediates   aggregates of β-strands   amyloid fibrils

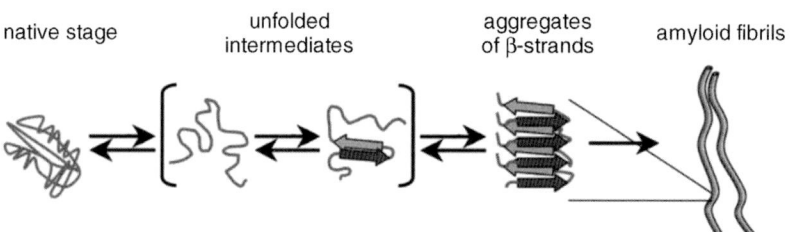

Figure 7.11 Schematic illustration of the formation of amyloid fibers by misfolding and assembling of soluble proteins. A natively folded monomer undergoes a conformational transition into a β-sheet-rich state. Self-assembly of these intermediates into ladders of β-strands results in the formation of ordered protofilaments. At least two winding protofilaments form the amyloid fibril. (Reproduced with permission from Cherny and Gazit (2008). Copyright 2008, Wiley-VCH Verlag GmbH.)

Table 7.2 Comparison between mechanical properties of amyloid fibrils, diphenylalanine nanotubes, and silk (Cherny and Gazit, 2008).

	Amyloid fibrils	Phe–Phe nanotubes[a]	Silk
Strength (GPa)	0.2–1.0		0.9–1
Shear modulus (GPa)	0.26–0.30		2.38
Young's modulus (GPa)	2.9–3.7	About 19	11–13
Thermal stability (°C)	$\geq$130	$\geq$150	230

a) PNTs made by self-assembling of very short aromatic peptides (hexapeptides and shorter) with the Alzheimer's β-amyloid diphenylalanine structural motif (for more details, see Section 7.3.1).

tissues assembles in the misfolded state into amyloid fibrils that lead to plaque deposition on the exterior of nervous tissue. This leads to the destruction of cells in the affected tissue after a few years. A diphenylalanine core motif has been identified as the shortest amyloidogenic peptide fragment of the Alzheimer's disease β-amyloid peptide. If dissolved in hexafluoro-2-propanol and diluted in water to a final concentration of 2 mg/ml, it assembles as a multiwall nanotube with an internal diameter of about 20 nm, an external diameter in the range of 80–300 nm, and micrometer length with a very high persistence length in the same range (see also Section 7.3.1).

A large group of diseases, such as type 2 diabetes, Alzheimer's disease, Huntington's disease, and Parkinson's disease, summarized as amyloidoses, are caused by the formation of amyloid fibrils. However more recently, it has been revealed that amyloids are not only pathological structures but are also involved in some physiological processes. For instance, amyloid fibrils are components of biofilms formed by *Escherichia coli* bacteria, and they have been observed in aerial hyphae of *Streptomyces coelicolor* bacteria. Interestingly, nature has duplicated essential structure elements of the amyloids in spider and worm silk. Silk proteins are composed of repetitive sequence of amino acids with a higher content of glycine, alanine, and proline. Probably, the hydrophobic domains of the proteins form β-structures and crystal-like dispersions that reinforce the fiber, whereas the glycine regions cause its flexibility. The regions of silk enriched with β-sheet structures correspond to similar structures in amyloid fibrils. They have a similar diameter (10 nm). X-ray studies have shown that the β-sheets in silk are oriented perpendicular to the fiber axis, as in amyloids. The structural similarity of the two material groups is also reflected in the remarkable mechanical and thermal properties (Table 7.2) that may lead to the development of new concepts of bio-inspired materials.

7.2.4
Self-Assembly of Two-Dimensional Protein Lattices: the Bacterial Surface Layers (S-Layers)

As already explained in Section 7.1, S-layers are an example of self-assembly of highly ordered two-dimensional structures. For several reasons, they have become a model system for systematic studies of the self-assembly process of protein sheets:

- widespread presence in cell walls of archaea and eubacteria,
- robustness,
- weak bonds (hydrophobic, electrostatic, and hydrogen bonds) between the monomers, and
- bonds between S-layer and underlying cell envelope or membrane are weaker than those between S-layer subunits.

After detachment from the cell surface by means of an ultrasonic homogenizer, the S-layer patches can be disintegrated with hydrogen bond breaking agents (e.g., guanidine hydrochloride). When the disintegrating agent is removed, the monomers can reassemble into the same lattice structure as the native one. Under *in vitro* conditions, four paths of recrystallization can be distinguished: (i) homogeneous nucleation and growth in an aqueous solution, (ii) heterogeneous nucleation and growth on a solid substrate, (iii) assembling at the air–water interface, and (iv) assembling on lipid films (Figure 7.12).

Self-assembly in solution can produce sheets or tubes. The shape and size depend on the structure of the particular S-layer protein as well as on the processing conditions such as monomer concentration, temperature, pH, and presence of additional ions in the solution (composition and ionic strength). Examples of typical

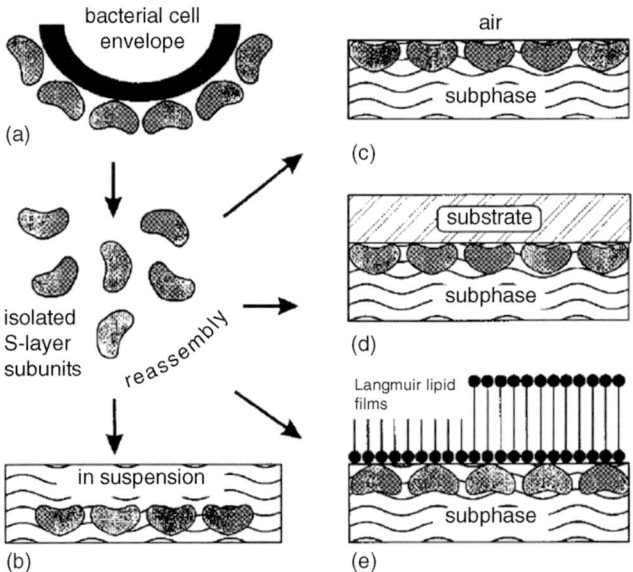

Figure 7.12 Schematic illustration of recrystallization of isolated S-layer subunits into crystalline arrays. After formation of monomers by detachment from the cell surface (a), the self-assembly can occur in suspension (b), at the water–air interface (c), on solid supports (d), and on Langmuir lipid films (e). (Reproduced with permission from Sleytr *et al.* (1999b). Copyright 1999, Wiley-VCH Verlag GmbH.)

(a)   (b)

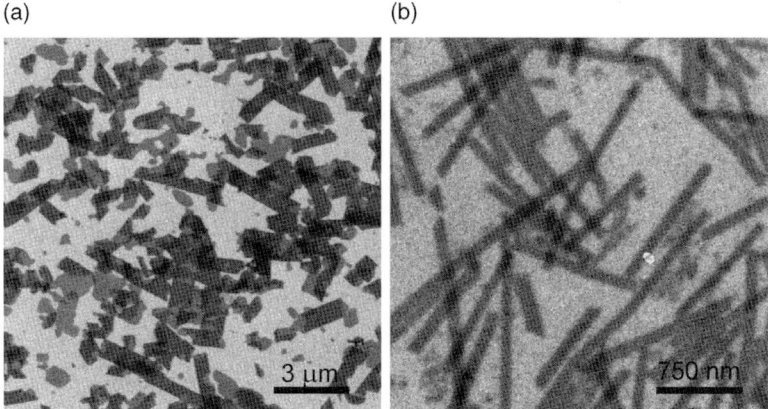

Figure 7.13 (a) SEM image of S-layers of *B. sphaericus* NCTC 9602 adsorbed on silicon substrate from a solution with concentration of $c \sim 1\,mg/ml$. (b) SEM image of tube-like S-layer structures of *G. stearothermophilus* ATTC 12980 adsorbed on silicon substrate from a solution with concentration of $c \sim 1\,mg/ml$. (Reproduced with permission from Bobeth *et al.* (2011). Copyright 2011, American Chemical Society.)

growth modes of S-layers after homogeneous nucleation in an aqueous solution are shown in Figure 7.13. Planar monolayers, folded bilayers, and tubes have been obtained with *Bacillus sphaericus*. Tube-like aggregates are preferentially obtained with *Geobacillus stearothermophilus*.

Self-Assembly of S-Layer Patches The formation of the well-organized lattice structures of the S-layer proteins is the result of *reversible association–dissociation processes* of protein monomers. The interplay of the kinetics of both mechanisms leading to such structures can be described in an idealized growth model. For simplicity, we consider the growth of a square-like patch, corresponding to S-layers with *p*4 lattice symmetry, and represent the monomers by small squares, thereby ignoring the pore structure of the S-layer (Figure 7.14).

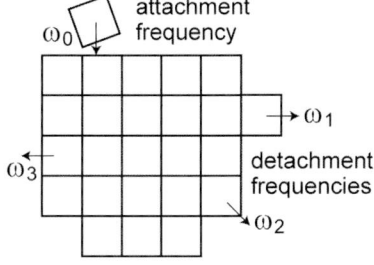

Figure 7.14 S-layer assembly modeled by kinetic Monte Carlo simulation. The attachment frequency ω_0 is assumed to be equal for any position, but the detachment frequencies are assumed to be unequal: $\omega_1 \gg \omega_2 > \omega_3$. (Reproduced with permission from Bobeth *et al.* (2011). Copyright 2011, American Chemical Society.)

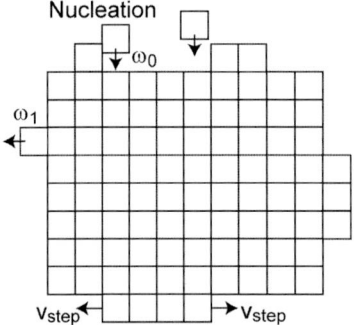

Figure 7.15 Patch growth by nucleation and lateral growth of monomer rows. (Reproduced with permission from Bobeth *et al.* (2011). Copyright 2011, American Chemical Society.)

The representation of the odd-shaped monomer by a small square in this model is justified by the fact that every four monomers taken together fit into the square unit cell of the S-layer lattice with *p4* symmetry. Change of patch size results from competition between random attachment and detachment of monomers, as illustrated in Figure 7.14. The attachment rate of monomers per lattice site at the patch edges, ω_0, is assumed to be proportional to the monomer density in the solution, n_1:

$$\omega_0 = D a_m n_1, \tag{7.18}$$

where the kinetic coefficient D has the dimension of a diffusion constant. It can be assumed that D is much smaller than the diffusion coefficient of the monomers in the solution since the attachment is essentially governed by conformational fitting. It is also possible to derive a rough analytical approximation for the growth velocity. It is easily imaginable that the detachment frequencies from the different positions differ much. With the simplifying assumptions of $\omega_1 \gg \omega_0$ and $\omega_2 = \omega_3 = 0$, patches grow by addition of monomers to incomplete rows as soon as a row nucleus consisting of two adjacent monomers has settled (Figure 7.15).

Under this assumption, the velocity of step motion v_{step} due to monomers attaching at a step site is given by $v_{step} = a_m \omega_0$. Since a monomer row grows on both sides, the time τ_{RM} to grow a complete row of length a is estimated as

$$\tau_{RM} = \frac{a}{2 v_{step}} = \frac{a}{2 a_m \omega_0}. \tag{7.19}$$

For the analysis of row nucleation at a patch edge, we introduce the number of single ad-monomers per edge length x_1. For the considered case $\omega_1 \gg \omega_0$, an attachment–detachment equilibrium of monomers will establish. The density of ad-monomers along a "facet" of a square-like patch is then given by

$$x_1 = \frac{\omega_0}{\omega_1} \frac{1}{a_m}. \tag{7.20}$$

The attachment of a monomer next to an ad-monomer makes an adsorbed dimer. The corresponding dimer density formation rate reads $\dot{x}_{dim} = 2 x_1 \omega_0$. The time τ_{dim}

for formation of one dimer on a patch facet of length a follows with Eq. (7.20):

$$\tau_{dim} = \frac{1}{\dot{x}_{dim}a} = \frac{\omega_1}{2\omega_0^2}\frac{a_m}{a}. \tag{7.21}$$

For small patches, nucleation of only one dimer is typically sufficient to grow a complete row of length a. For larger patches, several independently nucleated rows will grow and coalesce on a patch facet. A monomer row of length a is completed by the simultaneous growth of n_{mr} monomer rows on one patch facet. Thus, the corresponding time is given by

$$\tau_{RM} = \frac{a}{n_{mr} \cdot 2v_{step}}. \tag{7.22}$$

The number n_{mr} of simultaneously growing rows can roughly be estimated from the dimer nucleation rate $n_{mr} = \dot{x}_{dim}a\tau_{RM}$. Together with Eq. (7.20), we get

$$\tau_{RM} = \frac{1}{2\omega_0}\sqrt{\frac{\omega_1}{\omega_0}}, \tag{7.23}$$

which leads to the constant patch growth velocity v:

$$v = \dot{a}_{MRG} = \frac{2a_m}{\tau_{RM}} = 4\sqrt{\frac{\omega_0}{\omega_1}}\omega_0 a_m = v_P \cdot (n_1/n_{1S})^{3/2}, \tag{7.24}$$

where n_{1S} is the monomer solubility and

$$v_P = 4\sqrt{D/(\omega_1 a_m^2)}\left(n_{1S}a_m^3\right)^{3/2}D/a_m. \tag{7.25}$$

In the early stage of patch growth, single row growth can be observed. It is characterized by $\tau_{RM} < \tau_{dim}$. From Eqs. (7.21) and (7.23), one finds the corresponding condition for the patch size:

$$a < \sqrt{\omega_1/\omega_0}a_m. \tag{7.26}$$

The patch growth velocity during this early stage is given by

$$\dot{a}_{SRG} = 2a_m/\tau_{dim} = \beta a, \quad \beta = 4\omega_0^2/\omega_1, \tag{7.27}$$

and the patch size enlarges exponentially with time:

$$a(t) = a_0 \exp(\beta t). \tag{7.28}$$

The transition time t_C from exponential to linear growth is defined by the equation $\dot{a}_{SRG} = \dot{a}_{MRG}$, leading to

$$t_C = \frac{1}{8}\frac{\omega_1}{\omega_0^2}\left\{\ln\left(\frac{\omega_1}{\omega_0}\right) - 2\ln\left(\frac{a_0}{a_m}\right)\right\} \quad \text{and} \quad a(t_C) = \sqrt{\omega_1/\omega_0}a_m. \tag{7.29}$$

A more accurate analysis of patch growth without the restriction $\omega_1 \gg \omega_0$ has been performed by kinetic Monte Carlo simulations (Bobeth et al., 2011). In Figure 7.16, the patch size is plotted as a function of time for four simulation runs with various frequency ratios. For each run, the initial nucleus was a square with edge size $a_0 = 3a_m$.

(a)

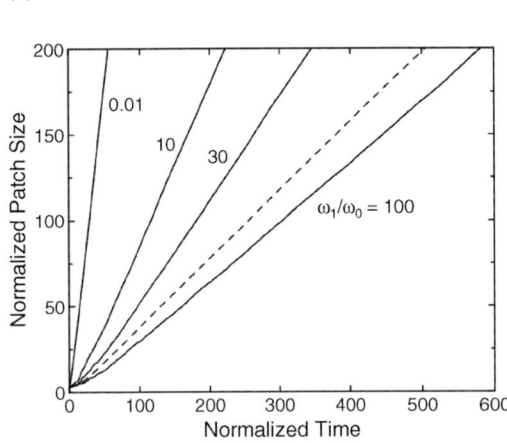

(b)

Figure 7.16 Monte Carlo simulations of self-assembly of S-layer patches. (a) Normalized patch size = square root of the number of monomers within a patch versus normalized time $\omega_0 t$. Kinetic parameters of the Monte Carlo simulations for the four curves are $\{\omega_1/\omega_0, \omega_2/\omega_0, \omega_3/\omega_0\} = \{10^{-2}, 10^{-4}, 10^{-6}\}$, $\{10, 0.1, 10^{-3}\}$, $\{30, 0.03, 3 \times 10^{-5}\}$, and $\{100, 10^{-2},$ $10^{-6}\}$. Dashed line: analytical estimate for $\omega_1/\omega_0 = 100$ for comparison. (b) Snapshots of patch shapes obtained from Monte Carlo simulations whose results are plotted for cluster size $\sqrt{z} = 200$. The numbers at the images are the ratio ω_1/ω_0. (Reproduced with permission from Bobeth *et al.* (2011). Copyright 2011, American Chemical Society.)

Since the arising patches are not really squares (Figure 7.16b), we introduce the normalized patch size $\bar{a}/a_m$ as the square root of the number of monomers in the patch $\bar{a}/a_m = \sqrt{z}$. After a transient stage, all curves show a roughly linear increase with time. The snapshots of patches in Figure 7.16b (corresponding to the parameters of the simulations given in Figure 7.16a) demonstrate that with higher frequency ratio ω_1/ω_0, the patch shape tends to change from rugged circle to near square. The time dependence of the patch size according to the analytical estimate is shown in Figure 7.16a (dashed line) in comparison with a Monte Carlo run for $\omega_1/\omega_0 = 100$. The slopes of the two curves agree well in the nearly linear range. The differences are larger for low ω_1/ω_0. Thus, the analytical estimate (Eq. (7.24)) serves as an approximation for high ratios ω_1/ω_0, that is, for low supersaturation.

Size and Shape of S-Layer Assemblies The final size of self-assembled S-layer patches is determined by the initial monomer concentration n_{10} and the nucleation rate J of stably growing patches. With increasing growth time, the pool of monomers is depleted, which determines the final patch size. The monomer density $n_1(t)$ initially decreases with time:

$$n_1(t) \approx n_{10} - \frac{1}{3}\left(\frac{v_0 t}{a_m}\right)^2 J_0 t, \tag{7.30}$$

where $v_0 = v(n_1 = n_{10})$ and J_0 are the initial patch growth velocity and nucleation rate, respectively (Bobeth *et al.*, 2011). Extrapolation of Eq. (7.30) to longer times suggests that monomers are depleted after the time

$$t_{\mathrm{MD}} \approx \left(\frac{3 n_{10} a_{\mathrm{m}}^2}{J_0 v_0^2} \right)^{1/3}.$$

(7.31)

The nucleation rate can be estimated by applying the models derived in Section 6.2.1.2 for homogeneous nucleation. The nucleation rate J_0 is given by

$$J_0 = f_0 \cdot n_c = f_Z \cdot \omega_c \cdot n_{10} \cdot \exp\left(-\frac{\Delta G_c}{k_B T} \right),$$

(7.32)

where the kinetic factor ω_c is the incorporation frequency of a monomer at the patch of critical size and n_c is the volume density of patches with critical size a_c. ΔG_c notes the maximum of the total free enthalpy change $\Delta G(a)$ due to monomer agglomeration. $\Delta G(a)$ consists of a term proportional to the patch size (decrease of the free enthalpy due to incorporation of the free monomers in the S-layer lattice) and a contribution caused by unsaturated bonds along the edges of the patch (with energy per length, κ_e):

$$\Delta G(a) = -\Delta g \cdot \frac{a^2}{a_m^2} + \kappa_e \cdot 4a$$

(7.33)

The free enthalpy change per monomer, Δg, can be expressed by the supersaturation of the solution, n_1/n_{1S}:

$$\Delta g = k_B T \ln (n_1/n_{1S}).$$

(7.34)

The maximum of $\Delta G(a)$ represents the critical patch size a_C and the critical free enthalpy change ΔG_C for nucleation:

$$a_C = \frac{2\kappa_e a_m^2}{\Delta g},$$

(7.35)

$$\Delta G_C = \frac{(2\kappa_e a_m)^2}{\Delta g} = 2\kappa_e a_c.$$

(7.36)

Patches grow stably for $a > a_C$. The critical patch size a_c and free enthalpy change ΔG_c decrease with increasing supersaturation (Figure 7.17).

The family of curves is generated by variation of the presently unknown edge energy. With an assumed supersaturation of $n_1/n_{1S} = 10$ and an S-layer "model unit cell" of size $a_m = 6$ nm, a critical patch size a_c of about $2-4a_m$ and a critical free enthalpy change ΔG_c of about $5-20 k_B T_r$ result for edge energies of $10-20$ meV/nm (with $T_r = $ room temperature). The frequency ω_c can be expressed by the growth velocity of stable patches $\dot{a}$:

$$\omega_c = \frac{d}{dt} \frac{a^2}{a_m^2} \bigg|_{a=a_c} = \frac{2a_c}{a_m^2} \dot{a} \bigg|_{a=a_c} = \frac{2a_c}{a_m^2} \cdot v.$$

(7.37)

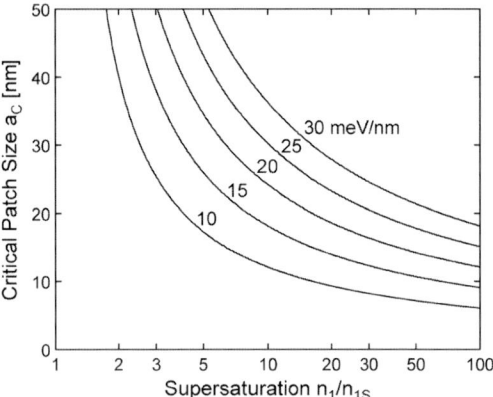

Figure 7.17 Critical patch size a_c versus supersaturation of monomers, with edge energy κ_e as parameter ($a_m = 6\,\text{nm}$, $T = 300\,\text{K}$).

According to observations, the patch growth velocity is typically on the order of 0.1 nm/s. Following classical nucleation theory (Haasen, 1991; Schmelzer *et al.*, 1999), an additional so-called Zeldovich factor f_Z with

$$f_Z = \sqrt{\frac{a_m^2}{2\pi k_B T}\left(-\frac{\partial^2 \Delta G(a)}{\partial a^2}\right)_{a=a_c}} = \sqrt{\frac{\kappa_e a_m}{2\pi k_B T}}\left(\frac{k_B T \ln\left(n_1/n_{1S}\right)}{2\kappa_e a_m}\right)^{3/2} \quad (7.38)$$

has to be applied that describes the influence of the shape of nucleation barrier on the nucleation rate. Equation (7.32) yields a steep increase of the nucleation rate with the supersaturation n_1/n_{1S} (Figure 7.18).

This means that the final patch size can be controlled by an appropriate choice of the supersaturation. The many small patches grown at high supersaturation tend to

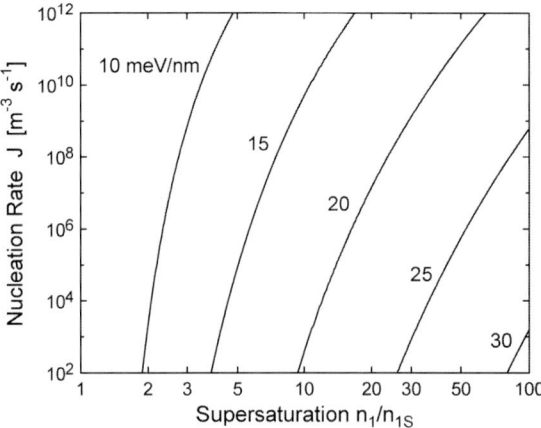

Figure 7.18 Nucleation rate versus supersaturation for various edge energies κ_e ($a_0 = 6\,\text{nm}$, $T = 300\,\text{K}$, $n_{1S} = 5 \times 10^{19}\,\text{m}^{-3}$, $D = 5 \times 10^{-15}\,\text{m}^2/\text{s}$, and $\omega_1 = 0.2\,\text{s}^{-1}$).

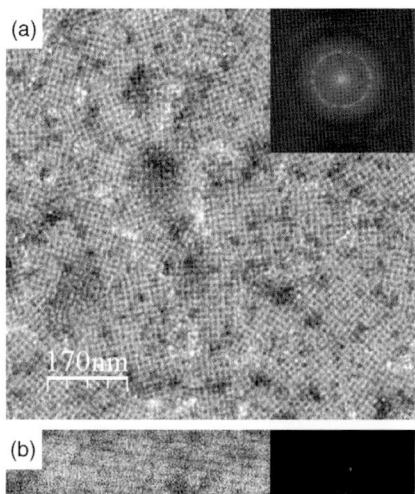

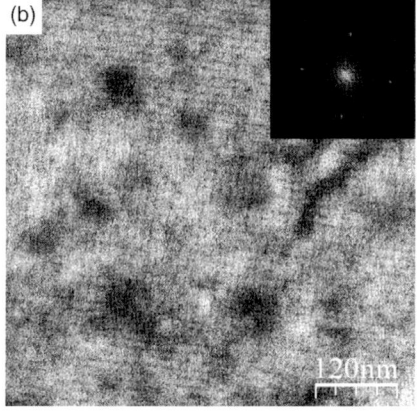

Figure 7.19 Electron micrographs of S-layers of *B. sphaericus* NCTC 9602 recrystallized on a carbon-coated copper substrate (TEM grids) (Nirschl *et al.*, 2009). (a) Polycrystalline structure deposited from high initial monomer concentration (1 mg/ml). The inset shows a circle-like distribution of reflexes in the power spectrum. (b) Monocrystalline structure deposited from low initial monomer concentration (0.1 mg/ml). The inset shows the corresponding power spectrum. (Reproduced with permission from Bobeth *et al.* (2011). Copyright 2011, American Chemical Society.)

associate into larger aggregates, resulting in a polycrystalline structure with high defect density. At low supersaturation, the nucleation rate is low, so the few patches grown from the few nuclei will be large. Self-assembled structures of the S-layer of *B. sphaericus* NCTC 9602 are shown in Figure 7.19. With high monomer concentration of 1 mg/ml, a polycrystalline structure with nanopores is formed (Figure 7.19a). Numerous small crystalline domains cause a circle-like distribution of reflexes in the power spectrum, corresponding to a lattice constant of 13 nm. At a concentration of 0.1 mg/ml, the monomers self-assemble into a large monocrystalline S-layer (Figure 7.19b). Four sharp reflexes in the power spectrum (inset) indicate *p*4 lattice symmetry with a lattice constant of 13 nm. In both cases, S-layer formation was triggered by addition of 1 mM $MgCl_2$.

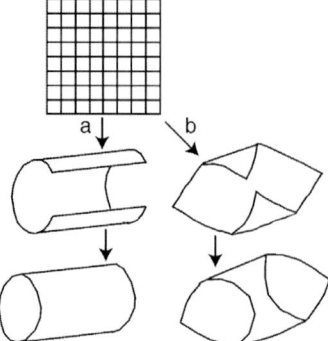

Figure 7.20 Possible tube formation mechanisms for larger S-layer patches with *p4* crystal symmetry as proposed in Pum *et al.* (1989). Depending on the binding between the proteins, the axis of curvature can be directed parallel to the edge (a) or diagonal (b).

S-layers formed by self-assembly of monomers are intrinsically curved, which can lead to the formation of tubes when the edges meet (Figure 7.20) (Pum *et al.*, 1989). The curvature depends on the molecular structure of the protein monomers and on the environment via interaction with surface charges. Since the S-layer structure is not symmetric with respect to its two faces, the two surface tensions are not equal. Hence, variations of pH, temperature, ionic strength, or presence of divalent cations affect the two sides differently, which results in changes of curvature. Tubes formed from self-assembled S-layer proteins of *B. sphaericus* and of *G. stearothermophilus* (Figure 7.12) correspond to the two closure modes. Figure 7.21 shows a tube formed by an S-layer of *B. sphaericus* shortly after closure, with the axis parallel to the diagonal of the *p4* lattice.

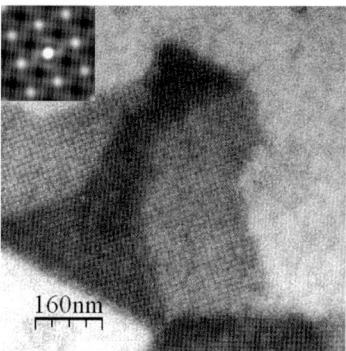

Figure 7.21 Electron micrograph of the early stage of an S-layer tube of *B. sphaericus* NCTC 9602 negatively stained. The left inset shows the periodicity of the lattice after an autocorrelation analysis applying the program WSxM for image analysis. The right inset shows the *p4* symmetry of the lattice power spectra obtained by FFT analysis of the electron micrograph revealing a lattice constant of about 13 nm. (Reproduced with permission from Bobeth *et al.* (2011). Copyright 2011, American Chemical Society.)

Tube formation by self-assembly of monomers is restricted to a limited region of supersaturation. The lower limit for tube formation is given when the patch size a_T necessary for tube formation cannot be reached due to fast depletion of the limited initial monomer concentration during stable growth of few patches. An upper limit of supersaturation is reached if the nucleation rate is so high that too many patches are formed that cannot grow sufficiently large before the monomer solution is depleted due to the competitive adsorption at many patches. This condition can be expressed in a relation between the time t_{TN} needed for tube growth and the depletion time t_{MD}. Tubes can only be observed when at least the time t_T for a single tube is shorter than the monomer depletion time. A crude lower estimate of the tube formation time t_T can be obtained from the initial patch growth velocity v_0 with $t_T = a_T/v_0$, where the patch size a_T is related to the tube radius ($a_T = \sqrt{2\pi r_T}$ if axis is parallel to the patch diagonal). Here, we suppose $a_T \gg a_C$ and neglect a transient time needed to form a stable patch. In order to be observable, tubes should reach at least a certain minimum concentration n_T in the solution, for example, on the order of 10^{12} m^{-3}. An approximate value for the time t_{TN} to reach this density is given by $t_{TN} = t_T + n_T/J_0$. Thus, we finally find $t_{TN} < t_{MD}$ as a necessary condition for the observation of tubes. For the given tube radius and tube density, this relation provides upper and lower bounds for the initial monomer density n_{10} since the initial patch growth velocity $v_0 = v(n_1 = n_{10})$ and the initial nucleation rate J_0 depend on n_{10}. With $J_0(n_{10})$ from Eq. (7.32), $v_0(n_{10})$ from Eq. (7.24), and t_{MD} from Eq. (7.31), we get an estimate of the processing region to form tubes. As shown in Figure 7.22, the range of supersaturation leading to tube formation depends on the tube radius r_T.

Any modification of the amino acid sequence of an S-layer should be reflected in the tube radius. It has been shown that the radius can be changed in a large range by genetic modification of the natural S-layer protein (Jarosch et al., 2001; Blecha, 2005; Bobeth et al., 2011). For the natural protein of G. stearothermophilus ATTC

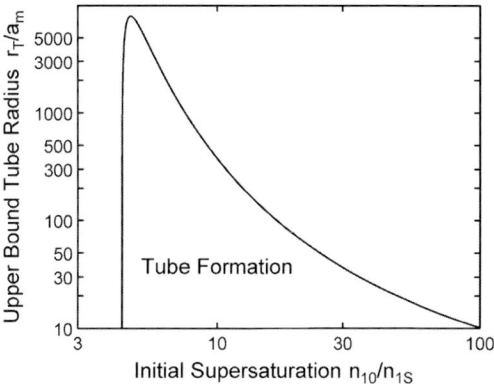

Figure 7.22 Normalized maximum tube radius r_T/a_m as a function of the initial monomer supersaturation n_{10}/n_{1S}. Parameters: $n_T = 10^{12}$ m^{-3}, $n_{1S} = 1\,\mu$M, $D = 10^{-15}$ m^2/s, $\omega = 0.1$ s^{-1}, $a_m = 6$ nm, $\kappa_e a_m/k_B T = 3.48$. (Reproduced with permission from Bobeth et al. (2011). Copyright 2011, American Chemical Society.)

12980, for instance, a tube radius of 35–50 nm has been observed (Blüher, 2008). By adding 12 histidine molecules at the C-terminus of the protein, only a negligible change in the tube radius (45 nm) has been measured, whereas the fusion of 121 amino acids (heat shock protein) at the C-terminus gave an average tube radius of 25 nm. With the fusion of a green fluorescent protein (239 amino acids) at the C-terminus, the radius decreased to 15–20 nm. Alternatively, the truncation of 219 amino acids from the C-terminus increased the tube radius to 125–145 nm. For proteins with larger truncations of the C-terminus, tube formation could not be observed.

7.2.5
Self-Organized Structures of Lipids

Lipids are a particular class of amphiphatic molecules. As explained in Chapter 1, they consist of a polar or negatively charged head group that is soluble in water and a nonpolar tail of one or two hydrocarbon chains that are insoluble in water. The nonpolar, hydrophobic regions of the molecule tend to minimize contact with the water. It means that the spontaneous formation of self-organized structures of lipids is driven by the interaction of the amphiphatic molecules with water. The nonpolar tails are organized to present a minimum of hydrophobic area to the aqueous surrounding, whereas the hydrophilic head groups form extended surfaces to surround the assembly of hydrophobic tails. It means that the forces that hold the tails together, called hydrophobic interactions, are the result of the thermodynamic equilibrium between the amphiphatic molecules and the surrounding water.

Depending on the shape of the molecules and their concentration, three types of macrostructure are observed (Figure 7.23):

i) *Micelle: spherical or cylindrical structure*
    The head group of the lipid is wider than the tail. Contact of the tails with water can be avoided if the head groups are arranged in a spherical or cylindrical shell with tails in the interior.
ii) *Bilayer: planar structure*
    Contact of the tails with water is avoided, except at the boundary where they contribute to a line tension along the circumference that tends to curve the bilayer into a vesicle. The hydrophobic tails inside the bilayer hinder transport of polar solutes across. The bilayer may contain proteins. Usually bilayers are composed of many different lipids.
iii) *Vesicle: spherical or cylindrical cavity*
    The closed bilayer constitutes a separate aqueous compartment. Vesicles are important structural elements in intracellular transport, exo- and endocytosis (out of and into cell), and biomineralization.

The different macrostructures of self-organized lipid structures are mainly controlled by the molecule shape. Molecules with one alkyl chain, for example, detergents such as lysophospholipids and polyphosphoinositides, form micelles (a), whereas cylindrical lipids, for example, phosphatidylcholine and sphingomyelin,

(a)

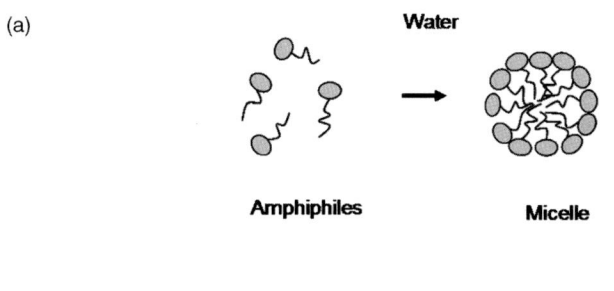

Water

Amphiphiles   Micelle

(b)

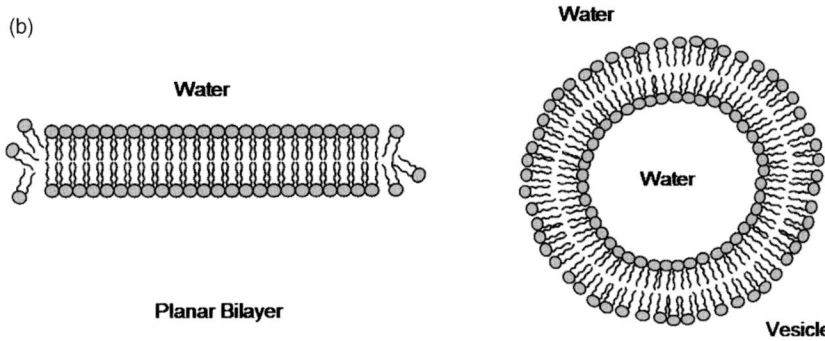

Water

Water

Water

Planar Bilayer

Vesicle

Figure 7.23 The three types of self-organized lipid structures.

form flat bilayers, cylinders, or vesicles (b). Lipids with tails wider than heads, for example, diacylglycerol and phosphatidylethanolamine, arrange themselves into structures with hydrophilic heads inside (inverted micelles or hydrophilic channels). The shape of lipids can be generally described by three parameters: S_0, the area of the polar head group; l, the maximum molecular length; and v, the molecular volume of the hydrophobic domain. The ratio $v/(lS_0)$ allows an estimation of the ratio of interfacial area to the hydrophobic and hydrophilic regions. The ratio is also called the critical packing parameter. It allows a rough classification of the shape of the lipid assemblies. If $v/(lS_0) \leq 1/3$, the lipids usually are arranged in a micellar-type shape. For $1/3 \leq v/(lS_0) \leq 1$, bilayers are observed, which form preferentially spheres or cylinders for $1/3 \leq v/(lS_0) \leq 1/2$. If $v/(lS_0) > 1$, hydrophilic channels or inverted micelles are formed.

Structure–Property Relations of Lipid Bilayers   With bio-inspired materials engineering in mind, we will try to understand the design principles of lipid bilayers realized in biological membranes in more detail. Lipid bilayers are the major constituent of the cell membrane, synonymously also often termed "cytoplasmic membrane," or "plasma membrane". The real structure of a lipid bilayer in a typical eukaryotic cell is very different from an ideal bilayer composed of two symmetric leaflets of lipid monolayers formed by the assembly of only one molecule type. Typically, the

(a)

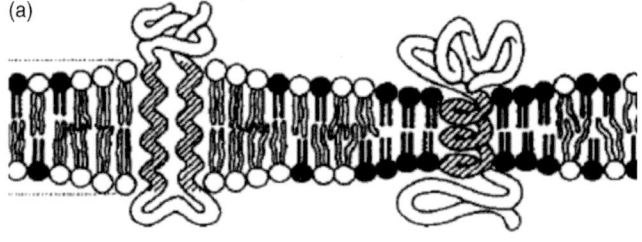

(b)

Figure 7.24 Mechanisms of selective lipid–protein interaction. (a) Interaction of the hydrophobic domains of proteins and the lipid bilayer. (b) Electrostatic selectivity by binding of filamentous protein (e.g., spectrin) to bilayers. Neutron surface scattering studies suggest that the flexible hinges between the condensed domains penetrate into the bilayer. (Reproduced with permission from Sackmann (1994). Copyright 1994, Elsevier.)

thickness of the bilayer is in the range of about 3–4 nm. In a cell membrane, there are usually hundreds of different lipid types assembled on continuous surface areas of hundreds of square micrometers (Janmey and Kinnunen, 2006). The chemical composition of the two leaflets is very different. In eukaryotic cells, for example, anionic lipids mainly face the cytoplasm, whereas lipids with large glycosylated head groups are assembled along the outer surface. Considerable amounts of other molecules, such as cholesterol, may also be present on either leaflet.

The chemical composition and the extent of asymmetry essentially determine the mechanical behavior and the biophysical and biochemical responses of the lipid bilayers. Proteins can be embedded in a lipid bilayer by appropriate coupling via lipid-like hydrophobic surface groups (Figure 7.24a). Globular proteins or ion channels with a belt of hydrophobic domains can be arranged in the bilayer with preferred orientation. A possible mismatch between the domain size and the extension of the lipid tails in the bilayer causes some localized structural relaxation of the bilayer. Furthermore, proteins can be tethered with a hydrophobic tail to the lipid bilayer surface. Alternatively, filamentous proteins can bind via electrostatic interaction selectively to the head groups of the bilayer (Figure 7.24b).

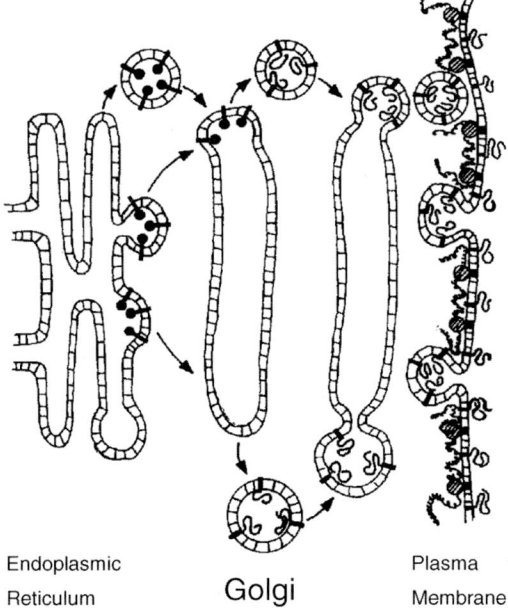

Endoplasmic

Reticulum   **Golgi**

Plasma

Membrane

Figure 7.25 Schematic drawing of one possible way of intracellular transport from the endoplasmic reticulum to the plasma membrane mediated by vesicles. (Reproduced with permission from Sackmann (1994). Copyright 1994, Elsevier.)

The lipid bilayer within the cell membrane mainly fulfills passive functions of an envelope. Its hydrophobic core makes an ideal bilayer almost impermeable to water and solutes. Lipid bilayers also realize some active functions via transmembrane channels and peripheral membrane-binding proteins. The gating of mechanically sensitive channels and the stress activation of enzymes are essentially controlled by mechanical forces exerted by the lipid bilayer. Furthermore, the self-organization capability of lipids enables the vesicle-mediated molecular transport among intra-cellular compartments and between these and the plasma membrane. The transport from the endoplasmic reticulum to the plasma membrane is schematically shown in Figure 7.25. It is based on a sequence of budding–fission–fusion events of vesicles. These processes demonstrate the high flexibility of the vesicle shape and an internal control of vesicle formation.

To get an understanding of the mechanical behavior of lipid membranes, it makes sense to consider first an ideal bilayer consisting of two leaflets of the same molecule type. Such a structure already realizes the basic physical properties of a real mem-brane. It is also an ideal starting system for biomimetic materials development. Owing to the very weak nonspecific interaction of the carbon chains, the individual molecules are highly mobile (self-diffusion coefficients of about $10^{-7}\,\mathrm{cm}^2/\mathrm{s}$) within the self-organized structures. The interaction between the two leaflets is weak compared to the interaction among the lipid molecules within the layer. The exchange of single molecules between a monolayer and the surrounding solution is negligible as their equilibrium concentration in water, $c_{equ.} = c_0 \cdot \exp - (\Delta\mu/k_B T)$ with $\Delta\mu \approx 30 k_B T$, is extremely low, less than 10^{-12} M. Therefore, the number of lipid molecules within a bilayer of a vesicle practically does not change. Hence, the volume V of the enclosed water and the surface area A of a vesicle also remain constant at given temperature. Also, the exchange of molecules in a bilayer from one monolayer to the other one (the so-called "flip-flop" process) proceeds at a very low rate because there is a large activation barrier for dragging a head group through the hydrophobic hydrocarbon region. Exchange processes of phospholipids typically occur in the timescale of hours up to days. However, other amphiphile components of the bilayer (e.g., cholesterol) can flip much faster. Furthermore, specific protein inclusions, the so-called flippases, floppases, and scramblases, can catalyze the transbilayer movement of lipids. The energy-dependent transport (controlled by ATP $\rightarrow$ ADP $+ P_i$ reactions) is much faster and enables the formation of a transbilayer asymmetry.

The high fluidity of the bilayer favors quick transport of proteins as, for instance, free integrins (see also Section 3.1) or transmembrane enzymes (diffusion coefficients in the range of about 10^{-9} cm^2/s) (Renner *et al.*, 2010). The high in-plane mobility of the molecules implies negligible shear strength and hence negligible resistance to uniaxial tension or compression. The high mobility also explains the observed self-healing of defect structures in lipid bilayers.

7.2.6
Liquid Crystals

The term "liquid crystal" (LC) refers to a state of condensed matter that combines fluidity with anisotropy. The anisotropy of the LCs reveals itself by their optical properties. It is found to be due to a textured microstructure with micrometer-sized domains. LCs are composed of rod- or disk-like molecules. LCs can also be found in biological materials. One example is the structure of spider silk (Magoshi *et al.*, 1985). Whereas the constituents of solid crystals – the atoms or molecules – are arranged in long-range positional and directional order, the molecules in LCs show only *long-range orientational order but no long-range positional order*. LCs are subdivided into thermotropic, lyotropic, and metallotropic LCs. Thermotropic LCs are formed by certain organic substances when a melting temperature is exceeded. Beyond another temperature threshold, they turn into "normal" liquids. Lyotropic LCs are highly concentrated liquid solutions of some organic substance. Dilution below a certain concentration turns them normal. The phase transition depends on temperature. Metallotropic LCs are composed of mixtures of organic and inorganic molecules. Their phase transition depends on the mixture ratio, in addition to temperature and concentration. Depending on the degree of positional disorder, *nematic and smectic LCs* are distinguished. As indicated in Figure 7.26, the orientation of the molecules in nematic LCs is rather uniform, while their other coordinates are randomly distributed. Smectic LCs are to a certain degree in positional long-range order with the molecules being assembled in layers. If the layers extend perpendicular to the direction of the molecules, the LC is called a smectic A phase. In

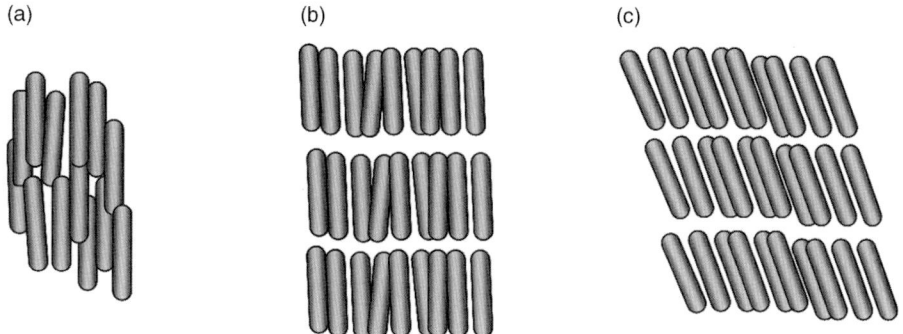

Figure 7.26 Basic phases of LC. (a) Nematic phase. (b) Smectic A phase. (c) Smectic C phase.

case of a tilt between the two directions, the LC is called a smectic C phase. In either case, there is no order within the layers.

Some rod-like biomolecules have chiral symmetry, which implies that there is no inversion symmetry. Typical examples are the helical structures of the coat proteins of rod-like viruses such as the M13 virus or the tobacco mosaic virus (TMV). The structure of LCs formed by chiral molecules reflects that chirality in the growth process. As shown schematically in Figure 7.27, chiral phases can grow from chiral molecules (see also the examples given in Section 7.3). Such structures are characterized by an additional twist of the molecule assembly, with the twist axis perpendicular to the longitudinal molecular axis (chiral nematic phase, often also called cholesteric nematic phase because it was first observed for cholesterol derivatives), or by spiral twisting of the molecular axis around the layer normal (smectic C* phase).

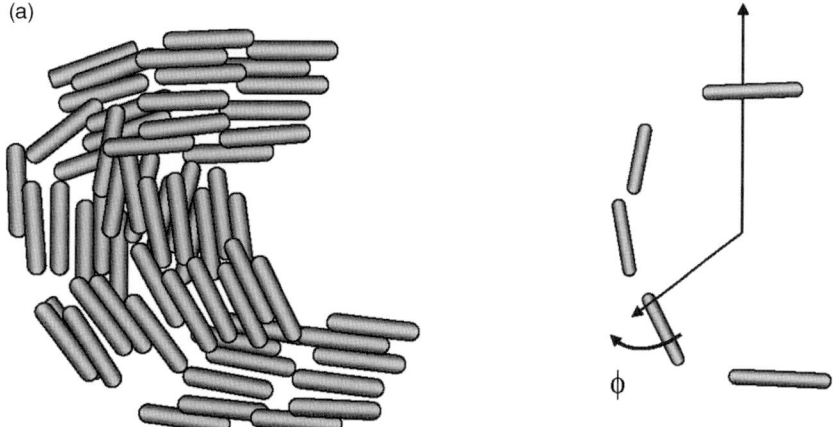

Figure 7.27 Schematic drawings of the microstructure of chiral phases. (a) Cholesteric nematic phase. (b) Smectic C* phase.

(b)

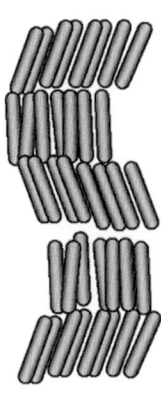
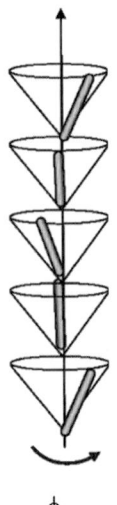

ϕ

Figure 7.27 *(Continued)*

7.3
Bioengineering

Self-assembly is one of the essential concepts in bionanotechnology. It facilitates the formation of complex structures from nanoparticles or other small units by means of a bottom-up process. Three options for such a process are listed here (Scheme 7.1):

First, complex nanostructures can be manufactured *in vitro* by self-assembly of large-scale nanostructured materials, for example, from natural biomolecules, genetically modified proteins, or supramolecular analogues such as tailored peptides. Second, biomolecules can be used for the generation of nanostructured templates that direct the assembly of artificial nanoparticles for manufacturing complex nanostructures. Third, the interaction of artificial nanoparticles can be modified by conjugation with biomolecules in a way that leads to a template-free directed self-assembly of the nanoparticles.[1]

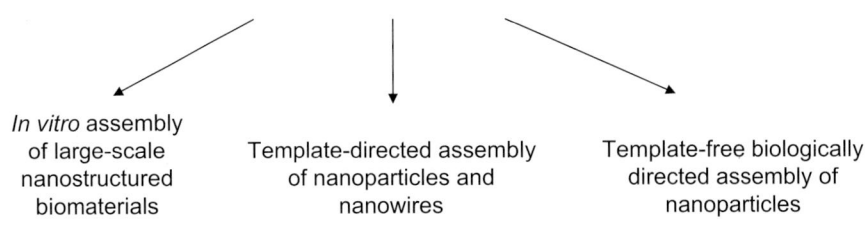

Biologically directed self-assembly of functional materials

In vitro assembly of large-scale nanostructured biomaterials

Template-directed assembly of nanoparticles and nanowires

Template-free biologically directed assembly of nanoparticles

Scheme 7.1 Processing routes for the generation of functional materials based on self-assembly of biomolecules.

1) With the notations "template-directed" and "directed self-assembly of nanoparticles", we follow a proposition made by Grzelczak *et al.* (2010).

The availability of *monodisperse nanosized structures capable of reversible adsorption with high selectivity of the binding sites* is the essential contribution of biology to the self-assembly in the processing routes mentioned above. Oligonucleotides, peptides, small proteins, antigen–antibody constructs, and viruses cover a wide range of applicable components. In addition, tailored properties of the inorganic nanoparticles (e.g., magnetization, polarizability, and particle shape) can be applied for controlling the assembly process by means of external fields (e.g., magnetic or electric field and hydrodynamic flow). Examples are discussed in the following.

7.3.1
In Vitro Self-Assembly of Large-Scale Nanostructured Biomaterials

Large-Scale DNA Assemblies   As mentioned in Section 1.2.1, the specific hybridization of base pairs is one of the most interesting properties of DNA molecules. Modification of the base sequences allows complex synthetic structures to be created. With the intention to generate large-scale DNA assemblies, the already mentioned "tile model", first proposed by Seeman in 1982 (Seeman, 1982), has turned out to be a powerful tool. The tiles formed by a small number of DNA strands are stiffened by crossing points and contain a finite number of free sticky ends protruding from the corners of the tiles. The base sequences of these sticky ends govern the self-assembly of the tiles into large-scale structures, such as 2D lattices, nanoribbons, nanotubes, and other complex 3D structures. For practical use, the so-called double crossover (DX) molecules and the triple crossover (TX) molecules have proven appropriate tile structures.

A relatively simple DX structure can be generated with two identical strands consisting of four self-complementary segments (Figure 7.28) (Liu *et al.*, 2006).

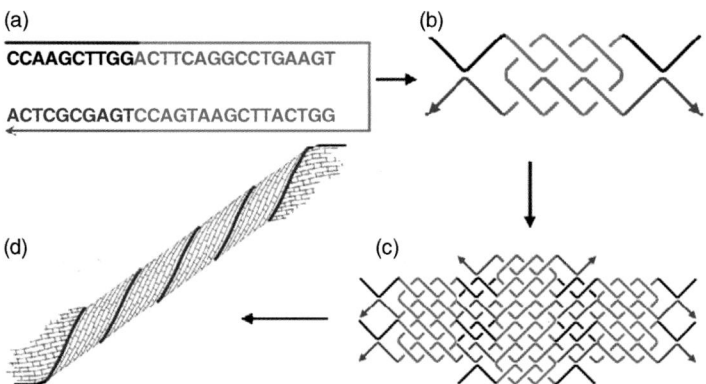

(a)

CCAAGCTTGGACTTCAGGCCTGAAGT

ACTCGCGAGTCCAGTAAGCTTACTGG

(b)

(d)

(c)

Figure 7.28   Schematic representation of the self-assembly of a DNA nanotube. DNA sequence of a single-strand containing four self-complementary segments (a). The arrow indicates the 3′-end of the DNA strand. As temperature decreases, the strands first form a two-stranded tile (b). The tiles associate with each other to form two-dimensional sheets (c) and eventually fold into nanotubes (d). (Reproduced with permission from Liu *et al.* (2006). Copyright 2006, Wiley-VCH Verlag GmbH.)

The DNA solutions with concentrations between 0.5 and 1.0 µM were slowly cooled from 95 °C to 4 °C over 48 h. First the single strands aggregate into individual tiles with two double-stranded regions and four single-stranded overhangs at the corners. The tiles can aggregate into a 2D lattice. The crossovers are connected with a number of nicks increasing the flexibility of the tiles. This can induce regions of localized curvature of the tile array. Such regions can act as nucleation sites for tube formation. Two growth modes compete in the growing aggregate: planar growth and tube growth. The outcome of this competition is kinetically controlled. For strands with 52 bases, it has been found that tubes, 20–45 nm in diameter and up to 60 µm long, are preferentially formed at low concentrations (around 1 µM). At higher concentrations, large-scale patch growth dominates. Tube formation is favored since it minimizes the number of unpaired sticky ends. It can be supported by some intrinsic curvature of the individual tiles. DNA nanotubes are useful components in nanotechnology as they are stiff and offer the possibility of various additional functionalizations. As in Section 1.3.2, targeted molecular species can be linked with the single-stranded DNA, either by covalent attachment of functional groups (as thiol and amino groups) or molecules (as biotin) or by extension of the assembling DNA with single-stranded or stem loops. Such additionally linked DNA segments can capture targets with complementary sequences.

Large functionalized DNA tiles can be used as a starting point for a technological platform on the submillimeter scale with the aim to generate arrays of quantum dots (QDs), proteins, or particular DNA targets. This idea that involves bottom-up and top-down techniques has been put forward by Lin *et al.* (2007) (Figure 7.29). In the bottom-up step, DNA nanotubes are self-assembled by using a 52-mer single-stranded oligomer. In the single strand, a biotinylated thymine is incorporated at such a position that it is always facing out of the nanotube. By molecular combing (see Section 1.3.2), the nanotubes are aligned onto a PDMS (polydimethylsiloxane) stamp patterned with regularly spaced wells. Subsequently, streptavidin (STV)-conjugated QDs are linked to the biotin binding sites on the nanotubes aligned on the PDMS stamp. The subsequent printing corresponds to the top-down step in the process. It allows a well-organized transfer of the QD pattern onto a substrate of at least a few hundred micrometers.

Peptide Nanotubes Peptide nanotubes (PNTs) are a new class of artificial nanomaterials with promising biochemical and physical properties for potential applications such as biosensors, drug delivery, molecular electronics, and templates for the preparation of inorganic nanomaterials. In comparison to DNA, they are stiffer and more thermally stable. Their design principles allow the well-controlled incorporation of specific binding sites. As they are built from natural amino acids, they are biocompatible *per se*. This favors these nanotubes over those manufactured from inorganic materials such as carbon. Man-made PNTs have been presented first by Ghadiri *et al.* in the early 1990s (Ghadiri *et al.*, 1993; Bong *et al.*, 2001). They developed the idea of nanotubes self-assembling from flat ring-shaped cyclic octapeptides consisting of an even number of alternating D- and L-conformation

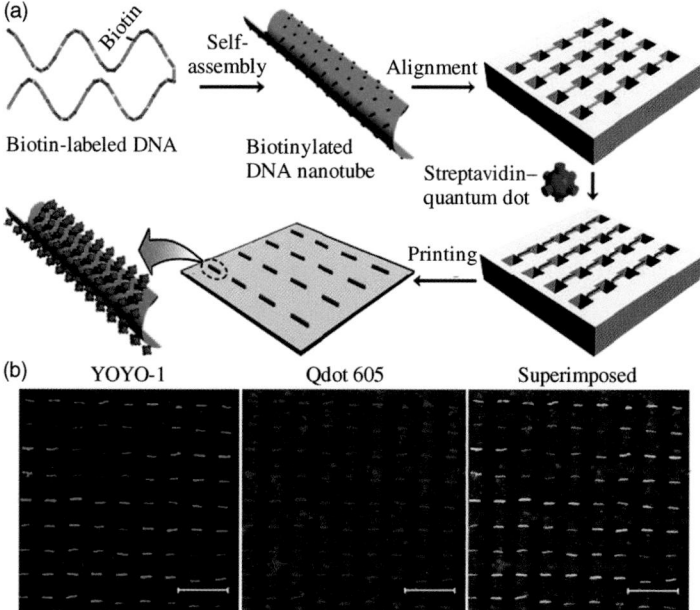

Figure 7.29 Alignment of STV–QD conjugate arrays templated by self-assembled biotinylated DNA nanotubes. (a) Schematic drawing of the procedure. (b) Confocal fluorescence microscopy images of the arrays on a glass surface. Images obtained in the green (left) and red (middle) channels show YOYO-1 stained DNA nanotubes and STV–QD (Qdot 605), respectively. A superimposed image is displayed on the right. Scale bar: 20 μm. (Reproduced with permission from Lin *et al.* (2007). Copyright 2007, Wiley-VCH Verlag GmbH.)

amino acids (Figure 7.30). Stacks of these rings, kept together by intermolecular hydrogen bonds, make a tube. The alternating isomers form β-sheet structures with their surface normal oriented parallel to the tube axis. For example, an eight-residue cyclic peptide with the sequence cyclo[—(L—Gln—D—Ala—L—Glu—D—Ala)₂—] when protonated, assembles into nanotubes hundreds of nanometers long and with internal diameters of 0.7–0.8 nm.

Later on, Zhang *et al.* (Santoso *et al.*, 2002; Zhang *et al.*, 2002) showed that nanotubes can also be assembled from linear amphiphilic peptides. Various hepta- and octapeptides had been chosen as building blocks. Similar to the assembly of lipids, they have a common design motif: a polar region consisting of one or two charged amino acids and a sequence of nonpolar residues made from hydrophobic amino acids. The simplest structures chosen as monomers were peptides (Gly-)$_n$(Asp-)$_2$ composed of n glycines ($n = 4,6,8,10$) forming a hydrophobic tail and two aspartic acids at the C-terminus forming the hydrophilic head group. At neutral pH, the head has a total of three negative charges from the carboxyl groups. The tail length of glycines varies, depending on the number of glycines, between 2.4 nm for $n = 4$ to 4.7 nm for $n = 10$. In aqueous solution, the monomers form a bilayer. The bilayers are folded into closed shapes. For $n = 4$, nanotubes have been observed with a diameter around 30–50 nm,

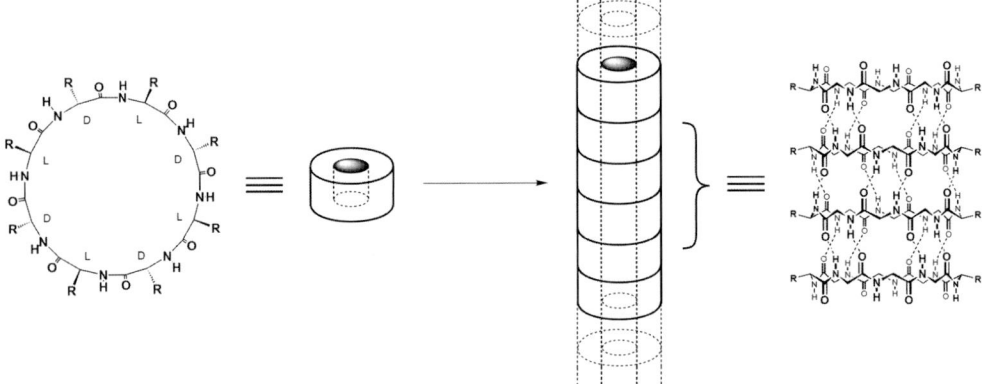

Figure 7.30 Self-assembled nanotube based on a cyclic peptide architecture. *Left*: Schematic drawing of the chemical structure of the peptide building blocks (D or L refers to the amino acid chirality). *Right*: The backbone structures of the assembled cycling blocks show the formation of a network of intermolecular hydrogen bonds between the blocks. (Reproduced with permission from Bong *et al.* (2001). Copyright 2001, Wiley-VCH Verlag GmbH.)

whereas for $n = 6$, vesicles are formed with a diameter of about 100 nm, along with nanotubes. For $n > 6$, interwoven nanotubes and membrane-like structures have been found. The formation of regular nanotubes for small peptides and the transition to larger vesicles can be interpreted as a result of the competition of the hydrophobic forces favoring closed shapes and the repulsive electrostatic forces inhibiting large curvature. Small peptides composed of glycine and aspartic acid are regarded as ancient examples of the prebiotic evolution on earth. It has been assumed that they were possibly the first amino acid sequences present in the prebiotic environment. It has been discussed that, among other options, the prebiotic molecular evolution may have begun in such isolated or semi-isolated environments. Similar to DNA, the peptide-based self-assemblies offer the advantage that the peptides can be tailored by adding additional functional groups by well-established techniques. For example, biotinylation of the monomers opens the wide range of coupling with other bio-molecules or inorganic nanoparticles.

For a generic approach to peptide-based nanotube synthesis, Gazit and coworkers tried to identify the simplest building blocks that can aggregate into a tube. With this motivation, they studied the pathological amyloid structures. As a remarkable characteristic of these structures, aromatic interactions play a central role in the molecular recognition of the folding domains and their stabilization. Reches and Gazit searched for the core recognition motif of the Alzheimer's disease, β-amyloid polypeptide (Aβ), which self-assembles into amyloid fibrils. When they considered the amino acid sequence of the β-amyloid polypeptide, they found a central aromatic core of diphenylalanine (-FF-) (Figure 7.31).

It can be shown that the diphenylalanine is indeed the core motif responsible for the amyloid structure. Monomers of the dipeptide $NH_2-Phe-Phe-COOH$ are

NH$_2$-DAEFRHDSGYEVHHQKLVFFAEDVGSNKGAIIGLMVGGVVIA-COOH

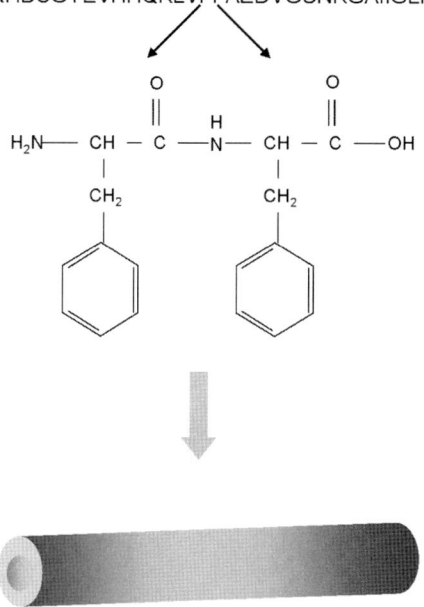

Figure 7.31 Self-assembly of PNTs by molecular recognition of motif derived from the β-amyloid polypeptide. The central aromatic core of the β-amyloid polypeptide involved in the molecular recognition process that leads to the formation of amyloid fibrils.

dissolved in the organic solvent hexafluoro-2-propanol (HFIP). When the concentrated HFIP solution is diluted with water down to a micromolar range, FF PNTs form rapidly. The monomers assemble as multiwall nanotubes with an internal diameter larger than 20 nm, external diameter in the range of 80–300 nm, and micrometer length. The PNTs are characterized by high Young's modulus and thermal stability (Table 7.2). The persistence length is around 1 μm. The tube-like shape of the assembly can be convincingly demonstrated by using the tubes as a template for growing metallic nanowires inside (see the following section). Interestingly, starting with concentrated HFIP dissolved in methanol at similar concentration as in the former aqueous solution leads to the formation of nanobeads with diameter of 2.1 nm and with a very narrow size distribution (~±0.1 nm) (Figure 7.32). The size of the nanobeads indicates that they are formed by the aggregation of two dipeptides.

Beads are formed in aqueous or methanol solutions with monomer concentrations above 1 mg/ml. The beads show optical fluorescence in the UV region at room temperature similar to inorganic semiconducting nanocrystals (Amdursky *et al.*, 2010b). A quantum effect, a particle size-dependent fluorescence, is known from inorganic semiconducting nanocrystals. Such so-called QDs have optical properties intermediate between those of bulk semiconductors and individual molecules. Quantum mechanical excitations of the semiconductor, called excitons, are nonlocal phenomena affecting the whole crystal. Hence, their optical properties depend strongly

Methanol   Aqueous solution

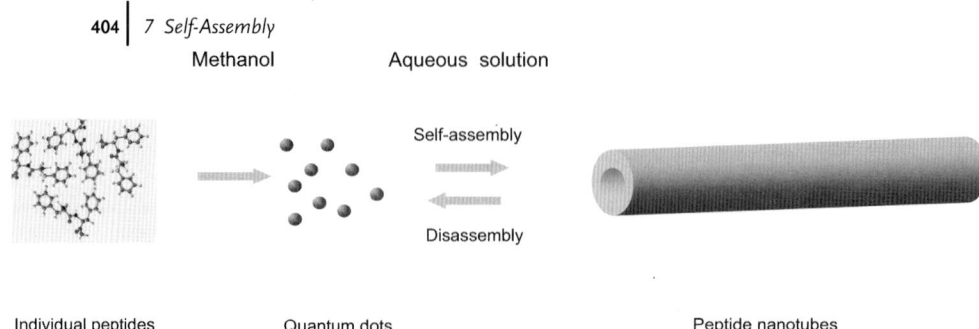

Individual peptides   Quantum dots   Peptide nanotubes

Figure 7.32 Schematic drawing of the self-assembly of diphenylalanine derived from studies of Gazit and coworkers. Formation of nanobeads composed of two diphenylalanine monomers in methanol. Reversible assembly of nanotubes in aqueous solution.

on the dimensionality and size of the nanostructures. Excitons are based on the Coulomb interaction of carriers of negative and positive charges – electrons and holes – in a crystalline structure. Inorganic QDs made of zinc sulfide (ZnS), cadmium selenide (CdSe), silicon, and germanium play a major role in applications for fluorescence markers, for example, in biomedicine, light-emitting devices, and solar cells.

Photoluminescence studies with the FF nanobeads revealed a broad adsorption peak in the range of 250–270 nm and a very sharp emission peak at 270 nm. The narrow photoexcitation peak is the result of the quantum confinement. Therefore, we speak of peptide quantum dots (PQDs). It has been shown that the synthesis of PQDs is not limited to the diphenylalanine system. Substituting one phenylalanine by tryptophan with a similar aromatic residue also allows the formation of QDs with a similar spectrum as for diphenylalanine. In this case, the exciton peak is located at 306 nm (Amdursky *et al.*, 2010a, 2010b). Peptide-based QDs offer some obvious advantages. First, as they are prepared from natural amino acids, there should be no problem with toxic side effects that often have to be handled carefully in case of inorganic QDs, especially those made of heavy metals. Second, the preparation from dipeptides is a very clean and cheap process.

There is another interesting observation concerning the structure relation between PNTs and PQDs made from diphenylalanine. Substituting water for methanol in the PQD solution causes the nanobeads to self-assemble into nano-tubes, a process described above. The transition between PQDs and PNTs is reversible. The optical properties of PQDs and PNTs are similar. For structures prepared from highly concentrated solutions, the photoluminescence excitation (PLE) peaks nearly coincide. The shift and broadening of the PLE peak of PQDs for lower concentration can be explained by the presence of additional free monomers. The optical behavior and the X-ray diffraction data indicate that the PNTs are regular assemblies of PQDs that probably are stabilized by hydrogen bonds between the backbone of the peptides and water molecules. Water molecules can enter through hydrophilic channels of the nanotube layer.

Amdursky *et al.* (2010a) have used the photoluminescence of the PNTs for the preparation of a luminescent hydrogel. By connecting the short FF motif with the

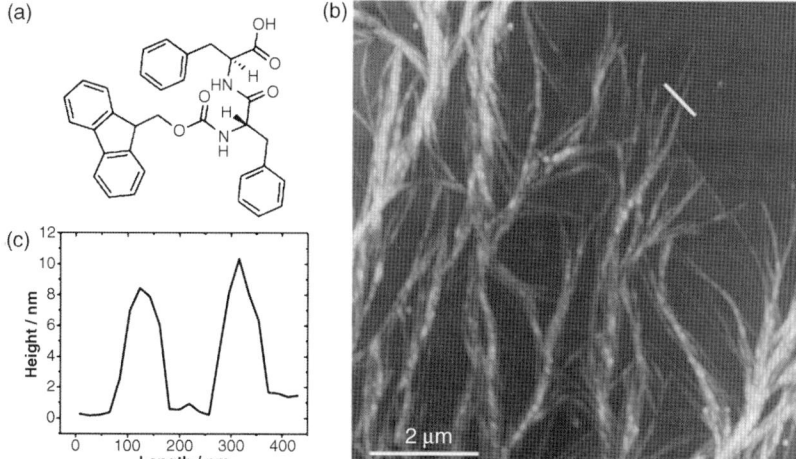

(a)

(b)

(c)

Figure 7.33 Morphology of the hydrogel. (a) Scheme of Fmoc—Phe—Phe—OH. (b) An AFM image of the hydrogel. The z-scale is 53 nm. (c) A cross section along the white line in part (b). (Reproduced with permission from Amdursky et al. (2010a). Copyright 2010, Wiley-VCH Verlag GmbH.)

protective group N-fluorenylmethoxycarbonyl (Fmoc), the hydrogel formation has been promoted (Figure 7.33). The hydrogel is composed of Fmoc-FF PNTs with diameters of 7–15 nm (Figure 7.33b and c). The photoluminescence is characterized by a narrow peak at $\lambda_{max} = 310$ nm with a full width at half maximum (FWHM) of 3.7 nm at room temperature. This narrow excitation peak indicates that the peptide is assembled in a crystalline structure. Therefore, the increase of that peak during the formation of the hydrogel is suitable for monitoring the growth kinetics of the crystalline structure, as shown in Figure 7.34. As proposed by Amdursky et al. (2010a), the exploration of the photoluminescence of FF peptide-based biomolecules offers various promising options for future development of novel types of biosensors, for monitoring of the level of amyloid fibrils, and also for ambitious development of new biophotonic devices, such as UV light-emitting devices, and biolasers.

With a minor change of the basic building block, the stability of the nanotube can be changed. Yan et al. (2007) modified the diphenylalanine in order to manufacture a carrier for intracellular delivery of DNA. For binding the negatively charged DNA at the peptide assembly, they used the cationic dipeptide (CDP) H—Phe—Phe—NH$_2$·HCl. Merely adding one positive charge to the diphenylalanine causes an additional topological transition of the cationic dipeptide nanotubes (CDPNTs). With monomer concentrations in aqueous solution below 7 mg/ml, the nanotubes transform gradually into vesicles with about 100 nm in diameter at physiological pH. Both structures coexist. Below 1 mg/ml, only vesicles are observed. Obviously, the repulsive electrostatic interaction (long range) overcompensates the tube stabilizing hydrogen bonding or π–π stacking interactions (short range) of the

(a)

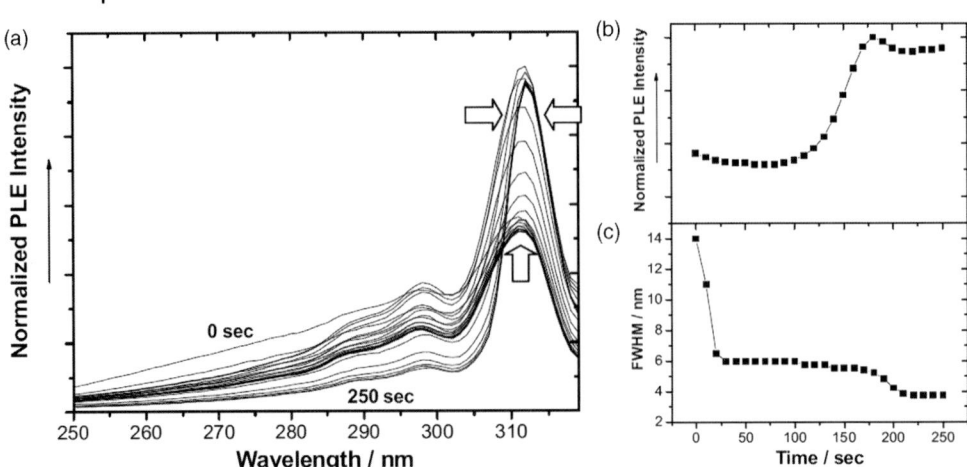

Figure 7.34 The dynamics of the hydrogel self-assembly process, followed by PLE spectrum. (a) Excitation spectra of Fmoc FF, concentration 3 mg/ml, taken at 10 s intervals during the first 250 s of the self-assembly process. The vertical arrow emphasizes the increased luminescence at the exciton wavelength. The horizontal arrows emphasize the narrowing of the exciton excitation peak. (b) Peak intensity versus time. (c) Optical width versus time, FWHM, of the peak at $\lambda = 310$ nm. (Reproduced with permission from Amdursky *et al.* (2010a). Copyright 2010, Wiley-VCH Verlag GmbH.)

peptide molecules at lower monomer concentrations. Probably, vesicles are favored over tubes here because of a shorter decay length of the electrostatic force. This transition behavior has been used for delivering oligonucleotides into the interior of cells (Figure 7.35). Single-stranded DNAs are immobilized at the cationic dipeptide nanotubes via electrostatic interaction. Afterwards, cells are incubated in a solution of CDPNT (concentration ≈ 0.5 mg/ml) loaded with oligonucleotides in a cell growth medium at 37 °C for 24 h. The CDPNT–oligonucleotide complexes enter the cells by converting into vesicles and accumulate in the cytoplasma of the cells. No cell death

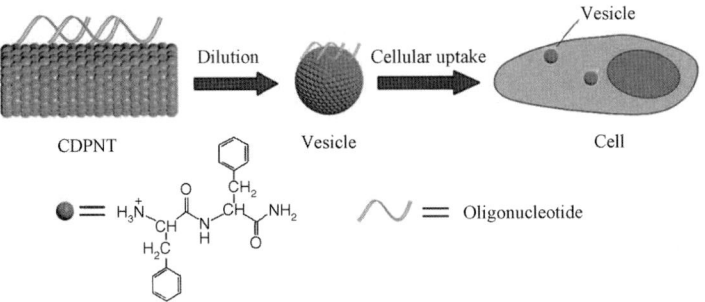

Figure 7.35 Schematic drawing of the transition of CDPNTs into vesicles during intracellular oligonucleotide delivery. (Reproduced with permission from Yan *et al.* (2007). Copyright 2007, Wiley-VCH Verlag GmbH.)

was observed in an experiment with HeLa cells. The experiment shows the high potential of CDPNTs as biocompatible nanotransporters for delivery of drugs, genes, and proteins.

The spatially aligned aromatic systems in diphenylalanine nanotubes feature another remarkable property, a *direct electron transfer along the nanotubes* (Yemini et al., 2005a). Therefore, they are promising candidates for nanostructured electrochemical biosensors. Similar to the inorganic counterpart, the carbon nanotubes (CNTs), it can be assumed that macroelectrodes structured with an array of electroconductive PNTs should show high electrochemical sensitivity. This idea has proved successfully, as first demonstrated by the Gazit group (Yemini et al., 2005b). Their electrochemical cell contains three electrodes: (i) a gold disk working electrode, (ii) a platinum wire counterelectrode, and (iii) a saturated calomel electrode (SCE) for reference. A design option for the working electrode for detection of β-D-glucose is shown in Figure 7.36. This sensor is based on the electroenzymatic detection of glucose, similar to the glucose sensor explained in Section 5.3.2. Glucose is oxidized, catalyzed by glucose oxidase (GOX), to gluconic acid and hydrogen peroxide. Thiol-modified PNTs are mixed with GOX in the

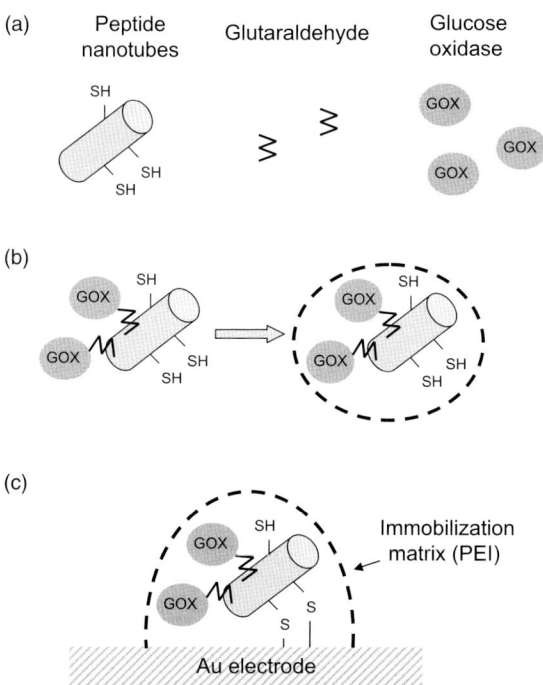

Figure 7.36 Scheme of the preparation of the PNT-based enzymatic electrode. (a) Thiol-modified PNTs are mixed with 1 μM GOX in the presence of 0.25% glutaraldehyde (GA). (b) 0.05% PEI is added to the solution. (c) The enzyme-modified PNTs are immobilized on the gold electrode surface via thiol bonds (S) and dried at room temperature. (Adapted with permission from Yemini et al. (2005a). Copyright 2005, American Chemical Society.)

presence of glutaraldehyde, which leads to stable covalent bonds between the GOX and the PNTs. In the next step, the functionalized PNTs are coated with a monolayer of polyethylenimine (PEI). The PEI monolayer improves the performance of the GOX. However, its structure is yet open enough to allow the formation of thiol bonds between the PNTs and the gold electrodes.

The amperometric response of the electrochemical cell at constant applied voltage is directly proportional to the hydrogen peroxide concentration:

$$glucose + O_2 \rightarrow gluconic\ acid + H_2O_2 \tag{7.39}$$

$$H_2O_2 \rightarrow 2H^+ + O_2 + 2e^- \tag{7.40}$$

The measured current increases with the number of electrons transferred to the gold macroelectrode. Gold electrodes are highly sensitized by a deposit of PNTs. A PNT-based electrode in a 10 mM H_2O_2 solution responds to applied voltages of 0.4 and 0.6 V with currents higher by factors of 15 and 3.5, respectively, in comparison to a bare gold electrode. Figure 7.37 shows the time dependence of the current as 0.2 mM glucose is successively added to the GOX-modified PNT-based gold electrode (A) and to the bare gold electrode (B). The anodic current of the PNT-based electrode, measured at 0.4 V versus SCE, reaches a steady state after a few seconds. The control electrode covered with the same immobilization matrix (GOX in PEI) but without PNTs shows no response. The high sensitivity and fast response of the PNT-based gold electrode could be explained by the increased electrode surface and the direct electron transfer between the spatially aligned aromatic systems in the PNTs.

The first successful use of the electronic properties of the PNTs justifies a comparison with the CNTs. Diameter and length are similar. There are obvious differences in chemical and thermal stability that favor the CNTs for application in harsh environments. As an essential advantage of PNTs, they assemble as individual

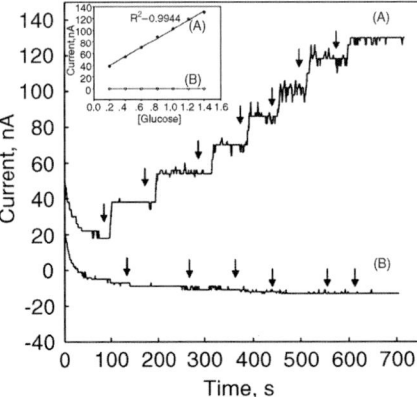

Figure 7.37 Amperometric response to successive additions of 0.2 mM β-D-glucose, measured at 0.6 V versus a SCE. (a) GOX and PNT-coated electrode. (b) GOX on a bare gold electrode, 0.1 M KCl (pH 7.5). Arrows indicate the successive addition of glucose. (Adapted with permission from Yemini *et al.* (2005a). Copyright 2005, American Chemical Society.)

units without getting tangled in aqueous solutions as CNTs do. The well-elaborated peptide chemistry facilitates many ways of additional functionalization of the PNTs. Finally, manufacturing and purification of the PNTs are simple and cost-efficient technologies.

Recently, Malvankar *et al.* (2011) reported metallic-like conductivity in microbial nanowire networks in films of the bacterium *Geobacter sulfurreducens* and also in pilin nanofilaments. Pilin nanofilaments are fibrous proteins in pilus structures of bacteria. Pili are cellular extensions of many bacteria. They are used for exchange of genetic material during bacteria conjugation. Short pili called fimbria support cell adhesion. The conducting structures were explored in connection with the manufacturing of microbial fuel cells working with 10 mM acetate as the electron donor. Biofilms of the *G. sulfurreducens* were grown on a gold electrode that acted as an electron acceptor. The current between anode and cathode increased over time with cell growth on the electrode. The conductance of the biofilm increased with film thickness. Moreover, conducting networks have been prepared from pilin nano-filamtes extracted from these bacteria. These filament networks showed electronic conductivities of ~5 mS/cm. Thus, they are comparable with synthetic organo-metallic nanostructures such as polyaniline. Up to now the molecular structure of pilin filaments is not known. Malvankar *et al.* (2011) hypothesized that, in view of the short length of the pilin protein monomers, aromatic groups in pilins might tilt with respect to peptide structures, minimizing their intermolecular distance. Similar to polyaniline or to the above-discussed PNTs, a face-to-face stacking of aromatic rings could lead to a π-orbital overlap that could explain the observed metallic behavior.

Large-Scale Membranes from Bacterial Surface Layer Proteins S-layer proteins have given rise to extensive activity aiming at the development of innovative nano-materials. There are at least three reasons for trying such approach. First, S-layer proteins are distinguished by structural robustness and simplicity, as already mentioned in Section 7.2.4. Second, there is a rich variety of bacteria and archaea that express S-layer proteins. They are found in a wide range of habitats differing in the chemical composition of nutrients and pollutants, in the pH that may vary between highly acidic (pH 2–3) to extremely alkaline (pH 9–10), and in the temperature range up to 130 °C. This offers a large variety of structures of appropriate S-layer proteins to be selected for different use in bioengineering. Finally, and maybe most important for future development, the remarkable progress in the genetics of S-layer proteins has provided several artificial ones manufactured by recombinant technologies. Based on the knowledge of the complete DNA code of the S-layer protein of particular bacteria, the protein can be expressed in host systems, for example, in the biotechnological "work horses" *E. coli* bacteria or *Saccharomyces cerevisiae* yeast cells. It has been proven repeatedly that such recombinant S-layer proteins behave like native ones. Furthermore, also proteins truncated in a defined way assemble into S-layers with remarkable structure similarity to the lattice structures of the native proteins. Finally, the genetic approach opens the rich world of chimeric fusion proteins. In this way, additional functionalities can be incorporated into the structure. Thus, S-layer proteins and their genetically modified

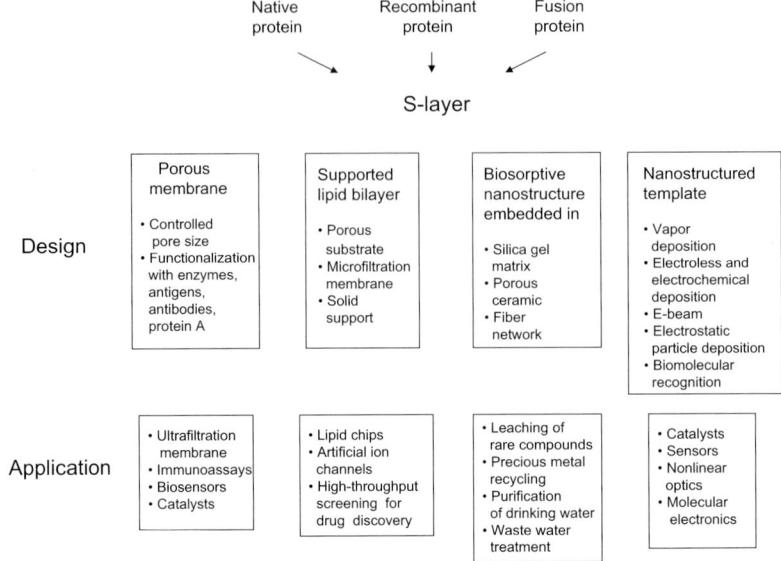

Figure 7.38 Design of S-layer constructs and their potential applications in bioengineering.

derivatives form a large platform for the bio-inspired manufacturing of nano-materials. Successful demonstrations and potential applications of the bio-inspired manufacturing of nanomaterials are compiled in Figure 7.38.

As one of the first examples making use of assembled protein membranes, ultrafiltration membranes (SUMs) with pores of equal size had been prepared from S-layers (Sára and Sleytr, 1987; Sleytr et al., 1999b). The flux through the membrane and the adsorption behavior of the ultrafiltration membrane can be tuned with respect to a specific target protein. By selective chemical modification, for example, by adding or converting carboxyl or amino groups at the surface and in the nanopores, the charge density can be modified in a controlled way (Weigert and Sára, 1995). Surface charge and hydrophilicity are two essential parameters to avoid the clogging of the nanopores by small proteins that would contribute to biofouling of ultrafiltration membranes.

The mentioned generation of functional monomolecular protein lattices, consist-ing of S-layer fusion proteins, facilitates the controlled attachment of functional proteins at the S-layer membrane, such as biotin, avidin, enzymes, antibodies, and antigens (Pleschberger et al., 2003). As evidence for the generic power of this approach Moll et al. (2002) have incorporated Streptavidin (STV) in an S-layer with the intention to open the way for many applications by exploiting the STV–biotin interaction for biomolecular coupling constructs. An S-layer array, displaying STV in defined repetitive spacing, was prepared in a two-step process: (i) STV monomers were fused with S-layer monomers to a chimeric protein retaining the self-assembly property of natural S-layer proteins. (ii) By mixing chimeric proteins with STV monomers in the ratio of 1 : 3, heterotetramers consisting of one chain of the fusion protein and three chains of STV were refolded.

In case of successful folding, the tetrameric STV subunit offers biotin binding sites that are not sterically blocked by the large S-layer protein. For the application, it is essential that the self-assembly behavior of the S-layer does not interfere with the fused construct. In the study of Moll *et al.* (2002), the S-layer protein SbsB of *G. stearothermophilus* PV72/p2 has been selected as the basic protein for the gene expression of the fused protein in *E. coli* cells. The S-layer protein SbsB of *G. stearothermophilus* PV72/p2 crystallizes in a *p*1 lattice ($a = 10.4$ nm, $b = 7.9$ nm, $\gamma = 81°$). It was known that truncated forms of SbsB lacking the N-terminal S-layer homology (SLH) domain that causes the binding of the inner face of the S-layer to a distinct secondary cell wall polymer (SCWP) self-assemble into an S-layer with same lattice parameters as the native protein. At the C-terminus, placed at the outer face of the S-layer, only the first 14 amino acids can be deleted to preserve the self-assembly behavior. Various fusion constructs were prepared using the N-terminus as well as the C-terminus for fusion of a STV monomer. Figure 7.39 shows an example of the successful attachment of a STV monomer at the N-terminus of SbsB and the formation of the heterotetrameric construct. The comparison of the S-layer lattice of the pure SbsB protein with that of the chimeric protein shows that the ability of self-assembly was retained. Moreover, a digital image reconstruction provided evidence for the formation of the heterotetrameric STV. The heterotetrameric fusion protein crystallized in suspension, on lipid layers, as well as on silicon wafers.

The fusion construct based on linking the STV monomer with the C-terminus showed a higher sensitivity of the self-assembly behavior to the structural changes. The heterotetrameric construct crystallized in an S-layer lattice only on an appropriate support formed with cell fragments of *G. stearothermophilus* PV72/p2. The periodically arranged STV tetramers replicating the SbsB lattice were capable of binding biotinylated ferritin. The coupling of biotinylated nanostructures with STV–S-layer fusion provides manifold options for application. As schematically shown in Figure 7.40, it enables the controlled arrangements of enzymes, antibodies, DNA, QDs, catalysts, and magnetic beads on various carriers (for further examples, see also Sleytr *et al.* (2003a)).

The concept of supporting protein membranes also facilitated a breakthrough in the use of lipid bilayers for biosensor application and drug discovery assays (Schuster *et al.*, 2001). Lipid bilayers can be regarded as the ideal structure to reconstitute proteins under nearly "natural" conditions or to mimic cell interfaces with embedded ion channels. The great advantage of the lipid bilayers is their in-plane liquid-like behavior combined with sufficient resistance against transmembrane transport of any kind of agents. However, handling a finite patch of lipid bilayer requires some kind of mechanical support that can be realized either by suspension along the edges or by lying on a planar substrate. Handling freestanding bilayers is problematic because of their low mechanical stability. Supporting by a planar solid substrate restricts the in-plane mobility of the bilayer and changes its

(a) (b)

(c) (d)

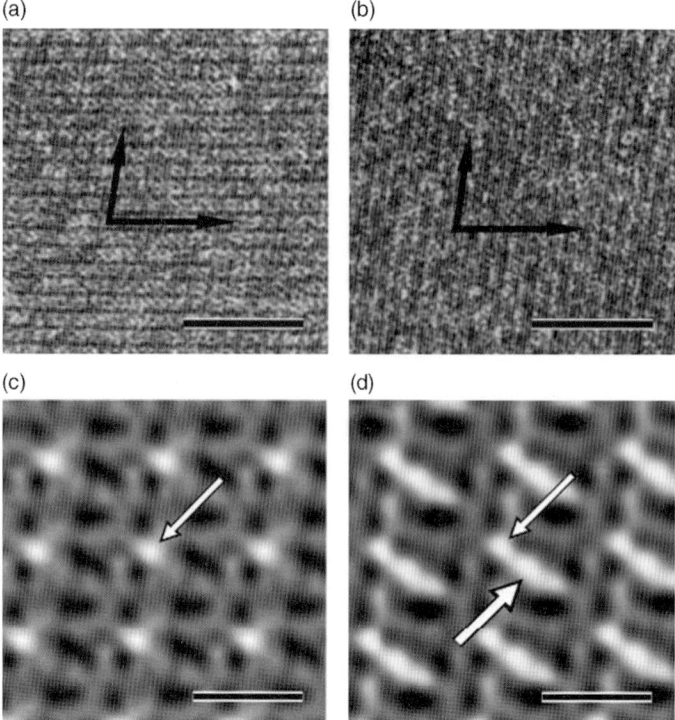

Figure 7.39 S-layer-based self-assembly of heterotetrameric STV lattices. Self-assembly products and digital reconstructions of SbsB (a and c) and the heterotetrameric fusion protein (N-terminus modification) (b and d). The crystalline sheets were formed in suspension and negatively stained with uranyl acetate for TEM. The dark arrows indicate the base vectors of the oblique $p1$ lattice (bars = 50 nm) (a and b). The digital image reconstructions were made by Fourier processing of electron micrographs [not identical to those shown in parts (a) and (b)] (c and d). The region of highest protein mass in the SbsB lattice is the SLH domain [part (c), bright arrow]. In the lattice of the fusion protein, STV showed up as additional protein mass [part (d), thick bright arrow] attached to the SLH domain (bars = 10 nm). (Reproduced with permission from Moll *et al.* (2002). Copyright 2002, National Academy of Sciences, USA.)

thermodynamic equilibrium conditions. Therefore, there was a large effort to develop a "soft" support that could lead to a compromise between "free" and "stable" bilayers. Hybrid bilayers were prepared by depositing a lipid monolayer on top of a hydrophobic layer, such as alkane silane or alkane thiol on silicon or gold substrates, respectively. An alternative approach consists in the use of the S-layer membrane to create a "semiliquid" bilayer. The combination of mechanical stability with nanoporosity of the S-layer seems to offer sufficient degrees of freedom to stabilize the out-of-plane properties of the lipid bilayer without large influence on its in-plane physical properties (especially the fluidity). Another advantage of the S-layer support is the practically unlimited transport of ions, which can be essential for the

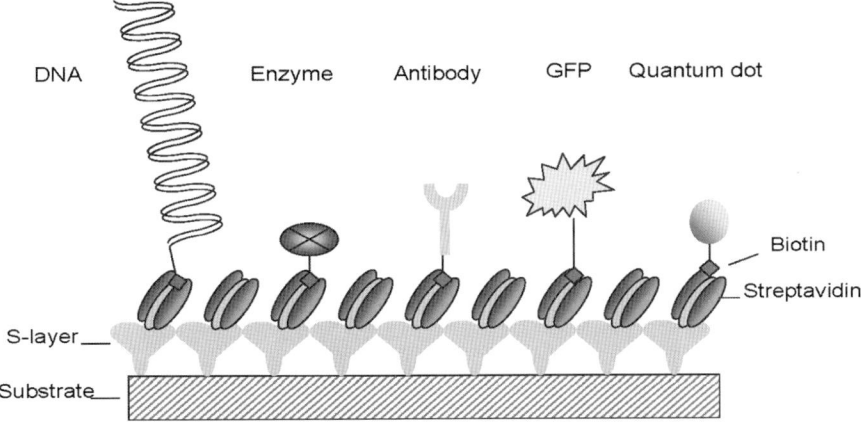

Figure 7.40 Schematic drawing illustrating the potential application of STV–fusion protein constructs.

above-mentioned sensor applications. Three options of S-layer-supported lipid bilayers have been experimentally realized (Schuster and Sleytr, 2000):

i) A tetraether lipid monolayer can be prepared on a Langmuir trough. A small patch of this tetraether membrane can be clamped by a glass micropipette and, subsequently, S-layer subunits can be crystallized on the lipid monolayer.

ii) A lipid bilayer can be prepared by dip-coating an object with an aperture in an aqueous solution covered with a lipid monolayer. Then, a freestanding lipid bilayer spanning the aperture is formed. In a second step, S-layer monomers are injected into the solution that can self-assemble along the array of hydrophilic head groups of the lipid film.

iii) Alternatively, a microfiltration membrane can be used as a support for an S-layer ultrafiltration membrane. This composite membrane can be used for the deposition of a lipid bilayer by means of adsorption of the two corresponding lipid monolayers. The lipid monolayers can be generated by the well-known Langmuir–Blodgett technique compressing lipids to an organized dense layer at the water–air interface using the so-called Langmuir balance.

Lipid membranes with solid support can be stabilized with S-layer membranes (Figure 7.41). S-layers are proved most suitable as they provided bilayers with high mobility of lipids and high stability against crack formation.

7.3.2
Template-Directed Assembly of Artificial Nanoparticles and Nanowires

Metallized DNA Networks   As shown in Section 1.3.2, the metallization of double-stranded DNA opens a promising processing route for the manufacturing of conducting nanowires. As discussed, there are specific binding sites of the DNA:

(a)

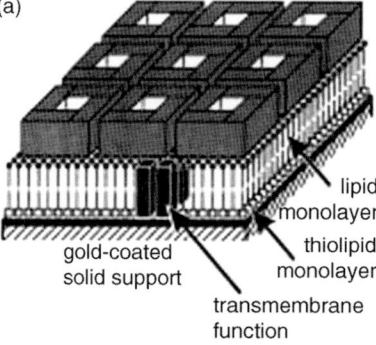

lipid
monolayer

thiolipid
monolayer

gold-coated
solid support

transmembrane
function

(b)

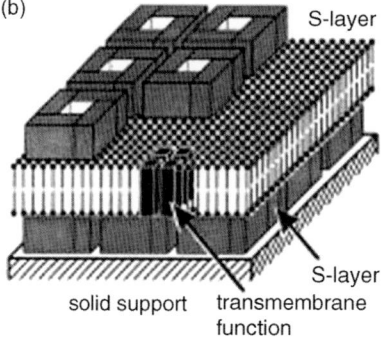

S-layer

S-layer

solid support   transmembrane
function

Figure 7.41 Schematic drawing illustrating the concept of solid-supported lipid membranes stabilized by S-layers. In part (a), the generation of the lipid bilayer makes use of the strong chemisorption of thiolipids to gold. The second leaflet and the S-layer that had recrystallized at the lipid film before in a Langmuir trough were transferred onto the thiolipid-coated solid support by the Langmuir–Schaefer technique. Integrated functional molecules allow the investigation of transmembrane functions. (b) As an alternative to soft polymer cushions, an S-layer is located between the solid support and the lipid layer. Optionally, the external leaflet of the lipid bilayer can be stabilized by the attachment of an S-layer. (Reproduced with permission from Sleytr *et al.* (1999b). Copyright 1999, Wiley-VCH Verlag GmbH.)

- the negatively charged phosphate backbone for directed adsorption of positively charged metal complexes,
- the electron donor centers at the nucleotides, and
- the base pairs by release of the imino proton in each base (for divalent metal ions).

The self-assembly of networks of double-stranded DNA and the option of the controlled incorporation of additional functional molecules are particular advantages of such concept. However, there is a disadvantage of double-stranded DNA if used as templates for manufacturing metallized nanostructures. Its stiffness as an individual molecule is too low to prevent distortions caused by the reorientation of the randomly nucleated metal clusters when they come into contact while growing. This reorientation driven by grain boundary energies leads to a roughening of the shape of a formerly straight DNA stretched between given contact points (see, for instance, Figure 1.83). Stiffer templates can be obtained by assembling DNA tiles as nanoribbons or nanotubes. Also, such nanoribbons or nanotubes can act as "pinboards" for a defined organization of additional chemical groups, proteins, QDs, CNTs, or other nanocomponents. This allows the design of nanoelectronic circuitries with a richer functionality.

The following examples describe two completely different approaches for the templated synthesis of DNA-nanoparticle hybrids. In the first approach, Liu et al. (2006) have used DNA tubes self-assembled from self-complementary DNA strands as templates for metallic nanowires (see Section 7.3.1). They applied the two-step metallization protocol already explained for double-stranded DNA in Section 1.3.2. An aqueous solution of palladium acetate $Pd(CH_3COO)_2$ was used for the activation of the nanotubes. In the second process step, a reducing buffer of sodium citrate, lactic acid, and dimethylaminoborane was added to drive the deposition of a continuous palladium nanowire on the nanotube. Figure 7.42 shows atomic force micrographs of such nanowires. The high regularity of the shape is obvious. The straight wires were 30–80 nm wide and up to 30 μm in length, which means a little

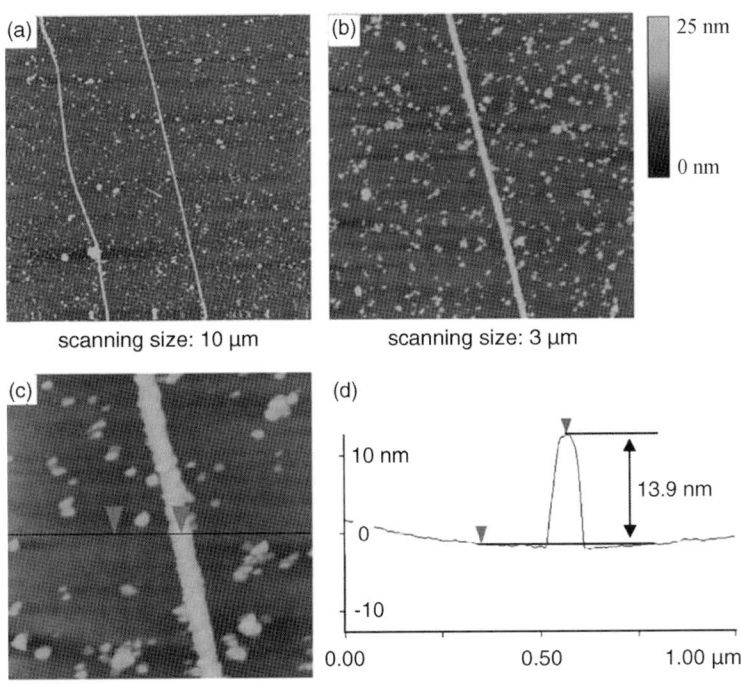

Figure 7.42 AFM images of palladium nanowires obtained by metallization of DNA nanotubes. From parts (a–c), the magnification of the images increases. All images have the same height scale, shown on the right. (d) Cross section profile of the Pd nanowire in part (c) taken along the black line, and the gray arrows in parts (c) and (d) marking the same positions. (Reproduced with permission from Liu et al. (2006). Copyright 2006, Wiley-VCH Verlag GmbH.)

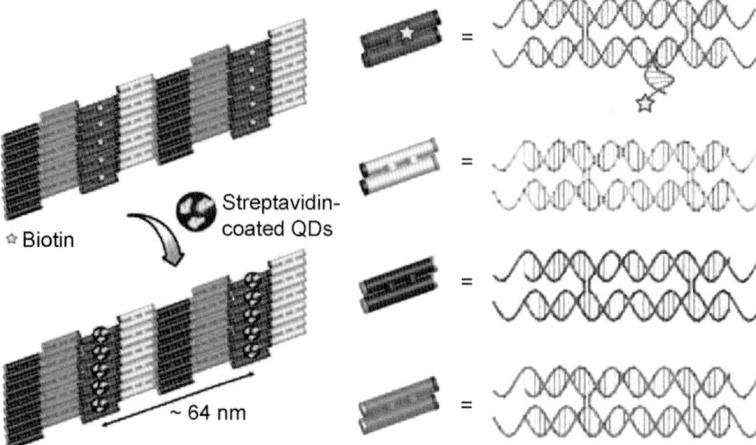

Figure 7.43 Process of DNA tile-directed self-assembly of QD arrays. (Reproduced with permission from Sharma *et al.* (2008). Copyright 2008, Wiley-VCH Verlag GmbH.)

bit wider than the initial DNA tubes and with about the same length. The thickness was in the range of 10–18 nm, compared to about 6 nm of bare DNA. Hence, the nanotubes were covered with a continuous palladium coating of 2–6 nm thickness.

The second approach has been chosen by Sharma *et al.* (2008, 2009). The goal of their design was the manufacturing of arrays of gold or silver nanoparticles as well as CdSe/ZnS QDs periodically arranged on a tubular template (Figures 7.43 and 7.44). Similar to the procedure of Liu *et al.* (2006), DNA tiles have been used as basic units for the template preparation.

The plane tile lattice was assembled with four DX tiles α, β, γ, and δ. Tiles of every type were aligned in parallel rows. The central strands of the α-tiles were conjugated with thiol groups at the upper face that facilitated the binding of 5 nm gold nanoparticles. The repulsive Coulomb interaction of the close-packed gold particles (their diameter is of similar size as the tile width) causes a curvature and twist of the tile lattice. The bottom surface of the γ-tiles was additionally modified with DNA stem loops. Placed on the side opposite to the gold-modified side, the stem loops partially counterbalance the action of the gold rows. Thus, the resulting deformation (curvature and winding) of the tile lattice can be controlled. The effective surface forces can wind the tile lattice into a nanotube. If the lattice forms a tube without winding, a stack of rings of gold particles is formed. Otherwise, the left- or right-handed single, double, or nested gold spirals are formed, depending on the pitch of the helical structure. Periodic metal structures of well-defined chirality are of high interest for the fabrication of components with novel optical properties. Collective oscillations of electrons (plasmons) excited by the light wave lead to an enhanced extinction at the plasmon frequencies. Near these frequencies, the real part of the dielectric constant changes its sign. The plasmon resonances of gold and silver nanoparticles are particularly intense, which results from the high electric

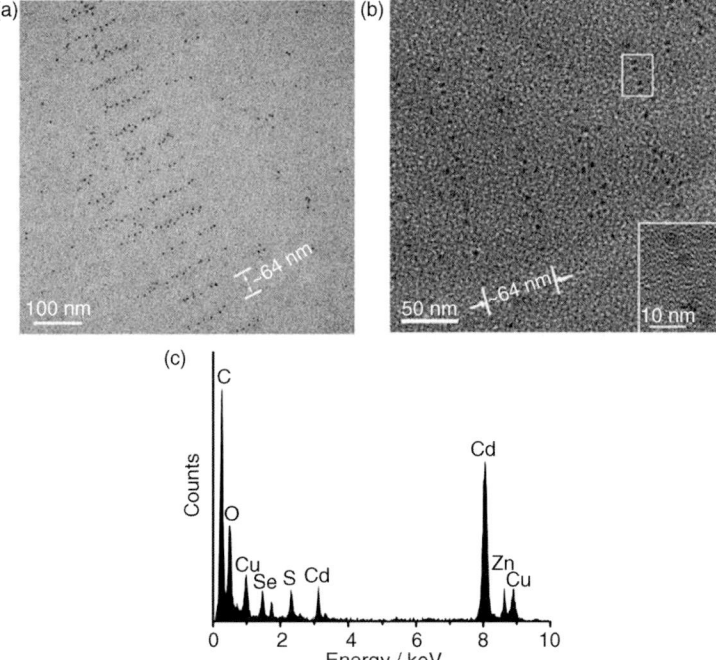

Figure 7.44 (a) TEM image of the periodic pattern of the organized QD arrays. (b) High-resolution TEM image (with a close-up inset) that reveals the crystalline structure of the QDs. (c) EDX spectrum that verifies the composition of the CdSe/ZnS QDs. (Reproduced with permission from Sharma *et al.* (2008). Copyright 2008, Wiley-VCH Verlag GmbH.)

conductivity. By variation of particle size and shape, the plasmon frequency can be tuned such that it gets into the wavelength range of visible and infrared light. The plasmon resonances also cause enhancement of the external electromagnetic field near sharp corners of the nanoparticles with nanometer-scale radii of curvature. This mechanism has found widespread application in particle-enhanced Raman spectroscopy. In optical technologies, the field enhancement at the nanoscale and the controllable extinction and reflection behavior facilitate the development of optical devices that can control electromagnetic energy on a length scale below the diffraction limit. Optical information can be guided around corners with nanometer-size radii of curvature. This peculiarity can be exploited for the development of plasmonic waveguides and nanoscale optical switches and also for the preparation of so-called metamaterials with negative refractive index that can make things virtually invisible. So it appears that patterns of biotemplated gold or silver nanoparticles represent promising components for the emerging field of plasmonics. As proposed by Sharma *et al.* (2008), the combination of QD assemblies with periodically arranged metallic nanoparticles can open the way to nanophotonic materials with well-controlled optoelectronic properties.

Nanoscale Patterns of Metal Clusters on S-Layer Templates The high regularity of the nanostructure of S-layer membranes makes them an ideal template for manufacturing artificial nanostructures by deposition of metallic or semiconducting nanoclusters. In the following, the term nanocluster refers to nearly monodisperse particles with diameters below 10 nm. The properties of particles of such size are intermediate between those of bulk material and single atoms. This opens fields of applications such as fluorescent QDs, light-emitting diodes, solar energy panels with enhanced energetic efficiency, chemical sensors, new catalysts with higher activity and improved selectivity, and visionary quantum computers. S-layer membranes can fulfill two tasks in corresponding routes of materials processing. First, their high regularity on the nanoscale is ideal for directing the growth of monodisperse nanoclusters. Second, defect-free S-layer assemblies enable close-packed periodic arrangements of the nanoclusters to be formed. There are three features of the periodic S-layer structure that facilitate the generation of nanoscale patterns: the existence of a highly regular distribution of nanopores, the charged amino acids, and the reaction sites at specific amino acid residues. As topography, charge distribution, and reaction sites are correlated in space for every S-layer protein, often more than one structural feature is involved in a particular processing route for the preparation of metallic or semiconducting nano-structures by biotemplating. Among the routes compiled in Figure 7.38, there are techniques based on vapor deposition, heterogeneous nucleation and growth in aqueous solution, and deposition of preformed nanoclusters onto the S-layer.

Based on metal vapor deposition, first examples of nanometer lithography had already been worked out by Douglas *et al.* in 1986 (Figure 7.45) (Douglas *et al.*, 1986). The S-layer was assembled onto a smooth substrate (amorphous carbon film). Afterward, a thin metal film [Ta/W (Douglas *et al.*, 1986) or Ti subsequently oxidized in air to TiO_x (Douglas *et al.*, 1992), chosen because of their small grain sizes], was deposited by electron beam evaporation. The film thickness was typically in the range of 1.2–3.5 nm. The evaporation was performed under an angle of $40°$ from the normal to the substrate surface to get a better replica of the S-layer structure by shadowing effects. The final process step was argon ion milling (2 keV ion energy) at normal incidence. The milling removed the metal preferentially at the holes of the underlying S-layer (Figure 7.46). The final metal–protein heterostructure was a fairly good replica of the S-layer. It can serve, for instance, as a metallic mask during further structuring of an underlying amorphous carbon film with the ion beam at the pore sites.

Another interesting example of the potential of this technique has been given by Panhorst *et al.* (2001): the patterning of ferromagnetic nanodot arrays that are very promising for application in magnetic storage media. In the experiment, thin films of 2.5 nm thickness of various ferromagnetic metals and alloys (Co, Fe, FeCo, CoNi, and NiFe) were sputtered onto an S-layer of the bacteria *Deinococcus radiodurans* (p6 symmetry, lattice constant 18 nm, and pore size 6 nm). After argon ion etching with ion energies of 50–300 eV, periodic nanodot patterns on top of the S-layer have been obtained (Figure 7.47). With each dot presenting a storage bit, the storage density of such structure would be about 10^{16} bits/m^2.

If S-layers are used as structure-directing templates for the growth of metallic or semiconducting nanoclusters in aqueous solution by wet chemical synthesis,

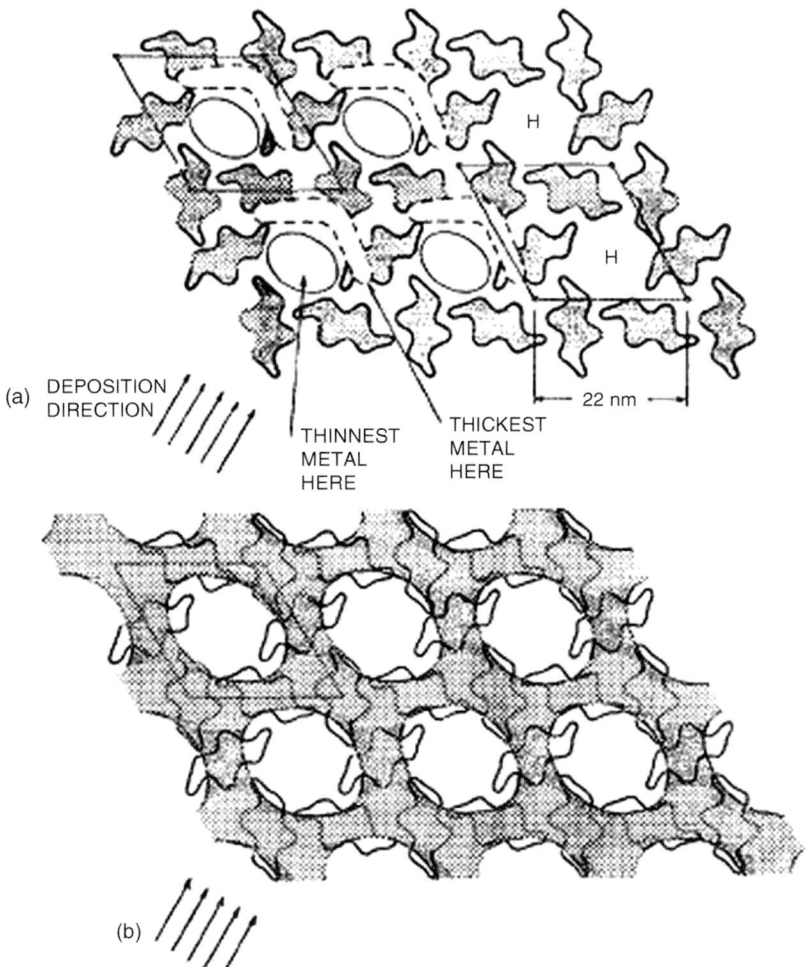

(a) DEPOSITION
DIRECTION

H

H

|—— 22 nm ——|

THINNEST
METAL
HERE

THICKEST
METAL
HERE

(b)

Figure 7.45 (a) Schematic structure of the 2D crystalline protein S-layer of *Sulfolobus acidocaldarius*. The shaded areas represent the regions of largest electron density projected onto the layer plane. Each shaded unit is a protein dimer. The structure, a triangle lattice with a basis of three protein dimers, exhibits an array of sixfold symmetric holes, labeled *H*. The metallization beam (Ta/W) is incident at an angle of 40° from the normal to the S-layer plane in the direction of the solid arrows. The unmilled metallized membranes exhibit maximum metal thickness in the regions of the dashed lines, in particular where the edges of the sixfold holes are exposed to the metal flux. The minimum metal thickness appears at the hole sites, probably because of shadowing by the front edges of the holes. (b) Schematic of the average metal structure and its relationship with the protein after milling. Metal is removed from the holes leaving the shaded region coated with metal grains. The holes in the metal film are on the average oval shaped, being narrower in the deposition direction, with a sharp outline formed by chains of grains. The coated region has the form of a pair of overlapping sets of periodic strips, spaced by 19 nm, running along the molecular rows more heavily coated in the oblique evaporation. (Reproduced with permission from Douglas *et al.* (1986). Copyright 1986, American Physical Society.)

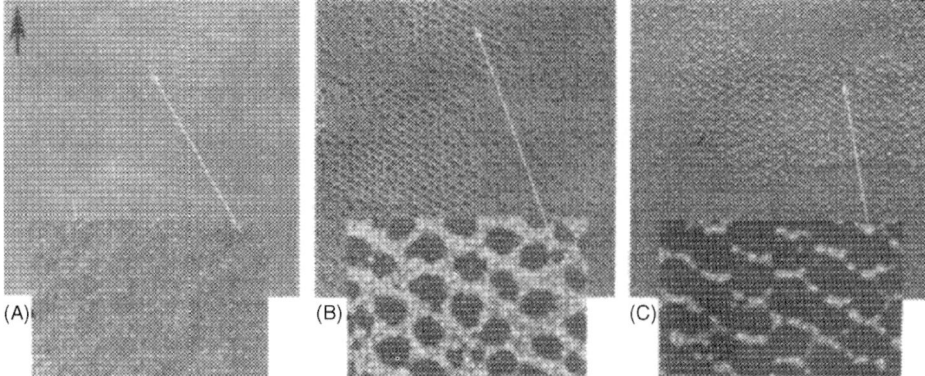

(A)   (B)   (C)

Figure 7.46 Transmission electron micrographs of typical (A) unmilled and (B and C) milled Ta/W films. The white bar indicates 100 nm in all pictures at low magnification and 20 nm in the insets. The black arrow indicates the Ta/W deposition direction in all pictures. The white arrows indicate the locations of the insets. (A) Typical unmilled Ta/W film (average thickness 1.2 nm) coated on an S-layer fragment at 40° from normal incidence. *Thicker* metal deposition corresponds to a *lighter* gray film. (A) Metal coats all areas with some thickness variation, as discussed in Figure 7.45. (B and C) Films such as in part (A) after ion milling. The metal films are perforated with holes located at the sixfold holes of the underlying S-layer. The holes are metal-free and on the average oval shaped, being narrower along the deposition direction, creating two periodic sets of metal strips. The holes are bounded by pearl chains of metal grains [inset in part (B)], which are not present before milling. Increased milling produces larger holes. Milling times: (B) 25 s, (C) 30 s. (Reproduced with permission from Douglas *et al.* (1986). Copyright 1986, American Physical Society.)

homogeneous nucleation of the clusters in the solvent has to be avoided. Therefore, the synthesis is performed in two steps, similar to the electroless metallization of DNA. In the first activation step, the S-layer assembly is incubated in a metal–salt solution in order to generate metal complexes at nucleation sites of the S-layer. The cluster formation is initiated by the subsequent addition of a reducing agent. The first wet chemical nanoparticle synthesis on S-layers has been demonstrated by

15 nm

0 nm

(a)   (b)

200 nm

Figure 7.47 Ferromagnetic nanodot pattern of permalloy ($Fe_{19}Ni_{81}$), dot size 10 nm, lattice parameter 18 nm. AFM (a) and SEM (b) images after argon etching with 150 eV (Panhorst *et al.*, 2001). (Reproduced with permission from Panhorst *et al.* (2001). Copyright 2001, American Physical Society.)

Table 7.3 Substance combinations applied for the formation of transition metal nanoparticles.

Metal salts	Reducing agents
$PdCl_2$	$(CH_3)_2HN\text{-}BH_3$ (DMAB)
$Pd(CH_3COO)_2$	H_2
Na_2PdCl_4	NaN_3
K_2PtCl_4	DMBA, NaN_3
$(NH_3)_2PtCl_2$	$NaBH_4$, $h\nu^{a)}$
$HAuCl_4$	Citrate, tannin, $NaBH_4$, N_2H_4, H_2O_2
$KAuCl_4$	Electron beam[a]

a) The electrons required for chemical reduction can also be provided by irradiation with ultraviolet light or electron beam.

Shenton *et al.* (1997) with the deposition of CdS nanoparticles. They obtained CdS by reducing $CdCl_2$ with H_2S. In view of the interesting catalytic properties of transition metal nanoparticles, much effort has been focused on their manufacture (Table 7.3).

The activation of the biotemplate is initiated by a hydrolysis reaction of the salt complex, as in the case of K_2PtCl_4, for instance (see also the discussion on bio-Pd in Section 6.3.1.1):

$$[PtCl_4]^{2-} + H_2O \leftrightarrow [PtCl_3 H_2O]^- + Cl^- \tag{7.41}$$

$$[PtCl_3 H_2O]^- + H_2O \leftrightarrow [PtCl_2(H_2O)_2]^0 + Cl^- \tag{7.42}$$

It has been observed that the subsequent reduction reaction proceeds significantly faster if the hydrolysis reaction has already reached its equilibrium (Mertig *et al.*, 2002; Seidel *et al.*, 2004). Typically, it needs about 10–20 h. Only twofold hydrolyzed platinum complexes possess a positive electroaffinity that is a necessary condition for the reduction of the metal complex. Activation of the S-layer assembly means binding of the metal complexes at the proteins. Spectroscopy studies revealed that the process reaches saturation after about 2–3 h when about 200–300 complexes per monomer are bound (Wahl, 2003). It is assumed that the complexes probably create covalent bonds with reactive groups of the protein.

Clusters grow only if the positively charged metal ions are reduced. In principle, metal binding amino acids of the S-layer protein could act as potential reducing agents. However, the fraction of amino acids that could act as electron donors (histidine, cysteine, and tryptophan) is not sufficient to reduce a significant part of the metal complexes bound during activation (in *Sporosarcina ureae*, for example, there are only 0.6% histidine, 0% cysteine, and 0.9% tryptophan). Therefore, cluster growth starts only after adding an additional reducing agent. Molecular simulation of the first nucleation events has shown that already a single atom of zero-valent platinum Pt(0) or monovalent platinum Pt(I) provides a critical nucleus (Colombi Ciacchi *et al.*, 2001). The further growth is an autocatalytic process in which newly reduced metal complexes are aggregating. An estimate shows that about 30% of metal complexes, needed for the formation of the final nanoparticle distribution, are already bound on the S-layer surface during activation (Wahl *et al.*, 2005). The clusters grow preferentially at the edges of the nanopores or inside of them. In comparison to planar regions on the protein surfaces, these sites are energetically favored owing to a larger number of binding sites and a larger spatial charge density. It seems that there is an analogy with the behavior of adatoms on surfaces of a solid crystal. Step ledges are preferred nucleation sites for adatoms due to the larger coordination with respect to terrace sites.

In order to facilitate an adaptation of the growing clusters to the structure variation of the S-layer, the sticking probability of small nuclei at the S-layer has to be small. Then by local dissolution–redeposition processes, the metal complexes deposited far from nanopores can also contribute to the particle growth at the energetically favored positions. The growth rate of the nanoparticles depends on the strength of the reducing agent. NaN_3 as a weak reducing agent leads to a low growth rate and small particle size. Figure 7.48 shows the deposition of Pt nanoparticles on an S-layer of *S. ureae* (Mertig *et al.*, 1999).

After about 24 h, almost all pores of the S-layer are filled with monodisperse Pt nanoparticles. The comparison of the high-resolution transmission electron microscopy (HRTEM) images of the metallized S-layer (Figure 7.49) and the native, negatively stained S-layer (Figure 7.4) demonstrates that the particles grow preferentially inside the nanopores.

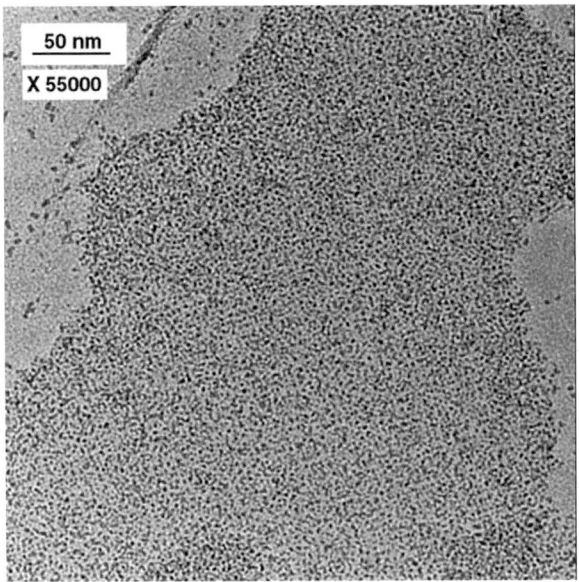

Figure 7.48 TEM micrograph of platinum clusters chemically deposited onto an S-layer sheet of *S. ureae*. The S-layer face was activated by treating 16.5 ml S-layer suspension (2 mg protein per 1 ml) with 1 ml of a 3 mM K_2PtCl_4. (Reproduced with permission from Mertig *et al.* (1999). Copyright 1999, with kind permission from Springer Science+Business Media.)

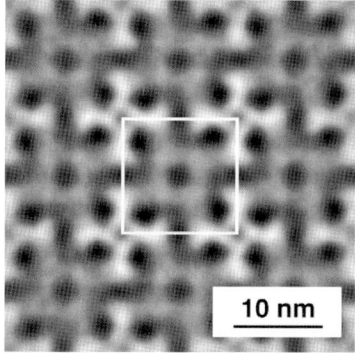

Figure 7.49 Spatial distribution of metal clusters (black) on the protein crystal revealed by image processing. The size of the quadratic unit cell of the cluster lattice (white frame) is 13.2 nm. (Reproduced with permission from Mertig *et al.* (1999). Copyright 1999, Springer Science+Business Media.)

The metallic character of the clusters has been proven by HRTEM. Figure 7.50 shows lattice fringes of the clusters with periodicities of 0.22, 0.20, and 0.14 nm. These distances perfectly agree with the lattice spacings of the (111), (002), and (202) planes, respectively, of an fcc-platinum crystal with a lattice constant of 0.39 nm. In addition, the purity of the nanoparticles was verified by energy-dispersive X-ray spectroscopy (EDX). The narrow size distribution of the particles has its maximum at about 2.0 nm (Figure 7.51). This shows clearly that not only the nucleation sites but also the particle sizes are controlled by the nanopores of the S-layer. Alternatively, DMAB as a strong reducing agent leads to a short growth time. The growth rate is

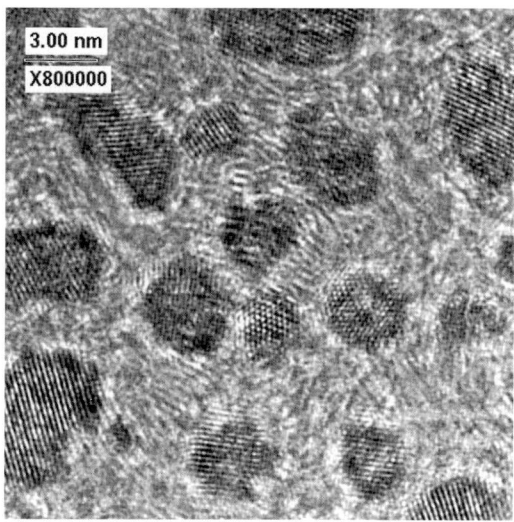

Figure 7.50 HRTEM micrograph of the metallized S-layer. (Reproduced with permission from Mertig *et al.* (1999). Copyright 1999, with kind permission from Springer Science+Business Media.)

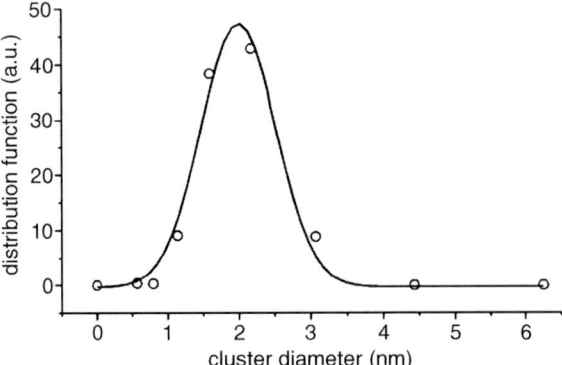

Figure 7.51 Size distribution of platinum clusters deposited on *S. ureae* S-layers by using NaN₃ as reducing agent. The distribution of cluster diameters was determined by analyzing about 30 000 metal clusters using the image analyzer QUANTIMET 570 (Leica, Bensheim, Germany). (Reproduced with permission from Mertig *et al.* (1999). Copyright 1999, Springer Science+Business Media.)

nearly independent of the presence of an S-layer. The cluster array on top of an S-layer shows the symmetry of the underlying S-layer, but there is a stronger variation of the cluster size than in case of the weaker reducing agent NaN₃. Obviously, the strongly reducing agent DMAB also causes homogeneous nucleation in the liquid in addition to heterogeneous nucleation.

To sum up, it can be stated that an optimized growth process of metal nano-particles on an S-layer assembly is characterized by the following four features:

i) Activation of the S-layer with an equilibrated aqueous solution of metal complexes.
ii) Sticking probability of reduced metal complexes on the S-layer smaller than 1.
iii) Autocatalytic growth of partially reduced metal complexes.
iv) Enhanced binding probability of the metal clusters inside the pores.

The combination of the self-assembly behavior with the control of metal cluster growth by wet chemical synthesis facilitates interesting applications for patterned metallization of microelectronic structures by inverse microcontact printing by means of a microfluidic system (Figure 7.52).

The localized metallization of 10 gold pixels of a piezoelectric sensor array is shown in Figure 7.53. With a structured stamp made from PDMS, a system of microfluidic channels is generated that conduct the solution of S-layer monomers of *B. sphaericus* to every pixel. After successful deposition of S-layer proteins at every pixel region, Pt clusters with 2–3 nm in diameter are grown on top of the S-layer in a second step (Figure 7.54). The S-layer formation and the Pt cluster growth are performed in the same microfluidic setup without changing the PDMS position. The reaction containment was flushed between the two process steps.

There are alternative techniques for structured metallization of microelectronic circuitries via S-layer directed growth of metallic clusters. A patterned reduction of

(a)

(b)

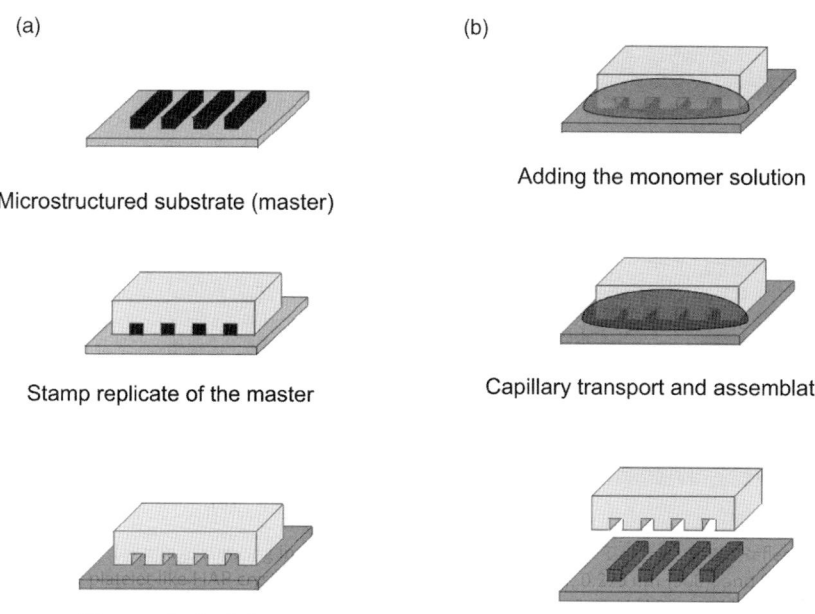

Microstructured substrate (master)

Adding the monomer solution

Stamp replicate of the master

Capillary transport and assemblation

Microcontact printing

Inverse microcontact printing

Figure 7.52 Microcontact and inverse microcontact printing. Stamp preparation and common microcontact printing. (a) as well as "inverse printing" (b) of a biomolecular pattern.

the metal complexes can be realized by means of generating the needed electrons by irradiation with an e-beam or UV light.

For e-beam-induced cluster growth, two different approches have been demonstrated. In the first approach, first metal complexes can be chemically bound to the S-layer surface, where they form a homogeneous ultrathin fine-grained metal film.

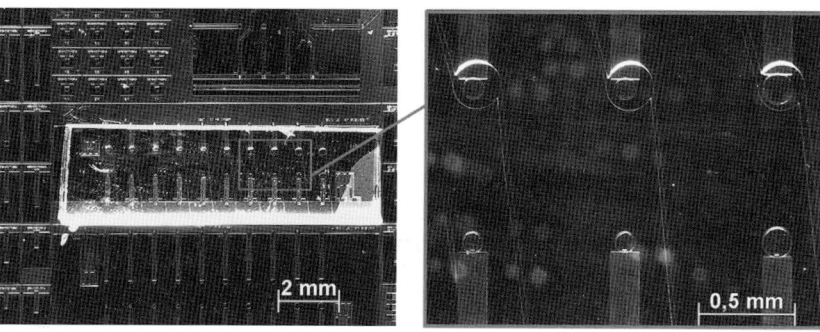

Figure 7.53 Localized metallization of a sensor array by inverse microcontact printing using self-assembled S-layers for biotemplating of nanosized Pt clusters. (a) A structured PDMS stamp is placed onto a region of 10 gold pixels on a patterned silicon wafer. (b) The enlarged image shows the channel systems and the microreaction volumes for three neighboring gold pixels of different size. (Reproduced with permission from Blüher (2008).)

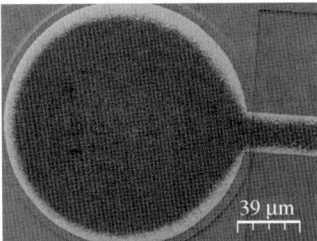

Figure 7.54 SEM image of the coated pixel region. The pixel surface is covered after the second deposition step with Pt clusters grown on the S-layer template. (Reproduced with permission from Blüher (2008).)

Second, the film is thermally decomposed into nanoclusters by electron beam irradiation. Dieluweit *et al.* (1998) reported the generation of gold nanoclusters on S-layers after chemical modification of the S-layer with thiol groups. A gold film had been deposited by exposing the modified S-layer to a solution of $HAuCl_4 \cdot 3H_2O$ for 4 days at $4\,^\circ C$. Afterwards, thermal destabilization of the gold film under electron beam exposure had led to an ordered array of monodisperse 4–5 nm-sized gold particles growing in the pores of the S-layer (Figure 7.55).

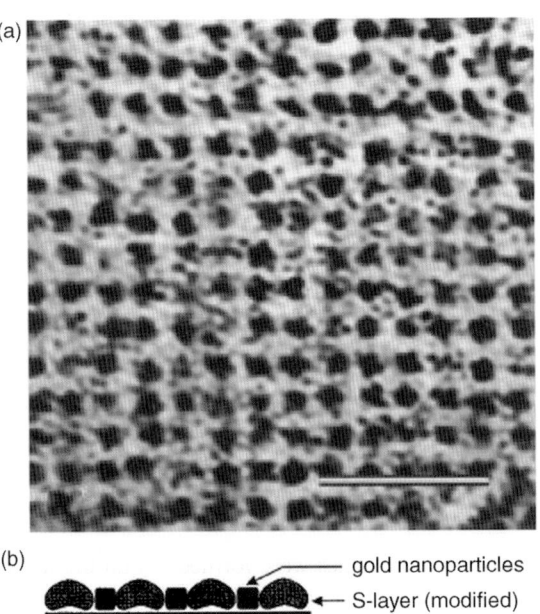

Figure 7.55 (a) Electron micrograph of a gold superlattice consisting of monodisperse gold nanoparticles with mean diameters of 4–5 nm. Bar = 50 nm. (b) Schematic drawing of a cross section through a support coated with S-layers, where the gold nanoparticles had formed in the pore region of the protein layer. (Reproduced with permission from Sleytr *et al.* (1999b). Copyright 1999, Wiley-VCH Verlag GmbH.)

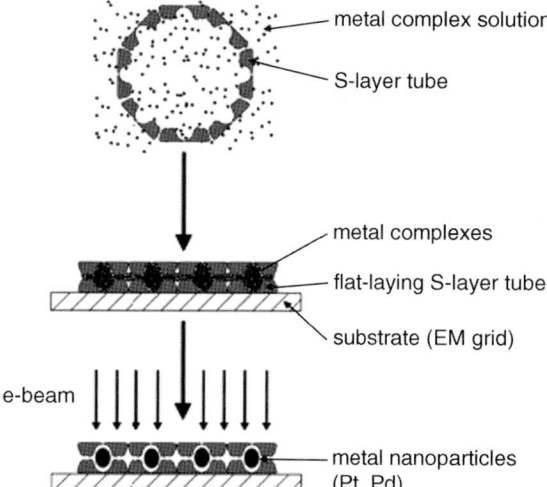

Figure 7.56 Schematic diagram of the electron beam-induced cluster formation. After treatment with a metal complex solution (Pt or Pd), the S-layer tubes were placed on a substrate where they collapsed to double layers. Exposure to an electron beam inside an electron microscope leads to the growth of metallic clusters in the S-layer cavities within a few minutes. (Reproduced with permission from Wahl *et al.* (2001). Copyright 2001, John Wiley & Sons, Inc.)

In the second approach, Wahl *et al.* (2001) used the electrons as a reducing agent of Pt and Pd clusters inside S-layer tubes. *B. sphaericus* were treated with a metal complex solution (Pt or Pd complex) for about 2 h. After activation, the S-layer tubes were placed on a flat substrate, where they collapsed. By a very short exposure to an electron beam inside an electron microscope, metal clusters growth was observed in the S-layer cavities of the flat tubes (Figure 7.56). The nanoparticles (5–7 nm) reproduced the $p4$ symmetry as well as the unit cell spacing of about 12.5 nm of the S-layer (Figure 7.57). The individual clusters had the lattice symmetry and lattice spacing of the bulk metal.

Peptide Templates for Metal Nanowire Growth As mentioned in the preceding section, monomers of the dipeptide NH_2—Phe—Phe—COOH can self-assemble into very regular tubes that can serve as templates for growing metallic nanowires inside (Reches and Gazit, 2003). When the tubes were incubated into a solution containing silver ions, high-resolution transmission electron microscopy followed by EDX analysis indicated that silver nanoparticles were formed within the tubes (Figure 7.58). By incubating the tubes into a boiling ionic silver solution and adding citric acid as a reducing agent, a uniform assembly of silver nanowires was obtained. After enzymatic digestion of the peptide coating, the diameter of the silver nanowires was found to be ~20 nm.

One-Pot Synthesis of Template-Directed Nanostructures The examples discussed above can be classified as two-step processes. First, the biomolecular template is

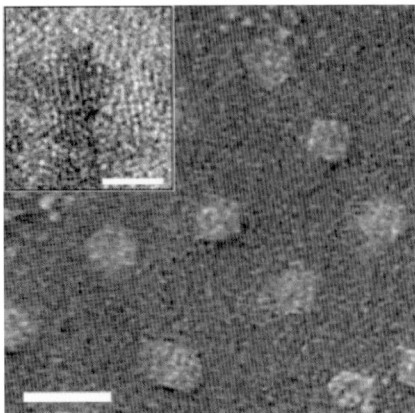

Figure 7.57 Electron beam-induced Pt cluster array in the S-layer tube of *B. sphaericus* (HRTEM). The micrograph is slightly defocused for better contrast. The *p4* symmetry and the 12.5 nm lattice constant of the S-layer are reproduced in the cluster pattern. Scale bar = 10 nm. *Insert*: Partial view of one cluster with lattice spacing visible at the correct focus in HRTEM. The spacing is the same as for bulk metallic Pt. Scale bar = 2 nm. (Reproduced with permission from Wahl *et al.* (2001). Copyright 2001, John Wiley & Sons, Inc.)

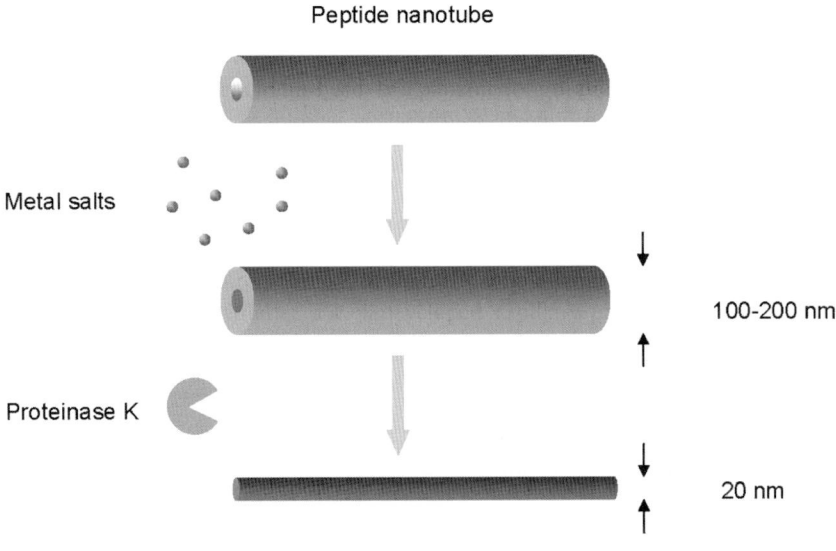

Figure 7.58 Formation of a silver nanowire inside a PNT. The nanowire is formed by the reduction of silver ions within the tube, followed by enzymatic degradation of the peptide mold. (Adapted with permission from Cherny and Gazit (2008). Copyright 2008, Wiley-VCH Verlag GmbH.)

generated by self-assembly. In a second step, the inorganic material is deposited onto the nanostructured pattern. However, a simultaneous formation of an extended biomolecular template and a mineral phase is also thinkable. We have already given an example of such a process in Section 6.3.2.2: the manufacture of mineralized collagen scaffolds by assembly of collagen and hydroxyapatite precipitation proceeding simultaneously.

For the so-called *one-pot synthesis*, an appropriate peptide has to be designed that combines peptide self-assembly and peptide-based biomineralization. Such concept has been realized for metallic nanostructures first by Rosin and coworkers (Chen *et al.*, 2008). They prepared highly ordered gold nanoparticles, assembled in double helices of micrometer length, by applying a one-step process. The design of the peptide began with the selection of a gold-binding amino acid sequence. With a phage display, the sequence AYSSGAPPMPPF, in the following referred to as (PEP$_{Au}$), was identified as highly affine to gold. In order to generate a peptide distinctly prone to self-assembly, an aliphatic chain was attached to the N-terminus of the hydrophilic PEP$_{Au}$. By coupling dodecanoic acid to the N-terminus, the amphiphilic molecule [C$_{12}$H$_{23}$CO]-PEP$_{Au}$, called C$_{12}$-PEP$_{Au}$ henceforth, was created. When the C$_{12}$-PEP$_{Au}$ monomers were dissolved in a buffer, they self-assembled into twisted ribbons. Individual fibers were micrometers in length ($>4\,\mu$m) and 6.1 ± 0.6 nm in width. The pitch of the twisted nanoribbon was 84.1 ± 4.2 nm with high regularity. In the proposed structure model, two C$_{12}$-PEP$_{Au}$ span the ribbon width, interacting via their aliphatic tails. The organization along the longitudinal axis of the ribbon is caused by a stack of parallel β-sheets and the hydrophobic interaction of the aliphatic tails. The amino acid sequence AYSS heading the N-terminus is the β-sheet forming substructure of the peptide. The ribbon-like shape is caused by the bulky outer subsequence of the peptide with the two proline dimers and one phenylalanine. The twisted arrangement of the PEP$_{Au}$ units leads to an effective shielding of the hydrophobic core of the ribbon.

If chloroauric acid (HAuCl$_4$) is added to the C$_{12}$-PEP$_{Au}$ solution at the very onset of the self-assembly process, small precipitates can be observed in the solution after about 30 min. TEM revealed double-helical assemblies of monodisperse gold particles with diameters of 8.2 ± 1.0 nm. There are about 22 nanoparticles per pitch distance. The edge-to-edge spacing between the particles along the helix is 1.5 ± 0.8 nm. The length of the helices is in the micrometer range. The regularity of the gold nanostructures promises interesting plasmonic properties.

7.3.3
Template-Free Directed Self-Assembly of Nanoparticles

One of the most advanced examples of practical use of template-free directed self-assembly is the *DNA-based assembly of nanoparticles*. This includes all kinds of manufacturing of "nanocrystal molecules" prepared by attaching single-stranded DNA oligonucleotides of defined length and sequences to individual particles. The recognition structures of complementary sequences of oligonucleotides are used for the assembly of those "nanocrystal molecules" composed of nanoparticles. This wide field pioneered by the Mirkin group has already been discussed in detail in Section 1.3.2. As mentioned there, the term "nanocrystal" covers various inorganic materials such as metallic nanoparticles, semiconducting QDs, CNTs, and so on. In the following, we will show that the concept of template-free directed self-assembly can be extended by making use of other biomolecular recognition structures.

Self-Assembly by Peptide–Metal Coordination   There are proteins with metals as an essential component governing their functional properties. Metals such as zinc (e.g., in alcohol dehydrogenase), iron (e.g., in the heme group of myoglobin or hemoglobin), calcium (e.g., in calmodulin), molybdenum (e.g., in dinitrogenase), and copper (e.g., in plastocyanin) are incorporated in so-called prosthetic groups. The interaction with the amino acid residues is usually realized by coordination bonds. Natural proteins containing metals inspired the manufacture of metal–peptide complexes specially designed for self-assembly of nanoparticles (Si *et al.*, 2008). The nanoparticles have to be functionalized with the carboxylated peptides. Induced by metal–ion coordination of the prosthetic groups, the peptides form a network with incorporated nanoparticles (Figure 7.59).

As the coordination bonds are weak, the nanoparticles can reach highly ordered equilibrium states of minimum free enthalpy in the result of self-assembly. Moreover, the assembly process is reversible. The bonds can be loosened with metal-chelating agents such as alkaline solutions of ethylenediaminetetraacetic acid (EDTA). The self-assembly of gold nanoparticles of size 8.7 ± 2.3 nm functionalized with a short carboxylated peptide NH_2—Leu—Aib—Tyr—COONa has been monitored by means of absorption measurements (Figure 7.60). Adding heavy metal ions (Pb^{2+}, Cd^{2+}, Cu^{2+}, and Zn^{2+}) to a colloidal solution of the Au nanoparticles leads to the formation of two- or three-dimensional nanostructures generated by the peptide–metal linkers (Figure 7.61). The structure change can be detected by a color change of the solution from red to blue. The optical extinction changing with

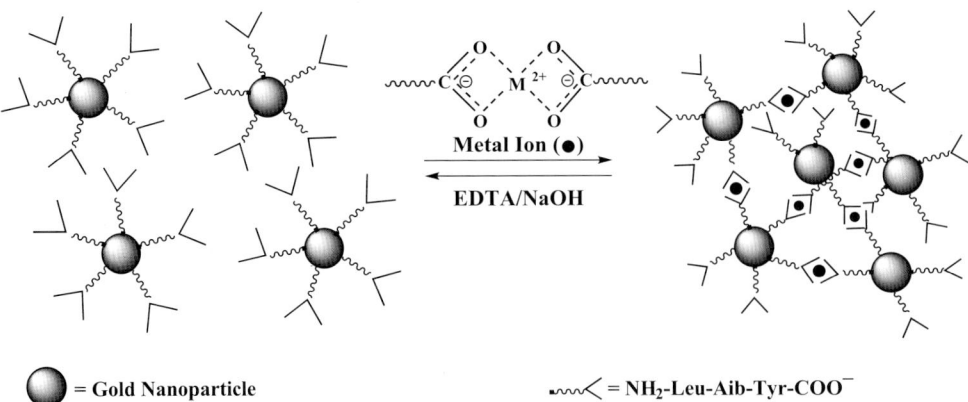

Figure 7.59 Metal–ion induced reversible self-assembly of gold nanoparticles. The solution changes its color from red to blue as a consequence of the aggregation of the gold nanoparticles. (Reproduced with permission from Si *et al.* (2008). Copyright 2008, Wiley-VCH Verlag GmbH.)

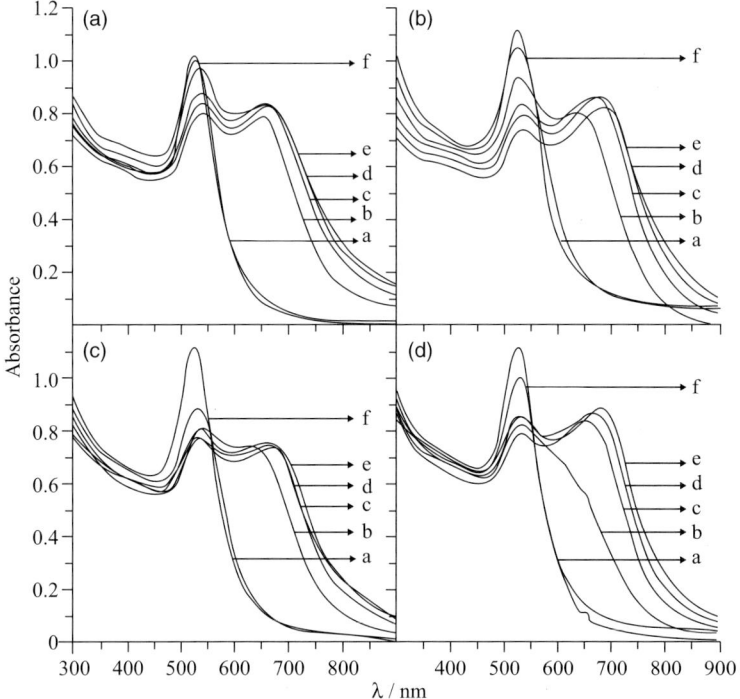

Figure 7.60 Variation of UV-VIS absorption spectra of a suspension of gold nanoparticles caused by metal-induced self-assembly. Influence of (a) Pb^{2+}, (b) Cd^{2+}, (c) Cu^{2+}, and (d) Zn^{2+} ions: spectra (a) before addition of metal ions, (b) 1, (c) 5, (d) 10, and (e) 15 min after addition of metal ions (0.6 mM), and (f) after addition of alkaline EDTA solution (0.6 mM). (Reproduced with permission from Si *et al.* (2008). Copyright 2008, Wiley-VCH Verlag GmbH.)

heavy metal content of the solution can be used for the development of colorimetric sensors for the detection of metal pollution in water.

Before addition of the metal ions, the UV-VIS spectrum of the gold nanoparticle solution shows a sharp extinction peak at 527 nm [curves (a) in Figure 7.60a–d]. It is related to the plasmon resonance (collective oscillation of free electrons) of the individual gold nanoparticles functionalized with the peptide. Addition of the metal ions leads to a slight redshift of that band at about 535 nm, and a second peak at higher wavelength [curves (b–e)]. The new peak is caused by the coupling of electron excitations (plasmons) in the neighboring gold particles after self-assembly into larger aggregates. The same mechanism has been used for the development of DNA-based colorimetric sensors, as already discussed in Section 1.3.2. The effect of the addition of EDTA can be seen in curve (f) showing the redispersion of the aggregates into separate gold particles.

Self-Assembly by Controlled Peptide Folding Controlled conformation changes of *de novo* designed peptides can be used for reversible self-assembly processes. An

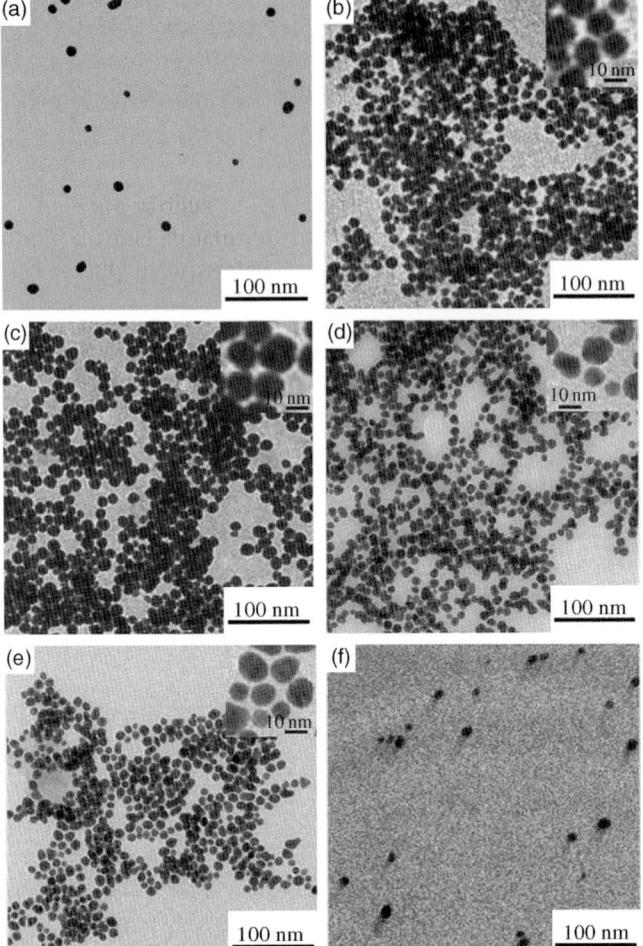

Figure 7.61 TEM images of a peptide–gold NPs suspension (a) before addition of metal ions, (b–e) 10 min after addition of Pb²⁺, Cd²⁺, Cu²⁺, and Zn²⁺ ions (0.6 mM), respectively, and (f) 5 min after addition of alkaline EDTA solution to the solution of peptide–gold NPs containing Pb²⁺ ions (0.6 mM). (Reproduced with permission from Si *et al.* (2008). Copyright 2008, Wiley-VCH Verlag GmbH.)

instructive example has been given by Aili *et al.* (2008a, 2008b) with the self-assembly of polypeptide-decorated gold nanoparticles. Two types of polypeptide linkers were designed, based on controlled dimerization and folding of polypeptides immobilized on gold nanoparticles:

  i) In the first case, the *coordination of Zn^{2+} ions to a de novo designed polypeptide* is used to induce dimerization and folding of the peptide (Figure 7.62). A glutamic acid-rich polypeptide was synthesized with 42 residues and a net charge -5 at neutral pH. Thus, a neutral pH dimerization and folding are

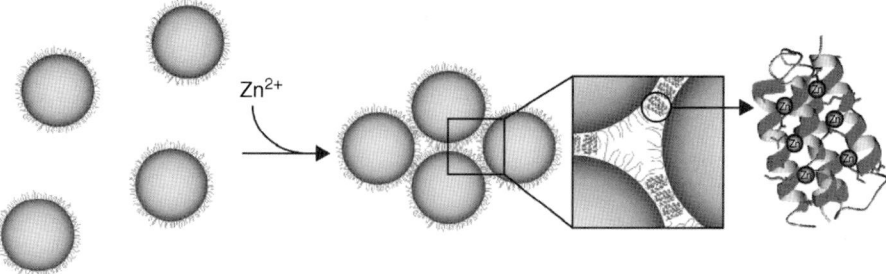

Figure 7.62 Association of Au nanoparticles by Zn^{2+}-induced dimerization and folding of peptides. The designed peptide dimerizes and folds into a four-helix bundle in the presence of Zn^{2+} ions. Immobilization of the peptides on gold nanoparticles leads to dimerization between peptides located on separate particles. (Reproduced with permission from Aili *et al.* (2008b). Copyright 2008, American Chemical Society.)

prevented. Below pH 6 or in the presence of metal ions (Zn^{2+}) at neutral pH, the free peptide folds under dimerization into a four–helix bundle (See Figure 7.62). Gold nanoparticles (13 nm) were functionalized with the peptides. A cysteine at position 22 in the loop region allowed the site-specific thiol-dependent immobilization on the gold surface. In case of immobilized peptides, the dimerization occurred between peptides bound on separate particles. Lateral dimerization between peptides immobilized on the same particle was sterically impossible. In Figure 7.62, the self-assembly induced by addition of Zn^{2+} is shown. Large aggregates of gold clusters were generated by dimerization. Similar to the former example, the aggregates can be redispersed in a controlled way by adding EDTA. EDTA is a strong zinc chelator. It removes Zn^{2+} from the peptides. The formation of Zn chelate complexes leads to the disruption of the dimerization of the peptides.

ii) The reversible association and folding of three polypeptides are used for assembly of Au nanoparticles (Aili *et al.*, 2008a). Two identical (primary) polypeptides used in case (i) are associated with a charge–complementary linker polypeptide. The linker polypeptide consists of two identical monomers joined covalently through a disulfide bond. The three polypeptides form a heterotrimeric complex consisting of two disulfide-linked four-helix bundles. For assembly of gold nanoparticles, the particles are functionalized with the identical (primary) peptides. After addition of the linker polypeptides to the solution, their association with the immobilized polypeptides induces folding and bridging of the two four-helix bundles, which leads to large aggregates of gold nanoparticles. The aggregates could be dissolved by reducing the disulfide bridges that connect the four-helix bundles.

Virus-Based Self-Assembly Structures In highly concentrated solutions of viruses, the formation of liquid crystalline (LCs) structures can be observed. Rod-like viruses with pronounced shape anisotropy, for instance, the M13 virus ($\sim$900 nm long and 7 nm wide) or the TMV ($\sim$300 nm long and 18 nm wide), are favored structures for

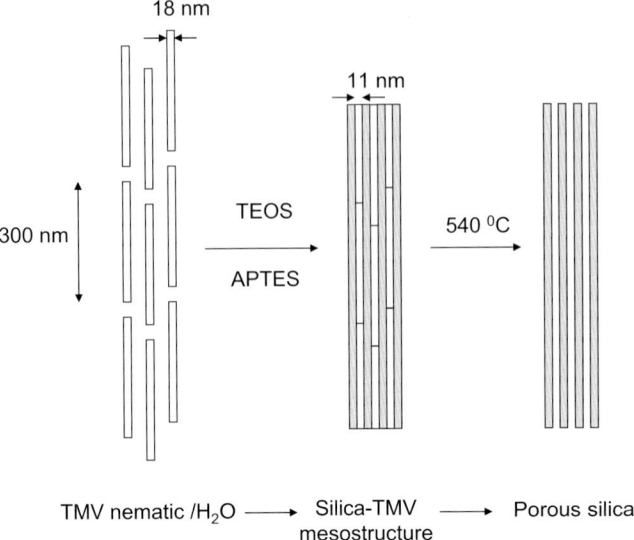

Figure 7.63 Synthesis of ordered mesostructured silica using TMV LC templates. (Adapted with permission from Fowler (2001). Copyright 2001, Wiley-VCH Verlag GmbH.)

self-assembly of mesostructured materials. As phages can be produced economically by infection of bacteria, large-scale application of such materials seems to be possible. The various liquid crystalline phases allow a large structural variability in processing.

A straightforward method for manufacturing mesoporous silica materials has been worked out by Mann and coworkers by using virus-based LCs (Fowler, 2001). From the concentrated aqueous solution of TMVs, a nematic liquid crystalline structure has been grown. In this structure the TMVs were arranged in orientational order with a cross-sectional periodicity of about 38 nm. A mixture of tetraethoxysilane (TEOS) and aminopropyltriethoxysilane (APTES) was given to that viscous gel. As schematically shown in Figure 7.63, the induced mineralization led to a silica gel consisting of mesoporous silica (between 55 and 60 wt%) and about 15 wt% TMV, depending on the amount of APTES. The native TMVs were rearranged into continuous chains and compressed during silicification. By thermal degradation of the TMVs, a hexagonal array of straight channels (11 nm in diameter) with 9–10 nm thick silica walls was produced (Figure 7.64). The final mesoporous silica was a replica of the nematic LC of TMVs. With a periodicity of about 20 nm, the silica structure exhibits an interesting geometry for application as a mesostructured scaffold or template for a secondary materials processing.

One essential advantage of using viruses in such processes is the additional degree of freedom in case of genetically modified coat proteins of the viruses. Thus, the properties of the LCs can be tailored to achieve specific physical or chemical properties. As discussed already in Section 2.3.1.2, the incorporation of binding sites for specific materials permits the self-assembly of patterned templates for inorganic nanoparticles, for instance, QDs in optoelectronic heterostructures (Whaley *et al.*, 2000).

(a)

(b)

(c)

(d)

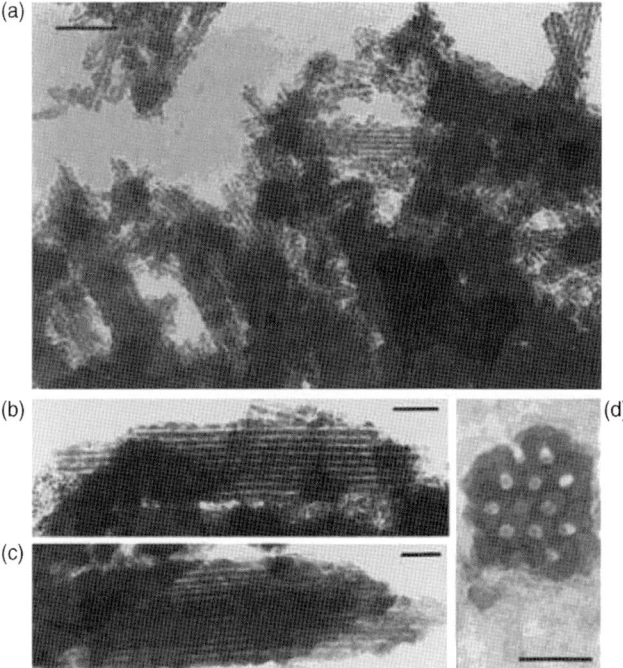

Figure 7.64 TEM images of ordered silica– TMV mesostructures. (a) Low magnification image, scale bar = 200 nm. (b) As-synthesized fragment showing well-ordered lattice fringes spaced at 18 nm, scale bar = 100 nm, (c) Calcined replica showing ordered linear channels, 11 nm in diameter, scale bar = 100 nm. (d) Small fragment viewed parallel to channel direction showing hexagonally ordered mesoporosity, scale bar = 50 nm. (Reproduced with permission from Fowler (2001). Copyright 2001, Wiley-VCH Verlag GmbH.)

The organization of the phages in liquid crystalline structures allows the formation of highly ordered hybrid inorganic–organic nanostructures. Lee *et al.* (2002) have given such example by ordering of ZnS QDs using genetically engineered viruses. The final material was a self-supporting hybrid film ordered at the nanoscale and also at the micrometer scale into ~72 μm domains, which were continuous over a centimeter length scale. Through screening of phage display libraries, an M13 phage was selected, which displayed peptides as part of the p3 minor coat protein (for the structure of the M13 phage, see Figure 1.19). The dominating peptide exhibited a ZnS binding motif (Cys—Asn—Asn—Pro—Met—His—Gln—Asn—Cys). The phages amplified to liquid crystalline concentrations were precipitated. Afterward, the bacteriophage pellets were resuspended in ZnS precursor solution (mixture of 1 mM $ZnCl_2$, and 1 mM Na_2S). The nanocrystal hybrid films (typically, ~15 μm thick) were transparent. They showed smectic-like lamellar structure. Its periodicity length (895 nm) corresponded nearly to that of the phage length (880 nm) plus the ZnS nanocrystal (diameter ~20 nm). In order to generalize this method, in a second experiment an M13 virus was selected that preferentially binds STV at the p3 minor coat protein (Lee *et al.*, 2003). This gave the option to incorporate any material that has been covalently conjugated to STV. By applying this method with 10 nm Au beads, self-supporting Au virus films were prepared. Also the manufacturing of electrodes for a lithium ion battery by self-assembly of genetically modified M13 viruses discussed in Section 2.3.1.2 serves as an example of such approach. In this case, a bifunctional virus was applied. Hybrid gold–cobalt nanowires were deposited onto the LC films, which led to a higher capacity of lithium batteries (Nam *et al.*, 2006).

Remarkably, the physical properties of phages alone can be used to create promising new materials by self-assembly into liquid crystalline structures. The development of a phage-based piezoelectric generator that produces up to 6 nA of current and 400 mV of potential is an impressive example of such option. In a fundamental study Lee *et al.* (2012) have explored that the M13 phage exhibits intrinsic piezoelectric properties related to the p8 major coat protein. Each phage exhibits about 2700 copies of the p8 major coat protein and only few copies of minor coat proteins p3, p6, p7, and p9 at its end. The p8 protein has an α-helical structure. It exhibits an electric dipole moment directed from the N-terminal region to the C-terminal region (Figure 7.65). Densely packed negative charges near the N-terminus and densely packed positive charges near the C-terminus are

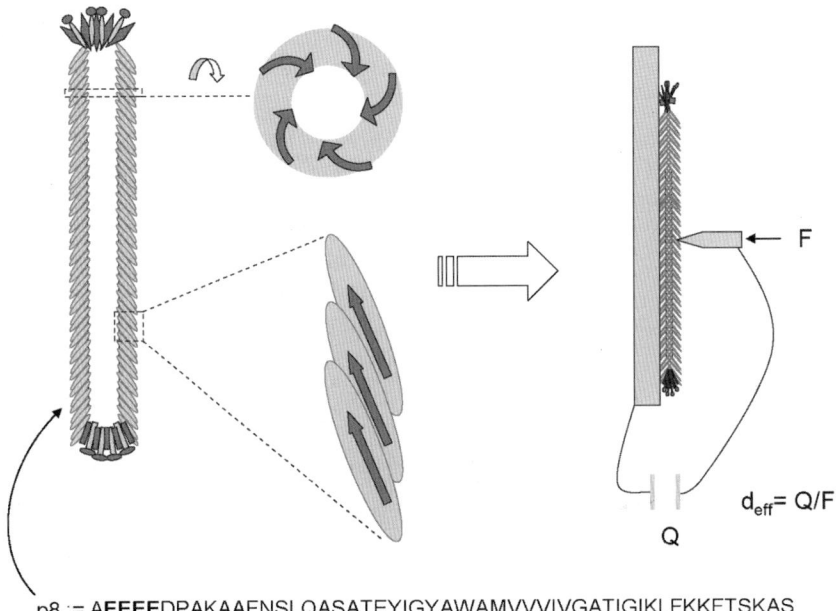

p8 := AEEEEDPAKAAFNSLQASATEYIGYAWAMVVVIVGATIGIKLFKKFTSKAS

Figure 7.65 Schematic of piezoelectric M13 phage structure. *Left*: M13 phage covered by ∼2700 p8 major coat proteins. The dipole moments are directed from the N-terminus to the C-terminus. The arrows indicate the dipole directions. The p8 coat proteins assemble with fivefold rotational and twofold screw symmetry. The p8 coat protein has ∼20° tilt angle with respect to the phage long axis. *Right*: Setup for measurement of the piezoelectric response under action of a point force. *Bottom*: Primary structure of the engineered major coat protein with the engineered four-glutamic acid (4E) sequence (Lee *et al.*, 2012).

responsible for the dipole moment of the p8 protein of the wild type. The dipole moment can be genetically enhanced by insertion of additional glutamate (E) molecules at the N-terminus between the alanine (A) and aspartate (D). In the wild-type phage, there are two glutamate (E) molecules between the first (Ala) and fifth (Asp) amino acid. In the study of Lee *et al.* (2012), phages with 1E, 2E, 3E, and 4E inserts have been prepared.

Phage films (monolayer and multilayer with a thickness of up to 300 nm) were deposited on gold-coated flexible substrates by dip- or dropcoating. Topography images by atomic force microscopy revealed a smectic structure (phages ordered in two directions and positions) and a band spacing of about 1 μm. The piezoelectric response was measured at monolayers of these phages. By applying a voltage at the metal-coated tip of an atomic force microscope, the vertical displacement of the film was measured. The slope of displacement versus voltage, the effective piezoelectric coefficient d_{eff}, is a measure for the piezoelectric response. It was found as $d_{eff} = 0.30 \pm 0.03$ pm/V for the wild type and $d_{eff} = 0.14 \pm 0.03$ pm/V and $d_{eff} = 0.70 \pm 0.05$ pm/V for monolayers of 1E and 4E phages, respectively. The piezoelectric response was increased by deposition of multilayer phage films. Films prepared with 4E phages reached a saturation value of $d_{eff} \sim 3.9$ pm/V for a thickness exceeding ~100 nm. Comparison with data measured at periodically poled lithium niobate (PPLN) films ($d_{eff} \sim 13.2$ pm/V) and collagen type 1 films ($d_{eff} \sim 1.1$ pm/V) shows that the piezoelectric properties of M13-based films are not far away from the relevant inorganic piezoelectric material PPLN. The low value of the collagen films can be understood by the piezoelectric response being mainly directed along the collagen fibril axis.

The 4E phage films have been used to fabricate a first simple phage-based piezoelectric power generator. Multilayered phage films with an area of ~1 cm^2 were placed between two gold electrodes by dropcasting. At 6 s intervals, a compressive load F was applied onto the electrodes. The resultant short-circuit current, the open-circuit voltage, and the charge were measured. At a maximum strain of ~0.14, a peak value of ~4 nA has been observed for the short-circuit current. A single current peak with $F = 34$ N peak load contained a charge of $Q \approx 380$ pC. The open-circuit voltage reached 400 mV. These experiments have demonstrated that the manufacturing of actuators and energy harvesters for miniature devices seems to be feasible by LC assays based on genetically engineered phages.

References

Aich, P., Labiuk, S.L. *et al.* (1999) M-DNA: A complex between divalent metal ions and DNA which behaves as a molecular wire. *J Mol Biol* **294** (2): 477–485.

Aili, D., Enander, K. *et al.* (2008a) Assembly of polypeptide-functionalized gold nanoparticles through a heteroassociation- and folding-dependent bridging. *Nano Letters*, **8** (8), 2473–2478.

Aili, D., Enander, K. *et al.* (2008b) Folding induced assembly of polypeptide decorated gold nanoparticles. *Journal of the American Chemical Society*, **130** (17), 5780–5788.

Amdursky, N., Gazit, E. et al. (2010a)
Quantum confinement in self-
assembled bioinspired peptide
hydrogels. *Advanced Materials*, **22** (21),
2311–2315.

Amdursky, N., Molotskii, M., Gazit, E., and
Rosenman, G. (2010b) Elementary
building blocks of self-assembled peptide
nanotubes. *Journal of the American
Chemical Society*, **132**, 15632–15636.

Blecha, A. (2005) *Gentechnisches Design
bakterieller Hüllproteine für die technische
Nutzung*. Technische Universität,
Dresden.

Blüher, A. (2008) *S-Schichtproteine als
molekulare Bausteine zur
Funktionalisierung mikroelektronischer
Sensorstrukturen*. Technische Universität,
Dresden.

Bobeth, M., Blecha, A. et al. (2011)
Formation of tubes during self-assembly
of bacterial surface layers. *Langmuir*,
27 (24), 15102–15111.

Bong, D.T., Clark, T.D. et al. (2001) Self-
assembling organic nanotubes.
*Angewandte Chemie – International
Edition in English*, **40** (6), 988–1011.

Chen, C., Zhang, P., and Rosi, N.L. (2008)
A new peptide-based method for the
design and synthesis of nanoparticle
superstructures: construction of highly
ordered gold nanoparticle double
helices. *Journal of the American Chemical
Society*, **130**, 13555–13557.

Cherny, I. and Gazit, E. (2008) Amyloids:
not only pathological agents but also
ordered nanomaterials. *Angewandte
Chemie – International Edition in English*,
47 (22), 4062–4069.

Colombi Ciacchi, L., Pompe, W. et al.
(2001) Initial nucleation of platinum
clusters after reduction of K(2)PtCl(4) in
aqueous solution: a first principles study.
Journal of the American Chemical Society,
123 (30), 7371–7380.

Dieluweit, S., Pum, D., and Sleytr, U.B.
(1998) Formation of a gold superlattice on
an S-layer with square lattice symmetry.
Supramolecular Science, **5**, 15–19.

Douglas, K., Clark, N.A., and Rothschild,
K.J. (1986) Nanometer molecular
lithography. *Applied Physics Letters*,
48 (10), 676–678.

Douglas, K., Devaud, G. et al. (1992)
Transfer of biologically derived
nanometer-scale patterns to smooth
substrates. *Science*, **257** (5070), 642–644.

Fowler, C.E. (2001) Tobacco mosaic virus
liquid crystals as templates for interior
designing of silica mesophases and
nanoparticles. *Advanced Materials*,
13 (16), 1266–1269.

Ghadiri, M.R., Granja, J.R., Milligan, R.A.,
McRee, D.E., and Khazanovich, N.
(1993) Self-assembling organic
nanotubes based on a cyclic peptide
architecture. *Nature*, **366**, 324–327.

Grzelczak, M., Vermant, J., Furst, E.M.,
and Liz-Marzan, L.M. (2010) Directed
self-assembly of nanoparticles. *ACS
Nano*, **4** (7), 3591–3605.

Haasen, P. (ed.) (1991) Homogeneous
second phase precipitation, in *Phase
Transformations in Materials*, Wiley-VCH
Verlag GmbH, Weinheim.

Hill, T.L. (1987) *Linear Aggregation Theory
in Cell Biology*. New York, Springer.

Horejs, C., Mitra, M.K. et al. (2011) Monte
Carlo study of the molecular
mechanisms of surface-layer protein
self-assembly. *Journal of Chemical
Physics*, **134** (12), 125103.

Howard, J. (2001) *Mechanics of Motor
Proteins and the Cytoskeleton*, Sinauer
Associates, Inc. Publishers, Sunderland,
MA.

Janmey, P.A. and Kinnunen, P.K. (2006)
Biophysical properties of lipids and
dynamic membranes. *Trends in Cell
Biology*, **16** (10), 538–546.

Jarosch, M., Egelseer, E.M. et al. (2001)
Analysis of the structure–function
relationship of the S-layer protein
SbsC of *Bacillus stearothermophilus*
ATCC 12980 by producing truncated
forms. *Microbiology*, **147** (Pt 5),
1353–1363.

Lee, S.-W., Lee, S.K., and Belcher, A.M.
(2003) Virus-based alignment of
inorganic, organic, and biological
nanosized materials. *Advanced Materials*,
15 (9), 689–692.

Lee, S.W., Mao, C. et al. (2002) Ordering of
quantum dots using genetically
engineered viruses. *Science*, **296** (5569),
892–895.

Lee, B.Y., Zhang, J., Zueger, C., Chung, W.-J., Yoo, S.Y., Wang, E., Meyer, J., Ramesh, R., and Lee, S.-W. (2012) Virus-based piezoelectric energy generation. *Nature Nanotechnology*, **7** (June), 351–356.

Lehn, J.-M. (1995) *Supramolecur Chemistry, Concepts and Perspectives*, Wiley-VCH Verlag GmbH, Weinheim.

Lin, C., Ke, Y. *et al.* (2007) Functional DNA nanotube arrays: bottom-up meets top-down. *Angewandte Chemie – International Edition in English*, **46** (32), 6089–6092.

Liu, H., Chen, Y., He, Y., Ribbe, A.E., and Mao, C. (2006) Approaching the limit: can one DNA oligonucleotide assemble into large nanostructures? *Angewandte Chemie – International Edition in English*, **45**, 1942–1945.

Magoshi, J., Magoshi, Y., and Nakamura, S. (1985) Physical properties and structure of silk: 9. Liquid crystal formation of silk fibroin. *Polymer Communications*, **26**, 60–61.

Malvankar, N.S., Vargas, M. *et al.* (2011) Tunable metallic-like conductivity in microbial nanowire networks. *Nature Nanotechnology*, **6** (9), 573–579

Mertig, M., Ciacchi, L.C., Seidel, R., Pompe, W., and De Vita, A. (2002) DNA as a selective metallization template. *Nano Letters*, **2**, 841–844.

Mertig, M., Kirsch, R., Pompe, W., and Engelhard, H. (1999) Fabrication of highly oriented nanocluster arrays by biomolecular templating. *European Physical Journal D*, **9**, 45–48.

Moll, D., Huber, C. *et al.* (2002) S-layer-streptavidin fusion proteins as template for nanopatterned molecular arrays. *Proceedings of the National Academy of Sciences of the United States of America*, **99** (23), 14646–14651.

Nam, K.T., Kim, D.W. *et al.* (2006) Virus-enabled synthesis and assembly of nanowires for lithium ion battery electrodes. *Science*, **312** (5775), 885–888.

Nirschl, M., Blüher, A., Erler, C., Katzschner, B., Vikholm-Lundin, I., Auer, S., Voros, J., Pompe, W., Schreiter, M., and Mertig, M. (2009) Film bulk acoustic resonators for DNA and protein detection and investigation of *in vitro* bacterial

S-layer formation. *Sensors and Actuators A: Physical*, **156** (1), 180–184.

Panhorst, M., Brückl, H., Kiefer, B., and Reiss, G. (2001) Formation of metallic surface structures by ion etching using a S-layer template. *Journal of Vacuum Science & Technology B*, **19** (3), 722–724.

Pleschberger, M., Neubauer, A. *et al.* (2003) Generation of a functional monomolecular protein lattice consisting of an S-layer fusion protein comprising the variable domain of a camel heavy chain antibody. *Bioconjugate Chemistry*, **14** (2), 440–448.

Pum, D., Sára, M., and Sleytr, U.B. (1989) Use of two-dimensional protein crystals from bacteria for nonbiological applications. *Journal of Vacuum Science & Technology*, **B7**, 1391–1397.

Reches, M. and Gazit, E. (2003) Casting metal nanowires within discrete self-assembled peptide nanotubes. *Science*, **300**, 625–627.

Renner, L., Pompe, T., Lemaitre, R., Drechsel, D., and Werner, C. (2010) Controlled enhancement of transmembrane enzyme activity in polymer cushioned supported bilayer membranes. *Soft Matter*, **6**, 5382–5389.

Sackmann, E. (1994) The seventh Datta Lecture. Membrane bending energy concept of vesicle- and cell-shapes and shape-transitions. *FEBS Letters*, **346** (1), 3–16.

Santoso, S., Hwang, W., Hartman, H., and Zhang, S. (2002) Self-assembly of surfactant-like peptides with variable glycine tails to form nanotubes and nanovesicles. *Nano Letters*, **2**, 687–691.

Sara, M. and Sleytr, U.B. (1987) Molecular sieving through S layers of *Bacillus stearothermophilus* strains. *Journal of Bacteriology*, **169** (9), 4092–4098.

Schmelzer, J., Röpke, G., and Mahnke, R. (1999) *Aggregation Phenomena in Complex Systems*, Wiley-VCH Verlag GmbH, Weinheim.

Schuster, B., Pum, D., Sara, M., Braha, O., Bayley, H., and Sleytr, U.B. (2001) S-layer ultrafiltration membranes: a new support for stabilizing functionalized lipid membranes. *Langmuir*, **17**, 499–503.

Schuster, B. and Sleytr, U.B. (2000) S-layer-supported lipid membranes. *Reviews in Molecular Biotechnology*, **74** (3), 233–254.

Seeman, N.C. (1982) Nucleic acid junctions and lattices. *Journal of Theoretical Biology*, **99** (2), 237–247.

Seidel, R., Ciacchi, L.C. *et al.* (2004) Synthesis of platinum cluster chains on DNA templates: conditions for a template-controlled cluster growth. *Journal of Physical Chemistry B* **108** (30), 10801–10811.

Sharma, J., Chhabra, R., Cheng, A., Brownell, J., Liu, Y., and Yan, H. (2009) Control of self-assembly of DNA tubules through integration of gold nanoparticles. *Science*, **323**, 112–116.

Sharma, J., Ke, Y. *et al.* (2008) DNA-tile-directed self-assembly of quantum dots into two-dimensional nanopatterns. *Angewandte Chemie – International Edition in English*, **47** (28), 5157–5159.

Shenton, W., Pum, D., Sleytr, U.B., and Mann, S. (1997) Synthesis of cadmium sulphide superlattices using self-assembled bacterial S-layers. *Nature*, **389** (9), 585–587.

Si, S., Raula, M. *et al.* (2008) Reversible self-assembly of carboxylated peptide-functionalized gold nanoparticles driven by metal-ion coordination. *ChemPhysChem*, **9** (11), 1578–1584.

Sleytr, U.B. and Beveridge, T.J. (1999a) Bacterial S-layers. *Trends in Microbiology*, **7** (6), 253–260.

Sleytr, U.B., Györvary, E., and Pum, D. (2003a) Crystallization of S-layer protein lattices on surfaces and interfaces. *Progress in Organic Coatings*, **47**, 279–287.

Sleytr, U.B., Messner, P., Pum, D., and Sára, M. (1999b) Crystalline bacterial cell surface layers (S layers): from supramolecular cell structure to biomimetics and nanotechnology. *Angewandte Chemie – International Edition in English*, **38**, 1034–1054.

Sleytr, U.B., Schuster, B. *et al.* (2003b) Nanotechnology and biomimetics with 2-D protein crystals. *IEEE Engineering in Medicine and Biology Magazine*, **22** (3), 140–150.

Wahl, R. (2003) *Reguläre bakterielle Zellhüllenproteine als biomolekulares Templat*. Technische Universität, Dresden.

Wahl, R., Engelhardt, H. *et al.* (2005) Multivariate statistical analysis of two-dimensional metal cluster arrays grown in vitro on a bacterial surface layer. *Chemistry of Materials*, **17** (7), 1887–1894.

Wahl, R., Mertig, M., Raff, J., Selenska-Pobell, S., and Pompe, W. (2001) Electron-beam induced formation of highly ordered palladium and platinum nanoparticle arrays on the s layer of *Bacillus sphaericus* NCTC 9602. *Advanced Materials*, **13** (10), 736–740.

Weigert, S. and Sára, M. (1995) Surface modification of an ultrafiltration membrane with crystalline structure and studies on interactions with selected protein molecules. *Journal of Membrane Science*, **106**, 147–159.

Whaley, S.R., English, D.S. *et al.* (2000) Selection of peptides with semiconductor binding specificity for directed nanocrystal assembly. *Nature*, **405** (6787), 665–668.

Yan, X., He, Q., Wang, K., Duan, L., Cui, Y., and Li, J. (2007) Transition of cationic dipeptide nanotubes into vesicles and oligonucleotide delivery. *Angewandte Chemie – International Edition in English*, **46**, 2431–2434.

Yemini, M., Reches, M. *et al.* (2005a) Peptide nanotube-modified electrodes for enzyme-biosensor applications. *Analytical Chemistry*, **77** (16), 5155–5159.

Yemini, M., Reches, M., Rishpon, J., and Gazit, E. (2005b) Novel electrochemical biosensing platform using self-assembled peptide nanotubes. *Nano Letters*, **5** (1), 183–186.

Zhang, S., Marini, D.M., Hwang, W., and Santoso, S. (2002) Design of nanostructured biological materials through self-assembly of peptides and proteins. *Current Opinion in Chemical Biology*, **6**, 865–871.

Appendix A
Constants, Units, and Magnitudes

A.1
Fundamental Constants

Constant	Symbol	Value	SI unit
Avogadro's number	N_a	6.022×10^{23}	molecules/mol
Boltzman's constant	k_B	1.381×10^{-23}	J/K
Elementary charge	e	1.602×10^{-19}	C
Electron mass	m	9.109×10^{-31}	kg
Faraday constant	$F = N_a e$	9.649×10^4	C/mol
Molar gas constant	$R = N_a k_B$	8.314	J/(K mol)
Permittivity of vacuum	ε_0	8.854×10^{-12}	C/Vm
Planck constant	h	6.626×10^{-34}	J s
Speed of light	c	2.998×10^8	m/s

A.2
Table of SI Base Units

Quantity	SI unit	Symbol
Length	Meter	m
Mass	Kilogram	kg
Time	Second	s
Electric current	Ampere	A
Temperature	Kelvin	K
Amount	Mole	mol

Bio-Nanomaterials: Designing Materials Inspired by Nature, First Edition. Wolfgang Pompe, Gerhard Rödel, Hans-Jürgen Weiss, and Michael Mertig.
© 2013 Wiley-VCH Verlag GmbH & Co. KGaA. Published 2013 by Wiley-VCH Verlag GmbH & Co. KGaA.

A.3
Table of Derived Units

Quantity	SI unit	Symbol	Composite symbol
Frequency	Hertz	Hz	s^{-1}
Force	Newton	N	$(kg\ m)/s^2$
Pressure	Pascal	Pa	N/m^2
Energy	Joule	J	Nm
Power	Watt	W	J/s
Charge	Coulomb	C	As
Potential difference	Volt	V	W/A

A.4
Magnitudes

A.4.1
Sizes

Hydrogen atom (radius), ≈ 0.05 nm.

Covalent bond length, ≈ 0.1 nm.

H-bond (distance between centers of atoms flanking H), ≈ 0.27 nm.

Au (face-centered cubic lattice constant), 0.408 nm.

Si (diamond lattice constant), 0.543 nm.

Sugar, amino acid, nucleotide (diameter), 0.5–1 nm.

Debye screening length (of physiological Ringer's solution), $\lambda_D \approx 0.7$ nm.

DNA (diameter), ≈ 2 nm.

Globular protein (diameter), 2–10 nm.

C60 fullerene (diameter), 1.1 nm.

Single-walled carbon nanotube (SWCNT) (diameter), ≈ 1 nm.

Bilayer membrane (thickness), ≈ 3 nm.

Bacterial surface layer (S-layer, thickness), ≈ 6 nm.

Amyloid protofibril (diameter), 2–5 nm.

Actin filament (diameter), ≈ 5 nm.

Microtubule (diameter), ≈ 25 nm.

M13 virus (diameter), ≈ 6 nm.

Tobacco virus (diameter), ≈ 18 nm.

Poliovirus (diameter), ≈ 25 nm.

Ribosome (diameter), ≈ 30 nm.

Wavelength of visible light, 400–650 nm.

Tobacco mosaic virus (length), ≈ 300 nm.

M13 virus (length), ≈880 nm.
Single-walled carbon nanotube (length), few micrometers–several millimeters.
Typical bacterium (diameter), ≈1 μm.
Typical human cell (diameter), ≈10 μm.
Lambda phage virus DNA (λ-DNA, contour length), ≈16.5 μm.
Osteoblast (diameter), ≈20 μm.
Yeast cell (diameter), 30–50 μm.
Osteoclast (diameter), ≈100 μm.
Human hair (diameter), ≈100 μm.
Escherichia coli genome (length if extended), ≈1.4 mm.
Human genome (total length), ≈1 m.

A.4.2
Energies

Thermal energy at room temperature, $k_B T_r = 4.1\,\text{pNnm} = 4.1 \times 10^{-21}$ J $= 0.025$ eV.
Double covalent bond (e.g., C=C), $240\,k_B T_r$.
Single covalent bond (e.g., C—C), $140\,k_B T_r$.
Green photon, $120\,k_B T_r$.
Streptavidin/biotin bond, $40\,k_B T_r$.
ATP hydrolysis, ≈$20\,k_B T_r$/molecule.
Hydrogen bond, 2–$10\,k_B T_r$.
Hydrophobic interaction, 3–$5\,k_B T_r$.
van der Waals interaction between atoms, 0.6–$1.6\,k_B T_r$.

A.4.3
Rates and Diffusion Constants

Turnover number of enzymes, $5 \times 10^{-2} - 1 \times 10^{7}\,\text{s}^{-1}$.
Diffusion constant of small molecules in water, $D \approx 1\,\mu\text{m}^2/\text{ms}$.
Diffusion constant of globular proteins in water, $D \approx 0.01\,\mu\text{m}^2/\text{ms}$.

Appendix B
Energy of a Bent Fiber

We consider a short segment of length ds of a bent thin fiber of radius r. The fiber is elastically deformed with small curvature $1/R \ll 1/r$. Then, the neutral axis agrees with the midplane. Tensile and compressive forces are distributed symmetrically to the neutral axis (Figure B.1).

At a distance y from the neutral axis, the strain ε is given as $\varepsilon = y/R$. The stress is proportional to the strain $\sigma = E \cdot \varepsilon$, where E is the Young's modulus. The elastic energy dW_{el} stored in the segment ds of the fiber is given as

$$dW_{el} = \iint\limits_{cross\ section} dxdy \int_0^{\varepsilon(y)} \sigma(\varepsilon')d\varepsilon' \cdot ds = \frac{1}{2}E \cdot \iint\limits_{cross\ section} dxdy \cdot \varepsilon(y)^2 \cdot ds$$

$$= \frac{E \cdot I}{2R^2}ds. \tag{B.1}$$

I is the geometrical moment of inertia of the cross section:

$$I = \iint\limits_{cross\ section} dxdy \cdot y^2. \tag{B.2}$$

The comparison of Eq. (B.1) with Eq. (1.2) yields $\beta^2 = (d\theta/ds)^2 = 1/R^2$ for the bend persistence length:

$$A = \frac{E \cdot I}{k_B T}. \tag{B.3}$$

Note that this relation is not only valid for fibers with circular cross section. The particular shape of the cross section is expressed by the value of I. We see that the persistence length describes the relation between stiffness of a semiflexible fiber and the thermal energy.

Bio-Nanomaterials: Designing Materials Inspired by Nature, First Edition. Wolfgang Pompe, Gerhard Rödel, Hans-Jürgen Weiss, and Michael Mertig.
© 2013 Wiley-VCH Verlag GmbH & Co. KGaA. Published 2013 by Wiley-VCH Verlag GmbH & Co. KGaA.

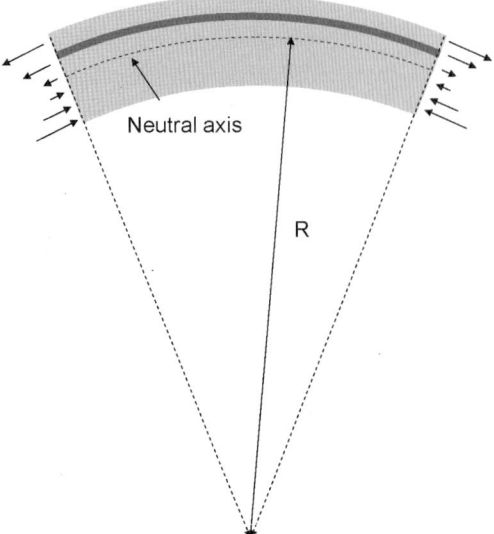

Figure B.1   Stress distribution within a bent fiber.

Appendix C
Circular Dichroism Spectroscopy

Circular dichroism is governed by the difference of the absorbances, $\Delta A = (\varepsilon_L - \varepsilon_R) \cdot c \cdot l$, which is known as the Beer–Lambert Law, where c is the molar concentration and l is the path length in the solution. Note that the molar extinction coefficients depend on the wavelength λ. The ellipticity θ is defined as the arc tangent of the ratio of the minor axis to the major axis, $\theta = arctan\,(b/a)$ (Figure C.1). For small differences of the absorbances, it scales with ΔA:

$$\theta\,(\text{degree}) = 2.303(A_L - A_R)180/(4\pi) = 33.0\Delta A. \tag{C.1}$$

The optical rotation is given as

$$\phi\,(\text{degree}) = 180l(n_L - n_R)/\lambda. \tag{C.2}$$

In order to compare results from different samples, molar rotation $[\phi]$ and molar ellipticity $[\theta]$ are introduced:

$$[\phi] = \frac{\tilde{M}}{100}\frac{\phi}{cl}, \quad [\theta] = \frac{\tilde{M}}{100}\frac{\theta}{cl} \quad \text{in} \quad \frac{\text{degree} \cdot \text{cm}^2}{\text{decimal}}, \tag{C.3}$$

where $\tilde{M}$ is the molecular weight in g, the concentration c is in g/cm^3, and the path length l is in decimeter. Changes of the secondary structure of proteins are measured usually by taking the CD spectrum of wavelengths between 190 and 250 nm. The absorption band of peptide bonds is in the region of 210–220 nm. In the unfolded state, CD of peptides is nearly zero.

For further studies, see Sheehan (2000).

Reference

Sheehan, D. (2000) *Physical Biochemistry:*
  Principles and Applications, John Wiley &
  Sons, Inc., New York.

Bio-Nanomaterials: Designing Materials Inspired by Nature, First Edition. Wolfgang Pompe, Gerhard Rödel,
Hans-Jürgen Weiss, and Michael Mertig.
© 2013 Wiley-VCH Verlag GmbH & Co. KGaA. Published 2013 by Wiley-VCH Verlag GmbH & Co. KGaA.

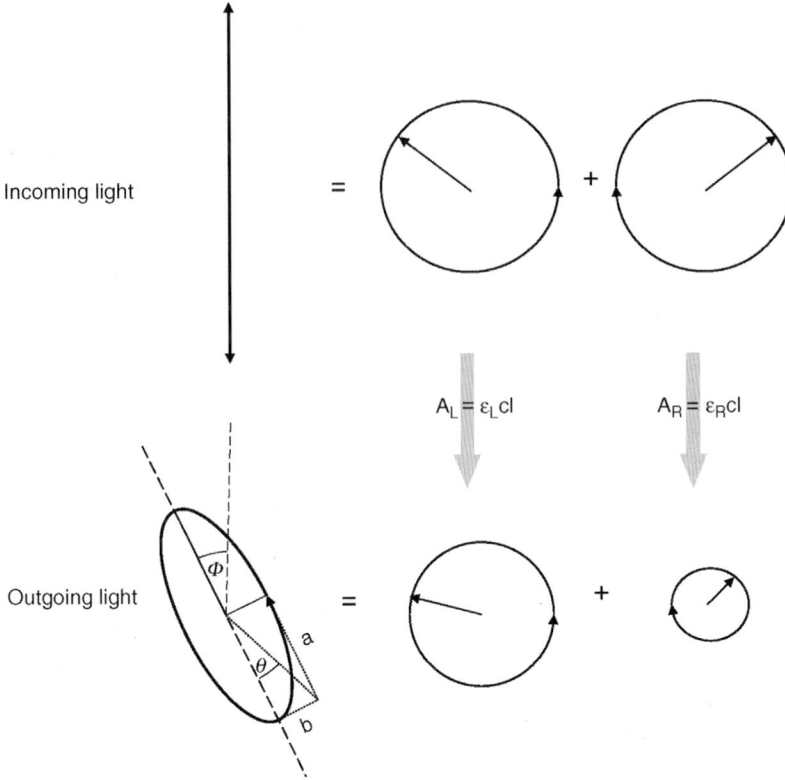

Incoming light

Outgoing light

$A_L = \varepsilon_L cl$

$A_R = \varepsilon_R cl$

Figure C.1 Change of polarization of the linearly polarized light by passing a solution of chiral molecules.

Appendix D
Task Solutions

Task 1.1: We consider the change of the total energy of a segment ds of a semielastic biopolymer that is loaded with a constant tensile force f. A tensile strain u causes an increase of the elastic energy of the polymer segment, dW_{el}, and a decrease $-uf\,ds$ of the potential energy of the loading system. In equilibrium, the total energy change $dW_{tot} = dW_{el} - fus$ approaches a minimum. As in the general case, the change of the elastic strain u is coupled with a change of the torsional deformation. Thus, we have to ask for the minimum of the total energy with respect to u and the twist density ω:

$$\frac{\delta dW_{tot}}{\delta u} = 0, \quad \frac{\delta dW_{tot}}{\delta \omega} = 0. \tag{D.1}$$

With Eq. (1.1), we get the coupled set of equations:

$$k_B T \left(Bu + \frac{1}{2} D\omega \right) - f = 0, \quad k_B T \left(C\omega + \frac{1}{2} Du \right) = 0, \tag{D.2}$$

with the solution

$$u = \frac{f}{k_B T B} \left(1 - \frac{D^2}{4CB} \right)^{-1}, \quad \omega = -\frac{Df}{2C k_B T B} \left(1 - \frac{D^2}{4CB} \right)^{-1}. \tag{D.3}$$

The elastic strain increases linearly with the force. The strain–twist coupling ($D \neq 0$) causes an increase of the elastic strain. The biopolymer can be approximately considered as inextensible if the following condition is fulfilled:

$$\frac{f}{k_B T B} \left(1 - \frac{D^2}{4CB} \right)^{-1} \leq 0.01. \tag{D.4}$$

The unknown constant B can be substituted by the Young's modulus and the fiber diameter by expressing the strain energy by $dW_{strain} = (\pi d^2/8) E u^2 ds$. Thus, we get for $D = 0$ from Eq. (D.4) the condition

$$\frac{4f}{\pi d^2 E} \leq 0.01. \tag{D.5}$$

Bio-Nanomaterials: Designing Materials Inspired by Nature, First Edition. Wolfgang Pompe, Gerhard Rödel, Hans-Jürgen Weiss, and Michael Mertig.
© 2013 Wiley-VCH Verlag GmbH & Co. KGaA. Published 2013 by Wiley-VCH Verlag GmbH & Co. KGaA.

For the particular choice of $f = 10\,\text{pN}$ and $E = 1\,\text{GPa}$, we get for the minimum value of the fiber diameter $d \geq 1.13\,\text{nm}$.

Task 1.2: As shown in Figure 1.34, the bend vector $\vec{\beta} = \mathrm{d}\vec{t}/\mathrm{d}s$ points inward. Its magnitude equals $\left| \vec{\beta} \right| = \mathrm{d}\theta/\mathrm{d}s = R^{-1}$. Thus, Eq. (1.2) yields

$$W_{\text{el}} = \frac{1}{2} k_{\text{B}} T \cdot A \int \vec{\beta}^{\,2}(s)\mathrm{d}s = \frac{1}{2} k_{\text{B}} T \cdot \frac{A \cdot l_0}{R^2}. \tag{D.6}$$

Task 1.3: With Eq. (1.4) and the experimental data given in Table 1.6, we get for the Young's modulus $E = A \cdot k_{\text{B}} T / I$, as summarized in Table D.1.

For comparison, some examples of Young's moduli of structural materials are given: carbon nanotube 1300 GPa, diamond 1200 GPa, stainless steel 211 GPa, glass 73 GPa, Plexiglas 3 GPa, and polypropylene 0.02 GPA.

Task 1.4: With the distance 0.34 nm between two base pairs along a double helix, the contour length of a λ-DNA follows as 16.49 μm. Together with the bend persistence length of about $A = 50\,\text{nm}$ (Table 1.6), Eq. (1.11) yields for the mean diameter of the coil $D_{\text{coil}} = 1.28\,\mu\text{m}$.

Task 1.5: The rate constants k are in the range between $0.06\,\text{s}^{-1}$ and $2 \times 10^{-5}\,\text{s}^{-1}$. One percent of the substrate is transformed in the time $t = k^{-1} \cdot \ln(100/99)$, which means between 0.168 and 502.5 s.

Task 3.1: For the distribution of the mean stress in the surface layer of the hydrogel, we make the simplifying assumption of an isotropic 2D stress state beneath single cells with

$$\bar{T}_{11} = \bar{T}_{22} = \bar{T}\text{max}, \quad \bar{T}_{33} \cong 0, \quad \bar{T}_{ik} \cong 0, \quad \text{for} \quad i \neq k. \tag{D.7}$$

Any interaction with the stress field of other cells can be neglected due to the low cell density in the experiments. The strain in the loaded layer region is given as

$$\bar{\varepsilon}_{11} = \bar{\varepsilon}_{22} = \frac{1-\nu}{E} \cdot \bar{T}_{\text{max}}, \quad \bar{\varepsilon}_{33} = -\frac{2\nu}{E} \bar{T}_{\text{max}}, \quad \bar{\varepsilon}_{ik} \cong 0 \quad \text{for} \quad i \neq k, \tag{D.8}$$

where E is the Young's modulus and ν is the Poisson's ratio of the hydrogel. The density of the elastic energy η follows as

$$\eta = \frac{1}{2} \sum \bar{T}_{ik} \bar{\varepsilon}_{ik} \geq \frac{1-\nu}{E} \bar{T}_{\text{max}}^2. \tag{D.9}$$

Table D.1 Second moments of inertia and the Young's modulus of fibrous biomolecules.

Biopolymer	Second moment of inertia (nm^4)	Young's modulus (GPa)
Double-stranded DNA	0.79	0.26
Amyloid protofibril:		
with 100% ordered β-sheets	1.63 for $r = 1.2\,\text{nm}$	10.1
with 50% ordered β-sheets	15.28 for $r = 2.1\,\text{nm}$	0.027
Actin filament	28.27	2.2
Microtubule	18 200	1.36

Equation (D.9) gives a lower bound for the energy density as we neglect the small contributions of $\bar{T}_{33}\bar{\varepsilon}_{33}/2 \cong 0$ and $\sum_{i\neq k}\bar{T}_{ik}\bar{\varepsilon}_{ik}/2 \cong 0$. For the estimate of the stored elastic energy per cell $\bar{U}$, we assume that the stress-influenced volume V approximately is given by $V \approx S \cdot h$, where S is the surface covered by a single cell and h is the depth of the deformed hydrogel layer. This leads to an upper bound for the thickness h with

$$h \leq \frac{\bar{U} \cdot E}{S(1 - v)\bar{T}_{\text{max}}^2}. \tag{D.10}$$

Taking the data for the intermediate hydrogel compliance $1/E = 0.2\,(\text{kPa})^{-1}$ from Figure 3.10, $\bar{U}_{\text{PSMA}} = 0.8\,\text{pJ}$, $\bar{U}_{\text{PMMA}} = 0.15\,\text{pJ}$, and for the cell diameter from Figure 3.9, $d \approx 45\,\mu\text{m}$, we get for the thickness h of the two systems nearly the same values, $h_{\text{PSMA}} \leq 10.2\,\mu\text{m}$, $h_{\text{PMMA}} \leq 10.4\,\mu\text{m}$. These numbers are in good agreement with experimental values for minimum thickness of the PAAm hydrogel layer where an impact of the stiff glass substrate can be excluded.

Task 5.1: In pure water at $25\,°\text{C}$, the concentration of water molecules is 55.5 M, which corresponds to the weight of 1 L in grams divided by its gram molecular weight (18.015 g/mol). Hundred water molecules occupy a volume $V = 100 \times 10^{-3}\,\text{m}^3/(55.5 \cdot N_a) = 2.99 \times 10^{-27}\,\text{m}^3$. The presence of two hydronium ions in this small volume corresponds to a concentration of hydrogen ions $[\text{H}^+] = [\text{H}_2\text{O}] \times 2/100 = 1.11\,\text{M}$. It leads to an extreme $\text{pH} = -\log([\text{H}^+]/\text{M}) = -0.045 \cong 0$. Nevertheless, the encaged enzyme is stable. Due to the small cavity, only two ions can influence the charge distribution of the enzyme.

Task 6.1: In comparison to large precipitates with dominating planar surfaces, for small precipitates there is an enhanced driving force for dissolution in a growth medium. For a spherical particle, the change of the chemical potential of dissolution is given by

$$\Delta\mu_r = \frac{2\gamma_{\text{sl}}v_s}{r}. \tag{D.11}$$

Therefore, an increase of the equilibrium concentration X_r is observed in a solution with small precipitates in comparison to the case of large precipitates (solute concentration X_∞):

$$X_r = X_\infty \exp\left(\frac{2\gamma_{\text{sl}}v_s}{k_B T r}\right) \approx X_\infty\left(1 + \frac{2\gamma_{\text{sl}}v_s}{k_B T r}\right). \tag{D.12}$$

For the given example, the concentration increases by 9.6%.

Task 6.2: For dilute solutions, very often it can be assumed that the activity coefficients are approximately independent of the concentration $\gamma_A \cong \gamma_{A_{eq}}$, $\gamma_B \cong \gamma_{B_{eq}}$:

$$\Delta\mu \cong k_B T \ln\left(\frac{[A^{q+}]^p \cdot [B^{p-}]^q}{\left[A_{\text{eq}}^{q+}\right]^p \cdot \left[B_{\text{eq}}^{p-}\right]^q}\right). \tag{D.13}$$

Task 6.3: From Eq. (6.34), we get

$$
\ln\left(\frac{f_0 n_0}{J_{hom}}\right) = \frac{\Delta G_{hom}^*}{k_B T} = \frac{2\pi}{3} \frac{r^{*3}}{v_s} \ln S_R.
\tag{D.14}
$$

With

$$
\ln S_R = \ln(n_0 / K_{sp} M)
\tag{D.15}
$$

and

$$
r^* = \frac{2\gamma_{sl} v_s}{k_B T \ln S_R},
\tag{D.16}
$$

Eq. (D.14) can be transformed to

$$
\ln\left(\frac{f_0 K_{sp} M}{J_{hom}}\right) + \ln S_R = \frac{16\pi}{3} \frac{\gamma_{sl}^3 v_s^2}{(k_B T)^3 (\ln S_R)^2}.
\tag{D.17}
$$

With $J_{hom} = f_0 K_{sp} M$, Eq. (D.17) yields

$$
\ln S_R = \left(\frac{16\pi}{3}\right)^{1/3} \frac{\gamma_{sl}}{k_B T} (v_s)^{2/3} \quad \text{and} \quad S_R = 58.7.
$$

Task 6.4: With $N_s = V_s / v_s$, volume and interfaces can be expressed by the radius r and the contact angle. Two interface energies can be eliminated by the equilibrium condition. This immediately leads to Eqs. (6.37) and (6.38).

Task 6.5:

$$
\text{Reaction equation:} \quad \frac{d\phi}{dt} = k_{ads}(1 - \phi)c_p - k_{des}\phi,
\tag{D.18}
$$

$$
\text{Equilibrium solution:} \quad \frac{d\phi}{dt} = 0 \rightarrow \phi_{eq} = \frac{K_L c_p}{1 + K_L c_p},
\tag{D.19}
$$

with $K_L = k_{ads}/k_{des}$.

Index

Bio-Nanomaterials: Designing Materials Inspired by Nature, First Edition. Wolfgang Pompe, Gerhard Rödel,
Hans-Jürgen Weiss, and Michael Mertig.
© 2013 Wiley-VCH Verlag GmbH & Co. KGaA. Published 2013 by Wiley-VCH Verlag GmbH & Co. KGaA.